AF765196

A Poem by Baba Tahir

به قبرستان گذر کردم کمابیش
بدیدم خفته دولتمند و درویش
نه درویش بی کفن خفته است در خاک
نه دولتمند دارد یک کفن بیش

Often I've seen walking by a cemetery
Lying people- the poor and also mighty
Neither with no shroud was there a poor one,
Nor a more one the mighty had to put on

NANOENCAPSULATION OF FOOD INGREDIENTS BY SPECIALIZED EQUIPMENT

Nanoencapsulation in the Food Industry

NANOENCAPSULATION OF FOOD INGREDIENTS BY SPECIALIZED EQUIPMENT

Volume 3

Series Editor:

PROF. SEID MAHDI JAFARI

Department of Food Materials and Process Design Engineering, Gorgan University of Agricultural Sciences and Natural Resources, Gorgan, Iran

ELSEVIER

ACADEMIC PRESS

An imprint of Elsevier

Academic Press is an imprint of Elsevier
125 London Wall, London EC2Y 5AS, United Kingdom
525 B Street, Suite 1650, San Diego, CA 92101, United States
50 Hampshire Street, 5th Floor, Cambridge, MA 02139, United States
The Boulevard, Langford Lane, Kidlington, Oxford OX5 1GB, United Kingdom

Library of Congress Cataloging-in-Publication Data
A catalog record for this book is available from the Library of Congress

British Library Cataloguing-in-Publication Data
A catalogue record for this book is available from the British Library

ISBN 978-0-12-815671-1

For information on all Academic Press publications
visit our website at https://www.elsevier.com/books-and-journals

Publisher: Charlotte Cockle
Acquisition Editor: Nina Rosa de Araujo Bandeira
Editorial Project Manager: Laura Okidi
Production Project Manager: Vignesh Tamil
Cover Designer: Miles Hitchen

Typeset by SPi Global, India

Contents

Contributors

M. Carmen Alfaro
Chemical Engineering Department, Escuela Politécnica Superior, C/Virgen de África, University of Seville, Seville, Spain

Cordin Arpagaus
NTB University of Applied Sciences of Technology Buchs, Institute for Energy Systems, Buchs, Switzerland

E. Arranz
Teagasc Food Research Centre, Moorepark, Fermoy Co. Cork, Ireland

Elham Assadpour
Department of Food Science and Technology, Baharan Institute of Higher Education, Gorgan, Iran

Talita Aline Comunian
Department of Food Engineering, School of Food Engineering, University of Campinas, Campinas, Brazil

F. Donsì
Department of Industrial Engineering, University of Salerno, Fisciano, Italy

C. Fernandez-Avila
Center for Innovation, Research and Knowledge Transfer in Food Technology (CIRTTA), XaRTA, TECNIO, Department of Animal and Food Science, Faculty of Veterinary Medicine, Universitat Autònoma de Barcelona, Bellaterra, Spain

Laura G. Gómez-Mascaraque
Teagasc Food Research Centre, Fermoy, Ireland

E. Hebishy
National Center for Food Manufacturing, College of Sciences, University of Lincoln, Holbeach Campus, Spalding, United Kingdom

Seid Mahdi Jafari
Department of Food Materials and Process Design Engineering, Faculty of Food Science and Technology, Gorgan University of Agricultural Sciences and Natural Resources, Gorgan, Iran

Roya Koshani
Department of Chemistry, Quebec Centre for Advanced Materials, Pulp and Paper Research Centre, McGill University, Montreal, QC, Canada; Department of Food Materials and Process Design Engineering, Gorgan University of Agricultural Sciences and Natural Resources, Gorgan, Iran

Amparo Lopez-Rubio
Food Preservation and Food Quality Department, IATA-CSIC, Valencia, Spain

M. José Martín-Piñero
Chemical Engineering Department, Facultad de Química, C/Profesor García Gonzalez, University of Seville, Seville, Spain

José Muñoz
Chemical Engineering Department, Facultad de Química, C/Profesor García Gonzalez, University of Seville, Seville, Spain

Jenifer Santos
Chemical Engineering Department, Facultad de Química, C/Profesor García Gonzalez, University of Seville, Seville, Spain

Hoda Shahiri Tabarestani
Department of Food Materials and Process Design Engineering, Faculty of Food Science and Technology, Gorgan University of Agricultural Sciences and Natural Resources, Gorgan, Iran

A.J. Trujillo
Center for Innovation, Research and Knowledge Transfer in Food Technology (CIRTTA), XaRTA, TECNIO, Department of Animal and Food Science, Faculty of Veterinary Medicine, Universitat Autònoma de Barcelona, Bellaterra, Spain

Luis A. Trujillo-Cayado
Chemical Engineering Department, Escuela Politécnica Superior, C/Virgen de África, University of Seville, Seville, Spain

Preface to the series

Enthusiasm for the consumption of healthy and functional food products has dramatically expanded with the growth of industrial life and obesity among people. Therefore, many researches have focused on novel topics, such as nanoencapsulation in preparing healthy cuisine. Nanoencapsulation is a new field of science combining different fields of technology in general and encapsulation in particular. Encapsulation can be defined as the technology of encasing bioactive compounds in solid, liquid, or gaseous states in matrices, which can be released under particular circumstances with a controlled rate. Recently, according to the perception of material properties and their reaction at the nanoscale research, the encapsulation area has moved to the nanoencapsulation field. The fabricated nanocarriers provide better opportunity for interaction, high bioavailability, solubility, and permeation due to their larger surface area. Also, nanoencapsulated ingredients enable targeted release plus high stability against harsh digestive steps, process conditions, and environment stresses.

Selecting the best method for nanoencapsulation of distinct food bioactive ingredients is the main step for designing an efficient delivery system in healthy food and functional products. Considering different techniques applied for fabricating nanoscale carriers, we have classified nanoencapsulation technologies into five groups based on the main mechanism/ingredient in our previous books (Elsevier, 2017). Due to substantial and overwhelming research activities on nanoencapsulation of food bioactive ingredients and nutraceuticals in recent years, it is necessary to work on specialized and in-depth book titles devoted to different groups of nanocarriers. On the other hand, release, bioavailability, characterization, safety, and application of nanoencapsulated ingredients in different food products are some other important topics, which deserve to have some relevant book titles to provide more detailed information and discussions. Therefore, the book series "Nanoencapsulation in the Food Industry" has been defined to address these emerging topics and cover the recent cutting-edge researchers in this field. Seven volumes defined in this series have the titles as follows:

- Vol. 1: Biopolymer Nanostructures for Food Encapsulation Purposes
- Vol. 2: Lipid-Based Nanostructures for Food Encapsulation Purposes
- Vol. 3: Nanoencapsulation of Food Ingredients by Specialized Equipment

- Vol. 4: Characterization of Nanoencapsulated Food Ingredients
- Vol. 5: Release and Bioavailability of Nanoencapsulated Food Ingredients
- Vol. 6: Application of Nano-/Microencapsulated Ingredients in Food Products
- Vol. 7: Safety and Regulatory Issues of Nanoencapsulated Food Ingredients

This book series would be useful for a diverse group of readers including food technologists, food engineers, nanotechnologists, nutritionists, food colloid experts, pharmacists, cosmetic experts, physicists, chemists, microbiologists, biotechnologists, engineers, and those who are interested in novel technologies in the area of food formulations, functional foods, and nutraceutical delivery systems. We hope this book series will stimulate further research in this rapidly growing area and will enable scientists to gain more practical knowledge about different nanocarriers and their properties to solve their particular problems.

Seid Mahdi Jafari

Preface to Vol. 3

The global market for bioactive ingredients is growing very rapidly, as well as manufacturers demand for natural food ingredients to be applied in the formulation of food products. The market for encapsulated bioactive and nutraceutical ingredients for functional foods and dietary supplements is largely driven by consumer interest to promote health, well-being, and prevention of diseases. People are becoming more aware of the type and source of food and beverages they need for a healthy diet. Due to the sensitivity of bioactive ingredients to undesirable process and storage environmental influences such as temperature, light, and oxidation, it is important to select an appropriate delivery system and a processing method to maintain their stability and thus their bioactivity. Also, delivery systems should be able to transport food bioactive ingredients safely and completely throughout the gastrointestinal tract and release them in the right place and right time. Nanotechnology and encapsulation are the two emerging technologies that enable food scientists to realize many innovations in the segment of functional food production. Nanocarriers can be prepared from different food-grade materials with different processing technologies. The production process has to be economical, reproducible, and robust.

Production of bioactive-loaded nanocarriers is more complex than generating microcapsules, and more advanced and sophisticated technologies are needed for the preparation of nanodelivery systems. In general, there are three main approaches for the preparation of nanostructures including top-down strategies, bottom-up strategies, and a combination of them. In top-down strategies, nanocarriers are formed by instrumental/high-energy processes, which is the main topic of this book. In bottom-up strategies, nanomaterials are provided by low-energy/formulation-based methods from initial ingredients like self-assembly of molecules and atoms to nanosize structures, formation of protein-polysaccharide coacervates, desolvation, precipitation, conjugation, layer-by-layer deposition, microemulsification, and templating.

The overall aim of the *Nanoencapsulation of Food Ingredients by Specialized Equipment* is to present conventional and cutting-edge technologies based on sophisticated equipment, which can be applied for the production of various food bioactive-loaded nanocarriers. This book is covering recent and applied researches in all disciplines of bioactive and nutrient delivery. All chapters emphasize original results relating to experimental, formulation, analysis, and/or applications of nanostructures for food encapsulation purposes. After presenting a brief overview of these specialized equipments in Chapter 1, nanofibers made via electrospinning have been covered in Chapter 2. Then, Chapters 3 and 4 have been devoted to nanocarriers prepared by electrospraying and nanospray drying, respectively. Micro-/nanofluidics as an emerging technique for the production of bioactive-loaded nanocarriers have been explained in Chapter 5. Another important group of equipment, that is, high-pressure homogenizers and micro-fluidizers and their application for the preparation of nanocarriers containing nutraceuticals have been discussed in Chapters 6 and 7, respectively. Finally, Chapter 8 deals with nanocarriers made through ultrasonication.

All who are engaged in micro-/nanoencapsulation of food, nutraceutical, pharmaceutical, and cosmetic ingredients worldwide can use this book as either a textbook or a reference, which will give the readers a good and recent knowledge and potentials of high-energy (top-down) approaches, as well as their novel applications in developing bioactive delivery systems. We hope this book will stimulate further research in this rapidly growing area and will enable scientists to get familiar with specialized equipment for the production of nanocarriers.

I really appreciate the great cooperation of all authors of the chapters for taking time from their busy schedules to contribute to this project. Also, it is necessary to express my sincere thanks to all the editorial staff at Elsevier for their help and support throughout the project. Finally, special acknowledgement goes to my family for their understanding and encouragement during the editing of this great project.

Seid Mahdi Jafari

CHAPTER ONE

An overview of specialized equipment for nanoencapsulation of food ingredients

Elham Assadpour[a], Seid Mahdi Jafari[b]
[a]Department of Food Science and Technology, Baharan Institute of Higher Education, Gorgan, Iran
[b]Department of Food Materials and Process Design Engineering, Faculty of Food Science and Technology, Gorgan University of Agricultural Sciences and Natural Resources, Gorgan, Iran

Contents

1 Introduction

The global bioactive ingredient market was worth USD 29.8 billion in 2017 according to Research and Markets (2017), and it is projected to grow with about 6.6% to USD 41.1 billion by 2022. The market for encapsulated bioactive and nutraceutical ingredients for functional foods and dietary supplements is largely driven by consumer interest to promote health, well-being, and prevention of disease. People are becoming more aware of the type and source of food and beverages they need for a healthy diet (Faridi Esfanjani, Assadpour, & Jafari, 2018; Jafari & McClements, 2017). Due to sensitivity of bioactive ingredients to undesirable environmental influences such as temperature, light, and oxidation, it is important to select an

Nanoencapsulation of Food Ingredients by Specialized Equipment
https://doi.org/10.1016/B978-0-12-815671-1.00001-9

appropriate delivery system and a processing method to maintain their stability and thus their bioactivity (Abaee, Mohammadian, & Jafari, 2017; Akhavan, Assadpour, Katouzian, & Jafari, 2018; Mokhtari, Jafari, & Assadpour, 2017). Delivery systems for bioactive food ingredients can be formed from different food-grade materials with different processing technologies (Assadpour & Jafari, 2019b). The production process has to be economical, reproducible, and robust (Garti & McClements, 2012; McClements, 2015). Nanotechnology and encapsulation are the two emerging technologies that enable food scientists to realize many innovations in the segment of functional food production (Faridi Esfanjani & Jafari, 2016; Katouzian, Faridi Esfanjani, Jafari, & Akhavan, 2017; Rostamabadi, Falsafi, & Jafari, 2019).

Production of bioactive-loaded nanocarriers is more complex than generating microcapsules, and more advanced and sophisticated technologies are needed for the preparation of nanodelivery systems (Assadpour, Jafari, & Esfanjani, 2017; Mehrnia, Jafari, Makhmal-Zadeh, & Maghsoudlou, 2017; Raei, Rajabzadeh, Zibaei, Jafari, & Sani, 2015; Rafiee & Jafari, 2018; Rezaei, Fathi, & Jafari, 2019). In general, there are three main approaches for the preparation of nanostructures including top-down strategies, bottom-up strategies, and a combination of them (Katouzian & Jafari, 2016). In top-down strategies, nanocarriers are formed by instrumental/high-energy processes, which are the main topics of this book. In bottom-up strategies, nanomaterials are provided by low-energy/formulation-based methods from initial ingredients like self-assembly of molecules and atoms to nanosize structures, formation of protein-polysaccharide coacervates, desolvation, precipitation, conjugation, layer-by-layer deposition, microemulsification, and templating.

Another evaluation of nanoencapsulation technologies is their classification into five groups based on the main mechanism/ingredient used to make nanocapsules for the food industry (Assadpour & Jafari, 2019a; Jafari, 2017). They include lipid-based techniques, nature-inspired techniques, specialized equipment techniques, biopolymer-based techniques, and other miscellaneous techniques as shown in Fig. 1. There could be some overlaps in this classification; for example, some biopolymers are applied in specialized equipment techniques, but this classification is based on the main mechanism of nanocapsule formation. Selection of an appropriate technology for the production of food bioactive-loaded nanocarriers depends on many factors such as desired release profile and delivery purposes, physicochemical properties of the final product, economic considerations, available equipment, and technical knowledge.

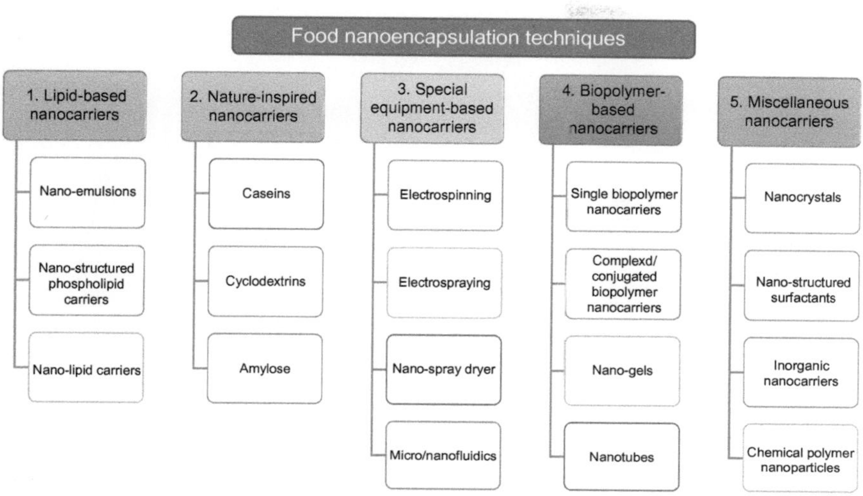

Fig. 1 A systematic classification of different nanoencapsulation technologies applicable to food bioactive ingredients and nutraceuticals. *(Reprinted with permission from Assadpour, E., & Jafari, S.M. (2019). Advances in spray-drying encapsulation of food bioactive ingredients: From microcapsules to nanocapsules. Annual Review of Food Science and Technology 10, 8.1–8.29).*

2 Specialized equipment for the production of food bioactive-loaded nanocarriers

Top-down techniques by applying specialized equipment involve size reduction to nanoscale by electrospinning (Chapter 2), electrospraying (Chapter 3), nanospray drying (Chapter 4), micro-/nanofluidics (Chapter 5), high-pressure homogenization (Chapter 6), microfluidization (Chapter 7), and ultrasonication (Chapter 8), along with some less common technologies such as milling and vortex fluidic system. A brief overview of these methods has been summarized in Fig. 2. In the following sections, a general background to these techniques has been provided, and more technical details and explanations will be presented in the relevant chapters.

2.1 Production of food bioactive-loaded nanofibers by electrospinning

Electrohydrodynamic processing refers to processing of electrically charged fluids, and it allows the production of dry nano- and microstructures by subjecting a polymeric fluid to a high-voltage electric field (Bhushani & Anandharamakrishnan, 2014). Usually, the fluid is pumped at a certain flow rate through a nozzle or conductive capillary (e.g., a stainless steel needle)

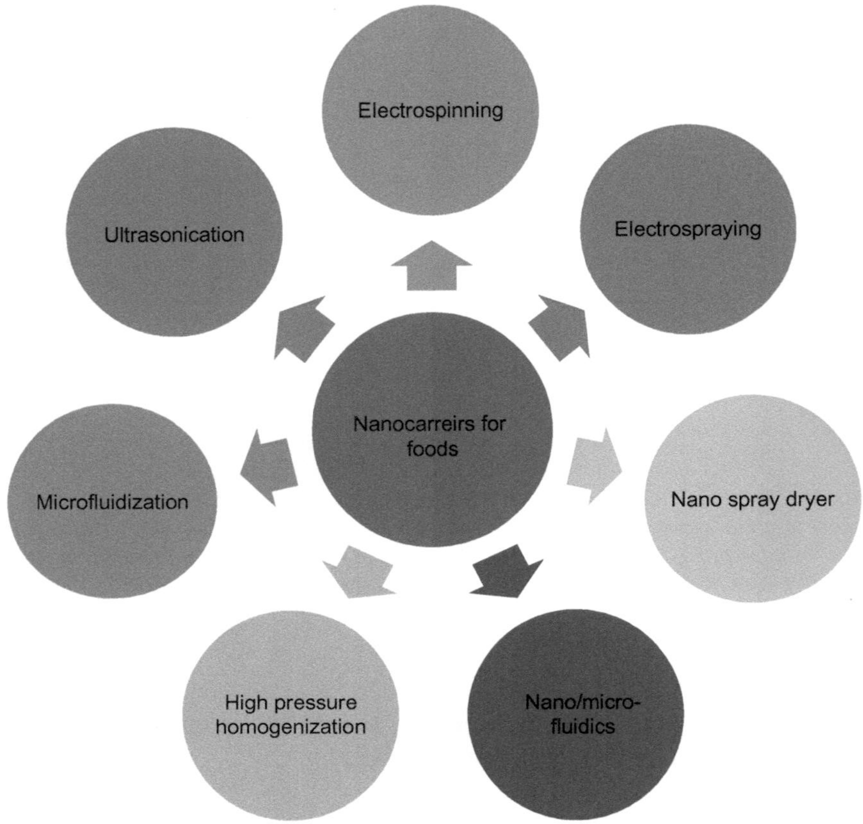

Fig. 2 Different specialized equipment for the production of nanocarriers for the encapsulation of food bioactive ingredients.

to which the voltage is applied. As a result, electrostatic (repulsive) interactions are generated within the liquid that is flowing out from the capillary, and when these electrical forces overcome the forces of surface tension, a charged polymer jet is ejected toward the opposite electrode, which usually consists of a grounded stainless steel collector. Due to the great surface/volume ratio and the repulsive electrical forces, the solvent in this polymeric jet is evaporated during its fly, so dry material is finally deposited on the collector (Bhardwaj & Kundu, 2010; Chakraborty, Liao, Adler, & Leong, 2009).

If the molecular cohesion between the polymeric chains in the polymer fluid being processed is high enough, the generated jet is elongated during its flight due to the balance of forces imposed on it, so that ultrathin fibers are produced upon drying (Kriegel, Arrechi, Kit, McClements, & Weiss, 2008).

In this case, the process is commonly referred to as "electrospinning," as described in Chapter 2. Conversely, if the intermolecular cohesion is sufficiently low, the jet breaks into fine droplets. Due to their surface tension, the jet fragments tend to acquire a spherical shape in the air, and thus they yield micro- or nanoparticles upon solvent evaporation. This phenomenon is generally known as "electrospraying" (Alehosseini, Ghorani, Sarabi-Jamab, & Tucker, 2017), which is discussed in Chapter 3. Therefore, the basic setup and general principles of both the electrospinning and the electrospraying techniques are the same, only differing in the morphology of the obtained nano- or microstructures, which in turn depend on the properties of the polymeric fluid to be processed and the process parameters. Similar setups can also be used to produce hydrogels through the electrostatic extrusion technique, also referred to as the electrosprayed-assisted or electrosprayed-aided extrusion (Zaeim, Sarabi-Jamab, Ghorani, Kadkhodaee, & Tromp, 2017). This technique is a variant of the conventional extrusion technique in which an external electric field is applied to break up the fluid droplets, which are extruded onto a gelling bath, exploiting the same principles of the electrospraying technology. It should be noted that electrostatic extrusion is a wet encapsulation technique not yielding dry encapsulation structures, although some authors classify it as part of electrospraying technologies (Tapia-Hernández, Rodríguez-Félix, & Katouzian, 2017).

A basic ordinary electrospinning setup (either in horizontal or vertical arrangement) consists of four main elements (Maftoonazad & Ramaswamy, 2018): (i) a high-voltage power supply (up to 50 kV) operated in direct current (DC) mode, which is applied on the needle containing the spinning solution and grounded collector by two electrodes; (ii) a syringe pump (flow control pump) for feeding spinning solution; (iii) a syringe with a flat-end metal needle (capillary tube or spinneret) to store a spinning solution; and (iv) an electrically conductive collector (target) presented in different geometries. When the high electric field reaches a critical value, the electrostatic forces surpass the surface tension of the optimum viscous polymer solution and a conical jet projection of fibers emerges (Taylor cone) on the surface of a drop at the tip of the spinneret. Further, in the space between the electrodes, the solvent gets evaporated during the process, rapidly transforming the jet into continuous thin fibers in a whipping motion (non-axisymmetric bending instability) in the collector (Jacobsen, Garcia-Moreno, Mendes, Mateiu, & Chronakis, 2018); thereby, solidification occurs before fibers are collected on the screen in the form of nonwoven mats. As explained in Chapter 3, electrospraying differs from electrospinning by the fact that when the polymer concentration is too low, the jet is destabilized due to varicose instability, forming highly charged droplets, which partially or fully solidify through solvent evaporation or cooling (Alehosseini et al., 2017; Jacobsen et al., 2018). In fact, the formation of electrospun fibers (Chapter 2) or electrosprayed particles is dependent on molecular chain entanglement of the polymer solution, which is generally a function of polymer chain topology and morphology.

Electrospinning is a very versatile procedure and can be generally divided into two strategies: melt electrospinning and solution electrospinning. Indeed, solution electrospinning requires a compatible solvent able to dissolve a given polymer and the bioactives to form a homogenous solution. Conversely, melt electrospinning is a solvent-free method (green technology) in which the polymer melt (composed of the polymer matrix and the bioactives) is heated up to its melting point and then it will be electrospun (Lian & Meng, 2017; Wen, Wen, Zong, Linhardt, & Wu, 2017). During the last few years, materials based on one-dimensional nanostructures, such as nanofibers, nanotubes, and nanowires, have created a subject of substantial interest due to their unique properties (Neo et al., 2014). Among them, electrospun nanofiber delivery systems have gained a wide interest compared with larger diameter fibers made of the same material, and their applications to the agriculture and food industries are relatively recent. Encapsulation for controlled release or immobilization of functional ingredient in nanofibers from biobased renewable materials might provide flexible tools to stabilize and preserve the quality of food or even to design healthier and more effective functional foods due to the special characteristics of the nanofibers (Fernandez et al., 2007).

Nanofibers are continuous filaments with a diameter ranging from several nanometers to 1 μm. These structures can be produced by electrospinning process, using specifically adapted apparatus. They are valued for their ultrahigh specific surface areas (i.e., surface-to-volume or surface-to-mass ratios), high porosity, and their easy use to fabricate different structures. Due to their unique properties, nanofibers have been found potentially useful in the field of food science and technology for nutraceutical delivery; enzyme immobilization; food packaging, reinforcement, and biosensing (Ghosal et al., 2018; Torres-Giner, Pérez-Masiá, & Lagaron, 2016); food-related separation applications such as filtration or clarification and solid-phase extraction; food indicator membranes and sensors; and energy harvesting and storage (Ghorani & Tucker, 2015).

The main factors that govern the morphology and properties of the nanofibers achieved by electrospinning are as follows (Ghorani & Tucker, 2015): (1) polymer properties (solubility, molecular weight, melting point, glass transition temperature, and crystallization velocity); (2) spinning solution properties (viscosity, viscoelasticity, concentration, dielectric constant, surface tension, solvent properties/vapor pressure, and electrical conductivity); (3) processing parameters in the electrospinning setup (applied voltage, needle to collector/nozzle-to-ground distance, collector composition and geometry, feed rate, and needle diameter); and (4) operating conditions (temperature, relative humidity, and atmospheric pressure). Also, different nozzle configurations (single, single with emulsion, side-by-side, or coaxial nozzles) can be used for obtaining various nanofibers, which together with electrospinning parameters will define the fiber features, and even overcome limitations in polymer processability (Neo et al., 2014).

Electrospinning technology has been increasingly developed and utilized for the production of functional food nanomaterials. This technology has been divided into needle, needleless, and core-shell methods (Vyslouzilova et al., 2017), which could be used for bioactive components entrapped into nanofibers. The classical approach for the preparation of nanofibers with incorporated bioactive molecules is blend electrospinning. In addition, the functionality of the nanostructures produced by electrospinning can be achieved through the utilization of coaxial system, emulsion electrospinning, inclusion of other functional molecules, and adsorption of functional components onto the surfaces of the nanostructures.

Electrospinning of most natural biopolymer solutions has proven to be challenging because of their broad molecular weight distribution, biovariations, and high processing costs, limiting solubility of biopolymers in most organic solvents due to their high crystallinity and polarity, polyelectrolyte nature in aqueous media, high tendency to form strong hydrogen bonds (high surface tension), and poor mechanical properties that make processability and handling of end product nanofibers difficult (Neo et al., 2014, 2013). One of the suggested approaches to overcome these technical issues is to blend natural biopolymers with biodegradable chemical polymers to make sufficient chain entanglements and/or other processing compatible aids (plasticizers and surfactants) for an effective electrospinning process with enhanced material properties such as a higher tensile strength and intact micro-/nanostructures (Mendes, Stephansen, & Chronakis, 2017).

Electrospun nanomicrostructures provide a good matrix for the encapsulation/immobilization of bioactive compounds without the loss of their activity or specificity (Jacobsen et al., 2018). It does not involve any severe conditions of temperature, pressure, and chemistry with capability for continuous fabrication, versatility, and facile operating processes (Ghorani & Tucker, 2015). Many bioactive compounds such as phenolics, antimicrobial agents, essential oils, fish oil, and enzymes have been encapsulated into electrospun nanofibers, which will be discussed in Chapter 2.

2.2 Production of food bioactive-loaded nanoparticles by electrospraying

The electrospraying conditions (both the properties of the feed formulation and the process parameters) generally need to be optimized for each polymer system or even for each polymer-bioactive combination if encapsulation is the purpose of the process, since the addition of small amounts of these ingredients and the procedure to disperse them may significantly alter the solution properties of the feed formulation (Gómez-Mascaraque & López-Rubio, 2016). The main properties of the feed formulation (usually a polymeric solution or dispersion), which have an impact on the performance of the hydrodynamic processing and the morphology of the resulting materials, are the rheological characteristics of the fluid, its surface tension, and conductivity. All these properties in turn depend on the type of polymer used, its molecular weight, concentration, the solvent used, and the addition of other substances (including the bioactive compounds to be encapsulated). Another critical factor is the volatility of the solvent used, which determines whether sufficient solvent evaporation can occur during the flight of the polymeric fluid from the tip of the capillary to the collector (Chakraborty et al., 2009). Also, electrospraying is affected by a number of process parameters, which include the applied voltage, the feed flow rate, the capillary diameter, the tip-to-collector distance, and environmental conditions such as the temperature, pressure, and relative humidity.

Compared with other technologies, electrospraying offers a series of advantages, which makes it a promising and versatile alternative for the nano- and microencapsulation of food ingredients. For instance, being a drying technique, it allows the production of dry nano-/microencapsulation structures in a one-step process from the feed formulations (Tapia-Hernández et al., 2015), without the need of a subsequent drying step. On the other hand, the droplet size obtained by atomizing through electrospraying is generally smaller than in conventional mechanical atomizers and in some cases with narrower size distributions (Jaworek & Sobczyk, 2008). Another advantage of electrospraying is the possibility of producing multilayer encapsulation structures in one step, without the need of applying subsequent coatings, by using a coaxial (or even multiaxial) configuration (Zhang et al., 2017; Zhang, Huang, Si, & Xu, 2012). From the operational point of view, electrospraying has also some benefits compared with other drying techniques. Once the process conditions are optimized, its operation is quite easy and cost-effective, and quality checks on the particles can be performed simply by interrupting the process briefly (Chakraborty et al., 2009), which is very convenient and cannot always be done with other encapsulation techniques requiring conditioning steps.

Despite all these advantages, there are still some challenges that need to be faced to apply electrospraying in the food industry. For instance, if the electrosprayed structures are to be used for food applications, the use of water as solvent is almost a must in order to avoid any residual traces of toxic solvents in the final materials (López-Rubio & Lagaron, 2012). But, the physical properties of water are not the most suitable for electrohydrodynamic processing. Also, the encapsulation matrices to be used for food applications must be food grade too, which restricts the range of available materials to edible biopolymers, that is, proteins and polysaccharides. In many cases, this represents an additional challenge, since some biopolymers are difficult to electrospray due to their complex molecular structures. Another challenge for electrospraying is to encapsulate lipophilic food ingredients, as they cannot be readily dissolved in the aqueous biopolymer formulations previously mentioned. One strategy to solve this problem is to prepare oil-in-water emulsions prior to electrospraying. This so-called emulsion electrospraying technique has been successfully applied to encapsulate a number of lipophilic bioactive ingredients.

Intensive research efforts have been made in the last decade to develop new edible electrosprayed materials based on biopolymers. For example, a number of proteins (zein, whey proteins, soy proteins, and gelatin) and polysaccharides (pullulan, maltodextrin, resistant starch, inulin/oligofructose, dextran, and chitosan) and their combinations have been applied for the encapsulation of different food bioactives and nutraceuticals by electrospraying, as described in Chapter 3. Also, Eltayeb, Stride, Edirisinghe, and Harker (2016) recently reported the production of lipid nanoparticles through electrohydrodynamic processing using stearic acid as wall material and ethanol as solvent.

2.3 Production of food bioactive-loaded nanoparticles by nanospray dryer

The recent rise in nanotechnological applications in foods particularly in the area of food functionality like enriched and fortified food products in particulate dosage form (Grumezescu, 2016; Oprea & Grumezescu, 2017) has increased the pressure on existing spray drying systems to produce nanoparticles with a high yield and a narrow size distribution. In 2009, the Swiss company Büchi Labortechnik AG launched the laboratory product Nano Spray Dryer B-90 for producing small quantities of submicron powders in the gram scale with very narrow distributions and high yields (Arpagaus, 2009; Arpagaus, Friess, & Schmid, 2009). A remarkable growth of the research activity in the field of nanospray drying can be recognized in the past 20 years, particularly in the fields of pharmaceuticals, materials technology, and bioactive food ingredients. Obviously, the potential of nanospray drying has not yet been fully exploited. Compared with traditional spray drying in the micrometer scale, the current level of nanospray drying is still in its infancy (Assadpour & Jafari, 2019c), but it is generally expected that more and more products will be developed in the coming years. Nanoscale powders by spray drying represent a new platform for many applications in food technology (Arpagaus, 2009).

The process steps of a nanospray dryer are essentially the same as in a traditional spray dryer, which includes heating of the drying gas, droplet formation by atomization of the fluid supply, drying of the droplets in the drying gas and formation of dry particles, and particle separation and collection of the dry particles from the drying gas. However, compared with a traditional spray dryer, some technological modifications on the experimental setup are necessary to produce and collect nanoscale particles, that is, (1) the nozzle system should be able to produce smaller droplets; (2) the drying gas flow needs to be gentle, laminar, and cocurrent with the sprayed droplets; and (3) the particle collector must be highly efficient in separating submicron particles. The droplet generation is based on vibration mesh technology, which has been adapted from nebulizers for inhalation therapy to the nanospray dryer application (Dhand, 2002; Knoch & Keller, 2005; Lass, Sant, & Knoch, 2006; Smart et al., 2002; Vecellio, 2006). Spray meshes are available with 4.0-, 5.5-, and 7.0-μm holes (first-generation spray head). The second generation of vibrating mesh technology in the Nano Spray Dryer B-90 HP launched in 2017 offers optimized productivity and better handling during installation and replacement. These nebulizers are available in small, medium, and large size. The droplet size depends on the mesh size and the physicochemical properties of the fluid, such as viscosity and surface tension.

The drying gas is heated up to the set inlet temperature in a compact heater at the top of the nanospray dryer. The heating unit consists of a porous metal foam with an embedded electrical heating coil. This coil ensures efficient heat transfer to the metal foam and uniform heat distribution in the entire heating volume (Arpagaus, Collenberg, Rütti, Assadpour, & Jafari, 2018). A highly efficient electrostatic particle collector separates the dried particles from the gas stream. The electrostatic particle collector consists of a stainless steel cylinder (anode = particle collection electrode) and a star-shaped counter electrode (cathode) inside the cylinder. During the nanospray drying process, a high voltage of approximately 15 kV is applied between the electrodes, and the dried particles are electrically charged deflected to the inner wall of the cylinder electrode. The electrostatic particle collector is able to capture submicron particles (<1 μm) at a separation efficiency >99% for small batches of 10 mg to 2.7 g powder (Arpagaus & Meuri, 2010; Arpagaus, Schafroth, & Meuri, 2010). After completion of the nanospray drying process, the fine dried powder particles are gently collected from the inner surface of the collection electrode cylinder using the particle scraper and the particle collection paper included in the scope of delivery of the laboratory instrument. Finally, the particles are filled into airtight glass vials and stored in a controlled and dry atmosphere (e.g., in a desiccator over silica gel at room temperature) until further usage and examination to prevent crystallization and moisture absorption (Schmid, Arpagaus, & Friess, 2011).

For nanospray drying of bioactive food ingredients, there are several process parameters that can be varied to optimize the yield, bioactive loading, encapsulation efficiency, particle size, release profile, stability, and morphology (Arpagaus, 2018a, 2018b, 2018c; Arpagaus et al., 2018; Arpagaus, John, Collenberg, & Rütti, 2017; Arpagaus, Rütti, & Meuri, 2013). The final particle size is directly related to the solid concentration in the feed solution and the droplet size (Maa, Nguyen, Sit, & Hsu, 1998). Submicron particle size is typically reached when using 4.0-μm spray cap and diluted solutions of about 0.1%–1% (*w*/*v*), as demonstrated in many studies. Other parameters to consider when nanospray drying are the inlet and outlet temperatures of the drying gas, the spray rate of the vibrating mesh atomizer, the drying gas type, and the drying gas flow rate and its humidity. The feed-related parameters, such as the feed composition (e.g., core-to-wall material ratio, bioactive content, surfactants, emulsifiers, stabilizers, and polymer glass transition temperature); the viscosity; the solid concentration; and the solvent type (e.g., boiling point, aqueous, organic, or mixture) will mainly affect the final powder properties. The yield of nanospray dried particles can be calculated from the total weight of the recovered particles and the original weight of the bioactive and polymer.

In general, the slow and gentle drying in a nanospray dryer yields spherical and compact particles. Small amounts of surface-active compounds (e.g., polysorbate) in the formulations are typically used to optimize the smoothness and sphericity of the particles, as shown by several researchers (Bürki, Jeon, Arpagaus, & Betz, 2011; Li, Anton, Arpagaus, Belleteix, & Vandamme, 2010; Schmid, 2011; Schmid et al., 2011; Schmid, Arpagaus, & Friess, 2009). Commonly applied analytical methods to characterize the nanospray dried powders are scanning electron microscopy (SEM) to determine particle size and morphology, laser diffraction to measure particle size (dynamic light scattering, DLS), and x-ray diffraction (XRD) or differential scanning calorimetry (DSC) to identify the amorphous/crystalline state.

A number of research studies have demonstrated the ability of nanospray drying to produce submicron powders from polymeric wall materials. The comparison of the morphologies reveals that the primary particles are almost spherical in shape and that the most part is submicron in size (Arpagaus et al., 2018). The wall materials used for the encapsulation of bioactive food ingredients by nanospray drying are mainly biopolymers such as sodium alginate, Arabic gum, whey proteins, chitosan, gelatin, modified starch, and maltodextrin. Also, nanospray drying has been applied for water-soluble vitamins, polyphenols, and plant extracts; nanoemulsions loaded with lipophilic bioactive compounds; nanogels made of egg yolk low-density lipoprotein; solid lipid nanoparticles containing lipophilic bioactives; and different salts as described in Chapter 4.

2.4 Production of food bioactive-loaded nanostructures by micro-/nanofluidics

Application of the nano-/microfluidic technique is very broad, covering research in the field of chemistry, medicine, and biology. Its application in the area of food science is still recent and is focused on the development of systems that promote the protection and controlled release of bioactive compounds. It is still a laboratory-scale technique but with great potential to scale up in the future (Maan, Nazir, Khan, Boom, & Schroën, 2015). The basic principle in nano-/microfluidics is the interfacial tension between the different fluids, which allows the formation of spherical droplets. Moreover, the particle size and the release of the encapsulated bioactive can be controlled very accurately according to the materials, type of device and flows used. Another important factor in this method is the production of highly monodisperse drops and, consequently, nano-/microcapsules that can be manipulated with an impressive degree of accuracy (Datta et al., 2014). The fact that this technique produces monodisperse drops is an important feature for the stability of the system since it is not necessary to use a large amount of emulsifier, besides resulting in a lower rate of creaming, flocculation, and sedimentation (Muijlwijk, Berton-Carabin, & Schroën, 2016).

Nano-/microfluidic devices consist of systems in which a channel must be in the range of micro-/nanometer; the disperse and continuous phases of emulsions flow inside these channels. The devices are set of channels molded or built on a base, which can be made of glass or polydimethylsiloxane (PDMS). Channels are interconnected to allow possible flow of materials in different directions in order to achieve a specific objective. Liquids or gases are injected and removed from the nano-/microfluidic devices using tubing, syringes, and pumps or hydrostatic pressure. Different materials are used in the manufacture of micro- and nanofluidic devices. Jena, Yue, and Lam (2012) used cyclic olefin copolymer (COC) in the manufacture of microfluidic devices in order to study various process parameters to develop a low-cost alternative system. These authors used a hot embossing technique to construct the devices and evaluated the embossing temperature, load, holding time, and demolding temperature and demonstrated a new type of system, with the optimal parameters: embossing temperature for COC of 10°C above its Tg (glass transition temperature), optimal embossing load of 2.94 kN, and holding time of 180s. In another interesting research, Eusner, Hale, and Hardt (2010) studied the use of polymethyl methacrylate (PMMA) in the fabrication of microfluidic devices. These authors also used the hot embossing technology to develop the system and showed that it is a repeatable process and capable of producing quality parts at different operating conditions and for features on the order of 50 μm. Nie, Xu, Seo, Lewis, and Kumacheva (2005) studied the production of polymer particles with various shapes and morphologies in microfluidic devices fabricated in polyurethane (PU) elastomer by using standard soft lithography. These authors reported the production of polymer spherical capsules and particles with nonspherical shapes using silicon oil and monomer in the oil and aqueous phases, respectively. Kawakatsu, Kikuchi, and Nakajima (1997) studied the production of regular-sized cell in microchannel produced with silicon designed and prepared by using semiconductor technology. These authors showed a method for both O/W and W/O emulsion cell production, showing the possibility for creating artificial biological cells.

Glass nano- and microfluidic devices are a class of devices composed of cylindrical capillaries inserted into a square capillary, whose inner dimension is slightly larger than the outer diameter of the cylindrical capillaries (Datta et al., 2014). The construction of these devices depends on the type of emulsion that will be produced: single, double, or multiple emulsions. The technique of nano-/microfluidic is a promising method based on flows of multiple fluids, making the elaboration of simple water-in-oil (W/O) or oil-in-water (O/W) emulsions, water-in-oil-in-water (W/O/W) or oil-in-water-in-oil (O/W/O) double emulsion, or multiple compartments emulsions (W/O/W/O or O/W/O/W) possible. The formation of single, double, or multiple emulsions can occur with different flow types: co-flow or flow focusing. In the co-flow system, all the phases of an emulsion flow in the same direction into the capillary(s) and between the internal capillary and the square; the faster the flow, the easier the internal droplet phase to break in droplets due to the formation of a thin film (Utada, Fernandez-Nieves, Stone, & Weitz, 2007), influencing the particle size of the systems. The system presented in flow-focusing devices differs by direction in which the phases of an emulsion are injected. In the same way as in the co-flow system, the phases of the emulsion, whether the system is single, double, or multiple, are forced through a small orifice (Garstecki, Gañán-Calvo, & Whitesides, 2005). However, in flow-focusing devices, the phases are injected in opposite directions. That is, the dispersed phase flows through the internal capillary, and the continuous phase flows between the inner capillary and the square in the opposite direction to the dispersed phase (in the case of single emulsions).

According to Vladisavljevic et al. (2013), the manufacturing technique used in the production of the PDMS devices is soft lithography, which produced a master with positive relief features. Different types of PDMS devices can also be obtained. In this case, the cylindrical, planar, and three-dimensional channels deserve prominence. The formation of droplets in PDMS nano-/microfluidic devices is influenced by the geometry of the devices: T-, X-, and Y-junction. The simplest structure is T-junction, which the device is in the form of the letter T and the dispersed phase is injected through the orthogonal channel and flows in the same direction of the continuous phase. Moreover, T-junction can also be applied when the continuous phase is injected through the orthogonal channel and the dispersed phase injected into the main channel, the two streams flowing in the same direction (Shui, Mugele, Van den Berg, & Eijkel, 2008; Van der Graaf, Steegmans, Van der Sman, Schroën, & Boom, 2005; Vladisavljevic et al., 2013; Xu, Li, Tan, Wang, & Luo, 2006).

Various materials may be used for the formation of nano- and microstructures for the encapsulation of bioactive compounds by nano-/microfluidic devices, such as lipids of high and low melt point, polysaccharides, and proteins, depending on the main objective and the physicochemical properties of the bioactive compounds, which have been discussed with more details in Chapter 5.

2.5 Production of food bioactive-loaded nanostructures by high-pressure homogenization

Mechanical homogenization refers to the capability of producing a homogeneous size distribution of particles suspended in a liquid by forcing the liquid under the effect of high pressure through a disruption valve. High-pressure homogenization (HPH) or dynamic HPH enables pressures 10–15 times higher than traditional homogenizers and covers the range of 100–400 MPa within which the pressure range of 200–400 MPa has been referred to as ultrahigh pressure homogenization (UHPH). Pressure levels of up to 400 MPa can be reached and maintained by these UHPH systems, although the maximum pressure achieved depends on the engineering design. Moreover, UHPH has been documented to play a key role in the delivery of bioactive compounds (Dumay et al., 2013), as explained in Chapter 6. HPH technology should not be confused with high hydrostatic pressure (Trujillo, Capellas, Saldo, Gervilla, & Guamis, 2002), as HPH is a continuous process, which combines both homogenization and pressure leading to shear forces, turbulence, and cavitation, produced due to the pass of the fluid throughout the HPH valve. There are other alternatives within HPH equipment such as the Microfluidizer processor, which finely controls the level of shearing and offers an ideal solution for the production of liposomes and for achieving ambition particle size reduction targets by using less energy (Henry, Fryer, Frith, & Norton, 2010; Jafari, He, & Bhandari, 2007a; Jafari, He, & Bhandari, 2007b). It has been discussed thoroughly in Chapter 7.

HPH or UHPH are technologies that have been applied in the food, pharmaceutical, and cosmetic areas. Particularly, UHPH technology has been applied to different food systems such as milk and vegetable beverages to improve their physical stability and microbial quality (Donsì, Ferrari, & Maresca, 2009). UHPH has been suggested as an alternative to thermal pasteurization, due to its effect on inactivating pathogenic and other food spoilage microorganisms (Georget et al., 2014). UHPH technology has also been widely used to create stable nanoemulsion systems (Fernandez-Avila & Trujillo, 2016; Hebishy, Buffa, Guamis, Blasco-Moreno, & Trujillo, 2015; Hebishy, Buffa, Juan, Blasco-Moreno, & Trujillo, 2017; Hebishy, Zamora, Buffa, Blasco-Moreno, & Trujillo, 2017). The coarse solution containing shell and bioactive materials is exposed to a high shear force under high pressure in the range of 10–150 MPa for HPH and up to 400 MPa for UHPH. HPH and microfluidization techniques have also been used to improve functionalities of different biopolymers such as gums (Gulrez, Al-Assaf, Fang, Phillips, & Gunning, 2012; Harte & Venegas, 2010; Porto, Augusto, Terekhov, Hamaker, & Cristianini, 2015; Villay et al., 2012) and proteins (Dissanayake & Vasiljevic, 2009; Donsì, Senatore, Huang, & Ferrari, 2010; Hu, Zhao, Sun, Zhao, & Ren, 2011; Shen & Tang, 2012).

Different physical processes have been proposed as the cause of disruption of droplets and cells in UHPH, and each mechanism shows up at a different extent as a result of various factors such as the viscosity of the product and the operating parameters (pressure, temperature, and flow rate): (1) shear stress, which is defined as the viscous forces within particles through a material cross section. It is the only mechanism that contributes to droplet disruption in a laminar flow (Floury, Legrand, & Desrumaux, 2004); however, both laminar and turbulence regimes coexist in this type of homogenizer; (2) turbulence, which is a mechanism caused by the inertial forces generated in the highly turbulent region, located just at the exit of the valve (Donsì et al., 2009); (3) cavitation, defined as a phenomenon when liquid passes from low pressure in which cavities/air bubbles are formed to high-pressure region, when they will be collapsing between each other. Intensities of cavitation strongly increase with homogenizing pressure (Floury et al., 2004) and decrease with higher fluid viscosity (Diels, Callewaert, Wuytack, Masschalck, & Michiels, 2005); (4) impingement, which is the impact of the droplets or cells against the walls of the UHPH valve; and (5) extensional stress, which stretches the droplets/cells, which are disrupted by elongational stresses (Floury et al., 2004). The physical properties of suspensions treated by HPH/UHPH can be characterized by a combination of various analytical techniques, such as visual observation, rheological methods, ultrasound profiling, electroacoustic spectroscopy (zeta potential), surface protein concentration, microscopic analysis, nuclear magnetic resonance, and optical methods (dynamic light scattering [DLS], diffusing wave spectroscopy, and Turbiscan).

HPH or UHPH is a highly recommended technique for the production of nanostructures due to its efficiency to reduce particle size, processing highly concentrated suspensions with a high reproducibility, lack of organic solvents usage, less time consumption, and industrial scalability compared with other methods. HPH can be performed under hot (Silva et al., 2011) or cold (Wise, 2000) conditions. The cold mode is more applied for thermal labile compounds, while lipid nanoparticles are usually obtained by using the hot mode to achieve a more uniform morphology. In particular, HPH represents an easily scalable, reproducible, high throughout, and robust mechanical disruption process for the production of submicrometric lipid-based nanostructures (Donsì, Annunziata, & Ferrari, 2013). In the food industry, HPH at pressures up to 350 MPa is the most common equipment used to produce fine emulsions (Santana, Perrechil, & Cunha, 2013). When used to treat a pre-existing emulsion, if sufficient emulsifier is present to cover the newly created oil-water interface, HPH is effective in reducing the size of the droplets to the nanorange. Increasing the number of passes through the homogenizer could also result in lower droplet size (McClements, 2011). HPH is available for both lab and industrial scale (Setya, Talegaonkar, & Razdan, 2014). The three main mechanisms of the size reduction of droplets into smaller ones are turbulence, shear, and cavitation forces (Schultz, Wagner, Urban, & Ulrich, 2004).

In general, HPH has been successfully applied with lipids, proteins, and carbohydrate-based delivery systems for the nanoencapsulation of different food bioactive ingredients such as phenolics and antioxidants, vitamins, fatty acids, antimicrobials, food colorants, flavors and aromas, and minerals, which have been discussed with more details in Chapter 6.

2.6 Production of food bioactive-loaded nanostructures by microfluidization

Microfluidization, as a top-down methodology, is a state-of-the-art technology among high-energy methods that is used for the production of nanoemulsions, submicron emulsions, nanosuspensions, nanoliposomes, and nanoencapsulated materials and to promote deagglomeration of particle clusters and cell disruption (Jafari et al., 2007a). On the other hand, microfluidization can also be applied in a bottom-up methodology when used as a microfluidic reaction technology to yield crystals or precipitates.

Bench-scale Microfluidizer devices apply high pressures (3.4–275 MPa) with a pneumatic intensification pump, which oscillates between suction and compression strokes, following a trapezoidal profile. In this way, the set point pressure is kept for longer than in conventional high-pressure valve homogenizers, which usually apply pressures according to a loosely triangular profile. Production-scale Microfluidizer devices can use two intensifier pistons to reduce the cycle lag between suction and compression strokes. In addition, the continuous electronic control of pumps allows a nearly constant pressure profile to be delivered, which enhances the reproducibility of industrial batches (Microfluidics, 2018).

In Microfluidizer devices, the inlet fluid is forced to flow into specifically designed microchannels, which are available with different geometries and characteristic diameters. These microchannels are the so-called interaction chambers (IXC). Fluids undergo laminar flow when transported through pipes toward and out of the IXC, while undergoing turbulent flow and cavitation inside the microchannels (Villalobos-Castillejos et al., 2018). At present, two kinds of geometries are commercially available for IXCs: the "Y-type" and the "Z-type" interaction chambers. The high pressure applied and sudden reduction in cross-sectional area when fluids flow from internal transport pipes to the inlet of interaction chambers yield high-flow velocities. This along with the changes of flow direction promotes high shear and extensional deformations as well as turbulent flow and associated eddies. Products flowing through a "Z-type" IXC undergo huge impact forces when hitting the walls of the interaction chamber. On the other hand, collision with microchannel walls is not as important in "Y-type" IXC, so energy losses are minimized. "Y-type" IXCs are well suited to the production of O/W nanoemulsions, O/W submicron emulsions, and liposomes and for polymer encapsulation. On the other hand, the "Z-type" is used for deagglomeration processes, W/O emulsions, and cell disruption. The usual reference to the characteristic IXC size is not sufficient when selecting a suitable one for specific applications, such as the preparation of nanoemulsions. The reproducibility of batches produced with the microfluidization technique is quite good on account of the well-defined geometry of IXCs, uniform pressure profile, and limited amount (microliter order of magnitude) of samples pushed through the IXC after each compression stroke, which, in turn, promotes a higher energy-density transfer rate (Panagiotou & Fisher, 2012). These factors are also responsible for the narrow particle size distributions and low mean diameters, which can be achieved by using this technology.

Microfluidization can be carried out by either single-channel or dual-channel feeding. With regard to emulsification applications, the former requires that a coarse emulsion is previously formed, typically by using a rotor-stator homogenizer or a high-speed mixer. Subsequently, the coarse dispersion is poured into a single feeding hopper and is forced to pass through the Microfluidizer. Therefore, the role of a Microfluidizer is to further reduce the mean particle size to the submicron or nanoscales and to get a narrower particle size distribution (Rayner & Dejmek, 2015). In contrast, in the dual-channel homogenization method, the continuous phase and dispersed phase feed the homogenizer from two different reservoirs, so that the formation of the dispersion is carried out simultaneously, forcing the oil phase and aqueous phase through an air-driven high-pressure Microfluidizer.

One of the most important processing variables for microfluidization is the possibility of recycling products whenever the final quality specifications are not achieved. If the mean particle size is not low enough or the particle size distribution (PSD) is not narrow enough, multiple passes through the interaction chambers are currently made. The optimum performance of the microfluidization technique is to apply a sufficiently high pressure to avoid recycling products, since each pass through an interaction chamber implies an increase in energy consumption and temperature. This may be counterbalanced by cooling the output of the Microfluidizer by means of a shell and tube heat exchanger or a cooling coil placed inside a temperature controlled bath. This raises the point of outlet temperature as one of the processing variables to control when using this technique. Excessive increase in temperature may even alter the performance of emulsifiers and the viscosity ratio (Panagiotou & Fisher, 2013). Decreasing the outlet temperature will slow down collisions among particles when producing emulsions or suspensions, which in turn will decrease the risk of coalescence and flocculation, respectively. A further problem related to recycling operations is the risk of recoalescence when dealing with emulsions (Jafari, Assadpoor, He, & Bhandari, 2008). Another processing variable, which is often overlooked, is the inlet temperature. Low inlet temperatures may compensate for the unavoidable heating effects caused by energy dissipation within the interaction chamber and, in other instances, must be optimized to ensure the right handling properties of emulsifiers.

In general, microfluidization can generate very small droplet sizes and narrow droplet size distributions provided the energy input is optimized and the right emulsifiers are selected. In order to achieve better droplet sizes and physical stability, the optimization of the emulsification process by microfluidization covers two main classes of processing variables. The first one is finding the appropriate design of the Microfluidizer device and the second one is identifying an appropriate combination of processing variables, such as premixing, number of cycles, homogenization pressure, and temperature. These processing variables are not independent and often influence the final nanoemulsion properties. On the other hand, in microfluidization, the time scale inside the interaction chamber is very small. Therefore, to obtain nanoemulsions by using this technique not only is it necessary to use an emulsifier that decreases the interfacial tension sufficiently but also one that is adsorbed quickly at the interface of the fine droplets so that it can protect them against coalescence. It should be noted that low molar mass emulsifiers have a higher rate of adsorption (Danov, Kralchevsky, & Ivanov, 2001). For this reason, surfactants are usually more effective than other emulsifiers at forming nanoemulsions with small droplet sizes using high-energy methods (McClements & Rao, 2011; Yalçınöz & Erçelebi, 2018), and a smaller quantity of these emulsifiers is needed compared with the biopolymers (Öztürk, Argin, Ozilgen, & McClements, 2015).

Microfluidization has been successfully applied for the production of different nanostructures such as nanospheres, polymeric and biopolymeric nanocapsules, solid lipid nanoparticles, nanoliposomes, and nanofibers loaded with various food bioactives and nutraceuticals as discussed in Chapter 7.

2.7 Production of food bioactive-loaded nanostructures by ultrasonication

Ultrasound is a series of sound waves (also known as acoustic waves) with frequencies above the human hearing range (>16 kHz), which can be classified into different frequency and intensity ranges (Jafari et al., 2007b). Firstly, low-intensity (high-frequency) ultrasound passing through a medium without causing any destructive physicochemical changes. In spite of operating on high frequencies ranging at 1–10 MHz, its power intensity is too low (<1 W/cm^2) to alter the material. Secondly, high-intensity (low-frequency) ultrasound is characterized by high-power intensities of 10–1000 W/cm^2 and frequencies of 16–100 kHz (Gharibzahedi & Jafari, 2018). Therefore, low-intensity ultrasound is commonly employed for detection/diagnosis purposes in the medium (such as sonography), while high-intensity ultrasound is being increasingly used for such material-altering purposes as depolymerizing the macromolecules, downsizing particles to nanoscale, making emulsions (homogenization), deflocculating droplets, and so on, especially in laboratories and food industries (Bermúdez-Aguirre, Mobbs, & Barbosa-Cánovas, 2011; Lorimer & Mason, 1987).

Ultrasound technique is commonly used to form different types of nanostructures suitable for nanoencapsulation, which represents an outstanding example of the "top-down" approach and a potential trigger of the "bottom-up" approach. Ultrasonically generated cavitations are principally responsible for many physicochemical effects of ultrasound waves particularly size reduction of materials (Patil & Pandit, 2007). The role of cavitation is not only limited to synthesis but also expanded to control morphological features of nanomaterials and processing techniques such as nanoencapsulation, coating, and nanocomposites (Abbasi, Samadi, Jafari, Ramezanpour, & Shams Shargh, 2019). Among existing technologies, ultrasonication owing to intrinsic advantages including safety, being environmentally friendly, straightforward operation, energy efficiency, and simple implementation, as well as scale-up potential, has been receiving a great deal of attention for >100 years (Falsafi, Maghsoudlou, Aalami, Jafari, & Raeisi, 2019). However, the interest in the employment of this technique for nanomaterial production was born only recently (Manickam & Rana, 2011; Soria & Villamiel, 2010).

Although acoustic cavitations play a key role in the production of nanomaterials, efforts are being made to find out the driving forces responsible for nanoparticle generation when a solution containing reactants is irradiated with ultrasound. The sonochemical reactions can occur in three different regions of a media as discussed in Chapter 8. The first place is the central environment of the bubble where the temperature and pressure are extremely high because of implosive collapse. The second region is interfacial zone where the conditions are lower than that of the core of the bubble and enough to induce a chemical reaction. The third region is the bulk liquid including ambient conditions. One possible explanation of nanostructures formed by sonication is the fast kinetics associated with the rapid intense collapse conditions of the bubble, which impedes the growth of the nuclei and subsequently structural organization or crystallization (Gharibzahedi & Jafari, 2018). This may yield to the nanostructured products. The reaction can as well happen in the interfacial region when the precursor is nonvolatile and unable to enter into the bubble. Nanoamorphous or nanocrystalline products are formed depending on the temperature of this region. Formation of nanostructures is even feasible in the third region, albeit slowly (Manickam & Rana, 2011).

A variety of specialized apparatuses with different setups are commercially available for sonochemical reactions. Examples include ultrasonic cleaning baths, direct-immersion ultrasonic horns, and flow reactors for continuous processes. A typical ultrasonic equipment mainly consists of (i) the electrical generator producing electrical current with a specified power rating, (ii) a single/multiple piezoelectric transducer converting electrical energy into sound energy by vibrating mechanically at ultrasonic frequencies, and (iii) an emitter in shape of titanium horns or baths to convey the ultrasonic wave from the transducer into the medium (Bermúdez-Aguirre et al., 2011). Ultrasonic probe, which is also referred to ultrasound horn, and ultrasonic cleaning bath are extensively employed in research laboratories. Flow reactors can produce high ultrasonic intensities for large-scale and industrial irradiations (Suslick, 1995).

Several types of nanodelivery systems consisted of lipid and surfactant molecules such as nanoliposomes, nanoemulsions, nanostructured lipid carriers, niosomes, cubosomes, and hexosomes that are mostly employed for nanoencapsulation have been successfully produced by ultrasonication (Rafiee, Nejatian, Daeihamed, & Jafari, 2019; Rezaei et al., 2019; Rostamabadi et al., 2019). Biopolymeric and polymeric nanostructures have also been developed by sonication and used as nanosized carrier-based delivery systems for various food bioactive ingredients, as explained in Chapter 8. Nanocarriers can be produced from biopolymers (such as proteins and polysaccharides) and synthetic polymers (such as polylactic acid, polycaprolactone, polyacrylonitrile, $mPEG_{5000}$-b-p, and paclitaxel) using a simple and safe sonication technique. The size, shape, charge, and internal organization of structures and their corresponding nanoparticles depend considerably on the method and materials used to fabricate them, thereby enabling them to be tailored for intended applications (Van der Sman, 2012).

The major challenges of ultrasound application include issues of scale-up and energy efficiency (Gharibzahedi & Jafari, 2018). Although laboratory-scale ultrasonic devices are readily commercially available for scientific investigations, larger-scale equipment for industrial applications still remains relatively uncommon. The utilization of pilot-scale sonicators and the optimization of process variables can assist manufactures in designing and scaling up for industrial facilities. Continuous-flow high-power processor is one of the recent advancements in the ultrasound technologies to produce nanosized structures that have a great potential for commercial applications. A number of technological challenges still need to be examined for the optimized fabrication of nanostructure-based delivery systems using sonication. In the meanwhile, health and safety factors should be prioritized before their widespread application. While encapsulation of bioactive compounds in different nanostructures manufactured by sonication has been studied, there is not enough information on the toxicity, oxidative stability, and bioavailability rate of incorporated ingredients.

3 Conclusion and further remarks

High-energy approaches by specialized equipment (top-down technologies) are promising methods to produce bioactive-loaded nanostructures currently used in the food industry, especially for the protection of lipophilic ingredients. However, although these processes are quite simple to operate once their conditions have been optimized, the complexity of the physics involved in each process results in the need of individually optimizing the formulations and process parameters for every ingredient-encapsulation matrix combination of interest. Most of the encapsulation systems have been developed using one single type of biopolymer as wall material, but recent works have already started exploiting biopolymer blends, since they represent an opportunity for modulating the performance of the nano- and microstructures by adjusting the ratio between both components. Hence, future research is expected to exploit this strategy and provides a new range of edible nanocarriers based on biopolymeric blends. Another important aspect that should be addressed in future works is the performance of bioactive-loaded nanocarriers in real food products. To date, the stability, release, and bioactive properties of the encapsulated food ingredients have been mainly assessed in vitro, using buffers or solutions as food simulants and simplified processing or storage conditions in the absence of food matrices. However, food products are complex systems, and their different components may potentially interact with the encapsulation structures, affecting their performance. Therefore, the real benefits of encapsulating food ingredients should also be investigated after incorporating them within real food products, and the impact that these encapsulation structures may have on the organoleptic properties and other qualitative parameters of the final product should also be assessed. Lastly, efforts should be made to scale up the production of encapsulation structures through these specialized equipment such as electrospinning/spraying, nanospray drying, ultrasonication, and high-pressure homogenization in order to have commercially available technologies for bulk production of functional food products.

References

Abaee, A., Mohammadian, M., & Jafari, S. M. (2017). Whey and soy protein-based hydrogels and nano-hydrogels as bioactive delivery systems. *Trends in Food Science & Technology*, *70*, 69–81. Supplement C.

Abbasi, F., Samadi, F., Jafari, S. M., Ramezanpour, S., & Shams Shargh, M. (2019). Ultrasound-assisted preparation of flaxseed oil nanoemulsions coated with alginate-whey protein for targeted delivery of omega-3 fatty acids into the lower sections of gastrointestinal tract to enrich broiler meat. *Ultrasonics Sonochemistry*, *50*, 208–217.

Akhavan, S., Assadpour, E., Katouzian, I., & Jafari, S. M. (2018). Lipid nano scale cargos for the protection and delivery of food bioactive ingredients and nutraceuticals. *Trends in Food Science & Technology*, *74*, 132–146.

Alehosseini, A., Ghorani, B., Sarabi-Jamab, M., & Tucker, N. (2017). Principles of electrospraying: A new approach in protection ofbioactive compounds in foods. *Critical Reviews in Food Science and Nutrition*, 1–18.

Arpagaus, C. (2009). Food technology at the nano scale: Advances in spray drying. *Food Engineering Ingredients*, *34*, 26–27.

Arpagaus, C. (2018a). Nano spray drying of pharmaceuticals. In *IDS'2018—21st international drying symposium, Valencia, Spain, 11–14 September* (pp. 1–8).

Arpagaus, C. (2018b). Nano spray drying of pharmaceuticals. *American Journal of Nanotechnology & Nanomedicine*, *1*, 64–68.

Arpagaus, C. (2018c). A short review on nano spray drying of pharmaceuticals. *Nanomedicine and Nanoscience Research*, 1–5.

Arpagaus, C., Collenberg, A., Rütti, D., Assadpour, E., & Jafari, S. M. (2018). Nano spray drying for encapsulation of pharmaceuticals. *International Journal of Pharmaceutics*, *546*(1–2), 194–214.

Arpagaus, C., Friess, W., & Schmid, K. (2009). Nano spray drying in laboratory scale. *G.I.T Laboratory Journal*, *11–12*, 30–31.

Arpagaus, C., John, P., Collenberg, A., & Rütti, D. (2017). Chapter 10: Nanocapsules formation by nano spray drying. In S. M. Jafari (Ed.), *Nanoencapsulation technologies for the food and nutraceutical industries* (pp. 346–401). Elsevier Inc.

Arpagaus, C., & Meuri, M. (2010). Laboratory scale spray drying of inhalable particles: A review. *Respiratory Drug Delivery*, 469–476.

Arpagaus, C., Rütti, D., & Meuri, M. (2013). Chapter 18: Enhanced solubility of poorly soluble drugs via spray drying. In D. Douroumis & A. Fahr (Eds.), *Drug delivery strategies for poorly water-soluble drugs* (pp. 551–585). John Wiley & Sons, Ltd.

Arpagaus, C., Schafroth, N., Meuri, M. (2010). Laboratory scale spray drying of lactose: A review, best@buchi information bulletin, Number 57/2010.

Assadpour, E., & Jafari, S. M. (2019a). Chapter 3—Nanoencapsulation: Techniques and developments for food applications. In A. L. Rubio, M. J. F. Rovira, M. M. Sanz, & L. G. Gómez-Mascaraque (Eds.), *Nanomaterials for food applications* (pp. 35–61). Elsevier.

Assadpour, E., & Jafari, S. M. (2019b). A systematic review on nanoencapsulation of food bioactive ingredients and nutraceuticals by various nanocarriers. *Critical Reviews in Food Science and Nutrition*, 1–47.

Assadpour, E., & Jafari, S. M. (2019c). Advances in spray-drying encapsulation of food bioactive ingredients: From microcapsules to nanocapsules. *Annual Review of Food Science and Technology*, *10*(1), 103–131.

Assadpour, E., Jafari, S. M., & Esfanjani, A. F. (2017). Protection of phenolic compounds within nanocarriers. *CAB Reviews*, *12*(057), 1–8.
Bermúdez-Aguirre, D., Mobbs, T., & Barbosa-Cánovas, G. V. (2011). Ultrasound applications in food processing. In *Ultrasound technologies for food and bioprocessing* (pp. 65–105). New York, NY: Springer.
Bhardwaj, N., & Kundu, S. C. (2010). Electrospinning: A fascinating fiber fabrication technique. *Biotechnology Advances*, *28*(3), 325–347.
Bhushani, J. A., & Anandharamakrishnan, C. (2014). Electrospinning and electrospraying techniques: Potential food based applications. *Trends in Food Science & Technology*, *38*(1), 21–33.
Bürki, K., Jeon, I., Arpagaus, C., & Betz, G. (2011). New insights into respirable protein powder preparation using a nano spray dryer. *International Journal of Pharmaceutics*, *408*, 248–256.
Chakraborty, S., Liao, I. C., Adler, A., & Leong, K. W. (2009). Electrohydrodynamics: A facile technique to fabricate drug delivery systems. *Advanced Drug Delivery Reviews*, *61* (12), 1043–1054.
Danov, K. D., Kralchevsky, P. A., & Ivanov, I. B. (2001). Dynamic processes in surfactant-stabilized emulsions. In *Encyclopedic handbook of emulsion technology* (pp. 621–659).
Datta, S. S., Abbaspourrad, A., Amstad, E., Fan, J., Kim, S. -H., Romanowsky, M., et al. (2014). 25th anniversary article: Double emulsion templated solid microcapsules: Mechanics and controlled release. *Advanced Materials*, 1–14.
Dhand, R. (2002). Nebulizers that use a vibrating mesh or plate with multiple apertures to generate aerosol. *Respiratory Care*, *47*, 406–416.
Diels, A. M., Callewaert, L., Wuytack, E. Y., Masschalck, B., & Michiels, C. W. (2005). Inactivation of Escherichia coli by high-pressure homogenisation is influenced by fluid viscosity but not by water activity and product composition. *International Journal of Food Microbiology*, *101*(3), 281–291.
Dissanayake, M., & Vasiljevic, T. (2009). Functional properties of whey proteins affected by heat treatment and hydrodynamic high-pressure shearing. *Journal of Dairy Science*, *92*(4), 1387–1397.
Donsì, F., Annunziata, M., & Ferrari, G. (2013). Microbial inactivation by high pressure homogenization: Effect of the disruption valve geometry. *Journal of Food Engineering*, *115*(3), 362–370.
Donsì, F., Ferrari, G., & Maresca, P. (2009). High-pressure homogenization for food sanitization. In *Global issues in food science and technology* (pp. 309–352). .
Donsì, F., Senatore, B., Huang, Q., & Ferrari, G. (2010). Development of novel pea protein-based nanoemulsions for delivery of nutraceuticals. *Journal of Agricultural and Food Chemistry*, *58*(19), 10653–10660.
Dumay, E., Chevalier-Lucia, D., Picart-Palmade, L., Benzaria, A., Gràcia-Julià, A., & Blayo, C. (2013). Technological aspects and potential applications of (ultra) high-pressure homogenisation. *Trends in Food Science & Technology*, *31*(1), 13–26.
Eltayeb, M., Stride, E., Edirisinghe, M., & Harker, A. (2016). Electrosprayed nanoparticle delivery system for controlled release. *Materials Science and Engineering: C*, *66*, 138–146.
Eusner, T., Hale, M., & Hardt, D. E. (2010). Process robustness of hot embossing microfluidic devices. *Journal of Manufacturing Science and Engineering*, *132*. 030920–1–030920–8.
Falsafi, S. R., Maghsoudlou, Y., Aalami, M., Jafari, S. M., & Raeisi, M. (2019). Physicochemical and morphological properties of resistant starch type 4 prepared under ultrasound and conventional conditions and their in-vitro and in-vivo digestibilities. *Ultrasonics Sonochemistry*, *53*, 110–119.
Faridi Esfanjani, A., Assadpour, E., & Jafari, S. M. (2018). Improving the bioavailability of phenolic compounds by loading them within lipid-based nanocarriers. *Trends in Food Science & Technology*, *76*, 56–66.

Faridi Esfanjani, A., & Jafari, S. M. (2016). Biopolymer nano-particles and natural nano-carriers for nano-encapsulation of phenolic compounds. *Colloids and Surfaces B: Biointerfaces*, *146*, 532–543.

Fernandez, A., National, S., Torres-giner, S., National, S., Lagaron, J. M., & National, S. (2007). Encapsulation of the functional ingredient beta-carotene in electrospun PLA fibers of interest in bioactive food packaging and food processing applications. In *Conference proceeding: 23rd IAPRI symposium of packaging, Windsor, UK*.

Fernandez-Avila, C., & Trujillo, A. J. (2016). Ultra-high pressure homogenization improves oxidative stability and interfacial properties of soy protein isolate stabilized emulsions. *Food Chemistry*, *209*, 104–113.

Floury, J., Legrand, J., & Desrumaux, A. (2004). Analysis of a new type of high pressure homogeniser. Part B. study of droplet break-up and recoalescence phenomena. *Chemical Engineering Science*, *59*(6), 1285–1294.

Garstecki, P., Gañán-Calvo, A., & Whitesides, G. (2005). Formation of bubbles and droplets in microfluidic systems. *Bulletin of the Polish Academy of Sciences Technical Sciences*, *53*, 361–372.

Garti, N., & McClements, D. J. (2012). *Encapsulution technologies food ingredients and nutraceuticals*. Cambridge, UK: Woodhead Publishing.

Georget, E., Miller, B., Aganovic, K., Callanan, M., Heinz, V., & Mathys, A. (2014). Bacterial spore inactivation by ultra-high pressure homogenization. *Innovative Food Science & Emerging Technologies*, *26*, 116–123.

Gharibzahedi, S. M. T., & Jafari, S. M. (2018). Chapter 9—Fabrication of nanoemulsions by ultrasonication. In *Nanoemulsions* (pp. 233–285). Academic Press.

Ghorani, B., & Tucker, N. (2015). Fundamentals of electrospinning as a novel delivery vehicle for bioactive compounds in food nanotechnology. *Food Hydrocolloids*, *51*, 227–240.

Ghosal, K., Chandra, A., Praveen, G., Snigdha, S., Roy, S., Agatemor, C., et al. (2018). Electrospinning over solvent casting: Tuning of mechanical properties of membranes. *Scientific Reports*, *8*(1), 5058.

Gómez-Mascaraque, L. G., & López-Rubio, A. (2016). Protein-based emulsion electrosprayed micro- and submicroparticles for the encapsulation and stabilization of thermosensitive hydrophobic bioactives. *Journal of Colloid and Interface Science*, *465*, 259–270. Supplement C.

Grumezescu, A. M. (2016). *Encapsulations—Nanotechnology in the agri-food industry. Vol. 2*. Amsterdam: Elsevier.

Gulrez, S. K. H., Al-Assaf, S., Fang, Y., Phillips, G. O., & Gunning, A. P. (2012). Revisiting the conformation of xanthan and the effect of industrially relevant treatments. *Carbohydrate Polymers*, *90*(3), 1235–1243.

Harte, F., & Venegas, R. (2010). A model for viscosity reduction in polysaccharides subjected to high-pressure homogenization. *Journal of Texture Studies*, *41*(1), 49–61.

Hebishy, E., Buffa, M., Guamis, B., Blasco-Moreno, A., & Trujillo, A. J. (2015). Physical and oxidative stability of whey protein oil-in-water emulsions produced by conventional and ultra high-pressure homogenization: Effects of pressure and protein concentration on emulsion characteristics. *Innovative Food Science and Emerging Technologies*, *32*, 79–90.

Hebishy, E., Buffa, M., Juan, B., Blasco-Moreno, A., & Trujillo, A. J. (2017). Ultra high-pressure homogenized emulsions stabilized by sodium caseinate: Effects of protein concentration and pressure on emulsions structure and stability. *LWT - Food Science and Technology*, *76*, 57–66.

Hebishy, E., Zamora, A., Buffa, M., Blasco-Moreno, A., & Trujillo, A. J. (2017). Characterization of whey protein oil-in-water emulsions with different oil concentrations stabilized by ultra-high pressure homogenization. *Processes*, *5*(6), 1–19.

Henry, J. V. L., Fryer, P. J., Frith, W. J., & Norton, I. T. (2010). The influence of phospholipids and food proteins on the size and stability of model sub-micron emulsions. *Food Hydrocolloids*, *24*, 66–71.

Hu, X., Zhao, M., Sun, W., Zhao, G., & Ren, J. (2011). Effects of microfluidization treatment and transglutaminase cross-linking on physicochemical, functional, and conformational properties of peanut protein isolate. *Journal of Agricultural and Food Chemistry*, *59*(16), 8886–8894.

Jacobsen, C., Garcia-Moreno, P. J., Mendes, A. C., Mateiu, R. V., & Chronakis, I. S. (2018). Use of electrohydrodynamic processing for encapsulation of sensitive bioactive compounds and applications in food. *Annual Review of Food Science and Technology*, *9*, 525–549.

Jafari, S. M. (2017). Chapter 1—An introduction to nanoencapsulation techniques for the food bioactive ingredients. In *Nanoencapsulation of food bioactive ingredients* (pp. 1–62). Academic Press.

Jafari, S. M., Assadpoor, E., He, Y., & Bhandari, B. (2008). Re-coalescence of emulsion droplets during high-energy emulsification. *Food Hydrocolloids*, *22*(7), 1191–1202.

Jafari, S. M., He, Y., & Bhandari, B. (2007a). Optimization of nano-emulsions production by microfluidization. *European Food Research and Technology*, *225*(5–6), 733–741.

Jafari, S. M., He, Y., & Bhandari, B. (2007b). Production of sub-micron emulsions by ultrasound and microfluidization techniques. *Journal of Food Engineering*, *82*(4), 478–488.

Jafari, S. M., & McClements, D. J. (2017). Nanotechnology approaches for increasing nutrient bioavailability. In *Advances in food and nutrition research*. Academic Press.

Jaworek, A., & Sobczyk, A. T. (2008). Electrospraying route to nanotechnology: An overview. *Journal of Electrostatics*, *66*(3–4), 197–219.

Jena, R. K., Yue, C. Y., & Lam, Y. C. (2012). Micro fabrication of cyclic olefin copolymer (COC) based microfluidic devices. *Microsystem Technologies*, *18*, 159–166.

Katouzian, I., Faridi Esfanjani, A., Jafari, S. M., & Akhavan, S. (2017). Formulation and application of a new generation of lipid nano-carriers for the food bioactive ingredients. *Trends in Food Science & Technology*, *68*, 14–25. Suppl C.

Katouzian, I., & Jafari, S. M. (2016). Nano-encapsulation as a promising approach for targeted delivery and controlled release of vitamins. *Trends in Food Science & Technology*, *53*, 34–48.

Kawakatsu, T., Kikuchi, Y., & Nakajima, M. (1997). Regular-sized cell creation in microchannel emulsification by visual microprocessing method. *Journal of the American Oil Chemists' Society*, *74*(3), 317–321.

Knoch, M., & Keller, M. (2005). The customised electronic nebuliser: A new category of liquid aerosol drug delivery systems. *Expert Opinion on Drug Delivery*, *2*, 377–390.

Kriegel, C., Arrechi, A., Kit, K., McClements, D., & Weiss, J. (2008). Fabrication, functionalization, and application of electrospun biopolymer nanofibers. *Critical Reviews in Food Science and Nutrition*, *48*(8), 775–797.

Lass, J. S., Sant, A., & Knoch, M. (2006). New advances in aerosolised drug delivery: Vibrating membrane nebuliser technology. *Expert Opinion on Drug Delivery*, *3*, 693–702.

Li, X., Anton, N., Arpagaus, C., Belleteix, F., & Vandamme, T. F. (2010). Nanoparticles by spray drying using innovative new technology: The Büchi nano spray dryer B-90. *Journal of Controlled Release*, *147*, 304–310.

Lian, H., & Meng, Z. (2017). Melt electrospinning vs. solution electrospinning: A comparative study of drug-loaded poly (epsilon-caprolactone) fibres. *Materials Science & Engineering. C, Materials for Biological Applications*, *74*, 117–123.

López-Rubio, A., & Lagaron, J. M. (2012). Whey protein capsules obtained through electrospraying for the encapsulation of bioactives. *Innovative Food Science & Emerging Technologies*, *13*, 200–206.

Lorimer, J. P., & Mason, T. J. (1987). Sonochemistry. Part 1—The physical aspects. *Chemical Society Reviews*, *16*, 239–274.
Maa, Y. F., Nguyen, P. a., Sit, K., & Hsu, C. C. (1998). Spray-drying performance of a bench-top spray dryer for protein aerosol powder preparation. *Biotechnology and Bioengineering*, *60*, 301–309.
Maan, A. A., Nazir, A., Khan, M. K. I., Boom, R., & Schroën, K. (2015). Microfluidic emulsification in food processing. *Journal of Food Engineering*, *147*, 1–7.
Maftoonazad, N., & Ramaswamy, H. (2018). Novel techniques in food processing: Bionanocomposites. *Current Opinion in Food Science*, *23*, 49–56.
Manickam, S., & Rana, R. K. (2011). Production of nanomaterials using ultrasonic cavitation—A simple, energy efficient and technological approach. In *Ultrasound technologies for food and bioprocessing* (pp. 405–444). New York, NY: Springer.
McClements, D. J. (2011). Edible nanoemulsions: Fabrication, properties, and functional performance. *Soft Matter*, 7(2), 297–316.
McClements, D. J. (2015). *Nanoparticle- and microparticle-based delivery systems: Encapsulation, protection and release of active compounds*. Boca Raton, FL: CRC Press.
McClements, D. J., & Rao, J. (2011). Food-grade nanoemulsions: Formulation, fabrication, properties, performance, biological fate, and potential toxicity. *Critical Reviews in Food Science and Nutrition*, *51*(4), 285–330.
Mehrnia, M. -A., Jafari, S. -M., Makhmal-Zadeh, B. S., & Maghsoudlou, Y. (2017). Rheological and release properties of double nano-emulsions containing crocin prepared with Angum gum, Arabic gum and whey protein. *Food Hydrocolloids*, *66*, 259–267.
Mendes, A. C., Stephansen, K., & Chronakis, I. S. (2017). Electrospinning of food proteins and polysaccharides. *Food Hydrocolloids*, *68*, 53–68.
Microfluidics. (2018). *Microfluidizer processor user guide*. Newton, MA, USA: Microfluidics.
Mokhtari, S., Jafari, S. M., & Assadpour, E. (2017). Development of a nutraceutical nano-delivery system through emulsification/internal gelation of alginate. *Food Chemistry*, *229*, 286–295.
Muijlwijk, K., Berton-Carabin, C., & Schroën, K. (2016). Cross-flow microfluidic emulsification from a food perspective. *Trends in Food Science & Technology*, *49*, 51–63.
Neo, Y. P., Perera, C. O., Nieuwoudt, M. K., Zujovic, Z., Jin, J., Ray, S., et al. (2014). Influence of heat curing on structure and physicochemical properties of phenolic acid loaded proteinaceous electrospun fibers. *Journal of Agricultural and Food Chemistry*, *62*(22), 5163–5172.
Neo, Y. P., Ray, S., Jin, J., Gizdavic-Nikolaidis, M., Nieuwoudt, M. K., Liu, D., et al. (2013). Encapsulation of food grade antioxidant in natural biopolymer by electrospinning technique: A physicochemical study based on zein-gallic acid system. *Food Chemistry*, *136*(2), 1013–1021.
Nie, Z., Xu, S., Seo, M., Lewis, P. C., & Kumacheva, E. (2005). Polymer particles with various shapes and morphologies produced in continuous microfluidic reactors. *Journal of the American Chemical Society*, *127*, 8058–8063.
Oprea, A. E., & Grumezescu, A. M. (2017). *Nanotechnology applications in food—Flavor, stability, nutrition and safety*. Elsevier.
Öztürk, B., Argin, S., Ozilgen, M., & McClements, D. J. (2015). Formation and stabilization of nanoemulsion-based vitamin E delivery systems using natural biopolymers: Whey protein isolate and gum arabic. *Food Chemistry*, *188*, 256–263.
Panagiotou, T., & Fisher, R. (2012). Improving product quality with entrapped stable emulsions: From theory to industrial application. *Challenges*, *3*, 84–113.
Panagiotou, T., & Fisher, R. (2013). Producing micron- and nano-size formulations for functional foods applications. *Functional Foods in Health and Disease*, *3*(7), 274–289.
Patil, M. N., & Pandit, A. B. (2007). Cavitation—A novel technique for making stable nano-suspensions. *Ultrasonics Sonochemistry*, *14*(5), 519–530.

Porto, B. C., Augusto, P. E. D., Terekhov, A., Hamaker, B. R., & Cristianini, M. (2015). Effect of dynamic high pressure on technological properties of cashew tree gum (Anacardium occidentale L.). *Carbohydrate Polymers*, *129*, 187–193.

Raei, M., Rajabzadeh, G., Zibaei, S., Jafari, S. M., & Sani, A. M. (2015). Nanoencapsulation of isolated lactoferrin from camel milk by calcium alginate and evaluation of its release. *International Journal of Biological Macromolecules*, *79*, 669–673.

Rafiee, Z., & Jafari, S. M. (2018). Application of lipid nanocarriers for the food industry. In J. -M. Mérillon & K. G. Ramawat (Eds.), *Bioactive molecules in food* (pp. 1–43). Cham: Springer International Publishing.

Rafiee, Z., Nejatian, M., Daeihamed, M., & Jafari, S. M. (2019). Application of different nanocarriers for encapsulation of curcumin. *Critical Reviews in Food Science and Nutrition*, 1–77.

Rayner, M., & Dejmek, P. (2015). Engineering aspects of food emulsification and homogenization. In *CRC Press*. Boca Raton, FL, USA.

Research and Markets (2017). Global bioactive ingredients market—Forecasts from 2017 to 2022.

Rezaei, A., Fathi, M., & Jafari, S. M. (2019). Nanoencapsulation of hydrophobic and low-soluble food bioactive compounds within different nanocarriers. *Food Hydrocolloids*, *88*, 146–162.

Rostamabadi, H., Falsafi, S. R., & Jafari, S. M. (2019). Nanoencapsulation of carotenoids within lipid-based nanocarriers. *Journal of Controlled Release*, *298*, 38–67.

Santana, R. C., Perrechil, F. A., & Cunha, R. L. (2013). High- and low-energy emulsifications for food applications: A focus on process parameters. *Food Engineering Reviews*, *5*, 107–122.

Schmid, K., 2011. Spray drying of protein precipitates and evaluation of the nano spray dryer B-90. PhD Thesis, Ludwig-Maximilians-University, Munich.

Schmid, K., Arpagaus, C., & Friess, W. (2009). Evaluation of a vibrating mesh spray dryer for preparation of submicron particles. *Respiratory Drug Delivery*, 323–326.

Schmid, K., Arpagaus, C., & Friess, W. (2011). Evaluation of the nano spray dryer B-90 for pharmaceutical applications. *Pharmaceutical Development and Technology*, *16*, 287–294.

Schultz, S., Wagner, G., Urban, K., & Ulrich, J. (2004). High-pressure homogenization as a process for emulsion formation. *Chemical Engineering and Technology*, *27*, 361–368.

Setya, S., Talegaonkar, S., & Razdan, B. K. (2014). Nanoemulsions: Formulation methods and stability aspects. *World Journal of Pharmacy and Pharmaceutical Sciences*, *3*, 2214–2228.

Shen, L., & Tang, C. H. (2012). Microfluidization as a potential technique to modify surface properties of soy protein isolate. *Food Research International*, *48*(1), 108–118.

Shui, L., Mugele, F., Van den Berg, A., & Eijkel, J. C. T. (2008). Geometry-controlled droplet generation in head-on microfluidic devices. *Applied Physics Letters*, *93*. 153113–1–153113–3.

Silva, H. D., Cerqueira, M. A., Souza, B. W. S., Ribeiro, C., Avides, M. C., Quintas, M. A. C., et al. (2011). Nanoemulsions of β-carotene using a high-energy emulsification–evaporation technique. *Journal of Food Engineering*, *102*(2), 130–135.

Smart, J., Berg, E., Nerbrink, O., Zuban, R., Blakey, D., New, M., et al. (2002). TouchSpray technology: Comparison of the droplet size with cascade impaction and laser diffraction. *Respiratory Drug Delivery*, *VIII 2*, 525–528.

Soria, A. C., & Villamiel, M. (2010). Effect of ultrasound on the technological properties and bioactivity of food: A review. *Trends in Food Science & Technology*, *21*(7), 323–331.

Suslick, K. S. (1995). Applications of ultrasound to materials chemistry. *MRS Bulletin*, *20*(4), 29–34.

Tapia-Hernández, J. A., Rodríguez-Félix, F., & Katouzian, I. (2017). 9—Nanocapsule formation by electrospraying A2–Jafari, Seid Mahdi. In *Nanoencapsulation technologies for the food and nutraceutical industries* (pp. 320–345). Academic Press.

Tapia-Hernández, J. A., Torres-Chávez, P. I., Ramírez-Wong, B., Rascón-Chu, A., Plascencia-Jatomea, M., Barreras-Urbina, C. G., et al. (2015). Micro-and nanoparticles by electrospray: Advances and applications in foods. *Journal of Agricultural and Food Chemistry, 63*(19), 4699–4707.

Torres-Giner, S., Pérez-Masiá, R., & Lagaron, J. M. (2016). A review on electrospun polymer nanostructures as advanced bioactive platforms. *Polymer Engineering & Science, 56*(5), 500–527.

Trujillo, A. J., Capellas, M., Saldo, J., Gervilla, R., & Guamis, B. (2002). Applications of high-hydrostatic pressure on milk and dairy products: A review. *Innovative Food Science & Emerging Technologies, 3*(4), 295–307.

Utada, A. S., Fernandez-Nieves, A., Stone, H. A., & Weitz, D. A. (2007). Dripping to jetting transitions in coflowing liquid streams. *Physical Review Letters, 99*, 094502.

Van der Graaf, S., Steegmans, M. L. J., Van der Sman, R. G. M., Schroën, C. G. P. H., & Boom, R. M. (2005). Droplet formation in a T-shaped microchannel junction: A model system for membrane emulsification. *Colloids and Surfaces, A: Physicochemical and Engineering Aspects, 266*, 106–116.

Van der Sman, R. G. M. (2012). Soft matter approaches to food structuring. *Advances in Colloid and Interface Science, 176*, 18–30.

Vecellio, L. (2006). The mesh nebulizer: A recent technical innovation for aerosol delivery. *Breathe, 2*, 252–260.

Villalobos-Castillejos, F., Granillo-Guerrero, V. G., Leyva-Daniel, D. E., Alamilla-Beltrán, L., Gutiérrez-López, G. F., Monroy-Villagrana, A., et al. (2018). Fabrication of Nanoemulsions by microfluidization. In *Nanoemulsions. Formulation, applications, and characterization* (pp. 207–232). Academic Press [Elsevier. London].

Villay, A., Lakkis de Filippis, F., Picton, L., Le Cerf, D., Vial, C., & Michaud, P. (2012). Comparison of polysaccharide degradations by dynamic high-pressure homogenization. *Food Hydrocolloids, 27*(2), 278–286.

Vladisavljevic, G. T., Khalid, N., Neves, M. A., Kuroiwa, T., Nakajima, M., Uemura, K., et al. (2013). Industrial lab-on-a-chip: Design, application ans scale-up for drug discovery and delivery. *Advanced Drug Delivery Reviews, 65*, 1626–1663.

Vyslouzilova, L., Buzgo, M., Pokorny, P., Chvojka, J., Mickova, A., Rampichova, M., et al. (2017). Needleless coaxial electrospinning: A novel approach to mass production of coaxial nanofibers. *International Journal of Pharmaceutics, 516*(1–2), 293–300.

Wen, P., Wen, Y., Zong, M. H., Linhardt, R. J., & Wu, H. (2017). Encapsulation of bioactive compound in electrospun fibers and its potential application. *Journal of Agricultural and Food Chemistry, 65*(42), 9161–9179.

Wise, D. L. (2000). *Handbook of pharmaceutical controlled release technology*. New York: Taylor & Francis.

Xu, J. H., Li, S. W., Tan, J., Wang, Y. J., & Luo, G. S. (2006). Preparation of highly monodisperse droplet in a T-junction microfluidic device. *AIChE Journal, 52*, 3005–3010.

Yalçınöz, Ş., & Erçelebi, E. (2018). Potential applications of nano-emulsions in the food systems: An update. *Materials Research Express, 5*(6). 062001.

Zaeim, D., Sarabi-Jamab, M., Ghorani, B., Kadkhodaee, R., & Tromp, R. H. (2017). Electrospray assisted fabrication of hydrogel microcapsules by single-and double-stage procedures for encapsulation of probiotics. *Food and Bioproducts Processing, 102*, 250–259.

Zhang, L., Huang, J., Si, T., & Xu, R. X. (2012). Coaxial electrospray of microparticles and nanoparticles for biomedical applications. *Expert Review of Medical Devices, 9*(6), 595–612.

Zhang, C., Yao, Z. -C., Ding, Q., Choi, J. J., Ahmad, Z., Chang, M. -W., et al. (2017). Tri-needle coaxial electrospray engineering of magnetic polymer yolk-shell particles possessing dual-imaging modality, multiagent compartments, and trigger release potential. *ACS Applied Materials & Interfaces, 9*(25), 21485–21495.

CHAPTER TWO

Production of food bioactive-loaded nanofibers by electrospinning

Hoda Shahiri Tabarestani, Seid Mahdi Jafari
Department of Food Materials and Process Design Engineering, Faculty of Food Science and Technology, Gorgan University of Agricultural Sciences and Natural Resources, Gorgan, Iran

Contents

Nanoencapsulation of Food Ingredients by Specialized Equipment
https://doi.org/10.1016/B978-0-12-815671-1.00002-0

1 Introduction

The encapsulation of nutraceuticals and functional ingredients is an emerging area, which has gained increased popularity in technological investigations. Encapsulation is defined as the technology for incorporating core materials such as bioactive compounds, nutrients, and drugs, physically or molecularly into wall materials (shell and encapsulant) including different types of food-grade polymers (natural or synthetic) to achieve the desired properties (Assadpour & Jafari, 2018; Hosseini, Ghorbani, Jafari, and Sadeghi Mahoonak, 2019; Kakoria & Sinha-Ray, 2018; Mahdavi, Jafari, Ghorbani, & Assadpoor, 2014). The encapsulated contents will be released in response to a trigger (such as shear, osmotic force, pH, temperature, and enzymatic reaction) when it is required (Moreira, Morais, Morais, Vaz, & Costa, 2018; Wen, Wen, Zong, Linhardt, & Wu, 2017). Flavors, colors, preservatives, sweeteners, proteins, minerals, lipids, probiotics, acids, bases, buffers, antioxidants, polyphenols, and so on are most commonly used as core materials (Mahdavee Khazaei, Jafari, Ghorbani, & Hemmati Kakhki, 2014; Sarabandi, Sadeghi Mahoonak, Hamishekar, Ghorbani, & Jafari, 2018).

Encapsulation technology in the food area has several applications in terms of ease of handling process for some liquid or gaseous ingredients, controlled release, targeted delivery, masking the undesirable taste of certain nutrients, enhanced bioavailability, increased stability during severe environmental conditions, protection from interactions with other food ingredients or extreme processing conditions, and production of novel functional foods with potential health benefits, dietary supplements, and additional attractiveness (Assadpour & Jafari, 2018; Jafari, Assadpoor, He, & Bhandari, 2008; Rajabi, Ghorbani, Jafari, Sadeghi Mahoonak, & Rajabzadeh, 2015). These properties are highly related to surface-volume ratios, and hence among different sizes, micro- and nanocarriers are preferred. In particular, nanometric structures have increased the surface area-to-volume ratio, which might enhance the interaction of the encapsulated bioactive components with the surrounding media. Nanotechnology as an emerging technology presents new more efficient ways to overcome technological barriers within many fields of application especially for the production of nutritional and functional foods (Jafari, Fathi, & Mandala, 2015; Torres-Giner, Pérez-Masiá, & Lagaron, 2016). Nanoscale food materials will make a better solubility and bioavailability, enhancing controlled release with greater precision targeting as compared with micron-size particles (Jafari & McClements, 2017).

A number of nanoencapsulation studies have been conducted in food research for nutraceutical delivery, which include coacervation, emulsification-solvent evaporation, inclusion complexation, nanoprecipitation, supercritical fluid technique, nanospray-drying, and electrohydrodynamic (electrospinning and electrospraying) processes (Assadpour & Jafari, 2019a, 2019b; Jafari, 2017). Electrohydrodynamic processing (EHD) is one of the emerging technologies that uses a high-voltage supply to induce a charge of certain polarity onto a polymer solution or polymer melt that is then accelerated to a collector of opposite polarity (Faridi Esfanjani & Jafari, 2016; Katouzian & Jafari, 2016). Electrospinning is a very versatile procedure and can be generally divided into two strategies: melt electrospinning and solution electrospinning. Indeed, solution electrospinning requires a compatible solvent able to dissolve a given polymer and the bioactives to form a homogeneous solution. Conversely, melt electrospinning is a solvent-free method (green technology) in which the polymer melt (composed of the polymer matrix and the bioactives) is heated up to its melting point and then it will be electrospun (Lian & Meng, 2017; Wen, Wen, Zong, et al., 2017).

Despite the many disadvantages, solution electrospinning keeps the highest number of published data for loading bioactive compounds in electrospun fibers due to its simplicity, no specific limitation of polymers and bioactives, and no possibility of melt electrospinning for the most biological molecules and thermosensitive polymers (Lian & Meng, 2017). Through electrospinning and electrospraying (see Chapter 3), the liquid/melt draws into nanofibers and nanodroplets, respectively (Alehosseini, Ghorani, Sarabi-Jamab, & Tucker, 2017). Both methods have been recently applied especially in food applications as cost-effective and scalable technologies for protecting sensitive bioactive ingredients with a high encapsulation efficiency (often >80%), different tactile sensation and mouthfeel to food, and a high specific surface area. EHD techniques are nonthermal processes that produce high-performance nano-/microfibers or capsules with various sized particles and a narrow size distribution (Jacobsen, Garcia-Moreno, Mendes, Mateiu, & Chronakis, 2018). However, several factors should be considered when creating a delivery system for nutrients for food applications: (1) all ingredients must be nontoxic without economic or scale-up problems; (2) they must be designed to withstand the harsh environmental conditions that foods experience during their manufacture, storage, transport, and utilization, such as thermal processing, freezing, dehydration, and mechanical stresses; (3) the addition of a delivery system should not affect the organoleptic characteristics of food products, such as appearance, texture, flavor, and mouthfeel; (4) they should be designed to maintain or enhance the bioavailability of the nutraceuticals; and (5) they may be designed to control the release of functional bioactive substances at the site of action and allow localized and sustained applications (McClements, Decker, Park, & Weiss, 2009).

The mechanism of electrospinning involves the collaborative effects of electrostatic repulsions by the accumulated charges on the surface of polymer solution and the coulombic forces exerted by the external electric field (Ghorani, Alehosseini, & Tucker, 2017). The difference between electrospinning and electrospraying is based on the degree of molecular cohesion in the raw material, a property that is most readily controlled by variations in the concentration and viscosity of the polymeric solution (Assadpour & Jafari, 2018). A high degree of interaction and sufficient molecular cohesion to withstand the electric pull results in a fiber, and the process is referred as electrospinning and vice versa (Alehosseini et al., 2017). Even though electrospinning is one of the "top-down" approaches used in nanotechnology and does not change the structure of the molecules, the size reduction to nanoscale may change some properties of the materials (Neo et al., 2013). The bioactive material to be encapsulated through the electrospinning technique is dispersed in a carrier polymer solution that is atomized into fibers. The solvent is evaporated using high-voltage electric fields, and the resulting ultrathin polymeric structures are collected as dry continuous fiber structures. A summary of research on different kinds of food-grade biopolymers including proteins, polysaccharides, and their complexes and conjugates with recent applications in encapsulation by electrospinning technique has been discussed in this chapter. Electrospinning is therefore reported here as a technology capable of producing added-value biopolymer micro- and nanofibers, which can have a good potential in food and nutraceutical formulations and coatings, bioactive food packaging, and food processing industries.

2 Production of electrospun nanofibers as delivery systems

Generally the delivery systems for encapsulated bioactive components can be solid or liquid, spherical or nonspherical particles, fibers or tubes, and networks with gel-like or sponge-like characteristics (Wen, Zong,

Linhardt, Feng, & Wu, 2017). During the last few years, materials based on one-dimensional nanostructures, such as nanofibers, nanotubes, and nanowires, have created a subject of substantial interest due to their unique properties (Neo et al., 2014). Among them, electrospun nanofiber delivery systems have gained a wide interest compared with larger-diameter fibers made of the same material, and their applications to the agriculture and food industries are relatively recent. Encapsulation for controlled release or immobilization of functional ingredient in nanofibers from biobased renewable materials might provide flexible tools to stabilize and preserve the quality of food or even to design healthier and more effective functional foods due to the special characteristics of the nanofibers (Fernandez, Torres-Giner, & Lagaron, 2007).

Nanofibers are continuous filaments with a diameter ranging from several nanometers to 1 μm. These structures can be produced by electrospinning process, using specifically adapted apparatus. They are valued for their ultra-high specific surface areas (i.e., surface-to-volume or surface-to-mass ratios), high porosity, and easy use to fabricate different structures. Due to their unique properties, nanofibers have been found potentially useful in the field of food science and technology for nutraceutical delivery, enzyme immobilization, food packaging, reinforcement, and biosensing (Ghosal et al., 2018; Torres-Giner et al., 2016), food-related separation applications such as filtration or clarification and solid phase extraction, food indicator membranes and sensors, energy harvesting, and storage (Ghorani & Tucker, 2015). As a delivery system the release rate of encapsulated bioactive compounds can be controlled by modifying the structure, porosity, diameter, and composition of nanofibers, as well as loading dosage and incorporation method of bioactive components (Wen, Wen, Zong, et al., 2017). Overall the nanostructured fiber and particle morphologies produced by EDH techniques offer tunable release kinetics for encapsulated bioactives, pertinent to diverse applications in particular food production sectors (Aguilar-Vázquez, Loarca-Piña, Figueroa-Cárdenas, & Mendoza, 2018). Also, electrospun nanofibers exhibit many advantages over traditional fibers, which are made from drawing, template synthesis, self-assembly, phase separation, and melt-blowing techniques (Anu Bhushani & Anandharamakrishnan, 2014).

There are many ways to classify electrospinning processes as shown in Fig. 1. Fiber extrusion followed by mechanical stretching and branching and occurrence of undulated fibers are exclusive phenomena of the electrospinning (Ghorani et al., 2017). The ultrafine fibers are formed by the repulsion of electrostatic charges and elongation of the solution with

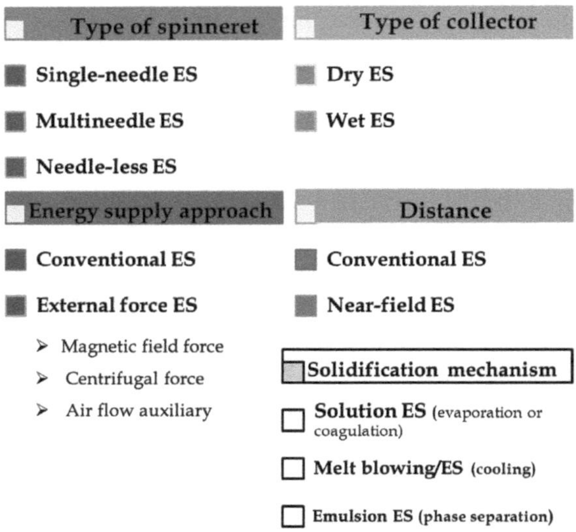

Fig. 1 Classification of different electrospinning (ES) setups.

different materials including synthetic and natural polymers, polymer blends, and polymers loaded with nanoparticles or bioactive agents (Neo, Ray, & Perera, 2018). This technique was started in 1897, patented as early as the 1930s, revitalized by Taylor during the 1960s through theoretical and experimental activities, and received much attention in the early 1990s with the advent of nanoscience and nanotechnology (Frenot & Henriksson, 2007). Many review articles and reports on electrospinning of nanofibers and related applications have been recently published. This technology has been developed due to high-output, consistent, and high molecular orientation, as well as good mechanical properties and low-diameter distributions.

2.1 Electrospinning setup

A basic ordinary electrospinning setup (either in horizontal or vertical arrangement) consists of four main elements (Maftoonazad & Ramaswamy, 2018) as depicted in Fig. 2: (i) a high-voltage power supply (up to 50 kV) operated in direct current (DC) mode, which is applied on the needle containing the spinning solution and grounded collector by two electrodes; (ii) a syringe pump (flow control pump) for feeding spinning solution; (iii) a syringe with a flat-end metal needle (capillary tube or spinneret) to store a spinning solution; and (iv) an electrically conductive collector (target) presented in different geometries. When the high electric field reaches a critical value, the electrostatic

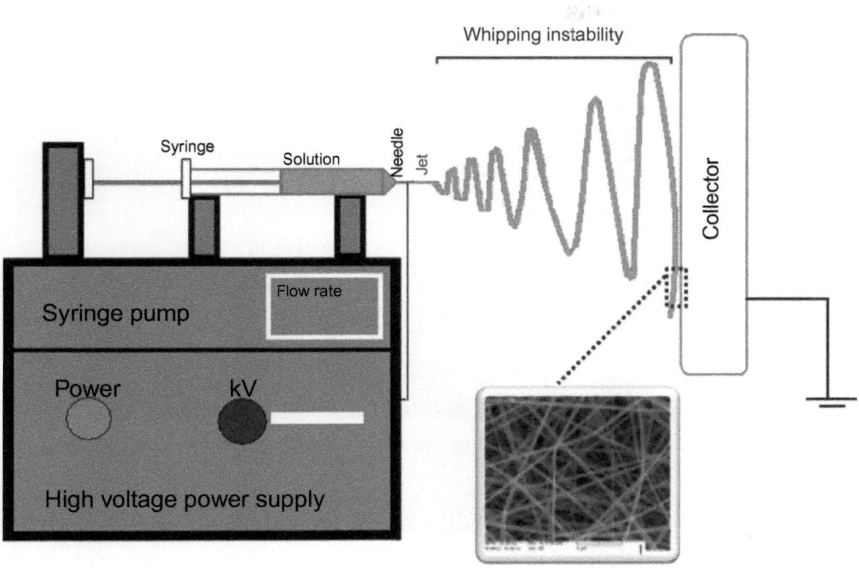

Fig. 2 Schematic representation of the electrospinning process.

forces surpass the surface tension of the optimum viscous polymer solution, and a conical jet projection of fibers emerges (Taylor cone) on the surface of a drop at the tip of the spinneret (Fig. 3). Further, in the space between the electrodes, the solvent gets evaporated during the process, rapidly transforming the jet into continuous thin fibers in a whipping motion (non-axisymmetric bending instability) in the collector (Jacobsen et al., 2018); thereby, solidification occurs before fibers are collected on the screen in the form of nonwoven mats. As explained in Chapter 3, electrospraying differs from electrospinning by the fact that when the polymer concentration is too low, the jet is destabilized due to varicose instability, forming highly charged droplets, which partially or fully solidify through solvent evaporation or cooling (Alehosseini et al., 2017; Jacobsen et al., 2018). In fact the formation of electrospun fibers or electrosprayed particles is dependent on molecular chain entanglement of the polymer solution, which is generally a function of polymer chain topology and morphology.

2.2 Electrospinning—Key experimental parameters

Electrospinning is a straightforward and versatile way to fabricate ultrafine fibers down to micro-/nanometer scale in various morphologies including nonwoven fibers, aligned fibers, beaded fibers (pearl necklace-like beaded fibers or

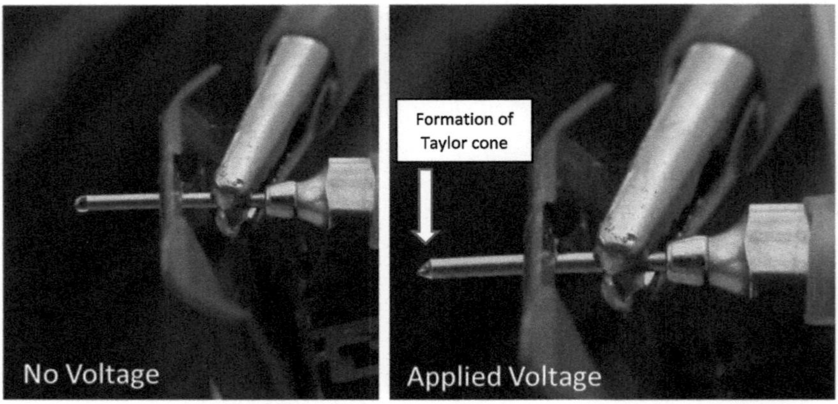

Fig. 3 Taylor cone formation by exerting the high voltage during electrospinning.

bead on string), hollow fibers, band-like fibers, porous fibers, core-shell fibers, wavy fibers, and spiral fibers and a great diversity of structures like grid, yarn, pattern, and twisted fibers (Mendes, Stephansen, & Chronakis, 2017). The main factors that govern the morphology and properties of the nanofibers achieved by electrospinning are (Ghorani & Tucker, 2015) as follows: (1) polymer properties (solubility, molecular weight, melting point, glass transition temperature, and crystallization velocity); (2) spinning solution properties (viscosity, viscoelasticity, concentration, dielectric constant, surface tension, solvent properties/vapor pressure, and electric conductivity); (3) processing parameters in the electrospinning setup (applied voltage, needle-to-collector/nozzle-to-ground distance, collector composition and geometry, feed rate, and needle diameter); and (4) operating conditions (temperature, relative humidity, and atmospheric pressure). Also, different nozzle configurations (single, single with emulsion, side-by-side, or coaxial nozzles) can be used for obtaining various nanofibers (Fig. 4), which together with electrospinning parameters will define the fiber features and even overcome limitations in polymer processability (Neo et al., 2014). The dependence of fiber morphologies, dimensions, and properties on different categories have been summarized by Ghorani et al. (2017). Most studies have suggested a stronger effect of solution parameters on the morphology of electrospun fibers as compared with processing or ambient parameters (Neo et al., 2013). The selection of a desirable solvent or solvent system as the carrier of a particular polymer is fundamental for the optimization of electrospinning. Solvent selection is vital in determining the critical minimum solution concentration to allow the transition from

Fig. 4 Different biphasic features of electrospun nanofibers.

electrospraying to electrospinning (Luo, Nangrejo, & Edirisinghe, 2010). In the following sections the influence of different parameters during electrospinning has been described.

2.2.1 Polymer molecular weight, concentration, and viscosity

Sufficient polymer molecular entanglements are the prerequisites for forming fibers. *Molecular weight* reflects the entanglement of polymer chains in solutions, indicating whether electrospraying (beads) or electrospinning (fibers) will take place. Low-molecular-mass solutions tend toward a "bead-on-string" morphology rather than pure fibers, while increased concentration of polymers induces alteration of spherical beads to rodlike, and lastly uniform fibers (Neo et al., 2013, 2014). By further increasing the molecular weight, microribbons will be obtained.

The *concentration* of polymer in a solution has an important effect on fiber morphology. The formation of good electrospun nanofibers (bead-free) is mainly due to significant increase of jet chain entanglement between polymer molecules because there is counterbalance to the high coulombic stretching force that causes jet stability (Colín-Orozco, Zapata-Torres, Rodríguez-Gattorno, & Pedroza-Islas, 2015). In relation to polymer concentration, three solution regimes including dilute, semidilute unentanglement, and semidilute entanglement are introduced. The critical chain overlap concentration c^*, which is the crossover concentration between the dilute and semidilute concentration regimes, is thus a critical parameter for electrospinning, which can be determined from intrinsic viscosity (Colín-Orozco et al., 2015). In a dilute regime, (nano)particles will be formed due to the low viscosity and high surface tensions of the solution. Further increase in biopolymer concentration results in a mixture of beads and fibers (Li & Wang, 2013) indicating the semidilute unentanglement regime, where the concentration of the polymer (c) is large enough to have some chains overlap ($c > c^*$) but not enough to have any significant degree of entanglement (Colín-Orozco et al., 2015). Therefore there is an appropriate concentration where smooth nanofibers can be obtained (Li & Wang, 2013). Furthermore the crossover of concentration from the semidilute unentanglement to semidilute entanglement regime is referred to as the critical entanglement concentration (Colín-Orozco et al., 2015). Above this critical point, fibers are intertwined and merged together due to the uneven evaporation of the solvent between the surface (skin) and core of the charged jet resulting in the formation of ribbon-like structures. According to Neo et al. (2013) the C_e value allowed the determination of the minimum polymer solution concentration required for electrospinning of bead-free zein fibers occurred at 25 wt% ($C = 2.1C_e$).

As polymer concentration increases, a mixture of beads and fibers will be obtained, and beyond a critical concentration, high viscosity of the solution will impede the continuous formation of fibers (Ghorani & Tucker, 2015). Rheological properties, especially *viscosity* of a solution as a key parameter, have a significant effect on the morphology of fibers and are related to polymeric concentration and molecular weight. Shekarforoush, Ajalloueian, Zeng, Mendes, and Chronakis (2018) reported a direct correlation between the apparent viscosity of chitosan-xanthan solutions and their optimal electrospinnability. High viscosity means high concentration of the polymer and higher viscoelastic forces, which act against the coulombic repulsion forces, which are the main force leading to the elongation of the jet after the Taylor cone apex (Ghorani et al., 2017). Therefore larger fibers or even hard ejection of the jet may result (Li & Wang, 2013). On the other hand, lower viscosity means lower viscoelastic forces, where surface tension is the dominant factor and just beads or beaded fibers are formed. Thus the viscosity of polymer solution should be high enough to prevent the jet from breaking apart into droplets and generate continuous and smooth fibers. It seems that there is a tendency confirming that low viscosities, for example, lower than 0.100 Pa s, do not favor fiber formation (Colín-Orozco et al., 2015).

Solution properties such as concentration, viscosity, molecular weight, conductivity, and surface tension are commonly dependent to each other. For instance the adjustment of polymer solution concentration or its average molecular weight will have an effect on its conductivity and surface tension. These parameters are described in Section 2.2.2.

2.2.2 Conductivity and surface tension

The *electric conductivity* of a solution reflects a charge density on a jet. Solutions with a higher conductivity can form a jet with a higher charge-carrying ability, resulting in stronger electrostatic forces, which yield nanofibers with a decreased fiber diameter and less bead formation (Liu, Xu, Zhao, & Yang, 2017). The solution electric conductivity is mainly determined by the polymer and solvent type, polymer concentration, temperature, and anionic salts (Li & Wang, 2013). The electric resistance (the inverse of conductivity) is the sum of resistances of all polymers and the sum of resistances of contacts between them (Colín-Orozco et al., 2015). It is necessary to increase the solution conductivity to a critical value in order to induce charges in the droplet surface; but conductivity should not be increased beyond a critical value, where there is a balance between the coulombic forces in the surface of fluids and the forces due to the external electric field (Ghorani et al., 2017; Neo et al., 2013). Large ionic conductivity can be obtained by adding ionic salts or polyelectrolytes and sometimes using organic acids as solvents (Li & Wang, 2013). In a composite structure, all conducting particles participate in the formation of the electric conductivity.

Another important factor in the process of electrospinning is the *surface tension*, which is defined as the amount of energy needed to increase its surface by unit of area and depends on the solvent composition, mass ratio of solvents, and temperature (Colín-Orozco et al., 2015). It is the primary force opposing applied voltage during electrospinning process and determines electrospinnability. In a constant concentration, feed solutions with a low surface tension produce fibers without beads. However, it does not mean all solutions with a low surface tension can be electrospun (Li & Wang, 2013). The minimum voltage for producing nanofibers increases with the surface tension of solutions but not always linearly.

The formation of beaded structure has been suggested to be either due to axisymmetric "Rayleigh instability" (dominated by surface tension) or axisymmetric "conducting instability" (electric competition between surface charge and electric fields), which are determined by the material properties of the fluid and the operating parameters of the process. Beaded structure can be eliminated through appropriate fluid viscoelasticity and polymer entanglement by suppressing "Rayleigh instability"; that is, the effect of surface tension is insignificant when the polymer solution preserves its viscoelasticity and undergoes solidification (Neo et al., 2018).

2.2.3 Applied electric voltage

The applied voltage is a critical element for electrospinning process because it provides surface charge on the electrospinning jet and affects nanofiber diameters. Generally, increasing applied voltages lead to decreasing nanofiber diameters with higher electrostatic repulsive forces on the fluid jet (Bhardwaj & Kundu, 2010). Besides, studies have shown an increase of fiber diameter with a higher applied voltage due to more polymer ejections or a larger mass flow. However, several studies have also shown that the applied voltage has an insignificant effect on the electrospun fiber diameters (Torres-Giner, Gimenez, & Lagaron, 2008).

2.2.4 Flow rate

The rate of feed solution influences jet velocity and transfer rate of the solution. For the evaporation of solvent and obtaining solid nanofibers, lower feeding rates are desirable (Bhardwaj & Kundu, 2010). Ideally, feeding rate must match the solution removing rate from the tip. Lower feeding rates can inhibit electrospinning (Alehosseini et al., 2017), and high feeding rates result in beaded large-diameter fibers due to unavailability of proper solvent evaporating time prior to reaching the collector (Bhardwaj & Kundu, 2010).

2.2.5 Working distance

Adjusting the needle tip-to-collector distance is a key factor for obtaining uniform fibers due to the sufficient flight time of the electrospinning jet. A larger distance generally produces fibers with a thinner diameter and less beaded structures, as the fiber receives more stretching and elongation time before it deposits on the collector. In contrast an inappropriately short distance is deemed to create a stronger electrostatic field that results in beaded structures, due to jet instability (Ghorani et al., 2017) or nanofiber fusion (Bhardwaj & Kundu, 2010). Nevertheless, a distance that is too large will also cause the reduction of electrostatic forces and effective voltage drop, thus resulting in less stretching of the fibers. Consequently an increase of the fiber diameter is also reported with the increase in the distance between spinneret tip and collector (Neo et al., 2013).

2.2.6 Operating conditions

Ambient parameters like temperature and relative humidity are found to affect the electrospun fiber morphologies by influencing the solution properties, such as viscosity and solvent evaporation rate (Ghorani et al., 2017). Temperature has binary opposing effects both of which create thinner fiber diameters: (1) low temperature will result in a longer stretching and elongation process of the solution jet due to slow solvent evaporation rate and slow jet solidification, and (2) high temperature will decrease the viscosity of the solution by moderating the rigidity of polymer chains and chain entanglement. Under the influence of different relative humidity and temperature values, porous, nonporous, and wrinkled electrospun fibers may be established. In addition, relative humidity has been demonstrated to decrease the fiber diameter due to the plasticizing effect or generating thicker fibers as a result of lower electrostatic fields that reduces stretching and elongation of the solution jet (Neo et al., 2018).

3 Encapsulation methods for bioactive compounds by electrospinning

Electrospinning technology has been increasingly developed and utilized for the production of functional nanomaterials. This technology has been divided into needle, needleless, and core-shell methods (Vyslouzilova et al., 2017), which could be used for bioactive components entrapped into nanofibers. The classical approach for the preparation of nanofibers with incorporated bioactive molecules is blend electrospinning. In addition, the functionality of the nanostructures produced by electrospinning can be achieved through the utilization of coaxial system, emulsion electrospinning, inclusion of other functional molecules, and the adsorption of functional components onto the surfaces of the nanostructures. Different electrospinning techniques have been depicted in Fig. 5.

3.1 Blend (single/uniaxial) electrospinning

Blend electrospinning is based on the direct incorporation of the bioactives into the polymeric solutions (solubilization of core and wall materials in the same solvent) prior to electrospinning process (Mendes et al., 2017). The bioactive component is dissolved or dispersed (if insoluble) in the polymeric matrix, and upon favorable conditions the solution jet solidifies, and the bioactive component is randomly distributed throughout the fiber matrix as

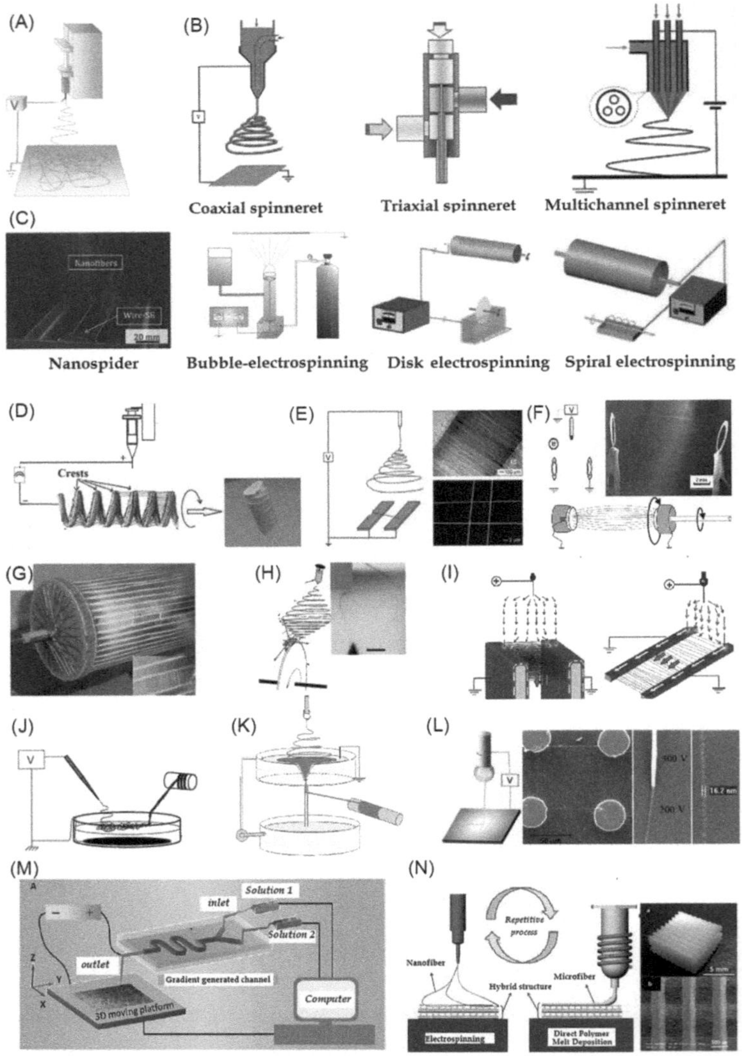

Fig. 5 Different electrospinning techniques: (A) the traditional setups of electrospinning, including a static high-voltage power supply, a spinneret, a microflow pump, and a collecting plate; (B) different spinnerets, including coaxial spinneret, triaxial spinneret, and multichannel spinneret; (C) the needleless electrospinning techniques, including the Nanospider with wire spinning electrode, bubble disturbing needleless electrospinning, disc needleless electrospinning, and spiral needleless electrospinning; (D) the helical spring collector; (E) the parallel electrodes and the collected aligned nanofibers; (F) the dual collection rings and the collected aligned nanofibers; (G) a rotating drum with parallel wires; (H) the rotating disc collector; (I) the automated parallel tracks; (J) the fiber collecting water bath; (K) a dynamic liquid support system for collecting nanofiber yarn; (L) the low-voltage near-field electrospinning; (M) a microfluidic assisted electrospinning; (N) combination of 3-D printing and electrospinning. *(Reprinted with permission from Yang, G., Li, X., He, Y., Ma, J., Ni, G., & Zhou, S. (2018). From nano to micro to macro: Electrospun hierarchically structured polymeric fibers for biomedical applications.* Progress in Polymer Science, 81, *80–113.)*

shown in Fig. 2. It has been reported that after electrospinning of the dispersion, the deposition of bioactive compounds on or near the surfaces result in their rapid diffusion into the release media (Wen, Zong, Linhardt, et al., 2017), causing this approach to be more attractive for fast-delivery (burst release) applications. Hence the release performance of the bioactive component from electrospun fibers is governed mainly by solubility, interaction, and compatibility between the encapsulated compounds and polymeric carrier/solvent at the molecular level (Wen, Wen, Zong, et al., 2017). Higher compatibility in the system means higher affinity of bioactive component to polymeric carrier than the solvent, and consequently, during the flash evaporation of solvent from the solution jet, the bioactive component will migrate into the inner rather than the surface of the fiber (Mendes et al., 2017). Thus the bioactive compounds are appropriately encapsulated inside the fiber and exhibit nearly zero-order kinetics of release.

Generally the release is either desorption/diffusion or dissolution/erosion controlled depending on the type of polymers (nonbiodegradable or biodegradable). The release of food bioactive compounds may be controlled strictly by erosion/degradation of polymeric matrix (Supra Type II) or diffusion through the matrix (Fickian Type I), polymer swelling diffusion mechanism (Type II), and non-Fickian diffusion (Type III), which depends both on polymer swelling and food bioactive compound diffusion. Estevez-Areco, Guz, Candal, and Goyanes (2018) have studied the release rate of rosemary extract from polyvinyl alcohol (PVA) electrospun fiber mats in different food simulant mediums and demonstrated that both polymer chain relaxation and Fick's diffusion were the leading mechanisms in acid medium, while polymer chain relaxation was the main mechanism in hydrophilic simulant, and burst release was observed in lipophilic medium. Blend electrospinning has also been tested for vitamin A in cress seed mucilage/PVA nanofibers as a carrier (Fahami & Fathi, 2018a) in which the release of vitamin in simulated intestinal fluid (SIF) was faster than simulated gastric fluids (SGF) and diffusion transport was the main release mechanism.

A wide range of food bioactives like phenolics, vitamins, and natural antimicrobial compounds have been encapsulated via blend electrospinning as further discussed in Section 5 of this chapter. Although blend electrospinning process is the most commonly used method in encapsulation of bioactive compounds, several major obstacles restrict its application in developing controlled delivery systems for food-based applications. First, a burst release trend of embedded bioactive molecules due to their high concentration distributed on the nanofiber surfaces is revealed (Neo et al., 2013). Second, the likely denaturation or inactivation of sensitive biological agents such as proteins, cells, and enzymes is possible due to their direct contact with organic solvents when preparing the polymer solution (Vyslouzilova et al., 2017; Wang, Yue, & Lee, 2015). Next, destruction of biological integrity of bioactives could happen after mechanical stirring, homogenization, or ultrasonication process. Finally, the traditional electrospinning method still faces enormous challenges to encapsulate water-soluble bioactive molecules into hydrophobic polymers (Wang et al., 2015). For these reasons, major advancements in this process such as coaxial electrospinning and emulsion electrospinning have gained the attention of researchers.

3.2 Coaxial electrospinning (coelectrospinning)

Coaxial electrospinning is a unique, sophisticated method to produce second generation of nanofibers, that is, composite functionalized nanofibers with an organized core/shell structure (Ghorani et al., 2017). The apparatus for coaxial electrospinning consists of a double-compartment syringe with two concentrically arranged nozzles (internal and outer capillary or needle), which are connected to individual syringe pumps for simultaneous electrospinning of versatile core/shell polymers and solvents (Vyslouzilova et al., 2017), as shown in Fig. 6. Traditional coaxial electrospinning is based on the creation of a compound droplet in the orifice of a dual-spinneret, which forms a composite Taylor cone that pulls up both the shell polymer and the core polymer. As a result the core-shell jet solidifies, and core–shell fibers are depositing on a grounded collector. In order to synthesize high-quality nanofibers using standard coaxial electrospinning, shell polymer selection should be based on electrospinnable solutions (Yang et al., 2017), and the inner solution could simply be nonspinnable polymers, non-polymeric materials, or even a powder (Shin & Lee, 2018; Vyslouzilova et al., 2017). Apart from the regular parameters that affect the quality of fibers being produced as described before, the miscibility of the solvents used in core and shell solutions and flow rate ratio of core and shell solutions are the factors that affect the coaxial electrospinning, significantly (Yang, Wen, et al., 2017).

Core-shell nanofiber structures provide greater encapsulation opportunities such as micro/nanolayering, bioactive loading efficiency (Wen, Wen, Huang, Zong, & Wu, 2017; Yang, Wen, et al., 2017), and controlled release, especially a greater control on delivery of delicate ingredients and

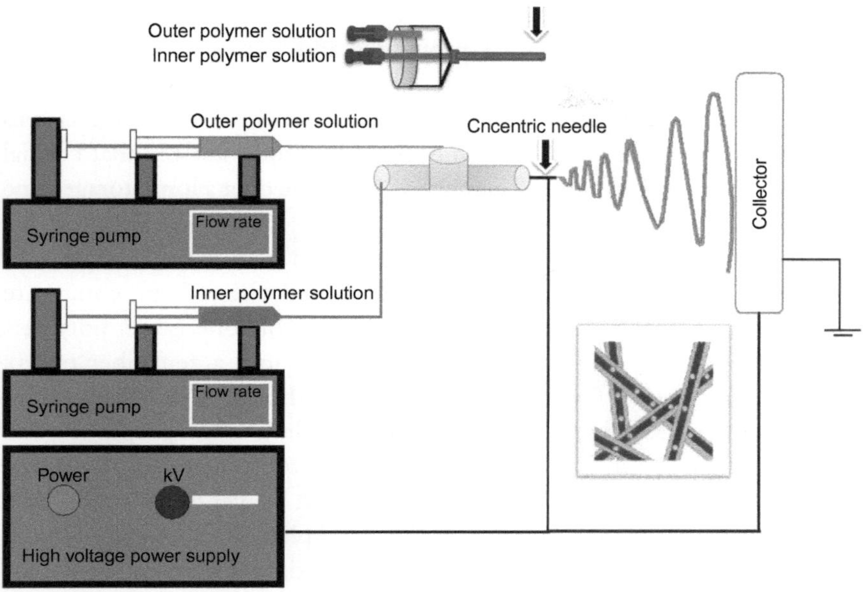

Fig. 6 Schematic illustration of coaxial electrospinning setup.

volatile material retention as compared with uniaxial electrospinning (Yao, Zhang, Lim, & Chen, 2017). Also, it is reported that immobilization of native enzymes in coaxial fibers increased their specific activity (Gabrielczyk, Duensing, Buchholz, Schwinges, & Jordening, 2018).

After its introduction, coaxial electrospinning has been rapidly implemented by many food research groups for the encapsulation of bioactive agents in the fiber core for different purposes and with a clear advantage over bulk materials due to their high surface-to-volume and surface-to-weight ratios and short diffusion path lengths (Korehei & Kadla, 2013). Coaxial electrospinning can generate fibers from various solution pairs, core-sheath, hollow, and functional fibers that may contain particles (Khalf & Madihally, 2017). An idea of using hollow ultrafine fibers was first proposed by Sakuldao, Yoovidhya, and Wongsasulak (2011) for cellulose acetate to encapsulate gelatin as a model protein compound and showed near zero-order release pattern of the protein in the gastrointestinal (GI) tract and anomalous diffusion mechanisms. Moreover a novel core-shell nanofilm structured for the delivery of protein to the colon was developed by coaxial electrospinning using bovine serum albumin (BSA) as a protein model (Wen, Wen, Huang, et al., 2017). Olive leaf extract was successfully encapsulated into coaxial silk fibroin/hyaluronic acid nanofibers, which exhibited favorable antibacterial properties and release stability (Dogan et al., 2016). Lopez-Rubio, Sanchez, Sanz, and Lagaron (2009) found that coaxial electrospinning process improves the viability of bifidobacteria cells compared with single electrospinning. A similar study showed that coaxial electrospinning maintained full activity of T4 phage over a long storage time (Korehei & Kadla, 2013).

Coaxial electrospinning has also been reported as an effective strategy to improve the oxidative stability of encapsulated fish oil in composite nanofibers based on zein (core) and polyvinylpyrrolidone (shell) polymers (Yang, Wen, et al., 2017) and rose hip seed oil into a zein fiber matrix for food preservation functions (Yao, Chang, Ahmad, & Li, 2016). This technique has been used to enhance the hydrophobic properties and improve tea polyphenol release properties in porous core-shell structured fibers of poly(L-lactic acid) (Wang & Xu, 2018). A bilayer fibrillar core-shell from zein/gelatin was used for encapsulation of phenolic antioxidants from *Momordica charantia* fruit, and the suitability of coaxial electrospinning to produce the film mat, which could be either consumed individually as a health supplement or as an added ingredient to food products, was studied by Torkamani, Syahariza, Norziah, Wan, and Juliano (2018). The coaxial encapsulation system proved to be an effective method for sour cherry polyphenol protection as quantified through digestion assays (Isik, Altay, & Capanoglu, 2018). Furthermore, Kim, Kim, Jang, and Shin (2018) improved the physical properties and maintained water content and bioactive release of spirulina extract using coaxial nanofibrous dressing composed of a polycaprolactone/alginate/spirulina. Investigation of isothiocyanate (ITC) release from electrospun modified poly(L-lactic acid)/mustard powder composite fibers showed a distinct longer ITC release behavior for coaxial fibers (14 days) than uniaxial fibers (8 days) (Yao et al., 2017).

Generally, in the coaxial electrospinning process, monitoring the existence of enough compatibility of core and shell fluids for different bioactive agents is essential. Also the main limitation of this technique is its low production rate (<300 mg/h) (Buzgo, Mickova, Rampichova, & Doupnik, 2018). In this regard a revolutionary idea for a needleless coaxial electrospinning method has been proposed, which can raise the number of Taylor cones and feed rates. This novel method exceeds the limited production capacity of current approaches, enabling coaxial electrospinning from free liquid surfaces, and leads to the formation of coaxial nanofibers with a higher and uniform shell/core ratio, which results in the possibility of better tuning of the degradation rate (Vyslouzilova et al., 2017). The throughput and quality increase favors the broader application of coaxial nanofibers from weir spinnerets as appropriate systems for controlled food bioactive delivery.

Besides, several other methods including modified coaxial/multiaxial electrospinning have been developed to enhance the productivity and alter the more complex structural features for use in various applications (Buzgo et al., 2018). These modifications also help in (i) producing multiaxially structured fibers with a high control on fiber properties, (ii) controlling the delivery approaches of multiple bioactives, (iii) improving the mechanical properties of the hollow fiber morphology, and (iv) altering the biological properties (Khalf & Madihally, 2017). A modified coaxial electrospinning process, characterized by the utilization of unspinnable liquids as the sheath working fluids, can also be implemented, which might comprise solvents, small molecule solutions, emulsions, and suspensions, and should lead to a range of novel structures (Yang, Wen, et al., 2017).

In triaxial electrospinning, three polymer solutions are supplied into a compound Taylor cone through a spinneret. Triaxial fibers could also be used to deliver two different bioactive compounds with a dual delivery system by using the intermediate layer that acts as an isolation layer between the sheath and core materials. Furthermore, triaxial fibers can be formed with a varying hydrophobicity and mechanical strength. In the triaxial electrospinning technique, polymers could also be loaded with nanoparticles. In particular, triaxial fibers can be successfully formed by ensuring that the outer sheath is formed first. For this purpose, it is necessary to adjust the volatility of the solvent in the core polymer to be lower than the level used for the sheath polymer. Thus selection of solvents with appropriate boiling points in different sheaths is critical; in some particular cases, outer sheaths should have a boiling point less than that in the inner cores (Khalf & Madihally, 2017). Triaxial electrospinning technology allows multiphasic delivery of bioactive molecules with complex release profiles.

3.3 Emulsion electrospinning

Emulsion electrospinning is a novel eco-friendly, clean, and safe production technique, which overcomes numerous disadvantages of blend and coaxial electrospinning. It can be easily accomplished using only an ordinary single-nozzle electrospinning setup, and there is no need to precisely control the process variables, and therefore, it can provide a simpler approach to decrease the burst release of bioactive compounds inside (Zhang & Zhang, 2018). Either water-in-oil (W/O) or oil-in-water (O/W) emulsions can be electrospun to directly encapsulate polar and hydrophilic molecules or hydrophobic compounds into core-shell fibers, respectively (Zhang, Feng, & Zhang, 2018). Besides these two systems, multiple emulsions, such as water-in-oil-in-water (W/O/W), are produced by additional emulsification of an O/W emulsion into water phase (Gharehbeglou, Jafari, Hamishekar, Homayouni, & Mirzaei, 2019; Gharehbeglou, Jafari, Homayouni, Hamishekar, & Mirzaei, 2019). Such multiple emulsions need higher requirements such as proper surfactant selection and emulsification protocol optimization (Assadpour, Maghsoudlou, Jafari, Ghorbani, & Aalami, 2016; Faridi Esfanjani, Jafari, & Assadpour, 2017).

Wang et al. (2015) investigated the production of BSA as a model protein by emulsion electrospinning of polystyrene fibers, which resulted in prevention of protein inactivation due to interaction with other food components or environmental factors. Additionally the emulsion electrospinning method provides an option to incorporate variety of bioactive agents into polymers. The bioactives are surrounded by suitable emulsifiers or surfactants in terms of HLB and impregnated into a polymeric carrier; the continuous phase is converted into shell of fibers, and the droplet phase is forming the fiber core (Zhang et al., 2018). Core/sheath structures are generated owing to the stretching and coalition of polymeric emulsions under the influence of electrostatic forces during the electrospinning process (Shin & Lee, 2018). The technology privileged is overcoming the poor solubility and instability of bioactive compounds in organic solvents and extends the release due to the encapsulated agent needing to pass through the core–shell fiber matrix before entering the medium (Kesici Güler, Cengiz Çallıoğlu, & Sesli Çetin, 2018).

The key points to achieve successful emulsion electrospinning are (i) electrospinnability of continuous dispersed phase polymers (sufficient conductivity, a suitable polymer concentration, and enough high molecular weight and surface tension), (ii) proper selection of surfactant, and (iii) droplet phase viscosity. In this regard, if the viscosity of droplet phase (core) is higher than that of continuous phase, the emulsion electrospinning process results in the formation of continuous core along the fiber axis (Mendes et al., 2017). In contrast, if the core has a lower viscosity than wall phase, the core phase breaks up into separated smaller droplets, and so, noncontinuous morphology in core will be observed (Moomand & Lim, 2014) due to the reduction of polymer chain entanglement. Likewise, there are a series of complex parameters including surfactant/emulsifier type (small molecule surfactants, particulate emulsifiers, and food hydrocolloids), additives (salts and thickening and gelling agents), the positional and interfacial structures of droplets, droplet volume fraction, and droplet size distribution that affect organization of core in emulsion electrospinning (Zhang et al., 2018).

In general the core morphology is among the most crucial variables affecting the release kinetics from emulsion nanofibers. In the case of emulsion electrospinning, the release of bioactive compounds from the surface of fibers or from the polymer matrix into the surrounding medium is mainly governed by the combination of diffusion or degradation mechanisms depending on dissolution or degradability of polymer structures.

In the case of typical W/O emulsion electrospinning, the shell of fibers is made of polymers stable in water (dissolved in an organic solvent), while the core is made from water-soluble material. The rate-limiting factor is therefore the diffusion rate of water and bioactives through polymer shell. The release patterns depend on the internal structure of fibers, the number of contact points available for the solvent to dissolve the core polymer, and the stability of the emulsion. The emulsion nanofibers have shown a more sustained release profile compared with coaxial nanofibers. However, the release rate from emulsion fibers is faster than blend spun fibers indicating improved availability of bioactive compounds due to their clustering with core droplets (Eng & Zhong, 2012). Ghitescu et al. (2018) attempted to encapsulate catechin within the core and poly(D,L-lactide-*co*-glycolide) (PLGA) as the fiber sheath via emulsion electrospinning and obtained tailored and sustained release rates for catechin. In vitro catechin release studies showed a burst release at the starting phase followed by a linear delivery at long term being driven by a diffusion-controlled mechanism.

O/W emulsion electrospinning has also been suggested by Kriegel, Kit, McClements, and Weiss (2010) with aqueous solution of PVA as the continuous phase and eugenol essential oil as the dispersed phase. A similar research in relation to PVA fibrous matrix using emulsion electrospinning was conducted by Camerlo, Vebert-Nardin, Rossi, and Popa (2013) to encapsulate a volatile fragrance (limonene). The encapsulation efficiency of limonene was $\geq$50%, and the release from PVA nanofibers continued over 15 days under ambient conditions. In another work, O/W emulsions based on PVA were applied for encapsulation of terpene-based volatile organic compounds, namely, *Chamaecyparis obtusa* phytoncide oil (Shin & Lee, 2018). The resulting nanofibers exhibited a well-aligned core/sheath structure with fiber diameters of 250–350 nm and 99.9% bacterial reduction against both *Staphylococcus aureus* and *Escherichia coli* with a sustained release over 14 days (Shin & Lee, 2018). Different studies with emulsion electrospinning have been achieved with various polymers and bioactive substances, which will be further described in Section 5.

Generally, using the emulsion electrospinning process, nanofibrous materials containing essential oils can be attractive because of the both unique and superior structure of essential oils and nanofibers (Kesici Güler et al., 2018). Potential food applications using emulsion electrospinning are not limited to the encapsulation of functional components and controlled release. Over the past years, there has been a growing interest in the development of emulsion electrospinning in accordance with enzyme immobilization purposes as a promising carrier to strengthen the functionality and the performance of enzymes due to improved stability under harsh environmental changes and a higher efficiency (Zhang et al., 2018). In terms of food packaging applications, the emulsion electrospinning has been utilized for the delivery of bioactive molecules through the fabrication of functionalized fiber-based packaging systems in which a wide variety of functional ingredients like essential oils are encapsulated to prevent food spoilage and foodborne illnesses resulting from microbial contamination (Zhang et al., 2018). To conclude, emulsion electrospinning is a suitable approach to create core-shell nanofibers loaded with various bioactive molecules. However, the main disadvantage regarding emulsion electrospinning technology is its full stop working in the case of low interfacial tension polymer matrices. Furthermore, obstacles to overcome include control on the methods used for creating emulsion systems since the structure of embedded molecules or polymers is affected (Zhang & Zhang, 2018).

3.4 High-throughput electrospinning technologies

There are two ways to produce nanofibers through the electrospinning technique: needleless and needle based. However, the throughput of classical needle electrospinning protocols is rather limited and reaches only 0.1–1.0 g/h for a single spinneret (Bazbouz & Russell, 2018). Methods for improving electrospinning throughput involve modified single-nozzle, multinozzle, and needleless systems. Multijet electrospinning technology has been developed by the multiplication of needles to obtain multiple jets and thus to enhance electrospinning productivity. The key disadvantage of multineedle electrospinning is the problem with finding optimal needle arrangements, applied voltage, and collector/electrode distance because multiple needles are forming nonhomogeneous electric fields resulting in the change of fiber jet properties.

It took >40 years to invent the needleless method of electrospinning overcoming the problems of multijet technology. The theory of needleless electrospinning is based on self-arrangement of fiber jets on the surface of liquid due to static wave of surface tension and electrostatic forces (Shaid, Wang, Padhye, & Jadhav, 2018). One example is the Nanospider in which the polymer solution is electrospun from either wire-based or roller electrodes, so that nanofiber production rate can be conveniently adjusted and fibers with diameters in the range of 50–500 nm can therefore be produced at a rate of about 1.5 $g\,min^{-1}$ per meter of roller length (Bazbouz & Russell, 2018). The first needleless electrospinning was based on a ring spinneret, and gradually, other shapes of spinneret were evolved. In needleless electrospinning method the needle is replaced with roller, ball, disc, cylinder, beaded wire, and so on. All these needleless electrospinning methods can be divided primarily into two groups: rotary and stationary spinneret needleless electrospinning. Rotary needleless electrospinning includes a rotary cylinder, disc, ball, spiral wire coil, splashing spinneret, cone, edge of metal plate, rotating roller, bowel edge, and rotary beaded chain. In stationary spinneret needleless electrospinning, the use of vertical rod, air bubble, conical wire coil, and metallic tube electrode has been reported (Shaid et al., 2018). Needleless electrospinning has also been used for the production of fibers from emulsions and opened a route in coaxial electrospinning. Alternatively, core-shell nanofibers could be produced at a high throughput by recently developed emulsion centrifugal spinning. This technology is based on the formation of ultrathin fibers by the application of high centrifugal forces on polymeric solutions. According to Kutzli, Gibis, Baier, and Weiss (2018), by the use of needleless electrospinning, commercial production of glycoconjugates could be possible.

3.5 Polymer-free (or free surface) electrospinning

Free surface electrospinning is a modified methodology of the electrospinning process, used for commercial production due to its higher yield and simpler plumbing than the spinneret approach (Moreira et al., 2018; Moreira, Terra, Costa, & Morais, 2018; Xiao & Lim, 2018). This process involves ejection of the spin dope from a continuous surface (conductive wire or drum) from which numerous polymer jets are formed. To obtain continuous fibers, polymers with a high molar mass are generally used. Their concentration has to be high enough to ensure chain entanglements in spin dope solution and to avoid breakage of the jet into droplets and stabilize the polymer jet via polymer chain entanglement (Xiao & Lim, 2018).

According to Moreira, Lim, et al. (2018) the high protein content of *Spirulina microalga* associated with free surface electrospinning can facilitate the scale-up of nanofibers for food potential applications. The possibility of the formation of uniform nanofibers with a low concentration of polyethylene oxide (0.8%, *w/w*) demonstrates the potential of the protein concentrate from *Spirulina* sp. *LEB* to develop nanofibers (Moreira, Lim, et al., 2018). Moreover, Xiao and Lim (2018) have used free surface electrospinning method to fabricate pullulan-alginate fibers with a diameter range from 87 to 57 nm. They found that the incorporation of $CaCl_2$ (up to 0.045%, *w/w*) and alginate (0.8%–1.6%, *w/w*) have key roles in promoting the formation of bead-free continuous ultrathin fibers with enhancing thermal stability and increasing polymer chain entanglement, respectively. Recently a few publications have demonstrated that nonpolymeric molecules such as phospholipids or Gemini surfactant can be electrospun without using a carrier polymer. These amphiphilic molecules exhibit a behavior similar to polymers in solution. When increasing their concentration, they form micelles, first spherical, then cylindrical, until these cylindrical micelles overlap and entangle at a high concentration (Allais et al., 2018).

Continuous fibers by electrospinning have also been produced with molecules exhibiting sufficient intermolecular forces in solution to self-assemble such as diphenylalanine and cyclodextrins (CDs) forming π–π interactions and hydrogen bonds, respectively, which are crucial to obtain homogeneous fibers. In this context, Allais et al. (2018) showed that tannic acid solubilized in a mixture of water/ethanol/pure water forms a supramolecular interconnected network strong enough to allow the electrospinning of a continuous and regular nanofiber. Furthermore, as opposed to the other small molecules for which polymer-free electrospinning was also demonstrated, tannic acid nanowebs can be efficiently cross-linked in water by either oxidative reactions or the formation of coordination complexes with a wide variety of metals. The proposed electrospinning and cross-linking strategy is easy, low cost, and scalable and uses nontoxic solvents and biocompatible and biofunctional molecules. Likewise, Shi and Yang (2016) fabricated porous, interconnected nonwoven nanofiber membranes by green electrospinning of PVA/citric acid and subsequent heat treatment (140°C for 2 h) for cross-linking that were insoluble in water. Bazbouz and Russell (2018) produced cellulose acetate/sodium-activated natural bentonite clay nanofibers by free surface electrospinning. In another study, Bazbouz, Liang, and Tronci (2018) investigated the scalable manufacture and structural stability in aqueous environments of UV-cured nanofibrous membranes fabricated by free surface electrospinning of aqueous solutions containing vinylbenzylated gelatin and poly(ε-caprolactone) dimethacrylate.

4 Electrospun nanofiber materials

Electrospinning is a very practical technique in order to obtain nanofibers from a wide range of materials involving natural biopolymers and biodegradable polymers, as well as their blends that have been reported for use in various applications, depending on the desired final characteristics of the product (Ghorani et al., 2017). There are some requirements for polymer matrices in electrospinning (Nieuwland et al., 2013): (1) the polymers should be highly soluble in solution; (2) the solutions must have the required conductivity, viscosity, and surface tension; (3) the concentration of polymers should be kept within a special range; (4) the polymers in the solution should have a random coil conformation. For the application of electrospun nanofibers in food products, edibility and safety are the priority (Deng, Li, Feng, & Zhang, 2019). Thus there is a growing interest for exploring wall materials (carrier) for food applications from naturally occurring biopolymers compared with fiber products prepared from synthetic polymer solutions. This is because of their nontoxicity, biocompatibility, biodegradability, low viscosity at high solid content, hydrophilicity, retention of volatiles, suitability for high-temperature applications, possibility of modification, ability to provide optimum structures (Wen, Zong, Linhardt, et al., 2017), and their availability as by-products of agrifood processing plants.

However, electrospinning of most natural biopolymer solutions has proven to be challenging because of their broad molecular weight distribution, biovariations, high processing costs, limiting solubility of biopolymers in most organic solvents due to their high crystallinity and polarity, the polyelectrolyte nature in aqueous media, high tendency to form strong hydrogen bonds (high surface tension), and poor mechanical properties that make processability and handling of end product nanofibers difficult (Neo et al., 2013, 2014). One of the suggested approaches to overcome these technical issues is to blend natural biopolymers with biodegradable chemical polymers to make sufficient chain entanglements and/or other processing compatible aids (plasticizers and surfactants) for an effective electrospinning process with enhanced material properties such as a higher tensile strength and intact micro/nanostructures (Mendes et al., 2017). In an excellent review by Mendes et al. (2017), the advances in the electrospinning of food proteins and polysaccharides, their applications, and comprehensive discussion on the morphology, physicochemical properties and bioactivity of biopolymeric nano-/microfibers have been well summarized. More recent developments on the potential of some natural and biodegradable polymers to construct delivery systems for controlled release via electrospinning are discussed in this section.

4.1 Biodegradable chemical polymers

Polyvinyl alcohol (PVA) is a famous semicrystalline, thermoplastic, biocompatible, electrospinnable, and nonhazardous polymer for applications in food science (Estevez-Areco et al., 2018), with a recommended maximum edible dose of 3 mg/kg body weight/day. PVA is soluble in water, but its water solubility depends on the degree of polymerization, hydrolysis, and solution temperature (Pusporini et al., 2018). Due to its high hydrophilic nature, almost all the PVA electrospinning studies have been conducted with its aqueous solutions. However, Mahmud, Perveen, Matin, and Arafat (2018) explored the electrospinning behavior of PVA using a binary solvent mixture to introduce a common solvent when PVA is blended with another polymer. PVA has been added into the biopolymers to facilitate their electrospinning process. In this regard, Islam and Rezaul (2010) fabricated PVA/alginate blend nanofibers by electrospinning. Furthermore, PVA electrospun mats have also been used by many authors to encapsulate different natural bioactive molecules and extracts. Interestingly, addition of PVA by an amount of only 1% (*w/w*) to *Aloe vera* solution made its electrospinning possible leading to the fabrication of *Aloe vera* nanofibers (99%) with an average diameter of around 80 nm (Isfahani, Tavanai, & Morshed, 2017).

Another "aid material" for easier electrospinning is *polyethylene oxide (PEO)*. It is a water-soluble polymer with low toxicity, which allows its potential use as a food additive, in particular, as a carrier polymer for fabricating nanofibers of several proteins and charged polysaccharides (Liu, Li, Tomasula, Sousa, & Liu, 2016). The addition of PEO improves the ability of electrospinning by modifying the physical properties of the blended solutions, especially the degree of the chain entanglements/associations and reduction of electric conductivity of the biopolymer solutions (Zhong et al., 2018). To function effectively as a carrier, PEO at higher molecular weights (>200 kDa or higher) is required. It has been reported that electrospinning of blends of a polyanionic polysaccharide with short chain PEO (35 kDa) resulted in sprayed beads (Liu et al., 2016).

Polyvinylpyrrolidone (PVP) is a FDA-approved polymer matrix, which is acceptable for food products because of its unique properties, such as low toxicity, high solubility in polar solvents, high processability, good adhesion, high hydrophilicity, great physiological adaptation and suitable biocompatibility. PVP nanofiber-based materials have important applications in delivery and release systems and bioactive packaging fields (Kesici Güler et al., 2018). Highly soluble polymers such as PVP are a suitable matrix for nanofibers with rapid release (Sriyanti et al., 2018) and often are used for solid dispersions containing amorphous bioactives in which amorphous-state bioactives show improved dissolution rates compared with their crystalline counterparts (Pusporini et al., 2018).

Polylactic acid (PLA) is another nontoxic, natural, and biodegradable linear aliphatic thermoplastic polyester, which is typically made from starch as a renewable plant resource (Liu, Liang, Wang, Qin, & Zhang, 2018). It can be solubilized readily in organic solvents to form spin dopes optimal for electrospinning. Some works have been reported on the use of PLA for electrospinning due to its good biocompatibility, tunable mechanical properties, and ease of fabrication into ultrafine fibers (Chuysinuan et al., 2017). According to Zhang, Huang, Kusmartseva, Thomas, and Mele (2017), essential oils of tea tree and manuka can be used for enhancing the properties of PLA electrospun fibers, in terms of both mechanical behavior and antibacterial activity.

Poly(ε-caprolactone) (PCL) is also a biocompatible aliphatic FDA-approved polyester. PCL has one of the lowest melting points among the hydrophobic polymers and decomposes at high temperatures, which make it suitable for melt encapsulating of many bioactive compounds in a broad temperature range (Lian & Meng, 2017). Some attempts have described the use of PCL as a bioactive delivery carrier. PCL has been blended with kafirin (KAF) in a mass ratio of 2:1 to obtain hybrid KAF/PCL fiber mats with desirable physical properties to open up new applications of underutilized cereal proteins in nutraceutical delivery. Compared with the hydrophobic surface of neat PCL fiber mats, KAF/PCL fiber mats showed a hydrophilic surface character (Xiao et al., 2016).

4.2 Natural biopolymers: Polysaccharides and gums

Food-grade polysaccharides are used extensively as wall material for the encapsulation of food bioactive compounds owing to their capability of forming amorphous glassy solids that provides structural support to the wall material of the delivery systems. Their ability to bind bioactive ingredients is complemented by their diversity in chemical structures; composition; molecular weights; and ionic characteristics, low cost, and widespread use in foods, thus making them the preferred choice for encapsulation (Labuschagne, 2018; Stijnman, Bodnar, & Hans Tromp, 2011). Typically the formation of electrospun polysaccharide fibers is dependent on the degree of their chain entanglements and the viscosity of the solution and requires weak shear-thinning properties to favor the breakdown of the liquid jet when pulled and extended by the electric field (Jacobsen et al., 2018). Stijnman et al. (2011) classified the polysaccharides based on their behavior during electrospinning into three groups: (1) jet and fiber forming, (2) jet and droplet forming, and (3) no jet or droplets. Polysaccharides containing anionic groups were found to belong to the last group. All nonspinning solutions have either a low-shear viscosity or a high-shear viscosity but strong shear thinning (Stijnman et al., 2011). Polysaccharides typically used for electrospinning process are algal origin (e.g., alginates), plant origin (e.g., pectin, guar gum, cellulose derivatives, and starch), microbial origin (e.g., dextran and xanthan gum), and animal origin (e.g., chitosan, chondroitin, and hyaluronic acid). Besides the polysaccharides traditionally used as a matrix in encapsulation, novel materials are being investigated via electrospinning.

Cellulose is the most abundant polysaccharide in nature and also the major structural material of plants. However, little research has been done on the use of pure cellulose for electrospinning due to its low solubility in both organic and aqueous solutions (Frenot & Henriksson, 2007). Rezaei, Nasirpour, and Fathi (2015) reviewed the electrospinning techniques used for producing cellulosic nanofibers regarding its application in food science. On the other hand, there are more studies involving the electrospinning of cellulose derivatives like nonionic cellulose ethers (Wali et al., 2018), hydroxypropyl cellulose, hydroxypropylmethyl cellulose, methyl cellulose, carboxymethyl cellulose (Esmaeili & Haseli, 2017), and cellulose acetate (Mendes et al., 2017; Rezaei et al., 2015). Among all cellulose derivatives, *cellulose acetate (CA)* is a potential candidate for being used in electrospinning process and controlled release because it is easy to be processed due to its good solubility in a wide variety of solvent systems, comparatively high modulus, and adequate flexural and tensile strengths (Rezaei et al., 2015). Another important cellulose derivative that has been used extensively is *carboxymethyl cellulose (CMC)* causing difficulty with electrospinning due to its insolubility in organic solvents and strong hydrogen bonding. However, a new degradable thermoplastic CMC (TCMC) has been fabricated, and its electrospun polymer blends with PEO have been suggested for food and pharmaceutical fields (Esmaeili & Haseli, 2017).

Chitin is one of the most abundant polysaccharide biopolymers in nature, particularly, in squid pens, shrimp or crab shells, and fungi (and yeasts) cell walls, which exhibits excellent properties like nontoxicity, biocompatibility, and biodegradability. Preparing chitin nanofibers (CNFs) facilitates its dispersal in water to create a colloid solution (Hsueh, Tsai, & Liu, 2017) and for enhancing saltiness perception (Hsueh et al., 2017; Jiang, Tsai, & Liu, 2017). CNFs may be produced by electrospinning method (Barber, Griggs, Bonner, & Rogers, 2013). However, chitin presents some limitations for electrospinning due to its poor solubility and high molecular weight of some chitin sources. The combination of microwave irradiation and electrospinning eliminates the need of problematical chemicals and reduces the energy and time needed (Salaberria, Labidi, & Fernandes, 2015).

Chitosan (CS), a biobased environmentally friendly material, is an amino linear polysaccharide obtained from chitin, through a deacetylation reaction with an alkali (Barzegari & Shariatinia, 2018). It is biodegradable and biocompatible and has a broad-spectrum antibacterial activity (Deng et al., 2018). It is well known that pure CS is difficult to electrospin. Clearly, because of polyelectrolyte nature of chitosan and its low charge density, repulsive forces arise between its ionic groups by exerting high electric fields during electrospinning, resulting in bead formation instead of continuous fibers (Divya & Jisha, 2017). Consequently, nanofiber creation is facilitated by combining CS with other commonly used macromolecules including PVA, PEO, polyethylene terephthalate (PET), PLA, poly(D,L-lactide-*co*-glycolide), and PCL (Jia et al., 2007). Moreover, Liu, Wang, Lan, and Qin (2019) conducted a study in which PLA, CS, and multiwalled carbon nanotube (MWCNT) composites were combined for antimicrobial packaging application. Since MWCNTs have an extremely high strength modulus, mechanical deficiencies in PLA/CS combinations were resolved (Liu et al., 2019). Electrospinning as a growing field of research in terms of chitin and chitosan has been comprehensively reviewed by Elsabee, Naguib, and Elsayed (2012).

Starch in fibrous form is very difficult to produce due to its branched amylopectin structure. Electrospinning fibrous form of starch is attempted by various researchers as reviewed recently by Hemamalini and Giri Dev (2018). It has been reported that blending of starch with other biopolymers is quite a successful strategy for the production of fibrous starch, in comparison with starch modification. Despite this, Fonseca et al. (2019) claimed that soluble potato starch with normal amylose content can be converted into ultrafine fibers by electrospinning like a neat polymer in a fiber-forming solution. According to Fonseca et al. (2019), aging time changes the solution viscosity, which results in fibers with distinct morphologies.

The enzymatic degradation of starch leads to acquiring *cyclodextrins (CDs)*, which are nontoxic cyclic oligosaccharides consisting of glucopyranose subunits bound through α-(1,4) links (Aytac & Uyar, 2016). The most common native CDs are α-, β-, and γ-CDs, containing 6, 7, and 8 D-(+)-glucopyranose units, respectively (Gharibzahedi & Jafari, 2017). In the crystalline state and in solution, CDs form a truncated cone-shaped molecule with the hydrophilic outer surface and a hydrophobic inner cavity, which are capable of forming noncovalent host-guest inclusion complexation with a variety of molecules such as antimicrobials and flavoring and therefore enhance water solubility and thermooxidative stability of such hydrophobic and volatile molecules (de Castro et al., 2018; Rezaei & Nasirpour, 2018). It is also possible to achieve electrospinning of nanofibers from CDs without using a polymeric matrix, thanks to the formation of sufficient aggregation in their highly concentrated solutions (Aytac, Ipek, Durgun, Tekinay, & Uyar, 2017; Aytac, Ipek, Durgun, & Uyar, 2017; Aytac, Keskin, Tekinay, & Uyar, 2017; Aytac, Yildiz, Kayaci-Senirmak, Tekinay, & Uyar, 2017). According to Kfoury, Auezova, Greige-Gerges, and Fourmentin (2018), the encapsulation in CDs can increase the aqueous solubility of essential oils up to 16-fold and reduce their photodegradation rates up to 44-fold while ensuring a gradual release. Some researchers have tried to use inclusion encapsulation by *chemically modified CDs* because of their much higher water solubility when compared with native CDs (Aytac, Ipek, Durgun, & Uyar, 2017), which makes it possible to extend their functionalities during storage and processing in the food. For instance, hydroxypropyl-β-CD (HPβCD) and hydroxypropyl-γ-CD (HPγCD CDs) formed stable inclusion complexes with the optimal molar ratio of 1:1 (menthol/CD) and inclusion complex formation enhanced the water solubility of menthol (Yildiz, Celebioglu, Kilic, Durgun, & Uyar, 2018).

Recently, complexes based on amylose analogue like *spiral dextrin (SD)* have attracted increasing scientific attention of food scientists (Rezaei, Fathi, & Jafari, 2019). SD subfraction (SD-40) was suggested as a new delivery system to protect vitamin E and isoflavone with a better stability and stronger antioxidant capacity for vitamin E (Wang, Luo, & Peng, 2018). In addition, *starch-guest inclusion complex* formation in electrospun starch fibers was introduced in which guest molecules were mixed with gelatinized starch before (dope mixing method) or during (bath mixing method) electrospinning into coagulation bath of ethanol solution (Kong & Ziegler, 2014). Surprisingly, this type of encapsulation is not efficient for large bioactive molecules due to limited size of amylose helices.

It has been reported that electrospinning of aqueous *pectin* solution is not possible due to its polyelectrolyte nature. However, electrospinning of pectin and pullulan blends from aqueous solutions resulted in the creation of fibrous mats by reducing surface tension and electric conductivity and so promoting molecular entanglements (Liu et al., 2016). Considering the great potential of pectin and the structural advantages of nanofibers, Cui et al. (2016) successfully obtained pectin electrospun nanofibers (100–500 nm) by introducing a low amount of PEO, Triton, and cosolvents. Cross-linking is highly desired for obtaining high water-resistant pectin nanofibers. Consequently, Cui et al. (2017) indicated that it is possible to customize the properties of pectin electrospun nanofibers via selecting proper types of pectin and cross-linking strategy. Moreover the source of pectin is considered as an important variable in creating electrospun blend fibrous mats with desired properties (Rockwell, Kiechel, Atchison, Toth, & Schauer, 2014).

Pullulan (PUL) is a food-approved, water-soluble, nonionic linear fungal (*Aureobasidium pullulans*) extracellular polysaccharide and is currently used extensively in the food industry due to its nontoxic, nonimmunogenic, nonmutagenic, and noncarcinogenic nature (Khanzadi et al., 2015; Rekha & Sharma, 2007). In electrospinning process, it has been reported that pullulan improves the solution properties of unspinable materials by increasing their viscosity and decreasing surface tension and electric conductivity, which favors the formation of fibers (Aguilar-Vázquez et al., 2018). Indeed, it is an interesting encapsulating material due to its high electrospinnability, oxygen barrier properties, and good moisture retention.

Gums have also been proposed as potential wall materials due to their unique functionalities, nontoxicity, biocompatibility, biodegradability, and particularly excellent capability to encapsulate food bioactive compounds with a limited digestion and absorption in the body, which enables their controlled release (Mendes et al., 2017). In this concept, Lubambo et al. (2013) observed that the precipitation and purification steps affect molar mass distribution of commercial *guar gum*, which leads to improved electrospun galactomannan fiber morphology. Shekarforoush, Faralli, Ndoni, Mendes, and Chronakis (2017) prepared uniform electrospun *xanthan* fibers with average diameters ranging from 128 to 240 nm depending on the polysaccharide concentration (0.5–2.5 wt/vol%) using formic acid as a solvent. For the xanthan solutions in formic acid, results obtained from rheological data revealed no observation in common "weak gel-like" and thixotropic properties with a small change in molecular weight based on size-exclusion chromatography.

Recently, *almond gum* has been introduced as an appropriate carrier for electrospinning and encapsulation of curcumin in order to enhance its dispersion in food or aqueous solutions (Rezaei & Nasirpour, 2018). Also, electrospun almond gum/PVA nanofibers (80:20; concentration, 7% *w/w*) have been explored to encapsulate vanillin to preserve it from unsuitable conditions (Rezaei, Nasirpour, Tavanai, & Fathi, 2016). Ranjbar-Mohammadi, Kargozar, Bahrami, and Joghataei (2015) fabricated curcumin-loaded *gum tragacanth*/PVA nanofibers with optimized electrospinning parameters, which showed good biological properties. Similarly, curcumin-loaded PCL/gum tragacanth nanofiber membranes were fabricated by electrospinning, provided the controlled release of curcumin for over 20 days (Ranjbar-Mohammadi & Bahrami, 2016).

Electrospinning process of novel basil seed mucilage (BSM)/PVA was investigated at different mass ratios of BSM/PVA and two voltages (Kurd, Fathi, & Shekarchizadeh, 2017). The authors reported that with a volume ratio of 20:80 and under voltage of 18 kV, ultrafine fibers (180 nm) with a high thermostability and 43% crystallinity index were obtained. Similarly, nanofibers from *cress (Lepidium sativum) seed mucilage (CSM)* with an anionic nature and average molecular weight of 540 kDa were considered as a novel source for electrospinning, which showed size distribution in the range of 95–278 nm depending on the electric field and CSM–PVA volume ratio (Fahami & Fathi, 2018b). Furthermore, Golkar, Allafchian, and Afshar (2018) optimized production conditions of *Alyssum lepidium mucilage* as a new source for electrospinning and found that the aqueous solution of *Alyssum lepidium*mucilage/PVA (80:20), voltage (18 kV), polymer concentration (50%), tip-to-collector distance (10 cm), and feed rate (0.125 mL/h) could be successfully used to obtain uniform nanofibers with a diameter of 139.9 nm. *Gum arabic (GA)*, *karaya gum (GK)*, and *kondagogu gum (KG)* nanofibers have been prepared via electrospinning with PVA as an aiding polymer in comparison with PEO. The optimum blend solution ratios of 70:30 to 90:10 for PVA/GA, PVA/GK, and PVA/KG resulted in uniform beadless nanofibers with an average diameter of 240, 220, and 210 nm, respectively (Padil, Senan, Waclawek, & Cernik, 2016).

4.3 Natural biopolymers: Proteins

Among natural biopolymers, proteins, particularly plant proteins, are preferred for the design of delivery systems over the polysaccharides (Jia, Dumont, & Orsat, 2016). This is because of their high nutritive values and low potential to be immunogenic and GRAS status. However, unlike polysaccharide matrices, the utilization of protein matrices to encapsulate phenolic compounds is not common (Jia et al., 2016) due to binding capacity of proteins to phenolics. Inevitably the amino and carboxyl functional groups of the proteins are able to form pH-dependent electrostatic complexes with the bioactive compounds (Aguilar-Vázquez et al., 2018). Fiber formation in the case of proteins is a result of chain entanglements and reversible junctions (Jacobsen et al., 2018). In fact, secondary and tertiary structures of protein complexes make them unfavorable for successful electrospinning (Fathi, Donsi, & McClements, 2018). The globular proteins cannot be electrospun into fibers in their native globular state due to their complex macromolecular and three-dimensional structures and the presence of active, intermolecular, and intramolecular forces of the molecules, which make the polymer more elastic, a property that prevents it to be spinnable (Soares, Siqueira, Prabhakaram, & Ramakrishna, 2018; Zhong et al., 2018). Consequently, globular proteins must be unfolded to be electrospinnable by a suitable solvent and the addition of denaturing agents and heat (Jacobsen et al., 2018) or mixed with a spinning-aid polymers like PEO and PVA (Fathi et al., 2018).

Successful attempts have been made to obtain food-grade nanofibers from *whey protein (WP)* by blending it with an electrospinnable carrier material such as PEO apparently at very low levels (1% *w/w* PEO in solution leading to ~3% wt in the final fiber) with diameters between 100 and 400 nm (Zhong et al., 2018). These findings suggest the considerable potential of WP fiber as a potential replacement for the fiber bundles that form the muscular tissue in meat (Zhong et al., 2018).

Zein, known as a prolamin protein with two major fractions of α- and β-zeins, is the most used plant protein because of its high electrospinnability without the need of another polymer or toxic solvent (Torres-Giner et al., 2008). It is well dissolved in aqueous ethanol or acetic acid solutions. The α-zein prolamin from corn has been the subject of previous studies for the encapsulation of bioactive compounds by electrospinning (Moomand & Lim, 2014). Yang et al. (2017) produced nanostructured preservatives with improved thermomechanical and antibacterial performances by combining sorghum extract with electrospun zein nanofibers. Erdogan, Demir, and Bayraktar (2015) indicated that reinforcing zein electrospun fiber structures by olive leaf extract (OLE) changed its morphological and structural properties due to the presence of oleuropein as natural cross-linker. OLE addition reduced fiber diameters up to 27% and enhanced their thermal stability.

Gelatin, as a commonly used FDA-approved biopolymer, has been long investigated for electrospinning due to its bioactivity, biodegradability, nontoxicity, and easy availability (Deng et al., 2019; Lin, Zhu, & Cui, 2018) with a widespread application in stabilizing food emulsions and inhibiting syneresis or oil separation in foods. However, the electrospinning of gelatin aqueous solutions can be achieved only when the process temperature is higher than the melting point of gelatin (Nieuwland et al., 2013) or in relatively mild solvents such as acetic acid/formic acid/water and ethanol/ethyl acetate/water solutions (Deng et al., 2019). In addition, gelatin nanofibers can be cross-linked by chemical, physical, and enzymatic methods to resolve their poor water resistance (Deng et al., 2019), improve their thermomechanical properties (Lin, Zhu, & Cui, 2018), and alter their conformation and interaction. The majority of researchers are focusing on nontoxic cross-linking methods for gelatin nanofibers for use in food products, edibility, and safety via incorporation of tannic acid (Tavassoli-Kafrani, Goli, & Fathi, 2017), procyanidin (Chen, Wang, & Jiang, 2012), reducing sugars and Millard reaction products (Deng et al., 2019), and biopolymer addition (Nieuwland et al., 2013). Recently, Banner et al. (2018) reported that the addition of citric acid and coconut oil into gelatin solutions modify its nanofiber diameters/morphology and create water-resistant electrospun gelatin nanofiber mats, respectively. Also, gelatin may be used as a hydrocolloid emulsifier in emulsion electrospinning. In this regard, Zhang and Zhang (2018) successfully encapsulated corn oil into electrospun nanofibers by emulsion electrospinning. For the gelatin-stabilized O/W emulsions, gelatin could rapidly diffuse to the newly formed water-oil interface and form the steric protein-based barrier. Steyaert, Rahier, Van Vlierberghe, Olijve, and De Clerck (2016) made use of electrospinning technique, to develop gelatin nanofibers used as an instant gelatin product, without the limitations of traditional amorphous instant gelatins including moisture sensitivity, low wettability, and low modulus of the cold gel.

4.4 Combined biopolymers

In the literature, there have been a few studies about nanofiber production by electrospinning of combined natural biopolymers, which are mainly proteins-polysaccharides. Recently, Aguilar-Vázquez et al. (2018) have produced electrospun fibers from blends of *pea (Pisum sativum) protein (PP)*; *PUL*; and Tween 80, a nonionic surfactant that is used to facilitate interaction between hydroxyl groups of PUL and the amino groups of PP. With PP/PUL 50:50 containing 20wt% Tween 80, fiber structures with a mean diameter of 149nm and high elasticity are obtained. In another study the electrospinning technique was used to prepare *PUL-CMC* nanofibers in which the formation of hydrogen bonds between the COO^{-1} groups of CMC and the OH groups of PUL may synergistically enhance electrospinnability (Shao, Niu, Chen, & Sun, 2018). Electrospinning of aqueous solutions of *calcium (CaCAS) or sodium caseinate (NaCAS) with PUL* was also reported for the first time by Tomasula et al. (2016). The authors showed that the PUL/CaCAS fibers tended to be finer than the PUL/NaCAS fibers due to sufficient chain entanglements resulting from cross-link formation with the phosphoserine residues and factors like surface tension and conductivity. Oguz, Tam, Aydogdu, Sumnu, and Sahin (2018) produced nanofibers containing *pea flour and HPMC* using electrospinning and evaluated the effects of pH, pea flour, and HPMC concentration on electrospinning solution properties and characteristics of nanofibers.

It has been reported that gums in combination with other encapsulating agents can offer an advantageous formulation for the encapsulation of bioactive compounds by using electrospinning. The combination of *xanthan gum with chitosan* as an encapsulating material provided a higher encapsulation efficiency of curcumin, physical stability in aqueous media, and long-term pH-stimulated release properties (Shekarforoush et al., 2018). A few reports have suggested blending gelatin with natural biopolymers such as cellulose or its derivatives (Mendes et al., 2017). Hence, electrospun ethyl cellulose/gelatin composite nanofibers with uniform highly porous structures and tunable physical properties have been fabricated based on the Hansen solubility parameters, which serve as a tool to calculate solution compatibility (Liu et al., 2018). Blending gelatin with unelectrospinnable globular proteins is also reported to yield electrospun fibers (Nieuwland et al., 2013). Shekarforoush et al. (2018) indicated that the optimal electrospinning process of *chitosan-xanthan* solutions is directed by the apparent viscosity properties and the electric conductivity. In another work, high concentrations (50wt%) of *chitin and cellulose nanocrystals loaded to chitosan/*

PEO have been regarded as an efficient way for better spinnability, but higher solubility (fibers contact angle of 18 degrees) was observed (Naseri, Mathew, & Oksman, 2016).

4.5 Lipid-based fibers

Phospholipids have been used for preparing nanofiber structures by electrospinning especially as delivery carriers of bioactive components. The fabrication of electrospun phospholipid fibers has been initially reported by Jørgensen, Qvortrup, and Chronakis (2015) who found that these fibers had an elastic modulus of 17.26 MPa and remained their elasticity up to 24 h. Shekarforoush, Mendes, Baj, Beeren, and Chronakis (2017) developed a type of phospholipids, namely, "asolectin" electrospun fiber as microencapsulation and antioxidant matrices, which permitted the release of both curcumin and vanillin, mainly due to the swelling of the phospholipid fiber matrix over time. Several studies have investigated the creation of *liposomal* nanofibers through electrospinning with loaded hydrophilic and lipophilic compounds. Traditional liposomes often encounter a low encapsulation efficiency and instability. Nevertheless, Cui, Yuan, Li, and Lin (2017) obtained better physicochemical properties and physical stability of liposomes by designing a supramolecular structure with colloidal SiO_2 nanoparticles as a core encircled by a lipid shell. Therefore a novel nanofibrous membrane of PEO containing SiO_2-eugenol liposomes was engineered, which exhibited an antioxidant activity on beef (Cui, Yuan, et al., 2017). It is interesting to note that, recently, a new generation of stimuli-responsive nanoliposomes has been engineered called "*proteoliposomes*" in which its phospholipid bilayer is inlaid with proteins containing bioactive agents. The main advantage of proteoliposomes is that released bioactive agents will exert their activity accurately fast to targeted sites (Lin, Dai, & Cui, 2017).

5 Encapsulation of different bioactive compounds within electrospun fibers

Encapsulation for controlled release or fixation of functional additives in nanofibers from biodegradable and renewable materials might provide flexible tools to stabilize and preserve the quality of food or even to design healthier and more effective functional foods due to the special characteristics of the nanofibers (Fernandez et al., 2007). Due to their unique properties, application of nanofibers for nutraceutical delivery and food science is getting popular recently. Utilization of nanotechnology on nutraceuticals can protect nutraceuticals during processing, storage, and gastrointestinal transit before they reach the target (Kesici Güler et al., 2018). Electrospun nano-/microstructures provide a good matrix for the encapsulation/immobilization of bioactive compounds without the loss of their activity or specificity (Jacobsen et al., 2018). It does not involve any severe conditions of temperature, pressure, and chemistry with capability for continuous fabrication, versatility, and facile operating processes (Ghorani & Tucker, 2015). This is especially interesting for the food industry, where the most widely used microencapsulation technology is spray-drying, and low-temperature alternatives are needed for thermosensitive ingredients. It combines the advantages of additive manufacturing (high encapsulation efficiencies, release performances, and improved bioavailability) with versatility of materials to be processed for a broad range of compounds (Gómez-Mascaraque, Tordera, Fabra, Martínez-Sanz, & Lopez-Rubio, 2019). Also the nanofiber mat has a high surface area and porous morphology to enable higher loading of bioactive compounds (Chen, Chen, Xu, & Yam, 2018). Encapsulation by electrospinning is not limited to the aforementioned methods described earlier in Section 3. Surface loading after electrospinning is also possible. In fact, encapsulation could also be accomplished after the electrospinning process. Loading of the bioactive components can be conducted through physical adsorption, chemical immobilization, or layer-by-layer (LbL) assembly, where the fiber mats serve only as the supporting structure (Neo et al., 2013). In this section a brief overview of the more recent researches on food bioactive-loaded natural or biobased electrospun fibers within the food sector has been summarized.

5.1 Phenolic compounds

Polyphenols (phenolic compounds) are currently the major micronutrient group of interest among plant-origin bioactive compounds with two important classes of *flavonoids and phenolic acids*; they have different bioactive functionalities, in particular antioxidant properties and their prospective favorable effects on human health (Barzegari & Shariatinia, 2018). However, their low solubility, stability, and bioavailability are limiting factors for their use as bioactive agents (Fuenmayor & Coiso, 2016; Jia et al., 2016). As a result, phenolic stabilization and micro-/nanoencapsulation by several valid strategies have been suggested (Assadpour, Jafari, & Esfanjani, 2017). Many researchers have reported electrospinning encapsulation matrices to improve and enhance polyphenol functionalities (Faridi Esfanjani & Jafari, 2016).

In this field, *chlorogenic acid* was effectively encapsulated into PCL nanofibers by electrospinning process and released in a sustained manner for 12 days (Ranjith, Balraj, Ganesh, & Milton, 2017b). The same result was reported for *naringin*-loaded PCL with a release rate dependency to the flow rate (Ranjith, Balraj, Ganesh, & Milton, 2017a). Neo et al. (2013) used zein nanofibers to produce packaging materials with nanoscale features embedded with *gallic acid*. In another work, the encapsulation of gallic acid/CD inclusion complex in electrospun PLA nanofibers improved antioxidant activity and controlled release of gallic acid during processing, storage, and usage (Aytac, Kusku, Durgun, & Uyar, 2016a). According to Yang et al. (2017), *tannic acid* (TA)-Fe^{3+} complexes incorporated into CA fibrous mats exhibited a high antioxidant capacity and increased the tensile strength to ~117% compared with pure CA. Moreover a much slower TA release was observed than its noncomplex form. Sunthornvarabhas et al. (2016) evaluated *tara tannin* as bioactive ingredients in PLA electrospun fibrous delivery systems. They found that the sheet of composite fibers exhibited antioxidation activity due to full release of galloylquinic acid as the smallest-molecular-weight composition. In the study conducted by de Oliveira Mori et al. (2014), the addition of *barbatimao bark tannins* increased the glass transition temperature of the nanofibers, along with a homogeneous fibrous structure, and the presence of zein crystals in the nanofiber. Parana, Bandhitsing, and Thitiwongsawet (2019) prepared electrospun PCL fiber mats as carriers for *resorcinol (RC, 1,3-dihydroxybenzene)*; an active ingredient with radical-scavenging activity equivalent to that of EGCG in green tea. Apparently, smaller fibers showed a higher water retention due to the porous structure and greater surface area per volume or mass ratio. According to the release study of RC from fiber mats (acetate buffer at 32°C for 48 h), the maximum amounts of released RC was about 75%–90%, and the rate constant (*k*) from the fibers with smaller diameters was higher than fibers with larger diameters.

Quercetin (QU) is a plant pigment (flavonoid) antioxidative agent that can scavenge free radicals due to the functional groups on its molecular structure. However, its main drawback is related to its poor aqueous solubility, low oral bioavailability, permeability, and fast oxygen/photo-induced degradation (Aytac, Ipek, Durgun, & Uyar, 2017; Aytac, Kusku, Durgun, & Uyar, 2016b). QU-loaded electrospun nanofibers have been produced by Vashisth et al. (2013) with poly(D,L-lactide-*co*-glycolide) (PLGA)-PCL nanofibers via core-shell electrospinning, and the antifungal activity of this nanofiber was investigated. Similarly, Li, Shi, Yu, Liao, and Wang (2014) use coaxial electrospinning to produce QU-loaded zein nanoribbons with a higher quality and improved functional performances. In the study performed by Aytac, Ipek, Durgun, and Uyar (2017), QU and QU/γ-CD inclusion complex (QU/γ-CD-IC)-loaded electrospun zein nanofibers were successfully obtained; the latter showed a fast and highly efficient antioxidant activity, enhanced water solubility, and stability. Related studies to the encapsulation of QU in a more efficient way by CD-IC incorporated electrospun nanofibers have already been reported by Aytac, Kusku, et al. (2016b). They conducted a study on encapsulating QU in the wall of PLA nanofibers incorporating β-CD inclusion complex (β-CD-IC). In this context a slow release profile of QU with a rather high antioxidant activity and photostability was achieved, due to its high surface area-to-volume ratio and highly porous structure of nanofiber. In addition, Aceituno-Medina, Mendoza, Rodríguez, Lagaron, and López-Rubio (2015) developed electrospun structures with bioactive protection for application in functional foods that was produced from hybrid amaranth protein isolate (API)/PUL ultrathin fibers with the addition of QU. The electrospun PUL/PVA/*rutin* nanofibrous mats with enhanced mechanical properties and potential applications in antiultraviolet packaging and dressing materials were introduced by Qian et al. (2016). Beadless fibers were observed when rutin is used in $<7.41\%$.

Electrospinning method has produced long and continuous 3% chitosan/2% PEO blend nanofibers ($d_{\text{average}} = 214\,\text{nm}$) containing uniform distribution of 1% *microalgal phenolic compounds* (Kuntzler, Costa, & Morais, 2018). The bioactive-loaded chitosan/PEO nanofibers showed a high temperature of maximum degradation (323°C), which is an important parameter for food packaging application as well as being able to maintain its antibacterial activity. Furthermore, Estevez-Areco et al. (2018) showed the potential application of electrospun PVA/*rosemary extract* fiber mats as active food packaging, specifically for hydrophilic and acid food products. Incorporation of rosemary extract into PVA electrospun mats resulted in the polyphenol retention of 88% and improved its thermal stability. Rosemary extract already has been added to WPI/PEO solutions for electrospinning (Colín-Orozco et al., 2015). According to the research conducted by Isik et al. (2018), electrospun encapsulation of *sour cherry (Prunus cerasus L.) concentrate*, in particular, coaxial method provided a high encapsulation efficiency, better protection of cyanidin-3-glucoside, and improved its bioaccessibility. Similar results were reported for encapsulated *M. charantia fruit* extract in coaxial shell-core zein/gelatin nanofibrillar bilayer structures with a significantly higher total antioxidant activity at 60–70 days of storage compared with nonencapsulated extract (Torkamani et al., 2018). Successful nanoencapsulation of *Moringa oleifera bioactive extract* within gelatin nanofibers (EE = 80%–85%) via electrospinning process was reported to be safe for oral consumption (Hani, Torkamani, Azarian, Mahmood, & Ngalim, 2017).

The encapsulation of *green tea extract (GTE)* by electrospinning has been addressed using PVP as a wall material by Pusporini et al. (2018). According to their results, the addition of GTE (6, 8, and 10 wt%) had a positive effect on the reduction of average fiber diameter (386, 379, and 344 nm), leading to the increase in antioxidant activity based on DPPH assay (26%, 29%, and 43%). This is due to the larger surface area of the GTE/PVP nanofiber mats, which contain higher amount of catechins (Pusporini et al., 2018). Furthermore, Chuysinuan et al. (2017) reported that *rice extract* incorporated in PLA nanofibers maintained its antioxidant activity when using an electrospinning process. Electrospinning, also, can have a potential interest for the *Aloe vera (AV) (Aloe barbadensis Miller)* encapsulation. In this respect, Torres-Giner, Wilkanowicz, Melendez-Rodriguez, and Lagaron (2017) showed that by AV leaf juice incorporated into either PVP or PVA, fiber-like structures formed, while barley starch (BS), whey protein concentrate (WPC), and maltodextrin generated capsules. AV stability against ultraviolet (UV) light exposure of PVP nanofibers was lower than WPC nanobeads due to positive interactions between WPC and the hydrophilic components of AV and inherent UV-blocking and oxygen barrier properties of WPC. In addition, PVP was explored as a hydrophilic matrix in the form of electrospun nanofiber mats (in a size range of 387–586 nm) to carry *α-mangostin* in mangosteen (*Garcinia mangostana L.*) pericarp extract with an enhanced antioxidant activity and release rate (Sriyanti et al., 2017).

Catechin is a nontoxic polyphenolic compound from flavonoid family with a relatively high molecular weight and low bioavailability (Pusporini et al., 2018) and its stability is influenced both by environmental pH and temperature conditions as well as exposure to UV light, which makes it a predestinated candidate for stabilization by encapsulation to obtain an enhanced and prolonged bioactivity (Ghitescu et al., 2018). Recently, PLGA submicron electrospun fibers containing catechin by use of W/O emulsions have been shown to successfully reduce levels of reactive oxygen species induced by MWCNT in vitro (Ghitescu et al., 2018). Electrospinning have been used to produce *(−)-epigallocatechin gallate (EGCG)*-loaded zein electrospun fibers with enhanced retention of EGCG due to hydrogen bonding and hydrophobic interactions of core and shell materials (Li, Lim, & Kakuda, 2009).

Curcumin is a natural bioactive ingredient that has been widely recognized for its high antioxidant capacity and antimicrobial properties (Rafiee, Nejatian, Daeihamed, & Jafari, 2019). It has been incorporated in new delivery carriers using electrospinning technology with the purpose of overcoming its poor water solubility, chemical instability in neutral or alkaline pH conditions, sensitivity to light, and thermal treatment and ensuring its stability and therapeutic effects (Rezaei & Nasirpour, 2018; Shekarforoush et al., 2018). In this regard, Blanco-Padilla, López-Rubio, Loarca-Piña, Gómez-Mascaraque, and Mendoza (2015) designed API-PUL electrospun fiber-based delivery systems to encapsulate curcumin. Round, uniform, and homogeneous fibers (224–248 nm) were obtained for loaded API/Pul 50:50 with an encapsulation efficiency of 73% and 93% at 0.05% and 0.075% loading, respectively. Encapsulated curcumin showed a controlled and sustained released behavior under digestion process and a superior antioxidant activity. In the work performed by Shekarforoush, Mendes, et al. (2017), they confirmed the efficacy of electrospun phospholipid (asolectin) microfibers as encapsulation matrices and antioxidant systems for curcumin. The antioxidant capacity of obtained electrospun fibers remained stable over time at different temperatures (refrigerated and ambient) and pressures (vacuum and ambient). Other researchers have reported that curcumin-loaded gelatin nanofibers with cationic cetyltrimethylammonium bromide (CTAB) surfactant exhibited effective inhibition against *S. aureus* (Deng, Kang, Liu, Feng, & Zhang, 2017). Regarding curcumin properties, Deng et al. (2017) proved a direct relation between release, antioxidant, and antimicrobial activities of the encapsulated curcumin and the type of surfactant used in the formation of electrospun gelatin nanofibers. Aytac and Uyar (2017) suggested that the electrospinning of core-shell nanofibers with CD-IC core could offer a slow release of curcumin and solubility enhancement. Likewise, Rezaei and Nasirpour (2018) evaluated curcumin and curcumin-βCDIC incorporated into the almond gum/PVA electrospun nanofibers. They showed an increase in solubility up to 70 and 160 times compared with pure curcumin, respectively. Also, curcumin-βCDIC produced nanofibers with a lower diameter and slightly higher thermal stability (Rezaei & Nasirpour, 2018). In a novel study, Alehosseini, Gómez-Mascaraque, Martínez-Sanz, and López-Rubio (2019) have used liposomes as a strategy to incorporate curcumin into gelatin and zein protein fibers with encapsulation efficiencies close to 90%. Besides, GTE was incorporated within formulations, which showed a high improvement in thermal stability, protective effect, and reduction in the release of curcumin in food simulants when gelatin was used as carrier. They suggested the selection of the zein and gelatin matrices for surfaces in contact with moisty and fatty food products, respectively.

One of the phenolic groups encapsulated through electrospinning are anthocyanins known as water-soluble pigments. Prietto et al. (2018) developed food-grade pH indicator membranes sensitive to pH variation by entrapping *red cabbage anthocyanins* (3%, 4%, or 5%, *w/v*) as a natural pH dye within zein (30%, *w/v*) electrospun ultrafine fibers (444–510 nm) for monitoring pH in colorimetric scale. Recently, Moreira, Terra, et al. (2018) developed a pH indicator from PLA/PEO ultrafine fibers (921 nm) containing *phycocyanin* (3%, *w/v*), pigment of microalgae, to be used as pH indicator membranes with best responses regarding the perception of color changes (ΔE). They concluded that pH indicators from phycocyanin-containing ultrafine fibers open a way as an innovative application for intelligent packaging. Moreover the *proanthocyanidin*-loaded zein electrospun fibers produced with an encapsulation efficiency of ~100% and a high antioxidant retention were reported (Wang et al., 2016).

5.2 Essential oils (EOs), food flavor additives/fragrances

EOs that are extracted from different parts of plants are composed of highly concentrated hydrophobic volatile aroma compounds (Celebioglu, Kayaci-Senirmak, İpek, Durgun, & Uyar, 2016) and have drawn attention as flavoring agents and natural preservatives in the food industry due to their inimitable specifications such as being antibacterial, antioxidant, antifungal, antiseptic, and aroma compounds (Celebioglu et al., 2016; Kesici Güler et al., 2018). However, their highly volatile nature, low water solubility, and little bioavailability limit their applications. Consequently, various encapsulation strategies are often applied to enhance the efficiency and long-term shelf life of EOs (Amiri & Rahimi, 2018). Several studies in the literature showed a high positive correlation between the incorporation of EOs in electrospun fibers and increase in thermal stability and sustained release property with its antioxidant and antimicrobial properties.

Ginger essential oil (GEO) has been encapsulated in electrospun ultrafine fibers of soy protein isolate, PEO, and zein. The packaging with 12% GEO-loaded ultrafine fibers showed a significant control of *Listeria monocytogenes* growth in Minas cheese during refrigerated storage of 12 days (Silva et al., 2018). In addition, Wang and Mele (2018) have demonstrated the electrospinning of *EOs from clary sage and black pepper* induced in a PLA matrix; they obtained nanofibers with bacterial inactivation efficiency of

76% and 100%, respectively. Electrospun gelatin and gelatin cross-linked tannic acid nanofibers were reported for the encapsulation of *orange Eos*, which provided an adequate controlled delivery and improved storage stability (Tavassoli-Kafrani, Goli, & Fathi, 2018). The same results were reported for *carvacrol* encapsulation in PLA (Altan, Aytac, & Uyar, 2018) and starch or PCL (Tampau, González-Martinez, & Chiralt, 2017) matrices by electrospinning.

Kesici Güler et al. (2018) produced *cinnamon essential oil (CEO)*/PVP nanofibers by emulsion electrospinning with good antibacterial properties and showed that spinnability and fiber smoothness were enhanced significantly with surfactant and CEO. According to Kfoury et al. (2018), encapsulation in CDs makes the applications of EOs broader and allows a delayed and controlled release of EOs under specific conditions of humidity and temperature. Moreover, fibrous CD matrices helps encapsulating a higher level of bioactive agents via polymer-free electrospinning (Celebioglu, Yildiz, & Uyar, 2018a, 2018b, 2018c). In this regard, various kinds of EO-loaded nanocarriers have been fabricated using CD inclusion complexes of *cineole and p-cymene* (Celebioglu et al., 2018a), *CEO* (Amiri & Rahimi, 2018), *camphor* (Celebioglu, Aytac, Kilic, Durgun, & Uyar, 2017), *thymol* (Aytac, Ipek, Durgun, Tekinay, & Uyar, 2017; Celebioglu et al., 2018c), *eugenol* (Celebioglu et al., 2018b), and *menthol* (Yildiz et al., 2018) in the form of self-standing electrospun nanofibrous webs. These nanostructured forms may have potentials for food and oral-care applications due to their particularly large surface area along with a fast-dissolving character, enhanced water solubility, high thermal stability, high antioxidant capacity, quite effective encapsulation efficiency, and controlled/delayed release behavior (Amiri & Rahimi, 2018; Celebioglu et al., 2018a, 2018b, 2018c). Moreover, *thyme EO*/β-CD ε-polylysine nanoparticles (TCPNs) showed an excellent antimicrobial activity against *Campylobacter jejuni* and its subsequent incorporation into gelatin nanofibrous matrix via electrospinning revealed as antimicrobial nanofibers with promising prospects in meat preservation without a negative impact on sensory evaluation (Lin, Zhu, & Cui, 2018). Previously, Lin et al. (2017) had proposed CEO/β-CD proteoliposomes with a high encapsulation efficiency and antibacterial activity. They indicated that CEO in proteoliposomes will not be released until the proteoliposomes meet the infective target bacteria, which could secrete protease from *Bacillus cereus*. This novel antibacterial packaging material significantly extends the shelf life of beef (Lin et al., 2017).

Hexanal and benzaldehyde are food flavor additives with a generally recognized as safe (GRAS) status from plants, which are potent antimicrobial compounds but susceptible to oxidative degradation (Jash & Lim, 2017). Encapsulated *hexanal and benzaldehyde* imidazolidine precursors within PLA fibers via electrospinning by 90:10 ethyl formate/dimethyl sulfoxide blend as a solvent exhibited steady volatile release behaviors from the precursor-loaded nonwovens, after being activated with a mild acid (Jash & Lim, 2017). The authors suggested aldehyde precursor-loaded nonwovens as favorable tools for active packaging applications and shelf-life extension of fresh crops. Additionally, Rieger and Schiffman (2014) produced *cinnamaldehyde* (0.5% and 5%)-loaded chitosan/PEO electrospun fiber mats (50 nm) to be used as functional delivery vehicles that potentially eradicate *Pseudomonas* infections. This is due to the intrinsic antibacterial activity of chitosan along with the quick release of cinnamaldehyde (Rieger & Schiffman, 2014).

Various studies have used electrospinning of nanofibers with CDs as an interesting nanoencapsulation method for food flavors. In this regard, Aytac et al. (2016a) run polymer-free electrospinning of inclusion complexes of three modified CDs (HPβCD, MβCD, and HPγCD) and *limonene* to fabricate functional nanofibers that contain fragrances/flavors with a high water solubility and thermal stability. CD/limonene-IC nanofibers showed a high antibacterial activity against *E. coli* and *S. aureus*, and greater amount of preserved limonene was observed for MβCD-IC nanofibers. In the same line, Aytac et al. (2016b) fabricated polymer-free CD/*geraniol*-inclusion complex nanofibers with a high antibacterial, enhanced antioxidant activity and slow release of geraniol. According to their findings, MβCD/geraniol-IC nanofibers released less geraniol compared with HPβCD/geraniol-IC and HPγCD/geraniol-IC samples, indicating the formation of much stronger inclusion complexation between MβCD and geraniol. Kayaci and Uyar (2012) formulated *vanillin*-CD complexes in PVA nanofibers using electrospinning technique. γ-CDs showed a prolonged and moderate release of vanillin along with higher thermal stability in comparison with α- and β-CDs. The same result was obtained for geraniol-CD inclusion complexes with PVA in that the loss of geraniol was minimal (~10%) for PVA/geraniol/γ-CD-IC nanofibers due to a more complexation efficiency between geraniol and γ-CD (Kayaci, Sen, Durgun, & Uyar, 2014). Rezaei et al. (2016) reported vanillin-loaded almond gum/PVA composite electrospun nanofibers exhibited a relatively good stability under dry ambient conditions and maintained around 58% of the incorporated vanillin for 90 days.

Lević et al. (2014) encapsulated *ethyl vanillin* in PVA electrospun fibers (100–1700 nm) and showed melting point of immobilized ethyl vanillin was lower (~55°C) compared with free flavor (~77 degrees). According to their results, films based on PVA/ethyl vanillin nanofibers are mechanically stable. The effectiveness of *butyrate hydroxyanisole (BHA)* as a volatile food antioxidant in gelatin electrospun fiber mats was evaluated for storage of strawberries in the work of Li et al. (2018). They reported that BHA was efficiently encapsulated (90%) and showed a good antioxidant capacity along with antibacterial (against *S. aureus*) and antifungal (against *Rhizopus* sp., *Mucor* sp., *Aspergillus* sp., and *Penicillium* sp.) activity.

5.3 Vitamins

Electrospinning has been considered as a useful technique for the encapsulation of vitamins. For instance, *retinyl acetate (RA)*, one of the most commonly used vitamin A derivatives, was effectively incorporated into electrospun nanofibers of PVA containing β-CD in order to form inclusion complexes for encapsulation to prolong its shelf life and thermal stability (Lemma et al., 2015). Additionally, Fahami and Fathi (2018a) reported *vitamin A*-loaded cress seed mucilage/PVA electrospun nanofibers with an enhanced thermal resistance, great stability against acidic conditions, encapsulation efficiency of 97%, and loading capacity of 29.51%. In a broader perspective, cress seed mucilage/PVA nanofibers might be suitable for the incorporation of hydrophobic ingredients (Fahami & Fathi, 2018a). In the same way, Alborzi, Lim, and Kakuda (2014) showed that the degradation processes of *folic acid (vitamin B9)* within the electrospun pectin/PEO polymers was reduced in low pH food materials. Electrospinning was also used as a strategy to diminish *folic acid* sensitivity to UV exposure for 2 h, allowing its better availability and enhancing its thermal stability in API/PUL-based ultrathin structures (Aceituno-Medina, Mendoza, Lagaron, & López-Rubio, 2015). Folic acid was also encapsulated in electrospun zein fibers, which was proposed as promising ingredients for application in foods that require thermal processing or exposure to irradiation (do Evangelho et al., 2018). Interestingly the results showed ~8.7% thermodegradation of fibers containing 1.5% folic acid at 180°C and a slight reduction in folic acid concentration, when exposed to 24 h of UVA irradiation.

The strategy of inclusion complexes (IC) between HPβCD and *vitamin E* has provided unique characteristics in terms of high loading of vitamin E (11%, *w/w*), improved UV photostability, enhanced water solubility, and prolonged shelf-life antioxidant activity. Moreover a 3-year-old vitamin E/HPβCD-IC nanofiber sample showed a very similar antioxidant efficiency when compared with freshly prepared vitamin E/HPβCD-IC samples (Celebioglu & Uyar, 2017). From a practical point of view, Fathi, Nasrabadi, and Varshosaz (2016) concluded that dextran electrospun ultrafine fibers (~400 nm) containing vitamin E showed a high potential for cheese fortification. Aytac and Uyar (2016) encapsulated *α-tocopherol/*β-CD inclusion complexes (αTC/β-CD-ICs) into electrospun nanofibers of PCL. As a result of stabilization and solubility increment of α-TC, they achieved delivery systems having a high antioxidant activity and photostability. According to Aytac, Keskin, Tekinay, and Uyar (2017), PLA/α-TC/γ-CD-IC nanofibers displayed a higher antioxidant activity than PLA/α-TC nanofibers and, when applied directly on beef, prolonged its shelf life. In another study the retention of α-TC in soluble dietary nanofibers (SDFs) and SDF films was reported to be higher than those without SDFs after heating at 40°C and UV irradiation exposure at room temperature (Li, Chotiko, Narcisse, & Sathivel, 2016).

β-Carotene (βC), which is the most commonly known type of carotenoids, is a nonpolar polyene pigment with provitamin A activity and widely used as a nutritional supplement due to its strong antioxidant activity (Peinado, Mason, Romano, Biasioli, & Scampicchio, 2016; Reksamunandar, Edikresnha, Munir, Damayanti, & Khairurrijal, 2017). However, high interaction with radicals and the photooxidation may reduce its bioavailability (Rostamabadi, Falsafi, & Jafari, 2019). Encapsulation is an effective way to stabilize carotenoids and improve their antioxidant activity and water solubility; therefore it is promising to use them in food processing, particularly in aqueous food matrices (Inanc Horuz & Belibagli, 2018). Polymeric nanofibers containing carotenoids have become the focus of much research. According to Reksamunandar et al. (2017), βC encapsulation into PVP through electrospinning showed promising results in terms of antioxidant properties. Besides, for light-sensitive food processes or food packaging strategies, Fernandez et al. (2007) suggested the use of electrospinning for encapsulation and enhanced stability of βC in ultrafine fibers of PLA. Similarly, Fernandez, Torres-Giner, and Lagaron (2009) demonstrated that βC encapsulation in ultrafine electrospun fibers of zein was useful when exposed to UV-Vis irradiation and βC was not observed to be oxidized or isomerized. Typically, Peinado et al. (2016) found that βC incorporation into electrospun nanofibers of PEO prolongs its shelf life and thermal stability. In another work, hybrid encapsulation structures based on βC-loaded nanoliposome into ultrathin fibers of PVA and PEO were reported by de Freitas Zompero, Lopez-Rubio, de Pinho, Lagaron, and de la Torre (2015) via electrospinning technique. The βC-loaded nanoliposomes with electrospun fibers showed a high UV photostability. PVA exhibited a better electrospinning performance in comparison with PEO. Inanc Horuz and Belibagli (2018) successfully produced composite zein nanofibers containing the carotenoids extracted from tomato peel with above 90% encapsulation efficiency, high thermal stability and better retention of lycopene, and 11-fold antioxidant activity during 14-day storage stability by electrospinning.

5.4 Fish oil and sterols

Encapsulation has been one of the strategies for protecting fish oil and other health-promoting omega-3 fatty acids in view of their high susceptibility to oxidative degradation and undesirable fishy flavor to meet consumer request for organoleptic quality and functionality (Esfahani, Jafari, Jafarpour, & Dehnad, 2019; Ghorbanzade, Jafari, Akhavan, & Hadavi, 2017). Regarding nanocarriers made by special equipment, several studies attempted to evaluate the omega-3 nanodelivery systems (García-Moreno et al., 2017; García-Moreno, Damberg, Chronakis, & Jacobsen, 2017; Moomand & Lim, 2014, 2015). In this perspective, Moomand and Lim (2015) achieved a higher antioxidant activity by encapsulating fish oil in electrospun fibers. Encapsulated fish oil in zein polymer (20%, *w/w*) by electrospinning has been reported to result in changes in ribbon-shaped fiber morphology (190–300 nm) and beaded fibers (up to 1 μm) under the effect of aqueous ethanol or aqueous isopropanol solvents, respectively (Moomand & Lim, 2015). At 30% (*w/w*) fish oil loading content, the encapsulation efficiency of the electrospun zein fibers reached 91% (ethanol-based fibers) and 96% (isopropanol- based fibers). In another study, Moomand and Lim (2014) found the release of fish oil was greater in the presence of pepsin under simulated gastric fluid (SGF) conditions. Furthermore, Yang, Wen, et al. (2017) reported that oxidative stability of encapsulated fish oil in the coaxial nanofibers (EE = 94%) was clearly improved compared with uniaxial obtained fibers, but the release behavior of encapsulated fish oil changed. Consequently, they proposed that the addition of ferulic acid into electrospun zein mats significantly enhanced the oxidative stability of encapsulated fish oil (EE = 94%) with no changes in the release property of fish oil (Yang, Wen, et al., 2017). In the same way, results from the literature have proved that both oil content and natural antioxidant addition are key factors to improve the oxidative stability of electrospun fibers.

For instance, PUL submicron fibers containing 10% wt fish oil showed the pure fiber morphology with a higher encapsulation efficiency (88.5%) and oxidative stability with antioxidants (500 ppm α-tocopherol and 500 ppm rosemary extract) when compared with 30 wt% oil-loaded fibers (García-Moreno, Damberg, et al., 2017). The same authors have electrospun the solutions of 200 kDa PUL (15 wt%) and showed a superior oxidative stability when neat fish oil (vs. emulsified oil) was incorporated into fibers due to a higher fish oil entrapment and the elimination of the homogenization process (García-Moreno, Özdemir, et al., 2017). Generally the unique structural features of electrospun fibers were suggested to play a relevant role in the preservation of fish oil. Nevertheless, García-Moreno et al. (2016) reported that fish oil encapsulation in PVA nanofibers, which was obtained via emulsion electrospinning, exhibited a much more rapid rate of oxidation than emulsified and unprotected fish oil. Higher content of oxidation products was mainly ascribed to the presence of metal traces in the polymer.

Paaver et al. (2016) enhanced the solubility and dissolution behavior of β-sitosterol by affecting the solid state of this bioactive substance. The β-sitosterol-loaded nanofibers were freely water soluble and exhibited a very short lag time in releasing the plant sterol.

5.5 Natural antimicrobial agents

Due to the unique properties of electrospinning technology, electrospun nanofibers containing antimicrobial agents show an excellent antimicrobial efficacy against a wide range of microorganisms and sustain the antibacterial ability over a long period of time. In this section, recent research activities on antimicrobial electrospun nanofibers incorporated with phytochemicals are presented. For more information regarding different metallic nanoparticles, carbon nanomaterials, antibiotics, and biopolymer-driven antimicrobial electrospun nanofibers are reviewed comprehensively by Song, Wu, Qi, and Kärki (2017).

Phytochemicals like phenolics, terpenoids, and alkaloids are found associated with an antimicrobial activity. In this regard, Kim et al. (2016) successfully fabricated a novel antibacterial nanofiber composite from PVA containing *Juniperus chinensis extract* with a good efficiency for antibacterial activity against both the Gram-positive/G-negative bacteria, even in lower amount of the extract. *ε-Polylysine (ε-PLY)*, a cationic biological metabolite with an excellent antibacterial activity, was selected to prepare (ε-PL)/chitosan nanofibers via electrospinning (Lin, Xue, Duraiarasan, &

Haiying, 2018). It behaved significantly as a potent novel antibacterial food packaging material (against *Salmonella*) without any losses on the sensory quality after treatment on chicken samples. Aytac, Yildiz, Kayaci-Senirmak, Tekinay, and Uyar (2017) produced polymer-free nanofibrous webs of CD/*linalool*-inclusion complexes, which showed fast-dissolving characteristics, high preservation, prolonged release, and great antibacterial activity against *E. coli* and *S. aureus*, making it a good candidate as fast-dissolving supplement material in food and pharmaceutical products. Nasir and Apriani (2017) have successfully synthesized *Ag/catechin*-gelatin nanofibers with a diameter around 399 nm by electrospinning with 40.4% inhibitory activity for *S. aureus*. Moreover, antimicrobial electrospun ultrafine fibers (against *S. aureus and* L. *monocytogenes*) from zein containing *eucalyptus EOs*/CD inclusion complexes were introduced as a promising material for use in antimicrobial applications (Dias Antunes et al., 2017). Likewise, Wang et al. (2017) reported that *curcumin*-loaded zein electrospun fibers (EE = 100%) had an antibacterial activity toward *S. aureus* and *E. coli*. The inhibition performance strongly depended on curcumin content.

Electrospun active nanofibres (~100–300 nm) were successfully fabricated from chitosan/PVA/β-CD solution with *cinnamon and oregano EOs*. The nanofibers potential to release EO volatiles and demonstration of antifungal activity against *Botrytis* sp. have been confirmed (Munhuweyi, Caleb, van Reenen, & Opara, 2018). Furthermore, Ceylan, Unal Sengor, Sağdıç, and Yilmaz (2017) indicated chitosan nanofibers (CNs) and *smoke*-loaded electrospun CNs (LSCNs) could be used as a functional coating on fish fillets to prevent microbial growth during storage. According to the microbiological stability assays, CNs and LSCNs were both effective with an almost 40%–50% microbial growth control, but the antimicrobial effect of CNs was more than that of LSCNs in terms of total mesophilic aerobic bacteria, psychrophilic bacteria, and yeast and mold count. In view of the well-established antimicrobial efficacy of *allyl isothiocyanate (AITC)* in many food systems, Vega-Lugo and Lim (2009) reported its successful encapsulation into PLA and soy protein/PEO blend electrospun fibers, which resulted in a controlled release of AITC triggered by an increase in the relative humidity. In another similar study, Dai and Lim (2015) showed that the electrospun PLA-PEO nonwoven matrix is a promising carrier for controlling the release of AITC from defatted mustard seed meal powder (MSMP). Aytac, Dogan, Tekinay, and Uyar (2014) demonstrated that PVA/AITC/β-CD-IC nanofibers have a higher antibacterial activity against *E. coli* and *S. aureus* compared with PVA/AITC nanofibers, due to the presence of higher amounts of AITC. Wang et al. (2015) developed an in vitro study on the efficiency of *pleurocidin*-incorporated PVA fibers to inhibit *E. coli* O157: H7 in apple cider and confirmed its higher inhibition potential compared with free pleurocidin.

The electrospinning technique has been used to prepare nanofibers for encapsulation and sustained release of the *bacteriocins* from probiotic bacteria during food processing and storage. For instance, the nisin-loaded gelatin nanofibers showed a high antimicrobial performance (Dheraprasart, Rengpipat, & Supaphol, 2009). The fiber mats were fragile, and cross-linking by glutaraldehyde strengthened the mat without affecting its antimicrobial activity. The potential use of API-PUL nanofibers when incorporating nisin as an antimicrobial compound was confirmed by Soto et al. (2016). Nisin-loaded nanofibers (20 mg/mL) showed a 95.41% EE and antibacterial activity against *Leuconostoc mesenteroides* in both solid and broth MRS media (1 log reduction). Han, Sherman, Filocamo, and Steckl (2017) evaluated the long-term antimicrobial (4 log reduction after 5 days of exposure) potential of trilayer coaxial electrospun fiber membranes containing nisin in the core with a hydrophobic PCL intermediate layer and a hygroscopic outer layer of CA. In comparison, nisin-incorporated/nisin-encapsulated fiber membranes by single fiber did not provide any antimicrobial activity, and coaxial fiber membranes showed a burst release of nisin with >99% bacterial kill (2 log reduction) after 1 day of exposure (Han et al., 2017).

5.6 Enzymes

Enzymes are biocatalysts with a practical application in the commercial and industrial sectors especially biosensors, bioreactors, food processing, and food additives (Rojas-Mercado, Moreno-Cortez, Lucio-Porto, & Pavón, 2018). Commercial application of biocatalysts depends on the efficiency of the immobilization method and residual enzyme activity (Gabrielczyk et al., 2018). Enzyme immobilization approaches can basically be divided into three categories: covalent or noncovalent attachment to a support, encapsulation, and carrier-free cross-linking (Gabrielczyk et al., 2018), with encapsulation as the promising method to retain the tertiary structure or active site of enzymes. Enzyme immobilization on electrospun fibers can promote and maintain the natural catalytic activity and the conformation of enzymes by multiple points of attachment onto the electrospun fiber support (Aldhahri et al., 2018) and allow enzyme separation from the reaction medium for recycling (Aldhahri et al., 2018; Rojas-Mercado et al., 2018).

Focusing on immobilizing enzymes on electrospun nanofibers, a key limitation is the choice of liquid system for dissolution of an enzyme and water-insoluble polymers (Gabrielczyk et al., 2018). However, fibrous materials are among the most desirable supports due to their intrinsically high specific surface area, interfiber porosity, easy handling, and good mechanical strength (Dos Santos, Zavareze, Dias, & Vanier, 2018).

Based on the currently available information, enzymes can be encapsulated within polymer fibers, coelectrospinning methods, or via postspinning immobilization by either covalent or noncovalent attachment to fibers (Aldhahri et al., 2018). Yet, covalent binding method requires fabricating nanofibers that contain reactive groups. In addition, reduction in enzyme activity has been observed over time, as well as limitation on enzyme loading due to the formation of a monolayer on the fiber surface (Dos Santos et al., 2018). Consequently, polymer encapsulation methods and coelectrospinning have been investigated more frequently. For instance, Rojas-Mercado et al. (2018) reported the encapsulation of ficin extract in PVA electrospun nanofibers. Glutaraldehyde vapor treatment (GAvt) was used as the cross-linker during immobilization process (1 h, pH = 8, 20% loading), which showed a high enzyme activity of the crude ficin extract (92%) that remained constant even after nine reuse cycles for 25 days. Previously, Moreno-Cortez et al. (2015) reported encapsulation and immobilization of *papain* in PVA electrospun fibrous membranes. The immobilization step was achieved through cross-linking by GAvt (24 h) in which the enzyme retained its catalytic activity after six cycles and maintained 40% of its initial activity after being stored for 14 days.

In addition, the coelectrospinning method offers an easy way for the encapsulation of enzyme into nanofibers with a good efficiency, high enzyme activity, and reusability, despite low enzyme loading (Dos Santos et al., 2018). In a research on the encapsulation of *fructosyltransferase* from *B. subtilis* (Gabrielczyk et al., 2018), the efficiency of core–shell immobilizates was at least 4.4-fold higher than one-dimensional fibers. The polymers used for enzyme immobilization may protect enzymes from pH and temperature instability. An alternative support for enzymatic immobilization is the use of CDs, which offer functional solutions by creating complex structures that exhibit unique properties, such as improved solubility and nontoxic nature of electrospun fibers. In this regard, immobilization of xylanase and xylanase-β CD complexes via electrospinning improved enzyme activity at a wide pH (4, 5, 7, and 8) and temperature range (Dos Santos et al., 2018). As a result of immobilization, optimum xylanase activity changed from 60°C to 70° C. Huang et al. (2017) developed a simple route based on electrospinning and LBL self-assembly processes to prepare controllable immobilization of *naringinase* on electrospun CA nanofibers with the purpose of grapefruit juice debittering. They demonstrated that naringin and limonin are the major bitter components removed by 22.72% and 60.71% via hydrolysis with naringinase and adsorption with CA nanofibers, respectively. Nunes et al. (2016) prepared enzyme entrapment fibrous mats by electrospinning where PVA was used for naringinase immobilization with a high retention (100%), and further cross-linking increased the thermal stability up to 121°C.

6 Bioactive-loaded electrospun fibers in food packaging

Recently, nanotechnology in food packaging has been an emerging area in which packaging materials can be manipulated for improving the barrier, mechanical and heat-resistance properties, biodegradability, and flame retardancy compared with normal polymers (Amna, Yang, Ryu, & Hwang, 2015). In comparison with the traditional films prepared by casting or coating methods, electrospun fibers are more responsive to surrounding atmosphere changes and enable a controllable release of encapsulated bioactives due to a high surface area and interconnected porous structures (Li et al., 2018; Shao, Yan, Chen, & Xiao, 2018). Furthermore, recent advances in this field have shown that multilayer systems containing electrospun ultrathin fibers can be applied as mono- and multilayers, LbL assemblies, and similar structures in packaging materials to improve their barrier performance (Cherpinski et al., 2018) and mechanical reinforcement of bipolymers (Cerqueira, Torres-Giner, & Lagaron, 2018; Cerqueira, Vicente, & Pastrana, 2018).

High barrier electrospun nanostructured interlayers with adhesive properties, made of different hydrocolloids (WPI, PUL, zein, and zein/WPI/PUL blends), have been used to enhance the barrier properties of polyhydroxyalkanoate (PHA) material in food packaging applications (Fabra, López-Rubio, & Lagaron, 2014). The same authors improved the barrier properties of thermoplastic corn starch-based films using bacterial cellulose nanowhiskers by means of PHA electrospun coatings (Fabra, López-Rubio, Ambrosio-Martín, & Lagaron, 2016). With respect to oxygen barrier properties, electrospun ultrathin zein fibers laminated in a sandwich-type structure improved PLA films (Cherpinski et al., 2018). Ghosal et al. (2018) showed the great potential of electrospinning over solvent casting method to fabricate polymeric membranes with a dramatic improvement in flexibility and plasticity of PCL-based nanofiber matrices. They suggested that electrospinning helps in plasticization and tuning of mechanical properties. Moreover, Deng et al. (2018) reported a hydrophobic surface (water contact angle of 118.0 degrees) for the gelatin/zein nanofibrous films, while the casted gelatin/zein film had a hydrophilic surface (water contact angle of 53.5 degrees). According to Cherpinski, Torres-Giner, Cabedo, Méndez, and Lagaron (2017), thermally postprocessed electrospun biopolymer coatings over fiber-based packaging materials are very promising systems since they self-adhere during annealing as a result of high surface-to-volume ratio of the fibers and furthermore provide enhanced barriers to gases and vapors when built with a sufficient thickness.

Additionally the use of nanocarriers to encapsulate bioactive molecules (such as EOs, peptides, antifungal and antibacterial agents, antioxidants, and oxygen scavengers) and include them in packaging materials will be possible not only during the production process (e.g., extrusion or solvent casting) but also through the formation of a layer (e.g., spray coating and electrospinning) leading to the formation of an active multilayer packaging (Vahedikia et al., 2019). Instead of mixing bioactive compounds directly with food, incorporation in packaging materials allows the functional effect at food surfaces where biological activity is localized (Amna et al., 2015; Cerqueira, Torres-Giner, et al., 2018; Cerqueira, Vicente, et al., 2018). This will also allow using responsive materials (an intelligent material with active properties) and, when triggered by an external influence (e.g., pH and temperature), will release the bioactive compound and signal microbiological and biochemical changes (Cerqueira, Torres-Giner, et al., 2018; Cerqueira, Vicente, et al., 2018).

Some studies involve the use of electrospinning as a novel method to develop antimicrobial functionalities in food packaging. A new class of antimicrobial hybrid (organic-inorganic) packaging mats, which potentially reduce the contamination of fresh/or processed meat and meat products, was developed by Amna et al. (2015) using electrospinning for the supplementation of virgin olive oil and zinc oxide with biodegradable polyurethane. Deng, Taxipalati, et al. (2018) fabricated electrospun chitosan/PEO/lauric arginate nanofibrous films with enhanced antimicrobial activity. They demonstrated that the antimicrobial mechanism of these nanofibrous films was cell membrane damage and, ideally, inhibition zone against *S. aureus* was larger than those against *E. coli*. Also, Li et al. (2018)

suggested that the functionalized gelatin-BHA electrospun fiber mats may have a great potential in active food packaging primarily for their high antioxidant, antibacterial, and broad-spectrum antifungal activity. Lin, Zhu, and Cui (2018) proposed electrospun gelatin-glycerin-ε-polylysine nanofibers as packaging materials due to their long-term anti-*Listeria* activity on beef.

To maintain greater CEO in the film and avoid too much loss of volatile substances, an antimicrobial packaging material was developed by incorporating CEO/β-CD inclusion complexes into PLA nanofibers via electrospinning (Wen et al., 2016). The same authors demonstrated the ability of electrospun PVA/β-CD films loaded with CEO to inhibit microbial growth of *S. aureus* and *E. coli*. They showed that electrospun films were able to contain higher EOs compared with conventionally casted films resulting in better antimicrobial properties and effectively prolonging the shelf life of strawberry (Wen et al., 2016). Besides, Shao, Yan, et al. (2018) fabricated electrospun ultrafine PVA/permutite $[(SiO_2)_xAl_2O_3]$/CEO fibrous films for active packaging applications, which can delay the rapid spoilage of strawberries during storage. According to their results, due to physical interaction between the CEO and the fibers, the release of CEO at low temperatures will achieve the minimum concentration required for preservation (Shao, Yan, et al., 2018). Liu, Liang, et al. (2018) produced electrospun antimicrobial PLA/tea polyphenol (TP) nanofibers for food packaging applications. The composite fibers were morphologically uniform, with fiber diameters that decreased with increasing TP content. The PLA/TP-3:1 composite fiber exhibited the strongest DPPH RSA (95.07%) and good antimicrobial activities against *E. coli* and *S. aureus*. In a study conducted by Altan et al. (2018), composite fibrous films were developed from zein and PLA by incorporating 20% carvacrol using electrospinning. Carvacrol-loaded electrospun fibers with antioxidant and antimicrobial properties had a sustained release property and inhibited 99.6 and 91.3% of the fungi growth. The potential use of curcumin-loaded zein nanofibers as edible and effective fungicidal coatings on apples was highlighted in the research of Yilmaz et al. (2016) in order to limit its postharvest decay (*Penicillium expansum and Botrytis cinerea*). α-Tocopherol loaded in hydrocolloid matrices of WPI, zein, and soy protein isolate nanofibers was directly electrospun as a coating onto wheat gluten films and showed good antimicrobial properties (Fabra, López-Rubio, & Lagaron, 2016). The authors reported no degradation in α-tocopherol during a usual sterilization process.

Electrospinning methods have also been proposed for stabilizing or controlling the release of the antioxidant compounds in food packaging. As an example, Fabra, López-Rubio, Sentandreu, and Lagaron (2016) fabricated multilayer corn starch-based food packaging containing β-carotene-loaded PCL matrices by electrospinning. Besides, Arrieta, López de Dicastillo, Garrido, Roa, and Galotto (2018) showed that plasticized PLA bio-nanocomposites reinforced with electrospun PVA containing *Durvillaea antarctica* extract are promising materials for sustainable antioxidant packaging applications. The electrospun fibers of PCL/vitamin E showed a high potential as functionalized coatings for extending the shelf life of food products by using electrospinning approach (Dumitriu, Mitchell, Davis, & Vasile, 2017). In another study, electrospun PCL nanofibers with *Urtica dioica* L. extract were incorporated into WPI to develop bioactive coatings, which improved rainbow trout fillet quality after 15 days of storage at 4°C (Alp Erbay, Dağtekin, Türe, Yeşilsu, & Torres-Giner, 2017). Similarly, chicken meat cuts covered by the PCL/microalgae biopeptide nanofibers showed a great stability ensuring the product quality maintenance during 12 days of preservation (Gonçalves et al., 2017). Tea polyphenols loaded in PUL-CMC electrospun nanofibrous films significantly decreased weight loss and maintained the firmness of strawberries and improved the quality of this fruit during storage (Shao, Niu, et al., 2018).

Indeed, the use of electrospinning in active food packaging by incorporating the fibers of both nanofillers or bioactive substances can add an extra value to the final product, changing the packed food conditions to extend the shelf life and improve safety and/or sensory properties (Torres-Giner et al., 2017). These results depict electrostatic spinning procedures as an emerging technology to design bioactive packaging with antimicrobial/antioxidant protection or delivery of nutraceuticals into food. However, further detailed evaluation should be conducted in terms of efficiency (Mousavi Khaneghah, Hashemi, & Limbo, 2018) and toxicity in food packaging applications.

7 Electrospun fibers loaded with phase change materials (PCMs)

Phase change materials (PCMs) are substances that undergo a phase transition at a specific temperature, and as a consequence, they are able to absorb and release latent heat with a very small temperature differential (Yuan, Zhang, Tao, Cao, & He, 2014). PCMs could be used during transfer, storage, and distribution stages to keep the cold chain of solid foods, beverages, cooked foods, and many other products in various sectors. Paraffin waxes, fatty acids, eutectics, and hydrated salts are among the most frequently used PCMs (Yuan et al., 2014; Chalco-Sandoval, Fabra, López-Rubio, & Lagaron, 2017).

Recently, PCMs have been encapsulated by electrospinning techniques, which can be of interest in the food industry in order to develop new smart packaging materials with the ability of temperature control (Pérez-Masiá, López-rubio, & Lagarón, 2013). However, single electrospinning is not suggested for PCM encapsulation due to insufficient encapsulation and partial leakage of PCMs through the fibers (Noyan, Onder, Sarier, & Arat, 2018). Consequently, coaxial electrospinning for encapsulating both hydrophilic and lipophilic PCMs in a variety of polymers has been considered as a suitable method to solve the aforementioned challenges. In this regard, Hu and Yu (2012) developed natural soy wax/polyurethane fibers by coaxial electrospinning with 100 heating-cooling cycles along with a uniform fiber morphology and homogeneous wax distribution. Furthermore, Noyan et al. (2018) reported a good encapsulation efficiency, enhanced mechanical properties, and remarkable thermal energy storages at broad melting temperatures (−1°C to 60°C) for polyacrylonitrile PCM fibers via coaxial electrospinning. PCMs from a mixture of plant oil-loaded PVA matrices have been developed via emulsion electrospinning (Zdraveva, Fang, Mijovic, & Lin, 2015). Such PCM fibers have shown a reliable heat-regulating performance, which can undergo at least 100 cycles of phase changes. Moreover, electrospun fibers containing PCMs based on lauric acid/PET (Chen, Wang, & Huang, 2008), PEG/CA (Chen, Zhao, & Liu, 2013), and dodecane/zein (Pérez-Masiá et al., 2013) exhibited remarkable heat management properties.

8 Modification of nanofiber surfaces

Modification of nanofiber surface through functionalization of polymeric nanofibers has been very attractive in the literature. This is due to the formation of functional groups on the nanofiber surface, which improves their biocompatibility and wettability and permits the immobilization of bioactive substances. Several methods including wet chemical methods, surface grafting, coelectrospinning of surface-active agents, and plasma treatment have been developed to modify nanofiber surfaces (Labuschagne, 2018). The latter has been found to be an efficient and green technology of enhancing the structural and physicochemical properties of the electrospun fibers.

In this regard, Cui, Bai, and Lin (2018) treated PEO nanofibers containing tea tree oil/β-CD inclusion complexes (TTO/β-CD-IC PEO) by plasma technology to enhance their release efficiency and antibacterial activity. The antibacterial assay of plasma-treated TTO/β-CD-IC PEO nanofibers exhibited a stable antibacterial effect on beef during 7-day storage. Moreover, electrospun nanofibers of gum arabic and karaya and kondagogu gums have been modified by methane plasma treatment (Padil et al., 2016). In some studies, heat treatments have been considered as a suitable technology for enhancing nanofibrous physicochemical properties. In this context, aqueous insoluble guar gum/PVA membranes were fabricated through postelectrospinning cross-linking by heat treatment (140°C for 2h); blends had 1wt% guar gum, 8wt% PVA (3:7 weight ratio), and 5wt% citric acid. Cross-linking of guar gum/PVA occurred through esterification reaction during heat treatment (Shi & Yang, 2016). Furthermore, Neo et al. (2014) demonstrated that heat treatment (150°C for 24h) remarkably increased the molecular weight of zein and zein-gallic acid electrospun fiber mats and reduced their release rate. The higher water-resistant nanofiber membranes based on defatted soy flour/gluten were also produced by cross-linking reaction without the use of any toxic cross-linkers (Lubasova, Mullerova, & Netravali, 2015). The authors applied oxidized sugar-containing aldehyde groups as green cross-linkers for the protein-based nanofiber membranes (Lubasova et al., 2015).

9 Conclusion and future trends

A great progress has been made in the last two decades in the field of electrospinning. It has turned out to be the most reliable, simple, and cheap technique for the fabrication of nanofibers, which can be used for many different applications. Electrospun polymer nanofibers have interesting characteristics resulting from their submicron diameter, fine interconnected porous network, and high surface-to-weight ratio compared with other fibrous structures. Thousands of research studies have been conducted by using both natural and synthetic polymers. Electrospinning provides wide options for loading bioactive compounds into the polymers. The field of food industry is notable, and the substantial applications of electrospinning within this field include stabilizing or controlling the release of bioactive compounds, food processing and packaging, and entangled mats simulating meat. In fact, by the applications of electrospinning and encapsulation in food science, healthier food products with desirable organoleptic properties and components with enhanced properties will be generated. Moreover, multilayer structures, in particular, electrospun fibrous layers, have been reported as an effective alternative for improving the barrier properties and mechanical performance of biopolymer-based films. Developments on environmentally friendly electrospun polymeric fibers, such as biocomposites and bionanocomposites, generally reinforced with natural additives have been reported. Even electrospun membranes of biobased and natural polymers are intended for fabricating into labels, sachets, or stickers, to be applied on the primary packaging, and add an intelligent or active function to food packages. In this connection, the combination of phase change materials with the electrospinning could provide new solutions for developing smart packaging systems with controlled barrier properties. Moreover, investigations relying on the construction of functional nanodroplet-loaded electrospun fibers with potential application for self-healing packaging might be an innovative concept. However, strategies to improve the mechanical properties of the fibers need to be addressed since stronger materials are needed to develop food packaging. In addition, further research on the interactions with the food matrix, shelf life, and toxicological studies is needed. Thanks to the exclusive properties of nanofibers, their use in food science will undoubtedly expand in the future. One of the main areas where electrospun nanofibers can be developed further in the future is three-dimensional (3-D) food printing in which the challenge is to integrate and manipulate electrospinning process in food printing platform to create on-demand foods. Recent advances have enabled electrospinning as an important technological alternative for the design of texture-modified foods to feed the elderly safely and nutritiously. Furthermore, electrospun nanofibers containing molecularly imprinted binding sites for the analysis of food ingredients and functional electrospun nanofibers for rapid and real-time recognizing of biogenic amines in food products, along with toxic compounds, would be interesting topics to study.

References

Aceituno-Medina, M., Mendoza, S., Lagaron, J. M., & López-Rubio, A. (2015). Photoprotection of folic acid upon encapsulation in food-grade amaranth (*Amaranthus hypochondriacus L.*) protein isolate—Pullulan electrospun fibers. *LWT—Food Science and Technology*, *62*(2), 970–975.

Aceituno-Medina, M., Mendoza, S., Rodríguez, B. A., Lagaron, J. M., & López-Rubio, A. (2015). Improved antioxidant capacity of quercetin and ferulic acid during in-vitro digestion through encapsulation within food-grade electrospun fibers. *Journal of Functional Foods*, *12*, 332–341.

Aguilar-Vázquez, G., Loarca-Piña, G., Figueroa-Cárdenas, J. D., & Mendoza, S. (2018). Electrospun fibers from blends of pea *(Pisum sativum)* protein and pullulan. *Food Hydrocolloids, 83*, 173–181.

Alborzi, S., Lim, L. -T., & Kakuda, Y. (2014). Release of folic acid from sodium alginate-pectin-poly(ethylene oxide) electrospun fibers under in vitro conditions. *LWT—Food Science and Technology, 59*(1), 383–388.

Aldhahri, M. M., Almulaiky, Y. Q., El-Shishtawy, R. M., Al-Shawafi, W., Alngadh, A., & Maghrabi, R. (2018). Facile immobilization of enzyme via co-electrospinning: A simple method for enhancing enzyme reusability and monitoring an activity-based organic semiconductor. *ACS Omega, 3*(6), 6346–6350.

Alehosseini, A., Ghorani, B., Sarabi-Jamab, M., & Tucker, N. (2017). Principles of electrospraying: A new approach in protection of bioactive compounds in foods. *Critical Reviews in Food Science and Nutrition, 8398*, 1–18.

Alehosseini, A., Gómez-Mascaraque, L. G., Martínez-Sanz, M., & López-Rubio, A. (2019). Electrospun curcumin-loaded protein nanofiber mats as active/bioactive coatings for food packaging applications. *Food Hydrocolloids, 87*, 758–771.

Allais, M., Mailley, D., Hebraud, P., Ihiawakrim, D., Ball, V., Meyer, F., et al. (2018). Polymer-free electrospinning of tannic acid and cross-linking in water for hybrid supramolecular nanofibres. *Nanoscale, 10*(19), 9164–9173.

Alp Erbay, E., Dağtekin, B. B., Türe, M., Yeşilsu, A. F., & Torres-Giner, S. (2017). Quality improvement of rainbow trout fillets by whey protein isolate coatings containing electrospun poly(ε-caprolactone) nanofibers with *Urtica dioica L.* extract during storage. *LWT, 78*, 340–351.

Altan, A., Aytac, Z., & Uyar, T. (2018). Carvacrol loaded electrospun fibrous films from zein and poly(lactic acid) for active food packaging. *Food Hydrocolloids, 81*, 48–59.

Amiri, S., & Rahimi, A. (2018). Poly(ε-caprolactone) electrospun nanofibers containing cinnamon essential oil nanocapsules: A promising technique for controlled release and high solubility. *Journal of Industrial Textiles*, (152808371876491).

Amna, T., Yang, J., Ryu, K. S., & Hwang, I. H. (2015). Electrospun antimicrobial hybrid mats: Innovative packaging material for meat and meat-products. *Journal of Food Science and Technology, 52*(7), 4600–4606.

Anu Bhushani, J., & Anandharamakrishnan, C. (2014). Electrospinning and electrospraying techniques: Potential food based applications. *Trends in Food Science & Technology, 38*(1), 21–33.

Arrieta, M. P., López de Dicastillo, C., Garrido, L., Roa, K., & Galotto, M. J. (2018). Electrospun PVA fibers loaded with antioxidant fillers extracted from *Durvillaea antarctica* algae and their effect on plasticized PLA bionanocomposites. *European Polymer Journal, 103*, 145–157.

Assadpour, E., & Jafari, S. M. (2018). A systematic review on nanoencapsulation of food bioactive ingredients and nutraceuticals by various nanocarriers. *Critical Reviews in Food Science and Nutrition, 8398*, 1–47.

Assadpour, E., & Jafari, S. M. (2019a). Advances in spray-drying encapsulation of food bioactive ingredients: From microcapsules to nanocapsules. *Annual Review of Food Science and Technology, 10*(1).

Assadpour, E., & Jafari, S. M. (2019b). Chapter 3—Nanoencapsulation: Techniques and developments for food applications. In A. L. Rubio, M. J. F. Rovira, M. M. Sanz, & L. G. Gómez-Mascaraque (Eds.), *Nanomaterials for food applications* (pp. 35–61). New York, USA: Elsevier.

Assadpour, E., Jafari, S. M., & Esfanjani, A. F. (2017). Protection of phenolic compounds within nanocarriers. *CAB Reviews, 12*(057), 1–8.

Assadpour, E., Maghsoudlou, Y., Jafari, S. -M., Ghorbani, M., & Aalami, M. (2016). Evaluation of folic acid nano-encapsulation by double emulsions. *Food and Bioprocess Technology, 9*(12), 2024–2032.

Aytac, Z., Dogan, S. Y., Tekinay, T., & Uyar, T. (2014). Release and antibacterial activity of allyl isothiocyanate/beta-cyclodextrin complex encapsulated in electrospun nanofibers. *Colloids and Surfaces. B, Biointerfaces*, *120*, 125–131.

Aytac, Z., Ipek, S., Durgun, E., Tekinay, T., & Uyar, T. (2017). Antibacterial electrospun zein nanofibrous web encapsulating thymol/cyclodextrin-inclusion complex for food packaging. *Food Chemistry*, *233*, 117–124.

Aytac, Z., Ipek, S., Durgun, E., & Uyar, T. (2017). Antioxidant electrospun zein nanofibrous web encapsulating quercetin/cyclodextrin inclusion complex. *Journal of Materials Science*, *53*(2), 1527–1539.

Aytac, Z., Keskin, N. O. S., Tekinay, T., & Uyar, T. (2017). Antioxidant α-tocopherol/γ-cyclodextrin-inclusion complex encapsulated poly(lactic acid) electrospun nanofibrous web for food packaging. *Journal of Applied Polymer Science*, *134*(21), 1–9.

Aytac, Z., Kusku, S. I., Durgun, E., & Uyar, T. (2016a). Encapsulation of gallic acid/cyclodextrin inclusion complex in electrospun polylactic acid nanofibers: Release behavior and antioxidant activity of gallic acid. *Materials Science & Engineering. C, Materials for Biological Applications*, *63*, 231–239.

Aytac, Z., Kusku, S. I., Durgun, E., & Uyar, T. (2016b). Quercetin/beta-cyclodextrin inclusion complex embedded nanofibres: Slow release and high solubility. *Food Chemistry*, *197*(Pt A), 864–871.

Aytac, Z., & Uyar, T. (2016). Antioxidant activity and photostability of α-tocopherol/β-cyclodextrin inclusion complex encapsulated electrospun polycaprolactone nanofibers. *European Polymer Journal*, *79*, 140–149.

Aytac, Z., & Uyar, T. (2017). Core-shell nanofibers of curcumin/cyclodextrin inclusion complex and polylactic acid: Enhanced water solubility and slow release of curcumin. *International Journal of Pharmaceutics*, *518*(1–2), 177–184.

Aytac, Z., Yildiz, Z. I., Kayaci-Senirmak, F., San Keskin, N. O., Kusku, S. I., Durgun, E., et al. (2016a). Fast-dissolving, prolonged release, and antibacterial cyclodextrin/limonene-inclusion complex nanofibrous webs via polymer-free electrospinning. *Journal of Agricultural and Food Chemistry*, *64*(39), 7325–7334.

Aytac, Z., Yildiz, Z. I., Kayaci-Senirmak, F., San Keskin, N. O., Tekinay, T., & Uyar, T. (2016b). Electrospinning of polymer-free cyclodextrin/geraniol–inclusion complex nanofibers: Enhanced shelf-life of geraniol with antibacterial and antioxidant properties. *RSC Advances*, *6*(52), 46089–46099.

Aytac, Z., Yildiz, Z. I., Kayaci-Senirmak, F., Tekinay, T., & Uyar, T. (2017). Electrospinning of cyclodextrin/linalool-inclusion complex nanofibers: Fast-dissolving nanofibrous web with prolonged release and antibacterial activity. *Food Chemistry*, *231*, 192–201.

Banner, J., Dautzenberg, M., Feldhans, T., Hofmann, J., Plümer, P., & Ehrmann, A. (2018). Water resistance and morphology of electrospun gelatine blended with citric acid and coconut oil. *Tekstilec*, *61*(2), 129–135.

Barber, P. S., Griggs, C. S., Bonner, J. R., & Rogers, R. D. (2013). Electrospinning of chitin nanofibers directly from an ionic liquid extract of shrimp shells. *Green Chemistry*, *15*(3), 601–607.

Barzegari, A., & Shariatinia, Z. (2018). Fabrication of chitosan-polyethylene oxide electrospun nanofibrous mats containing green tea extract. *International Journal of Chemical Engineering*, *15*(2), 65–77.

Bazbouz, M. B., Liang, H., & Tronci, G. (2018). A UV-cured nanofibrous membrane of vinylbenzylated gelatin-poly(ε-caprolactone) dimethacrylate co-network by scalable free surface electrospinning. *Materials Science and Engineering: C*, *91*, 541–555.

Bazbouz, M. B., & Russell, S. J. (2018). Cellulose acetate/sodium-activated natural bentonite clay nanofibres produced by free surface electrospinning. *Journal of Materials Science*, *53*(15), 10891–10909.

Bhardwaj, N., & Kundu, S. C. (2010). Electrospinning: A fascinating fiber fabrication technique. *Biotechnology Advances*, *28*(3), 325–347.

Blanco-Padilla, A., López-Rubio, A., Loarca-Piña, G., Gómez-Mascaraque, L. G., & Mendoza, S. (2015). Characterization, release and antioxidant activity of curcumin-loaded amaranth-pullulan electrospun fibers. *LWT—Food Science and Technology*, *63* (2), 1137–1144.

Buzgo, M., Mickova, A., Rampichova, M., & Doupnik, M. (2018). Chapter 11: Blend electrospinning, coaxial electrospinning, and emulsion electrospinning techniques. In M. L. Focarete & A. Tampieri (Eds.), *Core-shell nanostructures for drug delivery and theranostics* (pp. 325–347). Chennai, India: Woodhead Publishing, MPS Limited.

Camerlo, A., Vebert-Nardin, C., Rossi, R. M., & Popa, A. M. (2013). Fragrance encapsulation in polymeric matrices by emulsion electrospinning. *European Polymer Journal*, *49* (12), 3806–3813.

Celebioglu, A., Aytac, Z., Kilic, M. E., Durgun, E., & Uyar, T. (2017). Encapsulation of camphor in cyclodextrin inclusion complex nanofibers via polymer-free electrospinning: Enhanced water solubility, high temperature stability, and slow release of camphor. *Journal of Materials Science*, *53*(7), 5436–5449.

Celebioglu, A., Kayaci-Senirmak, F., İpek, S., Durgun, E., & Uyar, T. (2016). Polymer-free nanofibers from vanillin/cyclodextrin inclusion complexes: High thermal stability, enhanced solubility and antioxidant property. *Food & Function*, 7(7), 3141–3153.

Celebioglu, A., & Uyar, T. (2017). Antioxidant vitamin E/cyclodextrin inclusion complex electrospun nanofibers: Enhanced water solubility, prolonged shelf life, and photostability of vitamin E. *Journal of Agricultural and Food Chemistry*, *65*(26), 5404–5412.

Celebioglu, A., Yildiz, Z. I., & Uyar, T. (2018a). Electrospun nanofibers from cyclodextrin inclusion complexes with cineole and p-cymene: Enhanced water solubility and thermal stability. *International Journal of Food Science & Technology*, *53*(1), 112–120.

Celebioglu, A., Yildiz, Z. I., & Uyar, T. (2018b). Fabrication of electrospun eugenol/cyclodextrin inclusion complex nanofibrous webs for enhanced antioxidant property, water solubility, and high temperature stability. *Journal of Agricultural and Food Chemistry*, *66* (2), 457–466.

Celebioglu, A., Yildiz, Z. I., & Uyar, T. (2018c). Thymol/cyclodextrin inclusion complex nanofibrous webs: Enhanced water solubility, high thermal stability and antioxidant property of thymol. *Food Research International*, *106*, 280–290.

Cerqueira, M. A., Torres-Giner, S., & Lagaron, J. M. (2018). Chapter 6: Nanostructured multilayer films. In M. A. P. R. Cerqueira, J. M. Lagaron, L. M. Pastrana Castro, & A. A. Vicente (Eds.), *Nanotechnology for food packaging: materials, processing technologies, and safety issues* (pp. 147–171). New York, USA: Elsevier Inc.

Cerqueira, M. A., Vicente, A. A., & Pastrana, L. M. (2018). Chapter 1: Nanotechnology in food packaging: Opportunities and challenges. In M. A. P. R. Cerqueira, J. M. Lagaron, L. M. Pastrana Castro, & A. A. Vicente (Eds.), *Nanotechnology for food packaging: materials, processing technologies, and safety issues*. New York, USA: Elsevier Inc.

Ceylan, Z., Unal Sengor, G. F., Sağdıç, O., & Yilmaz, M. T. (2017). A novel approach to extend microbiological stability of sea bass (*Dicentrarchus labrax*) fillets coated with electrospun chitosan nanofibers. *LWT—Food Science and Technology*, *79*, 367–375.

Chalco-Sandoval, W., Fabra, M. J., López-Rubio, A., & Lagaron, J. M. (2017). Use of phase change materials to develop electrospun coatings of interest in food packaging applications. *Journal of Food Engineering*, *192*, 122–128.

Chen, X., Chen, M., Xu, C., & Yam, K. L. (2018). Critical review of controlled release packaging to improve food safety and quality. *Critical Reviews in Food Science and Nutrition*, *8398*, 1–14.

Chen, C., Wang, L., & Huang, Y. (2008). A novel shape-stabilized PCM: Electrospun ultra fine fibers based on lauric acid/polyethylene terephthalate composite. *Materials Letters*, *62*, 3515–3517.

Chen, Z., Wang, L., & Jiang, H. (2012). The effect of procyanidine crosslinking on the properties of the electrospun gelatin membranes. *Biofabrication*, *4*(3)035007.
Chen, C., Zhao, Y., & Liu, W. (2013). Electrospun polyethylene glycol/cellulose acetate phase change fibers with core–sheath structure for thermal energy storage. *Renewable Energy*, *60*, 222–225.
Cherpinski, A., Torres-Giner, S., Cabedo, L., Méndez, J. A., & Lagaron, J. M. (2017). Multilayer structures based on annealed electrospun biopolymer coatings of interest in water and aroma barrier fiber-based food packaging applications. *J Appl Polym Sci*, *45501*, 1–11.
Cherpinski, A., Torres-Giner, S., Vartiainen, J., Peresin, M. S., Lahtinen, P., & Lagaron, J. M. (2018). Improving the water resistance of nanocellulose-based films with polyhydroxyalkanoates processed by the electrospinning coating technique. *Cellulose*, *25*(2), 1291–1307.
Chuysinuan, P., Chimnoi, N., Pattani, L., Khlaychan, P., Nooeaid, P., & Techasakul, S. (2017). Poly(lactic acid) (PLA) electrospun fibers containing rice extract: Release characteristics and their antioxidant activity. *Key Engineering Materials*, *757*, 83–87.
Colín-Orozco, J., Zapata-Torres, M., Rodríguez-Gattorno, G., & Pedroza-Islas, R. (2015). Properties of poly (ethylene oxide)/whey protein isolate nanofibers prepared by electrospinning. *Food Biophysics*, *10*(2), 134–144.
Cui, H., Bai, M., & Lin, L. (2018). Plasma-treated poly(ethylene oxide) nanofibers containing tea tree oil/beta-cyclodextrin inclusion complex for antibacterial packaging. *Carbohydrate Polymers*, *179*, 360–369.
Cui, S., Yao, B., Gao, M., Sun, X., Gou, D., Hu, J., et al. (2017). Effects of pectin structure and crosslinking method on the properties of crosslinked pectin nanofibers. *Carbohydrate Polymers*, *157*, 766–774.
Cui, S., Yao, B., Sun, X., Hu, J., Zhou, Y., & Liu, Y. (2016). Reducing the content of carrier polymer in pectin nanofibers by electrospinning at low loading followed with selective washing. *Materials Science & Engineering. C, Materials for Biological Applications*, *59*, 885–893.
Cui, H., Yuan, L., Li, W., & Lin, L. (2017). Antioxidant property of SiO_2-eugenol liposome loaded nano fi brous membranes on beef. *Food Packaging and Shelf Life*, *11*, 49–57.
Dai, R., & Lim, L. -T. (2015). Release of allyl isothiocyanate from mustard seed meal powder entrapped in electrospun PLA–PEO nonwovens. *Food Research International*, *77*, 467–475.
de Castro, D. O., Tabary, N., Martel, B., Gandini, A., Belgacem, N., & Bras, J. (2018). Controlled release of carvacrol and curcumin: Bio-based food packaging by synergism action of TEMPO-oxidized cellulose nanocrystals and cyclodextrin. *Cellulose*, *25*(2), 1249–1263.
de Freitas Zompero, R. H., Lopez-Rubio, A., de Pinho, S. C., Lagaron, J. M., & de la Torre, L. G. (2015). Hybrid encapsulation structures based on beta-carotene-loaded nanoliposomes within electrospun fibers. *Colloids and Surfaces. B, Biointerfaces*, *134*, 475–482.
de Oliveira Mori, C. L. S., dos Passos, N. A., Oliveira, J. E., Mattoso, L. H. C., Mori, F. A., Carvalho, A. G., et al. (2014). Electrospinning of zein/tannin bio-nanofibers. *Industrial Crops and Products*, *52*, 298–304.
Deng, L., Kang, X., Liu, Y., Feng, F., & Zhang, H. (2017). Effects of surfactants on the formation of gelatin nanofibres for controlled release of curcumin. *Food Chemistry*, *231*, 70–77.
Deng, L., Li, Y., Feng, F., & Zhang, H. (2019). Study on wettability, mechanical property and biocompatibility of electrospun gelatin/zein nanofibers cross-linked by glucose. *Food Hydrocolloids*, *87*, 1–10.

Deng, L., Taxipalati, M., Zhang, A., Que, F., Wei, H., Feng, F., et al. (2018). Electrospun chitosan/poly(ethylene oxide)/lauric arginate nanofibrous film with enhanced antimicrobial activity. *Journal of Agricultural and Food Chemistry*, *66*(24), 6219–6226.
Deng, L., Zhang, X., Li, Y., Que, F., Kang, X., Liu, Y., et al. (2018). Characterization of gelatin/zein nanofibers by hybrid electrospinning. *Food Hydrocolloids*, *75*, 72–80.
Dheraprasart, C., Rengpipat, S., & Supaphol, P. (2009). Morphology, release characteristics, and antimicrobial effect of nisin-loaded electrospun gelatin fiber mat. *Journal of Food Protection*, *72*(11), 2293–2300.
Dias Antunes, M., da Silva Dannenberg, G., Fiorentini, Â. M., Pinto, V. Z., Lim, L. -T., da Rosa Zavareze, E., et al. (2017). Antimicrobial electrospun ultrafine fibers from zein containing eucalyptus essential oil/cyclodextrin inclusion complex. *International Journal of Biological Macromolecules*, *104*, 874–882.
Divya, K., & Jisha, M. S. (2017). Chitosan nanoparticles preparation and applications. *Environmental Chemistry Letters*, *16*(1), 101–112.
do Evangelho, J. A., Crizel, R. L., Chaves, F. C., Prietto, L., Pinto, V. Z., Miranda, M. Z. d., et al. (2018). Thermal and irradiation resistance of folic acid encapsulated in zein ultrafine fibers or nanocapsules produced by electrospinning and electrospraying. *Food Research International*. https://doi.org/10.1016/j.foodres.2018.08.019.
Dogan, G., Basal, G., Bayraktar, O., Ozyildiz, F., Uzel, A., & Erdogan, I. (2016). Bioactive sheath/core nanofibers containing olive leaf extract. *Microscopy Research and Technique*, *79*(1), 38–49.
Dumitriu, R. P., Mitchell, G. R., Davis, F. J., & Vasile, C. (2017). Functionalized coatings by electrospinning for anti-oxidant food packaging. *Procedia Manufacturing*, *12*, 59–65.
Dos Santos, J. P., Zavareze, E. D. R., Dias, A. R. G., & Vanier, N. L. (2018). Immobilization of xylanase and xylanase-beta-cyclodextrin complex in polyvinyl alcohol via electrospinning improves enzyme activity at a wide pH and temperature range. *International Journal of Biological Macromolecules*, *118*(Pt B), 1676–1684.
Elsabee, M. Z., Naguib, H. F., & Elsayed, R. (2012). Chitosan based nanofibers, review. *Materials Science and Engineering: C*, *32*, 1711–1726.
Eng, J. T. S., & Zhong, W. (2012). Electrospun nanofibers as carriers for bioactive molecules. *Journal of Textile Science & Engineering*, *2*(7).
Erdogan, I., Demir, M., & Bayraktar, O. (2015). Olive leaf extract as a crosslinking agent for the preparation of electrospun zein fibers. *Journal of Applied Polymer Science*, *132*(4).
Esfahani, R., Jafari, S. M., Jafarpour, A., & Dehnad, D. (2019). Loading of fish oil into nanocarriers prepared through gelatin-gum Arabic complexation. *Food Hydrocolloids*, *90*, 291–298.
Esmaeili, A., & Haseli, M. (2017). Optimization, synthesis, and characterization of coaxial electrospun sodium carboxymethyl cellulose-graft-methyl acrylate/poly(ethylene oxide) nanofibers for potential drug-delivery applications. *Carbohydrate Polymers*, *173*, 645–653.
Estevez-Areco, S., Guz, L., Candal, R., & Goyanes, S. (2018). Release kinetics of rosemary (*Rosmarinus officinalis*) polyphenols from polyvinyl alcohol (PVA) electrospun nanofibers in several food simulants. *Food Packaging and Shelf Life*, *18*, 42–50.
Fabra, M. J., López-Rubio, A., Ambrosio-Martín, J., & Lagaron, J. M. (2016). Improving the barrier properties of thermoplastic corn starch-based films containing bacterial cellulose nanowhiskers by means of PHA electrospun coatings of interest in food packaging. *Food Hydrocolloids*, *61*, 261–268.
Fabra, M. J., López-Rubio, A., & Lagaron, J. M. (2014). On the use of different hydrocolloids as electrospun adhesive interlayers to enhance the barrier properties of polyhydroxyalkanoates of interest in fully renewable food packaging concepts. *Food Hydrocolloids*, *39*, 77–84.

Fabra, M. J., López-Rubio, A., & Lagaron, J. M. (2016). Use of the electrohydrodynamic process to develop active/bioactive bilayer films for food packaging applications. *Food Hydrocolloids*, *55*, 11–18.
Fabra, M. J., López-Rubio, A., Sentandreu, E., & Lagaron, J. M. (2016). Development of multilayer corn starch-based food packaging structures containing β-carotene by means of the electro-hydrodynamic processing. *Starch/Stärke*, *68*(7–8), 603–610.
Fahami, A., & Fathi, M. (2018a). Development of cress seed mucilage/PVA nanofibers as a novel carrier for vitamin A delivery. *Food Hydrocolloids*, *81*, 31–38.
Fahami, A., & Fathi, M. (2018b). Fabrication and characterization of novel nanofibers from cress seed mucilage for food applications. *Journal of Applied Polymer Science*, *135*(6). 45811.
Faridi Esfanjani, A., & Jafari, S. M. (2016). Biopolymer nano-particles and natural nano-carriers for nano-encapsulation of phenolic compounds. *Colloids and Surfaces B: Bio-interfaces*, *146*, 532–543.
Faridi Esfanjani, A., Jafari, S. M., & Assadpour, E. (2017). Preparation of a multiple emulsion based on pectin-whey protein complex for encapsulation of saffron extract nanodroplets. *Food Chemistry*, *221*, 1962–1969.
Fathi, M., Donsi, F., & McClements, D. J. (2018). Protein-Based Delivery Systems for the Nanoencapsulation of Food Ingredients. *Comprehensive Reviews in Food Science and Food Safety*, *17*(4), 920–936.
Fathi, M., Nasrabadi, M. N., & Varshosaz, J. (2016). Characteristics of vitamin E-loaded nanofibres from dextran. *International Journal of Food Properties*, *20*(11), 2665–2674.
Fernandez, A., Torres-Giner, S., & Lagaron, J. M. (2007). Encapsulation of the functional ingredient beta-carotene in electrospun PLA fibers of interest in bioactive food packaging and food processing applications. In *Conference proceeding: 23 rd IAPRI Symposium of Packaging*. UK: Windsor.
Fernandez, A., Torres-Giner, S., & Lagaron, J. M. (2009). Food hydrocolloids novel route to stabilization of bioactive antioxidants by encapsulation in electrospun fibers of zein prolamine. *Food Hydrocolloids*, *23*(5), 1427–1432.
Fonseca, L. M., da Silva, F. T., Antunes, M. D., Mello El Halal, S. L., Lim, L. -T., & Dias, A. R. G. (2019). Aging time of soluble potato starch solutions for ultrafine fibers formation by electrospinning. *Starch/Stärke*, *71*(1–2). 1800089.
Frenot, A., Henriksson, M. W., & Walkenstrom, P. (2007). Electrospinning of cellulose-based nanofibers. *Journal of Applied Polymer Science*, *103*, 1473–1482.
Fuenmayor, C. A., & Coiso, M. S. (2016). Encapsulation of antioxidant of antioxidant phenolic compounds in zein ultrathin fibers via electrospinning. In *Conference proceedings*. .
Gabrielczyk, J., Duensing, T., Buchholz, S., Schwinges, A., & Jordening, H. J. (2018). A comparative study on immobilization of fructosyltransferase in biodegradable polymers by electrospinning. *Applied Biochemistry and Biotechnology*, *185*(3), 847–862.
García-Moreno, P. J., Damberg, C., Chronakis, I. S., & Jacobsen, C. (2017). Oxidative stability of pullulan electrospun fibers containing fish oil: Effect of oil content and natural antioxidants addition. *European Journal of Lipid Science and Technology*, *119*(12)1600305.
García-Moreno, P. J., Özdemir, N., Stephansen, K., Mateiu, R. V., Echegoyen, Y., Lagaron, J. M., et al. (2017). Development of carbohydrate-based nano-microstructures loaded with fish oil by using electrohydrodynamic processing. *Food Hydrocolloids*, *69*, 273–285.
García-Moreno, P. J., Stephansen, K., van der Kruijs, J., Guadix, A., Guadix, E. M., Chronakis, I. S., et al. (2016). Encapsulation of fish oil in nanofibers by emulsion electrospinning: Physical characterization and oxidative stability. *Journal of Food Engineering*, *183*, 39–49.
Gharehbeglou, P., Jafari, S. M., Hamishekar, H., Homayouni, A., & Mirzaei, H. (2019). Pectin-whey protein complexes vs. small molecule surfactants for stabilization of double

nano-emulsions as novel bioactive delivery systems. *Journal of Food Engineering*, *245*, 139–148.

Gharehbeglou, P., Jafari, S. M., Homayouni, A., Hamishekar, H., & Mirzaei, H. (2019). Fabrication of double W1/O/W2 nano-emulsions loaded with oleuropein in the internal phase (W1) and evaluation of their release rate. *Food Hydrocolloids*, *89*, 44–55.

Gharibzahedi, S. M. T., & Jafari, S. M. (2017). 7. Nanocapsule formation by cyclodextrins. In *Nanoencapsulation technologies for the food and nutraceutical industries* (pp. 187–261). San Diego, USA: Academic Press.

Ghitescu, R. E., Popa, A. M., Schipanski, A., Hirsch, C., Yazgan, G., Popa, V. I., et al. (2018). Catechin loaded PLGA submicron-sized fibers reduce levels of reactive oxygen species induced by MWCNT in vitro. *European Journal of Pharmaceutics and Biopharmaceutics*, *122*, 78–86.

Ghorani, B., Alehosseini, A., & Tucker, N. (2017). 8. Nanocapsule formation by electrospinning. In S. M. Jafari (Ed.), *Nanoencapsulation technologies for the food and nutraceutical industries* (pp. 264–319). San Diego, USA: Academic Press.

Ghorani, B., & Tucker, N. (2015). Fundamentals of electrospinning as a novel delivery vehicle for bioactive compounds in food nanotechnology. *Food Hydrocolloids*, *51*, 227–240.

Ghorbanzade, T., Jafari, S. M., Akhavan, S., & Hadavi, R. (2017). Nano-encapsulation of fish oil in nano-liposomes and its application in fortification of yogurt. *Food Chemistry*, *216*, 146–152.

Ghosal, K., Chandra, A., Praveen, G., Snigdha, S., Roy, S., Agatemor, C., et al. (2018). Electrospinning over solvent casting: Tuning of mechanical properties of membranes. *Scientific Reports*, *8*(1), 5058.

Golkar, P., Allafchian, A., & Afshar, B. (2018). Alyssum lepidium mucilage as a new source for electrospinning: Production and physicochemical characterisation. *IET Nanobiotechnology*, *12*(3), 259–263.

Gómez-Mascaraque, L. G., Tordera, F., Fabra, M. J., Martínez-Sanz, M., & Lopez-Rubio, A. (2019). Coaxial electrospraying of biopolymers as a strategy to improve protection of bioactive food ingredients. *Innovative Food Science & Emerging Technologies*, *51*, 2–11.

Gonçalves, C. F., Schmatz, D. A., Uebel, S., Kuntzler, S. G., Alberto, J., Costa, V., et al. (2017). Microalgae biopeptides applied in nanofibers for the development of active packaging. *Polímeros*, *27*(4), 290–297.

Han, D., Sherman, S., Filocamo, S., & Steckl, A. J. (2017). Long-term antimicrobial effect of nisin released from electrospun triaxial fiber membranes. *Acta Biomaterialia*, *53*, 242–249.

Hani, N. M., Torkamani, A. E., Azarian, M. H., Mahmood, K. W., & Ngalim, S. H. (2017). Characterisation of electrospun gelatine nanofibres encapsulated with Moringa oleifera bioactive extract. *Journal of the Science of Food and Agriculture*, *97*(10), 3348–3358.

Hemamalini, T., & Giri Dev, V. R. (2018). Comprehensive review on electrospinning of starch polymer for biomedical applications. *International Journal of Biological Macromolecules*, *106*, 712–718.

Hosseini, H., Ghorbani, M., Jafari, S. M., & Sadeghi Mahoonak, A. (2019). Encapsulation of EPA and DHA concentrate from Kilka fish oil by milk proteins and evaluation of its oxidative stability. *Journal of Food Science and Technology*, *56*(1), 59–70.

Hsueh, C. -Y., Tsai, M. -L., & Liu, T. (2017). Enhancing saltiness perception using chitin nanofibers when curing tilapia fillets. *LWT*, *86*, 93–98.

Hu, W., & Yu, X. (2012). Encapsulation of bio-based PCM with coaxial electrospun ultrafine fibers. *RSC Advances*, *2*(13), 5580–5584.

Huang, W., Zhan, Y., Shi, X., Chen, J., Deng, H., & Du, Y. (2017). Controllable immobilization of naringinase on electrospun cellulose acetate nanofibers and their application to juice debittering. *International Journal of Biological Macromolecules*, *98*, 630–636.

Inanc Horuz, T., & Belibagli, K. B. (2018). Nanoencapsulation by electrospinning to improve stability and water solubility of carotenoids extracted from tomato peels. *Food Chemistry*, *268*, 86–93.

Isfahani, F. R., Tavanai, H., & Morshed, M. (2017). Release of aloe vera from electrospun aloe vera-PVA nanofibrous pad. *Fibers and Polymers*, *18*(2), 264–271.

Isik, B. S., Altay, F., & Capanoglu, E. (2018). The uniaxial and coaxial encapsulations of sour cherry (*Prunus cerasus* L.) concentrate by electrospinning and their in vitro bioaccessibility. *Food Chemistry*, *265*, 260–273.

Islam, S., & Rezaul, M. (2010). Fabrication and characterization of poly (vinyl alcohol)/alginate blend nanofibers by electrospinning method. *Colloids and Surfaces A: Physicochemical and Engineering Aspects*, *366*, 135–140.

Jacobsen, C., Garcia-Moreno, P. J., Mendes, A. C., Mateiu, R. V., & Chronakis, I. S. (2018). Use of electrohydrodynamic processing for encapsulation of sensitive bioactive compounds and applications in food. *Annual Review of Food Science and Technology*, *9*, 525–549.

Jafari, S. M. (2017). Chapter 1—An introduction to nanoencapsulation techniques for the food bioactive ingredients. In *Nanoencapsulation of food bioactive ingredients* (pp. 1–62). San Diego, USA: Academic Press.

Jafari, S. M., Assadpoor, E., He, Y., & Bhandari, B. (2008). Encapsulation efficiency of food flavours and oils during spray drying. *Drying Technology*, *26*(7), 816–835.

Jafari, S. M., Fathi, M., & Mandala, I. (2015). Chapter 13—Emerging product formation. In C. M. Galanakis (Ed.), *Food waste recovery* (pp. 293–317). San Diego: Academic Press.

Jafari, S. M., & McClements, D. J. (2017). Nanotechnology approaches for increasing nutrient bioavailability. In *Advances in food and nutrition research*. New York, USA: Academic Press.

Jash, A., & Lim, L. -T. (2017). Triggered release of hexanal from an imidazolidine precursor encapsulated in poly(lactic acid) and ethylcellulose carriers. *Journal of Materials Science*, *53* (3), 2221–2235.

Jia, Z., Dumont, M. -J., & Orsat, V. (2016). Encapsulation of phenolic compounds present in plants using protein matrices. *Food Bioscience*, *15*, 87–104.

Jia, Y. -T., Gong, J., Gu, X. -H., Kim, H. -Y., Dong, J., & Shen, X. -Y. (2007). Fabrication and characterization of poly (vinyl alcohol)/chitosan blend nanofibers produced by electrospinning method. *Carbohydrate Polymers*, *67*(3), 403–409.

Jiang, W. -J., Tsai, M. -L., & Liu, T. (2017). Chitin nanofiber as a promising candidate for improved salty taste. *LWT*, *75*, 65–71.

Jørgensen, L., Qvortrup, K., & Chronakis, I. S. (2015). Phospholipid electrospun nanofibers: Effect of solvents and co-axial processing on morphology and fiber diameter. *RSC Advances*, *5*(66), 53644–53652.

Kakoria, A., & Sinha-Ray, S. (2018). A review on biopolymer-based fibers via electrospinning and solution blowing and their applications. *Fibers*, *6*(3), 45.

Katouzian, I., & Jafari, S. M. (2016). Nano-encapsulation as a promising approach for targeted delivery and controlled release of vitamins. *Trends in Food Science & Technology*, *53*, 34–48.

Kayaci, F., Sen, H. S., Durgun, E., & Uyar, T. (2014). Functional electrospun polymeric nanofibers incorporating geraniol–cyclodextrin inclusion complexes: High thermal stability and enhanced durability of geraniol. *Food Research International*, *62*, 424–431.

Kayaci, F., & Uyar, T. (2012). Encapsulation of vanillin/cyclodextrin inclusion complex in electrospun polyvinyl alcohol (PVA) nanowebs: Prolonged shelf-life and high temperature stability of vanillin. *Food Chemistry*, *133*(3), 641–649.

Kesici Güler, H., Cengiz Çallıoğlu, F., & Sesli Çetin, E. (2018). Antibacterial PVP/cinnamon essential oil nanofibers by emulsion electrospinning. *Journal of the Textile Institute*, *110*(2), 302–310.

Kfoury, M., Auezova, L., Greige-Gerges, H., & Fourmentin, S. (2018). Encapsulation in cyclodextrins to widen the applications of essential oils. *Environmental Chemistry Letters*, *17*(1), 129–143.

Khalf, A., & Madihally, S. V. (2017). Recent advances in multiaxial electrospinning for drug delivery. *European Journal of Pharmaceutics and Biopharmaceutics*, *112*, 1–17.

Khanzadi, M., Jafari, S. M., Mirzaei, H., Chegini, F. K., Maghsoudlou, Y., & Dehnad, D. (2015). Physical and mechanical properties in biodegradable films of whey protein concentrate-pullulan by application of beeswax. *Carbohydrate Polymers*, *118*, 24–29.

Kim, M. S., Kim, H. J., Jang, J. Y., & Shin, H. S. (2018). Development of coaxial alginate-PCL nanofibrous dressing for controlled release of Spirulina extract. *Journal of Biomaterials Science. Polymer Edition*, *29*(12), 1389–1400.

Kim, J. H., Lee, H., Jatoi, A. W., Im, S. S., Lee, J. S., & Kim, I. -S. (2016). Juniperus chinensis extracts loaded PVA nanofiber: Enhanced antibacterial activity. *Materials Letters*, *181*, 367–370.

Kong, L., & Ziegler, G. R. (2014). Rheological aspects in fabricating pullulan fibers by electro-wet-spinning. *Food Hydrocolloids*, *38*, 220–226.

Korehei, R., & Kadla, J. (2013). Incorporation of T4 bacteriophage in electrospun fibres. *Journal of Applied Microbiology*, *114*(5), 1425–1434.

Kriegel, C., Kit, K., McClements, D., & Weiss, J. (2010). Nanofibers as carrier systems for antimicrobial microemulsions. II. Release characteristics and antimicrobial activity. *Journal of Applied Polymer Science*, *118*(5), 2859–2868.

Kuntzler, S. G., Costa, J. A. V., & Morais, M. G. (2018). Development of electrospun nanofibers containing chitosan/PEO blend and phenolic compounds with antibacterial activity. *International Journal of Biological Macromolecules*, *117*, 800–806.

Kurd, F., Fathi, M., & Shekarchizadeh, H. (2017). Basil seed mucilage as a new source for electrospinning: Production and physicochemical characterization. *International Journal of Biological Macromolecules*, *95*, 689–695.

Kutzli, I., Gibis, M., Baier, S. K., & Weiss, J. (2018). Formation of whey protein isolate (WPI)-maltodextrin conjugates in fibers produced by needleless electrospinning. *Journal of Agricultural and Food Chemistry*, *66*(39), 10283–10291.

Labuschagne, P. (2018). Impact of wall material physicochemical characteristics on the stability of encapsulated phytochemicals: A review. *Food Research International*, *107*(2017), 227–247.

Lemma, S. M., Scampicchio, M., Mahon, P. J., Sbarski, I., Wang, J., & Kingshott, P. (2015). Controlled release of retinyl acetate from beta-cyclodextrin functionalized poly(vinyl alcohol) electrospun nanofibers. *Journal of Agricultural and Food Chemistry*, *63*(13), 3481–3488.

Lević, S., Obradović, N., Pavlović, V., Isailović, B., Kostić, I., Mitrić, M., et al. (2014). Thermal, morphological, and mechanical properties of ethyl vanillin immobilized in polyvinyl alcohol by electrospinning process. *Journal of Thermal Analysis and Calorimetry*, *118*(2), 661–668.

Li, J., Chotiko, A., Narcisse, D. A., & Sathivel, S. (2016). Evaluation of alpha-tocopherol stability in soluble dietary fiber based nanofiber. *LWT—Food Science and Technology*, *68*, 485–490.

Li, Y., Lim, L. T., & Kakuda, Y. (2009). Electrospun zein fibers as carriers to stabilize (-)-epigallocatechin gallate. *Journal of Food Science*, *74*(3), C233–240.

Li, X. Y., Shi, C. J., Yu, D. G., Liao, Y. Z., & Wang, X. (2014). Electrospun quercetin-loaded zein nanoribbons. *Bio-medical Materials and Engineering*, *24*(6), 2015–2023.

Li, Z., & Wang, C. (2013). Effects of working parameters on electrospinning. In Z. Li & C. Wang (Eds.), *One-dimensional nanostructures: Electrospinning technique and unique nanofibers* (pp. 15–28). Berlin, Heidelberg: Springer Berlin Heidelberg.

Li, L., Wang, H., Chen, M., Jiang, S., Jiang, S., Li, X., et al. (2018). Butylated hydroxyanisole encapsulated in gelatin fiber mats: Volatile release kinetics, functional effectiveness and application to strawberry preservation. *Food Chemistry*, *269*, 142–149.

Lian, H., & Meng, Z. (2017). Melt electrospinning vs. solution electrospinning: A comparative study of drug-loaded poly (epsilon-caprolactone) fibres. *Materials Science & Engineering C, Materials for Biological Applications*, *74*, 117–123.

Lin, L., Dai, Y., & Cui, H. (2017). Antibacterial poly(ethylene oxide) electrospun nanofibers containing cinnamon essential oil/beta-cyclodextrin proteoliposomes. *Carbohydrate Polymers*, *178*, 131–140.

Lin, L., Xue, L., Duraiarasan, S., & Haiying, C. (2018). Preparation of ε-polylysine/chitosan nanofibers for food packaging against *Salmonella* on chicken. *Food Packaging and Shelf Life*, *17*, 134–141.

Lin, L., Zhu, Y., & Cui, H. (2018). Electrospun thyme essential oil/gelatin nanofibers for active packaging against *Campylobacter jejuni* in chicken. *LWT*, *97*, 711–718.

Liu, S., Li, R., Tomasula, P., Sousa, A., & Liu, L. (2016). Electrospun food-grade ultrafine fibers from pectin and pullulan blends. *Food and Nutrition Sciences*, *7*, 636–646.

Liu, Y., Liang, X., Wang, S., Qin, W., & Zhang, Q. (2018). Electrospun antimicrobial polylactic acid/tea polyphenol nanofibers for food-packaging applications. *Polymers*, *10*(5), 561.

Liu, Y., Wang, S., Lan, W., & Qin, W. (2019). Fabrication of polylactic acid/carbon nanotubes/chitosan composite fibers by electrospinning for strawberry preservation. *International Journal of Biological Macromolecules*, *121*, 1329–1336.

Liu, P., Xu, H., Zhao, Y., & Yang, Y. (2017). Rheological properties of soy protein isolate solution for fibers and films. *Food Hydrocolloids*, *64*, 149–156.

Lopez-Rubio, A., Sanchez, E., Sanz, Y., & Lagaron, J. M. (2009). Encapsulation of living bifidobacteria in ultrathin PVOH electrospun fibers. *Biomacromolecules*, *10*(10), 2823–2829.

Lubambo, A. F., Freitas, R. A. D., Sierakowski, M. -R., Lucyszyn, N., Sassaki, G. L., Serafim, B. M., et al. (2013). Electrospinning of commercial guar-gum: Effects of purification and filtration. *Carbohydrate Polymers*, *93*(2), 484–491.

Lubasova, D., Mullerova, J., & Netravali, A. N. (2015). Water-resistant plant protein-based nanofiber membranes. *Journal of Applied Polymer Science*, *132*(16).

Luo, C. J., Nangrejo, M., & Edirisinghe, M. (2010). A novel method of selecting solvents for polymer electrospinning. *Polymer*, *51*(7), 1654–1662.

Maftoonazad, N., & Ramaswamy, H. (2018). Novel techniques in food processing: Bionanocomposites. *Current Opinion in Food Science*, *23*, 49–56.

Mahdavee Khazaei, K., Jafari, S. M., Ghorbani, M., & Hemmati Kakhki, A. (2014). Application of maltodextrin and gum Arabic in microencapsulation of saffron petal's anthocyanins and evaluating their storage stability and color. *Carbohydrate Polymers*, *105*(1), 57–62.

Mahdavi, S. A., Jafari, S. M., Ghorbani, M., & Assadpoor, E. (2014). Spray-drying microencapsulation of anthocyanins by natural biopolymers: A review. *Drying Technology*, *32*(5), 509–518.

McClements, D. J., Decker, E. A., Park, Y., & Weiss, J. (2009). Structural design principles for delivery of bioactive components in nutraceuticals and functional foods. *Critical Reviews in Food Science and Nutrition*, *49*(6), 577–606.

Mahmud, M., Perveen, A., Matin, A., & Arafat, M. T. (2018). Effects of binary solvent mixtures on the electrospinning behavior of poly (vinyl alcohol). *Materials Research Express*, *5*(11).

Mendes, A. C., Stephansen, K., & Chronakis, I. S. (2017). Electrospinning of food proteins and polysaccharides. *Food Hydrocolloids*, *68*, 53–68.

Moomand, K., & Lim, L. -T. (2014). Oxidative stability of encapsulated fish oil in electrospun zein fibres. *Food Research International*, *62*, 523–532.

Moomand, K., & Lim, L. -T. (2015). Effects of solvent and n-3 rich fish oil on physicochemical properties of electrospun zein fibres. *Food Hydrocolloids*, *46*, 191–200.

Moreira, J. B., Lim, L. -T., Zavareze, E. d. R., Dias, A. R. G., Costa, J. A. V., & Morais, M. G. d. (2018). Microalgae protein heating in acid/basic solution for nanofibers production by free surface electrospinning. *Journal of Food Engineering*, *230*, 49–54.

Moreira, J. B., Morais, M. G. d., Morais, E. G. d., Vaz, B. d. S., & Costa, J. A. V. (2018). Chapter 14—Electrospun polymeric nanofibers in food packaging. In A. M. Grumezescu & A. M. Holban (Eds.), *Impact of nanoscience in the food industry* (pp. 387–417). Academic Press.

Moreira, J. B., Terra, A. L. M., Costa, J. A. V., & Morais, M. G. d. (2018). Development of pH indicator from PLA/PEO ultrafine fibers containing pigment of microalgae origin. *International Journal of Biological Macromolecules*, *118*, 1855–1862.

Moreno-Cortez, I. E., Romero-Garcia, J., Gonzalez-Gonzalez, V., Garcia-Gutierrez, D. I., Garza-Navarro, M. A., & Cruz-Silva, R. (2015). Encapsulation and immobilization of papain in electrospun nanofibrous membranes of PVA cross-linked with glutaraldehyde vapor. *Materials Science & Engineering. C, Materials for Biological Applications*, *52*, 306–314.

Mousavi Khaneghah, A., Hashemi, S. M. B., & Limbo, S. (2018). Antimicrobial agents and packaging systems in antimicrobial active food packaging: An overview of approaches and interactions. *Food and Bioproducts Processing*, *111*, 1–19.

Munhuweyi, K., Caleb, O. J., van Reenen, A. J., & Opara, U. L. (2018). Physical and antifungal properties of β-cyclodextrin microcapsules and nanofibre films containing cinnamon and oregano essential oils. *LWT*, *87*, 413–422.

Naseri, N., Mathew, A. P., & Oksman, K. (2016). Electrospinnability of bionanocomposites with high nanocrystal loadings: The effect of nanocrystal surface characteristics. *Carbohydrate Polymers*, *147*, 464–472.

Nasir, M., & Apriani, D. (2017). Synthesis and property of Ag(NP)/catechin/gelatin nanofiber. *IOP conference series: Materials science and engineering*, *293*, 012004.

Neo, Y. P., Perera, C. O., Nieuwoudt, M. K., Zujovic, Z., Jin, J., Ray, S., et al. (2014). Influence of heat curing on structure and physicochemical properties of phenolic acid loaded proteinaceous electrospun fibers. *Journal of Agricultural and Food Chemistry*, *62* (22), 5163–5172.

Neo, Y. P., Ray, S., Jin, J., Gizdavic-Nikolaidis, M., Nieuwoudt, M. K., Liu, D., et al. (2013). Encapsulation of food grade antioxidant in natural biopolymer by electrospinning technique: A physicochemical study based on zein-gallic acid system. *Food Chemistry*, *136*(2), 1013–1021.

Neo, Y. P., Ray, S., & Perera, C. O. (2018). *Nanostructures for food applications*. New York, USA: Elsevier Inc.

Nieuwland, M., Geerdink, P., Brier, P., Eijnden, P. V. D., Henket, J. T. M. M., Langelaan, M. L. P., et al. (2013). Food-grade electrospinning of proteins. *Innovative Food Science and Emerging Technologies*, *20*(2013), 269–275.

Noyan, E. C. B., Onder, E., Sarier, N., & Arat, R. (2018). Development of heat storing poly (acrylonitrile) nanofibers by coaxial electrospinning. *Thermochimica Acta*, *662*, 135–148.

Nunes, M. A. P., Martins, S., Rosa, M. E., Gois, P. M. P., Fernandes, P. C. B., & Ribeiro, M. H. L. (2016). Improved thermostable polyvinyl alcohol electrospun nanofibers with entangled naringinase used in a novel mini-packed bed reactor. *Bioresource Technology*, *213*, 208–215.

Oguz, S., Tam, N., Aydogdu, A., Sumnu, G., & Sahin, S. (2018). Development of novel pea flour-based nanofibres by electrospinning method. *International Journal of Food Science & Technology*, *53*(5), 1269–1277.

Paaver, U., Laidmäe, I., Santos, H. A., Yliruusi, J., Aruväli, J., Kogermann, K., et al. (2016). Development of a novel electrospun nanofibrous delivery system for poorly water-soluble β-sitosterol. *Asian Journal of Pharmaceutical Sciences, 11*(4), 500–506.
Padil, V. V., Senan, C., Waclawek, S., & Cernik, M. (2016). Electrospun fibers based on Arabic, karaya and kondagogu gums. *International Journal of Biological Macromolecules, 91*, 299–309.
Parana, P., Bandhitsing, T., & Thitiwongsawet, P. (2019). Electrospun polycaprolactone fiber mats as carriers for resorcinol. *Science & Technology Asia, 4*(1), 37–44.
Peinado, I., Mason, M., Romano, A., Biasioli, F., & Scampicchio, M. (2016). Stability of β-carotene in polyethylene oxide electrospun nanofibers. *Applied Surface Science, 370*, 111–116.
Pérez-Masiá, R., López-rubio, A., & Lagarón, J. M. (2013). Food hydrocolloids development of zein-based heat-management structures for smart food packaging. *Food Hydrocolloids, 30*(1), 182–191.
Pusporini, P., Edikresnha, D., Sriyanti, I., Suciati, T., Munir, M. M., & Khairurrijal, K. (2018). Electrospun polyvinylpyrrolidone (PVP)/green tea extract composite nanofiber mats and their antioxidant activities. *Materials Research Express, 5*(5). 054001.
Prietto, L., Pinto, V. Z., El Halal, S. L. M., de Morais, M. G., Costa, J. A. V., Lim, L. T., et al. (2018). Ultrafine fibers of zein and anthocyanins as natural pH indicator. *Journal of the Science of Food and Agriculture, 98*(7), 2735–2741.
Qian, Y., Qi, M., Zheng, L., King, M. W., Lv, L., & Ye, F. (2016). Incorporation of rutin in electrospun pullulan/PVA nanofibers for novel UV-resistant properties. *Materials (Basel), 9*(7), 1–8.
Rafiee, Z., Nejatian, M., Daeihamed, M., & Jafari, S. M. (2019). Application of different nanocarriers for encapsulation of curcumin. *Critical Reviews in Food Science and Nutrition*, 1–77.
Rajabi, H., Ghorbani, M., Jafari, S. M., Sadeghi Mahoonak, A., & Rajabzadeh, G. (2015). Retention of saffron bioactive components by spray drying encapsulation using maltodextrin, gum Arabic and gelatin as wall materials. *Food Hydrocolloids, 51*, 327–337.
Ranjbar-Mohammadi, M., & Bahrami, S. H. (2016). Electrospun curcumin loaded poly (epsilon-caprolactone)/gum tragacanth nanofibers for biomedical application. *International Journal of Biological Macromolecules, 84*, 448–456.
Ranjbar-Mohammadi, M., Kargozar, S., Bahrami, S. H., & Joghataei, M. T. (2015). Fabrication of curcumin-loaded gum tragacanth/poly(vinyl alcohol) nanofibers with optimized electrospinning parameters. *Journal of Industrial Textiles, 46*(5), 1170–1192.
Ranjith, R., Balraj, S., Ganesh, J., & Milton, M. C. J. (2017a). Electrospun polycaprolactone nanofibers for sustained release of Naringin. *International Journal of Scientific Research in Science and Technology, 3*(6), 235–238.
Ranjith, R., Balraj, S., Ganesh, J., & Milton, M. C. J. (2017b). Preparation and characterization of polycaprolactone nanofibers for the controlled release of chlorogenic acid. *International Journal for Research in Applied Science & Engineering Technology, 5*(Ix), 234–238.
Rekha, M. R., & Sharma, C. P. (2007). Pullulan as a promising biomaterial for biomedical applications: A perspective. *Trends in Biomaterials & Artificial Organs, 20*, 2.
Reksamunandar, R. P., Edikresnha, D., Munir, M. M., Damayanti, S., & Khairurrijal (2017). Encapsulation of β-carotene in poly(vinylpyrrolidone) (PVP) by Electrospinning Technique. *Procedia Engineering, 170*, 19–23.
Rezaei, A., Fathi, M., & Jafari, S. M. (2019). Nanoencapsulation of hydrophobic and low-soluble food bioactive compounds within different nanocarriers. *Food Hydrocolloids, 88*, 146–162.
Rezaei, A., & Nasirpour, A. (2018). Encapsulation of curcumin using electrospun almond gum nanofibers: Fabrication and characterization. *International Journal of Food Properties, 21*(1), 1608–1618.

Rezaei, A., Nasirpour, A., & Fathi, M. (2015). Application of cellulosic nanofibers in food science using electrospinning and its potential risk. *Comprehensive Reviews in Food Science and Food Safety*, *14*(3), 269–284.

Rezaei, A., Nasirpour, A., Tavanai, H., & Fathi, M. (2016). A study on the release kinetics and mechanisms of vanillin incorporated in almond gum/polyvinyl alcohol composite nanofibers in different aqueous food simulants and simulated saliva. *Flavour and Fragrance Journal*, *31*(6), 442–447.

Rieger, K. A., & Schiffman, J. D. (2014). Electrospinning an essential oil: Cinnamaldehyde enhances the antimicrobial efficacy of chitosan/poly(ethylene oxide) nanofibers. *Carbohydrate Polymers*, *113*, 561–568.

Rockwell, P. L., Kiechel, M. A., Atchison, J. S., Toth, L. J., & Schauer, C. L. (2014). Various-sourced pectin and polyethylene oxide electrospun fibers. *Carbohydrate Polymers*, *107*, 110–118.

Rojas-Mercado, A. S., Moreno-Cortez, I. E., Lucio-Porto, R., & Pavón, L. L. (2018). Encapsulation and immobilization of ficin extract in electrospun polymeric nanofibers. *International Journal of Biological Macromolecules*, *118*(Pt B), 2287–2295.

Rostamabadi, H., Falsafi, S. R., & Jafari, S. M. (2019). Nanoencapsulation of carotenoids within lipid-based nanocarriers. *Journal of Controlled Release*, *298*, 38–67.

Sakuldao, S., Yoovidhya, T., & Wongsasulak, S. (2011). Coaxial electrospinning and sustained release properties of gelatin-cellulose acetate core-shell ultrafine fibres. *ScienceAsia*, *37*(4), 335.

Salaberria, A. M., Labidi, J., & Fernandes, S. C. M. (2015). Different routes to turn chitin into stunning nano-objects. *European Polymer Journal*, *68*, 503–515.

Sarabandi, K., Sadeghi Mahoonak, A., Hamishekar, H., Ghorbani, M., & Jafari, S. M. (2018). Microencapsulation of casein hydrolysates: Physicochemical, antioxidant and microstructure properties. *Journal of Food Engineering*, *237*, 86–95.

Shaid, A., Wang, L., Padhye, R., & Jadhav, A. (2018). Needleless electrospinning and electrospraying of mixture of polymer and aerogel particles on textile. *Advances in Materials Science and Engineering*, 1–7.

Shao, P., Niu, B., Chen, H., & Sun, P. (2018). Fabrication and characterization of tea polyphenols loaded pullulan-CMC electrospun nanofiber for fruit preservation. *International Journal of Biological Macromolecules*, *107*(Pt B), 1908–1914.

Shao, P., Yan, Z., Chen, H., & Xiao, J. (2018). Electrospun poly(vinyl alcohol)/permutite fibrous film loaded with cinnamaldehyde for active food packaging. *Journal of Applied Polymer Science*, *135*(16). 46117.

Shekarforoush, E., Ajalloueian, F., Zeng, G., Mendes, A. C., & Chronakis, I. S. (2018). Electrospun xanthan gum-chitosan nanofibers as delivery carrier of hydrophobic bioactives. *Materials Letters*, *228*, 322–326.

Shekarforoush, E., Faralli, A., Ndoni, S., Mendes, A. C., & Chronakis, I. S. (2017). Electrospinning of xanthan polysaccharide. *Macromolecular Materials and Engineering*, *302*(8). 1700067.

Shekarforoush, E., Mendes, A. C., Baj, V., Beeren, S. R., & Chronakis, I. S. (2017). Electrospun phospholipid fibers as micro-encapsulation and antioxidant matrices. *Molecules*, *22*(10), 1–16.

Shi, J., & Yang, E. (2016). Post-electrospinning crosslinking of guar/polyvinyl alcohol membrane. *Thermal Science*, *20*(1), 1–5.

Shin, J., & Lee, S. (2018). Encapsulation of phytoncide in nanofibers by emulsion electrospinning and their antimicrobial assessment. *Fibers and Polymers*, *19*(3), 627–634.

Silva, F. T. D., Cunha, K. F. D., Fonseca, L. M., Antunes, M. D., Halal, S. L. M. E., Fiorentini, Â. M., et al. (2018). Action of ginger essential oil (Zingiber officinale) encapsulated in proteins ultrafine fibers on the antimicrobial control in situ. *International Journal of Biological Macromolecules*, *118*, 107–115.

Soares, R. M. D., Siqueira, N. M., Prabhakaram, M. P., & Ramakrishna, S. (2018). Electrospinning and electrospray of bio-based and natural polymers for biomaterials development. *Materials Science & Engineering. C, Materials for Biological Applications*, *92*, 969–982.

Song, K., Wu, Q., Qi, Y., & Kärki, T. (2017). 20. Electrospun nanofibers with antimicrobial properties. In M. Afshari (Ed.), *Electrospun nanofibers. Series in Textiles: Vol. 1*(pp. 551–569). Woodhead Publishing.

Soto, K. M., Hernandez-Iturriaga, M., Loarca-Pina, G., Luna-Barcenas, G., Gomez-Aldapa, C. A., & Mendoza, S. (2016). Stable nisin food-grade electrospun fibers. *Journal of Food Science and Technology*, *53*(10), 3787–3794.

Sriyanti, I., Edikresnha, D., Rahma, A., Munir, M. M., Rachmawati, H., & Khairurrijal, K. (2017). Correlation between structures and antioxidant activities of polyvinylpyrrolidone/ *Garcinia mangostana* L. extract composite nanofiber mats prepared using electrospinning. *Journal of Nanobiomaterial*, 1–10.

Sriyanti, I., Edikresnha, D., Rahma, A., Munir, M. M., Rachmawati, H., & Khairurrijal, K. (2018). Mangosteen pericarp extract embedded in electrospun PVP nanofiber mats: Physicochemical properties and release mechanism of α-mangostin. *International Journal of Nanomedicine*, *13*, 4927–4941.

Steyaert, I., Rahier, H., Van Vlierberghe, S., Olijve, J., & De Clerck, K. (2016). Gelatin nanofibers: Analysis of triple helix dissociation temperature and cold-water-solubility. *Food Hydrocolloids*, *57*, 200–208.

Stijnman, A. C., Bodnar, I., & Hans Tromp, R. (2011). Electrospinning of food-grade polysaccharides. *Food Hydrocolloids*, *25*(5), 1393–1398.

Sunthornvarabhas, J., Liengprayoon, S., Aouf, C., Rungjang, W., Sangseethong, K., Lecomte, J., et al. (2016). Tara tannin as active ingredient in electrospun fibrous delivery system. *Journal of Applied Polymer Science*, *133*(27), 1–7.

Tampau, A., González-Martinez, C., & Chiralt, A. (2017). Carvacrol encapsulation in starch or PCL based matrices by electrospinning. *Journal of Food Engineering*, *214*, 245–256.

Tavassoli-Kafrani, E., Goli, S. A. H., & Fathi, M. (2017). Fabrication and characterization of electrospun gelatin nanofibers crosslinked with oxidized phenolic compounds. *International Journal of Biological Macromolecules*, *103*(2013), 1062–1068.

Tavassoli-Kafrani, E., Goli, S. A. H., & Fathi, M. (2018). Encapsulation of orange essential oil using cross-linked electrospun gelatin nanofibers. *Food and Bioprocess Technology*, *11*(2), 427–438.

Tomasula, P. M., Sousa, A. M. M., Liou, S. C., Li, R., Bonnaillie, L. M., & Liu, L. S. (2016). Short communication: Electrospinning of casein/pullulan blends for food-grade applications. *Journal of Dairy Science*, *99*(3), 1837–1845.

Torkamani, A. E., Syahariza, Z. A., Norziah, M. H., Wan, A. K. M., & Juliano, P. (2018). Encapsulation of polyphenolic antioxidants obtained from *Momordica charantia* fruit within zein/gelatin shell core fibers via coaxial electrospinning. *Food Bioscience*, *21*, 60–71.

Torres-Giner, S., Gimenez, E., & Lagaron, J. M. (2008). Characterization of the morphology and thermal properties of Zein Prolamine nanostructures obtained by electrospinning. *Food Hydrocolloids*, *22*, 601–614.

Torres-Giner, S., Pérez-Masiá, R., & Lagaron, J. M. (2016). A review on electrospun polymer nanostructures as advanced bioactive platforms. *Polymer Engineering & Science*, *56*(5), 500–527.

Torres-Giner, S., Wilkanowicz, S., Melendez-Rodriguez, B., & Lagaron, J. M. (2017). Nanoencapsulation of *Aloe vera* in synthetic and naturally occurring polymers by electrohydrodynamic processing of interest in food technology and bioactive packaging. *Journal of Agricultural and Food Chemistry*, *65*(22), 4439–4448.

Vahedikia, N., Garavand, F., Tajeddin, B., Cacciotti, I., Jafari, S. M., Omidi, T., et al. (2019). Biodegradable zein film composites reinforced with chitosan nanoparticles and cinnamon essential oil: Physical, mechanical, structural and antimicrobial attributes. *Colloids and Surfaces B: Biointerfaces*, *177*, 25–32.

Vashisth, P., Nikhil, K., Pemmaraju, S. C., Pruthi, P. A., Mallick, V., Singh, H., et al. (2013). Antibiofilm activity of quercetin-encapsulated cytocompatible nanofibers against *Candida albicans*. *Journal of Bioactive and Compatible Polymers*, *28*(6), 652–665.

Vega-Lugo, A. -C., & Lim, L. -T. (2009). Controlled release of allyl isothiocyanate using soy protein and poly (lactic acid) electrospun fibers. *Food Research International*, *42*(8), 933–940.

Vyslouzilova, L., Buzgo, M., Pokorny, P., Chvojka, J., Mickova, A., Rampichova, M., et al. (2017). Needleless coaxial electrospinning: A novel approach to mass production of coaxial nanofibers. *International Journal of Pharmaceutics*, *516*(1–2), 293–300.

Wali, A., Zhang, Y., Sengupta, P., Higaki, Y., Takahara, A., & Badiger, M. V. (2018). Electrospinning of non-ionic cellulose ethers/polyvinyl alcohol nanofibers: Characterization and applications. *Carbohydrate Polymers*, *181*, 175–182.

Wang, H., Hao, L., Niu, B., Jiang, S., Cheng, J., & Jiang, S. (2016). Kinetics and antioxidant capacity of proanthocyanidins encapsulated in zein electrospun fibers by cyclic voltammetry. *Journal of Agricultural and Food Chemistry*, *64*(15), 3083–3090.

Wang, H., Hao, L., Wang, P., Chen, M., Jiang, S., & Jiang, S. (2017). Release kinetics and antibacterial activity of curcumin loaded zein fibers. *Food Hydrocolloids*, *63*, 437–446.

Wang, P. P., Luo, Z. G., & Peng, X. C. (2018). Encapsulation of vitamin E and soy isoflavone using spiral dextrin: Comparative structural characterization, release kinetics, and antioxidant capacity during simulated gastrointestinal tract. *Journal of Agricultural and Food Chemistry*, *66*(40), 10598–10607.

Wang, P., & Mele, E. (2018). Effect of antibacterial plant extracts on the morphology of electrospun poly(lactic acid) fibres. *Materials (Basel)*, *11*(6), 0–11.

Wang, Y., & Xu, L. (2018). Preparation and characterization of porous core-shell fibers for slow release of tea polyphenols. *Polymers (Basel)*, *10*(2), 144–157.

Wang, X., Yue, T., & Lee, T. -C. (2015). *Development of pleurocidin-poly(vinyl alcohol) electrospun antimicrobial nanofibers to retain antimicrobial activity in food system application.* New York, USA: Elsevier Ltd.

Wen, P., Wen, Y., Huang, X., Zong, M. H., & Wu, H. (2017). Preparation and characterization of protein-loaded electrospun fiber mat and its release kinetics. *Journal of Agricultural and Food Chemistry*, *65*(23), 4786–4796.

Wen, P., Wen, Y., Zong, M. H., Linhardt, R. J., & Wu, H. (2017). Encapsulation of bioactive compound in electrospun fibers and its potential application. *Journal of Agricultural and Food Chemistry*, *65*(42), 9161–9179.

Wen, P., Zhu, D. H., Feng, K., Liu, F. J., Lou, W. Y., Li, N., et al. (2016). Fabrication of electrospun polylactic acid nanofilm incorporating cinnamon essential oil/beta-cyclodextrin inclusion complex for antimicrobial packaging. *Food Chemistry*, *196*, 996–1004.

Wen, P., Zhu, D. -H., Wu, H., Zong, M. -H., Jing, Y. -R., & Han, S. -Y. (2016). Encapsulation of cinnamon essential oil in electrospun nanofibrous film for active food packaging. *Food Control*, *59*, 366–376.

Wen, P., Zong, M. -H., Linhardt, R. J., Feng, K., & Wu, H. (2017). Electrospinning: A novel nano-encapsulation approach for bioactive compounds. *Trends in Food Science & Technology*, *70*, 56–68.

Xiao, Q., & Lim, L. T. (2018). Pullulan-alginate fibers produced using free surface electrospinning. *International Journal of Biological Macromolecules*, *112*, 809–817.

Xiao, J., Shi, C., Zheng, H., Shi, Z., Jiang, D., Li, Y., et al. (2016). Kafirin protein based electrospun fibers with tunable mechanical property, wettability, and release profile. *Journal of Agricultural and Food Chemistry, 64*(16), 3226–3233.

Yang, W., Sousa, A. M. M., Fan, X., Jin, T., Li, X., Tomasula, P. M., et al. (2017). Electrospun ultra-fine cellulose acetate fibrous mats containing tannic acid-Fe(3+) complexes. *Carbohydrate Polymers, 157*, 1173–1179.

Yang, H., Wen, P., Feng, K., Zong, M. H., Lou, Y., & Wu, H. (2017). Encapsulation of fish oil in a coaxial electrospun nano fibrous mat and its properties. *RSC Advances*, 14939–14946.

Yang, S. B., Yoo, S. H., Rabbani, M. M., Kim, I. K., Oh, W., Han, S. I., et al. (2017). Incorporation of sorghum extract into electrospun zein nanofibers and their characterization. *Journal of Nanoscience and Nanotechnology, 17*(12), 9002–9008.

Yao, Z. -C., Chang, M. -W., Ahmad, Z., & Li, J. -S. (2016). Encapsulation of rose hip seed oil into fibrous zein films for ambient and on demand food preservation via coaxial electrospinning. *Journal of Food Engineering, 191*, 115–123.

Yao, J., Zhang, S., Lim, L. -T., & Chen, X. (2017). Investigation of isothiocyanate release from electrospun modified poly(L-lactic acid)/mustard powder composite fibers. *Polymer Journal, 49*(5), 449–456.

Yildiz, Z. I., Celebioglu, A., Kilic, M. E., Durgun, E., & Uyar, T. (2018). Menthol/cyclodextrin inclusion complex nanofibers: Enhanced water-solubility and high-temperature stability of menthol. *Journal of Food Engineering, 224*, 27–36.

Yilmaz, A., Bozkurt, F., Cicek, P. K., Dertli, E., Durak, M. Z., & Yilmaz, M. T. (2016). A novel antifungal surface-coating application to limit postharvest decay on coated apples: Molecular, thermal and morphological properties of electrospun zein–nanofiber mats loaded with curcumin. *Innovative Food Science & Emerging Technologies, 37*, 74–83.

Yuan, Y., Zhang, N., Tao, W., Cao, X., & He, Y. (2014). Fatty acids as phase change materials: A review. *Renewable and Sustainable Energy Reviews, 29*, 482–498.

Zdraveva, E., Fang, J., Mijovic, B., & Lin, T. (2015). Electrospun poly(vinyl alcohol)/phase change material fibers: Morphology, heat properties, and stability. *Industrial & Engineering Chemistry Research, 54*(35), 8706–8712.

Zhang, C., Feng, F., & Zhang, H. (2018). *Key laboratory for agro-products postharvest handling of ministry of agriculture.* New York, USA: Elsevier Ltd.

Zhang, W., Huang, C., Kusmartseva, O., Thomas, N. L., & Mele, E. (2017). Electrospinning of polylactic acid fibres containing tea tree and manuka oil. *Reactive and Functional Polymers, 117*, 106–111.

Zhang, C., & Zhang, H. (2018). Formation and stability of core–shell nanofibers by electrospinning of gel-like corn oil-in-water emulsions stabilized by gelatin. *Journal of Agricultural and Food Chemistry, 66*(44), 11681–11690.

Zhong, J., Mohan, S. D., Bell, A., Terry, A., Mitchell, G. R., & Davis, F. J. (2018). Electrospinning of food-grade nanofibers from whey protein. *International Journal of Biological Macromolecules*, 1–10.

Further reading

Liu, Y., Deng, L., Zhang, C., Chen, K., Feng, F., & Zhang, H. (2018). Comparison of ethyl cellulose-gelatin composite films fabricated by electrospinning versus solvent casting. *Journal of Applied Polymer Science, 135*(46).

Ozkan, G., Kamiloglu, S., & Capanoglu, E. (2018). Chapter 11: Use of nanotechnological methods for the analysis and stability of food antioxidants. In A. M. Grumezescu & A. M. Holban (Eds.), *Impact of nanoscience in the food industry.* New York, USA: Elsevier Inc.

Patel, A., Patra, F., Shah, N., & Khedkar, C. (2018). Chapter 1: Application of nanotechnology in the food industry: Present status and future prospects. In A. M. Grumezescu &

A. M. Holban (Eds.), *Impact of nanoscience in the food industry* (pp. 1–27). New York, USA: Academic Press.

Pisoschi, A. M., Pop, A., Cimpeanu, C., Turcus, V., Predoi, G., & Iordache, F. (2018). Nanoencapsulation techniques for compounds and products with antioxidant and antimicrobial activity—A critical view. *European Journal of Medicinal Chemistry*, *157*, 1326–1345.

Yang, G., Li, X., He, Y., Ma, J., Ni, G., & Zhou, S. (2018). From nano to micro to macro: Electrospun hierarchically structured polymeric fibers for biomedical applications. *Progress in Polymer Science*, *81*, 80–113.

CHAPTER THREE

Production of food bioactive-loaded nanoparticles by electrospraying

Laura G. Gómez-Mascaraque[a], **Amparo Lopez-Rubio**[b]
[a]Teagasc Food Research Centre, Fermoy, Ireland
[b]Food Preservation and Food Quality Department, IATA-CSIC, Valencia, Spain

Contents

1 Introduction

Among the numerous microencapsulation techniques that have already been developed, drying technologies (mainly spray-drying and freeze-drying) are the most commonly used for the formulation of food ingredients (Gharsallaoui, Roudaut, Chambin, Voilley, & Saurel, 2007; Jafari, Assadpoor, He, & Bhandari, 2008). One of the main reasons is that

Nanoencapsulation of Food Ingredients by Specialized Equipment
https://doi.org/10.1016/B978-0-12-815671-1.00003-2

drying generally increases the stability of the formulations compared with the solutions or suspensions obtained through other microencapsulation techniques, extending the shelf life of the formulated food ingredients (Kwak, 2014). In fact, many other microencapsulation techniques such as emulsification or coacervation are usually combined with a drying technique in order to obtain a convenient dry powder in the final step (Mahdavi, Jafari, Ghorbani, & Assadpoor, 2014; Silva, Favaro-Trindade, Rocha, & Thomazini, 2012).

Nedović, Kalušević, Manojlović, Petrović, and Bugarski (2013) estimated that about 90% of the produced microencapsulation structures in the food industry are prepared by spray-drying (Nedović et al., 2013), due to its ease of operation, high production rates, low operating costs, reproducibility of the technique, and the possibility of using a wide range of encapsulating materials (Đorđević et al., 2015). However, the high temperatures used in the drying step make this technique unsuitable for some ingredients such as labile bioactive compounds or sensitive probiotic microorganisms (Pérez-Masiá, Lagaron, & Lopez-Rubio, 2015). On the other hand, freeze-drying is also a very simple technique to entrap ingredients of interest within a protective matrix, which does not require the use of high temperatures (Jafari, Mahdavi-Khazaei, & Hemmati-Kakhki, 2016; Kwak, 2014). But, it is approximately 30–50 times more expensive than spray-drying (Gharsallaoui et al., 2007), its processing time is very long, and the obtained materials are highly porous, which generally results in a poor barrier between the active ingredient and the environment or the food matrix in which they are to be incorporated (Zuidam & Shimoni, 2010). Therefore, although the application of drying technologies is, in many cases, unavoidable for the formulation of microencapsulated ingredients, both spray-drying and freeze-drying involve important limitations, which justify the need for developing new alternatives.

Electrospraying is an alternative drying technique based on the application of high-voltage electric fields to polymer fluids and can be performed under mild conditions (López-Rubio & Lagaron, 2012). This technology, whose application for microencapsulation purposes was first proposed in the biomedical field (Bock, Dargaville, & Woodruff, 2012), has only recently been extended to the area of food technology (Bock et al., 2012), with particularly promising applications in the development of new functional foods (Gómez-Mascaraque, Lagarón, & López-Rubio, 2015; Pérez-Masiá et al., 2015; Pérez-Masiá, Lagaron, & Lopez-Rubio, 2015). Due to its novelty, some recent reviews on encapsulation for food purposes do not yet include this technology in their list of available microencapsulation methods (Đorđević et al., 2015; Zuidam & Shimoni, 2010). In this chapter, this pioneering encapsulation technique will be described, and its main advantages and current challenges will be addressed. Additionally, the latest scientific advances in the application of electrospraying to the nano- and microencapsulation of food ingredients will be discussed.

2 Fundamentals of the electrospraying technique

Electrospraying, also known as electrohydrodynamic atomization (EHDA), is a particular case of the more general concept of electrohydrodynamic processing techniques. Electrohydrodynamic processing refers to the processing of electrically charged fluids, and it allows the production of dry nano- and microstructures by subjecting a polymeric fluid to a high-voltage electric field (Bhushani & Anandharamakrishnan, 2014). A scheme of a simple setup for electrohydrodynamic processing is depicted in Fig. 1. Usually, the fluid is pumped at a certain flow rate through a nozzle or conductive capillary (e.g., a stainless steel needle) to which the voltage is applied. As a result, electrostatic (repulsive) interactions are generated within the liquid that is flowing out from the capillary, and when these electrical forces overcome the forces of surface tension, a charged polymer jet is ejected toward the opposite electrode, which usually consists of a grounded stainless steel collector. Due to the great surface/volume ratio and the repulsive electrical forces, the solvent in this polymeric jet is evaporated during its fly, so dry material is finally deposited on the collector (Bhardwaj & Kundu, 2010; Chakraborty, Liao, Adler, & Leong, 2009).

If the molecular cohesion between the polymeric chains in the polymer fluid being processed is high enough, the generated jet is elongated during its flight due to the balance of forces imposed on it, so that ultrathin fibers are produced upon drying (Kriegel, Arrechi, Kit, McClements, & Weiss, 2008). In this case, the process is commonly referred to as "electrospinning" (cf. Fig. 1A), as described in Chapter 2. Conversely, if the intermolecular cohesion is sufficiently low, the jet breaks into fine droplets. Due to their surface tension, the jet fragments tend to acquire a spherical shape in the air, and thus they yield micro- or nanoparticles upon solvent evaporation. This phenomenon is generally known as "electrospraying" (cf. Fig. 1B) (Alehosseini, Ghorani, Sarabi-Jamab, & Tucker, 2017). Therefore, the basic setup and general principles of both the electrospinning and the electrospraying techniques are the same, only differing in the morphology of the

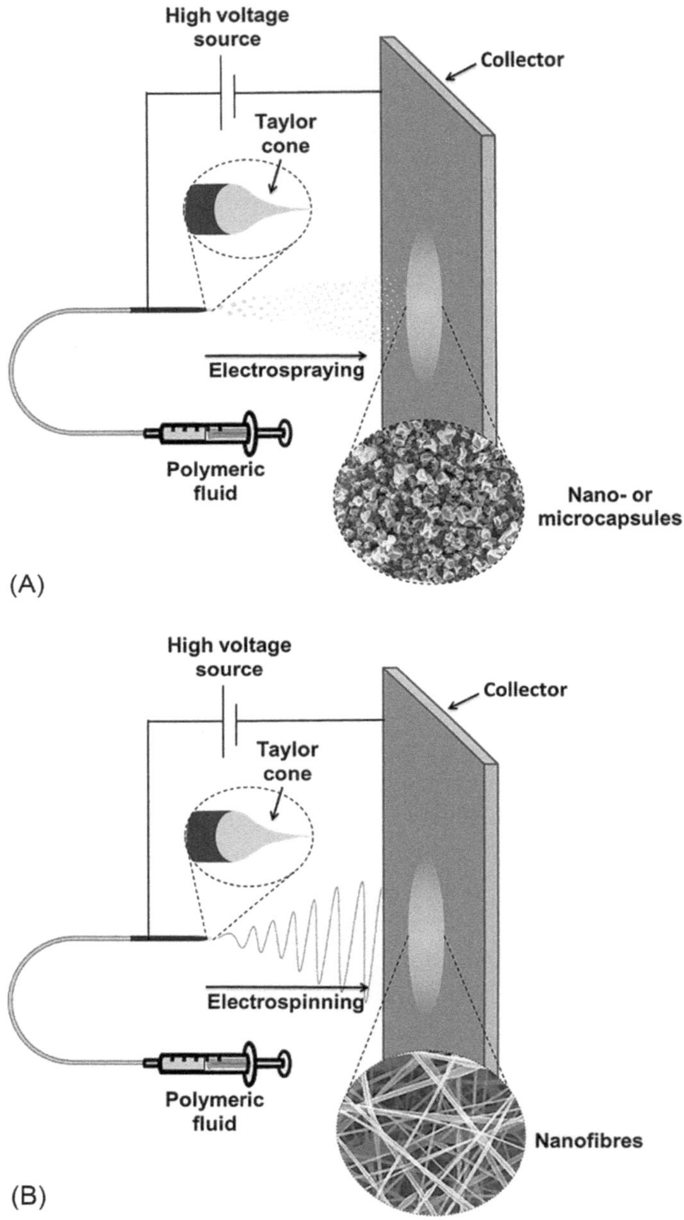

Fig. 1 Schematic representation of (A) electrospraying and (B) electrospinning processes and image of structures obtained thereof.

obtained nano- or microstructures, which in turn depend on the properties of the polymeric fluid to be processed and the process parameters (see Chapter 2). This provides the unique opportunity of tailoring the morphology of the encapsulation structures, being able to produce either fibers or particles, by adjusting the feed formulation and/or the processing conditions. However, for some applications in the food sector, particles rather than fibers are preferred, as powders are easier to handle and to disperse within the food matrices than fiber mats (Gómez-Mascaraque et al., 2015). Therefore, electrospraying conditions are usually pursued for this purpose, and they will be the focus of this chapter.

Similar setups can also be used to produce hydrogels through the electrostatic extrusion technique, also referred to as the electrospray-assisted or electrospray-aided extrusion (Zaeim, Sarabi-Jamab, Ghorani, Kadkhodaee, & Tromp, 2017). This technique is a variant of the conventional extrusion technique in which an external electric field is applied to break up the fluid droplets, which are extruded onto a gelling bath, exploiting the same principles of the electrospraying technology. It should be noted that electrostatic extrusion is a wet encapsulation technique not yielding dry encapsulation structures, although some authors classify it as part of electrospraying technologies (Tapia-Hernández, Rodríguez-Félix, & Katouzian, 2017); electrostatic extrusion will not be covered in this chapter.

Jaworek and Sobczyk (2008) and later on Rosell-Llompart, Grifoll, and Loscertales (2018) summarized the complex physics involved in the electrospraying process and provided equations for the balance of bulk forces and the balance of stress tensors to which the polymer jet is subjected upon electrospraying, which include electrodynamic, gravitational, inertial, and drag contributions. However, as they stated, no solutions to the general equations have been provided yet, and only simplified equations for particular spraying modes and specific considerations have been solved. This gives an idea of the complexity of the physics involved. As a consequence, the electrospraying conditions (both the properties of the feed formulation and the process parameters) generally need to be optimized for each polymer system or even for each polymer-bioactive combination if encapsulation is the purpose of the process, since the addition of small amounts of these ingredients as well as the procedure to disperse them may significantly alter the solution properties of the feed formulation (Gómez-Mascaraque & López-Rubio, 2016). The impact that the solution properties and the process parameters generally have on the electrosprayed process is explained in the succeeding text. Table 1 summarizes the solution properties of the formulations, which have

Table 1 Solution properties and process parameters of successfully electrosprayed formulations

Wall material	Solvent/ additives	Concentration	Surface tension (mN/m)	Electrical conductivity (μS/cm)	Viscosity (mPa*s)	Processing parameters (distance, voltage, flow rate)	References
Whey protein concentrate (WPC)	Water	20–40 w/v	42–47	2280–2753	11–49	7 cm, 14 kV, 0.15–0.3 mL/h	López-Rubio and Lagaron (2012), Pérez-Masiá, Lagaron, and López-Rubio (2014a)
	Water (+Span 20)	20% w/v	32	1750	5.6	9–11 cm, 10 kV, 0.15 mL/h	Pérez-Masiá, López-Nicolás, et al. (2015)
	Water (emulsion containing soy bean oil) (+Tween 20)	20% w/v	36	1102–1149	21–26	10 cm, 15–17 kV, 0.15 mL/h	Gómez-Mascaraque and López-Rubio (2016), Gómez-Mascaraque, Perez-Masiá, González-Barrio, Periago, and López-Rubio (2017)
	Water (containing liposome dispersion)	20% w/v	–	–	–	10 cm, 10 kV, 0.15 mL/h	Gómez-Mascaraque, Sipoli, de La Torre, and López-Rubio (2017a, 2017b)
	Water (+Tween 20)	30% w/v	35	2057	19	9–20 cm[a], 12–18 kV[a], 0.15 mL/h	Pérez-Masiá, Lagaron, and Lopez-Rubio (2015)

WPC	PBS or skimmed milk	30–40 wt%	–	–	–	7 cm, 12–14 kV, 0.3 mL/h	López-Rubio, Sanchez, Wilkanowicz, Sanz, and Lagaron (2012)
WPC (+resistant starch)	Water (+Tween 20)	30% w/v	–	–	–	10 cm, 10–14 kV, 0.15 mL/h	Gomez-Mascaraque, Morfin, Pérez-Masiá, Sanchez, and Lopez-Rubio (2016)
Zein	85 wt% aqueous ethanol	13%	–	–	–	15 cm, 12 kV, 0.2 mL/h	Torres-Giner, Martinez-Abad, Ocio, and Lagaron (2010)
		12% w/v	–	–	–	10 cm, 13 kV, 0.15 mL/h (uniaxial and coaxial)	Costamagna et al. (2017), Gómez-Mascaraque et al. (2017), Gómez-Mascaraque, Tordera, Fabra, Martínez-Sanz, and Lopez-Rubio (2019)
	80 wt% aqueous ethanol	5–12 wt%	–	700–1242	4–10	8 cm, 16 kV, 0.2 mL/h	Bhushani, Kurrey, and Anandharamakrishnan (2017)
		2.5–15 wt%	–	–	–	7 cm, 14 kV, 0.05–0.15 mL/h	Gómez-Estaca, Balaguer, Gavara, and Hernandez-Munoz (2012), Gómez-Estaca, Balaguer, López-Carballo, Gavara, and Hernández-Muñoz (2017)
	80 wt% aqueous ethanol (emulsion containing soy bean oil) (+Tween 20)	12% w/v	26	373–386	20	10 cm, 13 kV, 0.15 mL/h	Gómez-Mascaraque, Perez-Masiá, et al. (2017)

Continued

Table 1 Solution properties and process parameters of successfully electrosprayed formulations—cont'd

Wall material	Solvent/additives	Concentration	Surface tension (mN/m)	Electrical conductivity (μS/cm)	Viscosity (mPa*s)	Processing parameters (distance, voltage, flow rate)	References
	Zein dissolved in ethanol/acetic acid 75:25 (core) Pure ethanol pumped through the outer orifice of a coaxial nozzle	12% w/v	–	–	–	20 cm, 21 kV, 2 mL/h (total) (inner: 1.3–2 mL/h outer: 0–0.7 mL/h)	Liu, Zhang, Yu, Wu, and Li (2018)
Gelatin	Acetic acid 20% v/v	8% w/v	41	2420	16	10 cm, 15 kV, 0.15–0.2 mL/h	Gómez-Mascaraque et al. (2015), Gómez-Mascaraque, Hernández-Rojas, et al. (2017), Gómez-Mascaraque and López-Rubio (2016)
	Water/ethanol/acetic acid 1:2:1 v/v	5% w/v	–	–	–	10 cm, 14 kV, 0.15 mL/h	Gómez-Estaca, Gavara, and Hernández-Muñoz (2015)
Soy protein isolate (SPI)	Water (+ denaturation and/or Span 20)	10%	32–43	2493–2810	27–59	9–20 cm[a], 9–16 kV[a], 0.15 mL/h	Pérez-Masiá et al. (2014a)
	Water (+ denaturation and Tween 20)	10% w/v	–	–	–	10 cm, 17 kV, 0.15 mL/h	Gómez-Mascaraque and López-Rubio (2016)

Pullulan	Skimmed milk	15–20 wt%	–	–	–	7 cm, 12–14 kV, 0.3 mL/h	López-Rubio et al. (2012)
	Water	20% w/v	53	17	133	9–20 cm[a], 9–16 kV[a], 0.15 mL/h	Pérez-Masiá et al. (2014a)
Dextran	Water	40% w/v	59	23	94	9–20 cm[a], 9–16 kV[a], 0.15 mL/h	Pérez-Masiá et al. (2014a)
	Water	20–25 wt%	23–41	46–47	44–201	15 cm, 20 kV, 0.01 mL/min (1 needle) 10 cm, 42 kV, 5 mL/h (24 needles)	García-Moreno et al. (2017)
	Water (+Tween 20)	20% w/v	36	92	30	9–20 cm[a], 12–18 kV[a], 0.15 mL/h	Pérez-Masiá, Lagaron, and Lopez-Rubio (2015)
Resistant starch	Water (+surfactants)	20 wt%	26–35	25–136	2.2–5.6	9–11 cm, 9–10 kV, 0.15 mL/h	Pérez-Masiá et al. (2014a), Pérez-Masiá, Lagaron, and López-Rubio (2014b), Pérez-Masiá, López-Nicolás, et al. (2015)
Maltodextrin	Water (+surfactants)	20 wt%	25–35	681–843	2.2–5.8	10 cm, 9 kV, 0.15 mL/h	Pérez-Masiá et al. (2014a, 2014b)
Oligofructose	Water (+Span 20)	20 w/v	26	55	5.3	9–20 cm[a], 9–16 kV[a], 0.1 mL/h	Pérez-Masiá et al. (2014a)

Continued

Table 1 Solution properties and process parameters of successfully electrosprayed formulations—cont'd

Wall material	Solvent/ additives	Concentration	Surface tension (mN/m)	Electrical conductivity (μS/cm)	Viscosity (mPa*s)	Processing parameters (distance, voltage, flow rate)	References
Inulin	Water (+Span 20)	20 w/v	26	63	5.5	9–20 cm[a], 9–16 kV[a], 0.1 mL/h	Pérez-Masiá et al. (2014a)
	Water (+guar gum)	20 w/v	50	80	18	9–20 cm[a], 9–16 kV[a], 0.1 mL/h	Pérez-Masiá et al. (2014a)
Chitosan	Acetic acid 90% v/v	2% w/w	40	392	49	9–20 cm[a], 12–18 kV[a], 0.15 mL/h	Pérez-Masiá, Lagaron, and Lopez-Rubio (2015)
	Acetic acid 60–90% v/v	0.5%–5% w/v[b]	30–35	100–800	60–200	10 cm, 17 kV, 0.15 mL/h	Gómez-Mascaraque, Sanchez, and López-Rubio (2016)
Gum arabic	Water (+autoclave + Tween 80)	35 wt%	35	3145	455	10 cm, 19 kV, 0.2 mL/h	Zaeim, Sarabi-Jamab, Ghorani, Kadkhodaee, and Tromp (2018)
Alyssum homolocarpum seed gum	Water (+Tween 20) (emulsion by high speed homo-genizer)	0.5 wt%	41	130	450	15 cm, 20 kV, 0.1 mL/h	Khoshakhlagh, Koocheki, Mohebbi, and Allafchian (2017), Khoshakhlagh, Mohebbi, Koocheki, and Allafchian (2018)
Stearic acid	Ethanol	1–4 wt%	21–23	–	0.75–1.20	10 cm, 14.5 kV, 0.9 mL/h	Eltayeb, Stride, Edirisinghe, and Harker (2016)

[a]Exact conditions not specified.
[b]Depending on the molecular weight.

been successfully electrosprayed with potential applications in the nano- or microencapsulation of food ingredients, as well as the process parameters used to produce the encapsulation structures.

2.1 Influence of the polymer fluid properties

The main properties of the feed formulation (usually a polymeric solution or dispersion), which have an impact on the performance of the hydrodynamic processing and the morphology of the resulting materials, are the rheological characteristics of the fluid, its surface tension, and conductivity. All these properties in turn depend on the type of polymer used, its molecular weight, concentration, the solvent used, and the addition of other substances (including the bioactive compounds to be encapsulated).

The viscosity and rheological behavior of the feed formulations are key factors, as they are related to the extent of chain entanglements within the polymer fluid and, thus, have a decisive impact on the preferential formation of fibers or capsules (Ramakrishna, 2005). At lower viscosities (lower molecular weights and polymer concentrations), the polymer jet is more likely to break up and form droplets (Kriegel et al., 2008). Some authors have tried to generalize the ideal viscosity ranges for the formation of fibers or particles. For instance, Doshi and Reneker (1995) suggested that the ideal viscosity to achieve electrospinning ranges from 800 to 4000 mPas, considering solutions below 800 mPas too dilute to undergo chain entanglement (Doshi & Reneker, 1995). However, nanofibers have been successfully obtained from polymeric solutions with lower viscosities (Gómez-Mascaraque et al., 2015). Indeed, the ideal ranges greatly vary depending on the polymer type and entanglement behavior in the selected solvent (Kriegel et al., 2008). Moreover, the rheological properties are not the only factor involved. An increase in the surface tension usually favors jet break up, while an increase in the electrical conductivity tends to elongate the jet favoring fiber formation (Ramakrishna, 2005), although too high conductivities may lead to jet instability.

Another critical factor is the volatility of the solvent used, which determines whether sufficient solvent evaporation can occur during the flight of the polymeric fluid from the tip of the capillary to the collector (Chakraborty et al., 2009).

2.2 Influence of the process parameters

Electrospraying is affected by a number of process parameters, which include the applied voltage, the feed flow rate, the capillary diameter, the tip-to-collector distance, and environmental conditions such as the temperature, pressure, and relative humidity.

The applied voltage, which usually varies in the range of 6–20 kV, determines the electrostatic forces applied to the polymer jet (depending on the solution properties). A sufficient voltage must be applied to overcome the surface tension of the polymer at the tip of the capillary (Chakraborty et al., 2009). An increase in the applied voltage increases the charge density causing the jet to accelerate faster and to stretch to a greater extent. This usually results in a reduction of the particle diameter, although some exceptions have been observed (Kriegel et al., 2008) given that the resulting size and morphology depend on a number of factors. Thus, generalizations hold true only within certain ranges, which are often dictated by practical considerations (Jaworek & Sobczyk, 2008).

On the other hand an increase in the flow rate and/or the needle diameter, or a decrease in the tip-to-collector distance usually leads to an increase in the diameter of the obtained structures. Flow rates must be low enough to allow sufficient time for solvent evaporation in combination with an adequate flying distance, so the optimal ranges strongly depend on the solvent volatility (Kriegel et al., 2008).

Environmental factors also play a role in electrospraying, since they have an impact on the solution properties. For instance, higher temperatures increase the molecular mobility and, thus, increase the solution conductivity, while they would decrease the viscosity and surface tension. Moreover, the evaporation of the solvent would be faster. On the contrary, an increase in the relative humidity decreases the evaporation rate of the solvent, which causes an increase in the diameter of the generated structures (Jaworek & Sobczyk, 2008; Kriegel et al., 2008).

3 Advantages and challenges of electrospraying for food applications

3.1 Advantages of the technique for the food sector

Compared with other technologies, electrospraying offers a series of advantages, which makes it a promising and versatile alternative for the nano- and microencapsulation of food ingredients. For instance, being a drying technique, it allows the production of dry nano-/microencapsulation structures in a one-step process from the feed formulations (Tapia-Hernández et al., 2015), without the need of a subsequent drying step. As commented on earlier, drying generally increases the shelf stability of liquid formulations, and electrospraying allows the simultaneous encapsulation and drying of food ingredients, similarly to spray-drying or freeze-drying. As these two, electrohydrodynamic techniques can also achieve uniform dispersion of compounds within the polymeric matrices with a high loading capacity and minimal compound losses (Chakraborty et al., 2009). But unlike them, it does not require high temperatures nor freezing for drying, so it is suitable for both thermosensitive and cryosensitive bioactive food ingredients (Gómez-Mascaraque et al., 2016).

On the other hand, the droplet size obtained by atomizing through electrospraying is generally smaller than in conventional mechanical atomizers and in some cases with narrower size distributions (Jaworek & Sobczyk, 2008). Since particles in the nano- and submicron range are generally obtained (Gomez-Mascaraque et al., 2016; Gómez-Mascaraque et al., 2015), this is also advantageous to minimize their impact on the textural properties of the food products. Moreover, due to the mutual repulsion of the charged droplets generated, they self-disperse in space (Zuidam & Velikov, 2018), so aggregation is in principle prevented. And, since the deposition efficiency of a charged spray is, in general, considerably higher than for uncharged droplets (Jaworek & Sobczyk, 2008), the product yield is in many cases greater than for other atomizing techniques.

As already mentioned, although capsule-like nano-/microstructures are usually preferred for many food applications, electrohydrodynamic processing also allows the design of different nanosized and micron-sized morphologies (cf. Fig. 2) such as ultrafine fibers, which in turn allows tailoring their release properties. Another advantage of electrospraying is the possibility of producing multilayer encapsulation structures in one step, without the need of applying subsequent coatings, by using a coaxial (or even multiaxial) configuration (Zhang et al., 2017; Zhang, Huang, Si, & Xu, 2012). In this configuration, the nozzle consists of two or more concentric capillaries, so that the ingredient to be encapsulated flows through the inner one and the coating biopolymer solutions flow through the external ones from independent circuits (Gómez-Mascaraque et al., 2019). As a result, core-shell capsules can be produced being the ingredient of interest located in the core of the nano-/microparticles. A scheme of this coaxial configuration is shown in Fig. 3.

From the operational point of view, electrospraying has also some benefits compared with other drying techniques. Once the process conditions are optimized, its operation is quite easy and cost-effective, and quality checks on the particles can be performed simply by interrupting the process briefly (Chakraborty et al., 2009), which is very convenient and cannot

Fig. 2 Different types of morphologies of structures obtained through electrohydrodynamic processing.

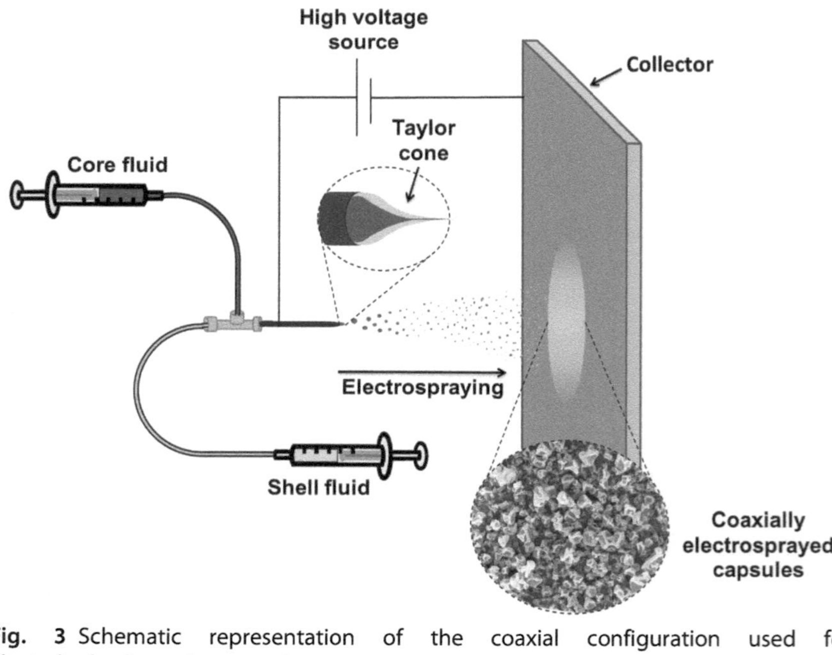

Fig. 3 Schematic representation of the coaxial configuration used for electrohydrodynamic processing.

always be done with other encapsulation techniques requiring conditioning steps. Regarding the scalability, as the application of electrospraying for encapsulation is still in its exploratory phase (Gómez-Mascaraque et al., 2016), most of the research works published or patented in this field have been conducted at lab scale, in equipment with considerably low production capacities. However, due to the increased interest in the electrohydrodynamic processing techniques for the production of different materials in various sectors, many technological solutions are already being proposed to scale up the production (Zhang et al., 2015). Some industrial-scale setups are already commercially available for the production of electrospun materials. For instance, Inovenso Ltd. supplies an apparatus with a production rate of about 0.2 kg/h, and the company Elmarco commercializes an electrospinning system with a claimed annual throughput of up to $50 \times 10^6\,m^2$ of nonwoven nanofibers (Persano, Camposeo, Tekmen, & Pisignano, 2013). Although these industrial solutions have not been applied for encapsulation purposes in the food sector yet, the industrialization perspectives of the technique are promising.

3.2 Challenges of electrospraying in the food sector

Despite all these advantages and the increasing interest in electrospraying for encapsulation of bioactive compounds, there are still some challenges that need to be faced to apply this technology in the food industry. First of all, if the electrosprayed structures are to be used for food applications, the use of water as solvent is almost a must in order to avoid any residual traces of toxic solvents in the final materials (López-Rubio & Lagaron, 2012).

But, the physical properties of water are not the most suitable for electrohydrodynamic processing. Water has a relatively high boiling point compared with many organic solvents, which implies that its evaporation from aqueous polymer solutions or suspensions is generally slower. Moreover, the surface tension of water is also considerably high (72 mN/m (Chakraborty et al., 2009)), so higher voltages would need to be applied in order to obtain stable jets; very high voltages cannot be used to process aqueous solutions since water also has a high conductivity, so ionization of the water molecules would occur causing corona discharges (Pérez-Masiá et al., 2014b). As a result, the feed formulations must be very carefully designed in order to obtain a stable electrospraying process and avoid dripping of the polymeric solutions (Gómez-Mascaraque et al., 2016). One approach to improve the physical properties of aqueous formulations prior to electrospraying is the addition of surfactants to reduce their surface tension. This strategy has successfully been employed to produce electrosprayed capsules from aqueous polysaccharide solutions, like starch or maltodextrin (Pérez-Masiá et al., 2014b), and aqueous protein dispersions such as soy protein isolate or whey protein concentrate (Pérez-Masiá et al., 2014a).

Naturally, the encapsulation matrices to be used for food applications must be food-grade too, which restricts the range of available materials to edible biopolymers, that is, proteins and polysaccharides. In many cases, this represents an additional challenge, since some biopolymers are difficult to electrospray due to their complex molecular structures. A number of them are polyelectrolytes, having numerous positively or negatively charged groups in their structure, which complicate the electrohydrodynamic process. Some others have limited chain flexibilities; their molecular weights are too high (which imply too high viscosities) or too low (which do not provide sufficient entanglements) or even are organized in the form of globular structures (this is the case of globular proteins), which do not provide sufficient chain entanglements either (Gómez-Mascaraque et al., 2016). Nevertheless, some strategies have already been proposed to overcome these limitations, which are mainly focused on manipulating the biopolymer chain conformation. For instance, denaturation of globular proteins has been employed as an effective approach to favor chain entanglements. By unfolding the proteins, their functional groups are exposed and available to form new intermolecular interactions (Gómez-Mascaraque et al., 2016). This approach has proven to be successful to improve the electrosprayability of soy proteins from aqueous dispersions (Pérez-Masiá et al., 2014a). The inverse problem, that is, the presence of too many chain entanglements, can also be encountered when attempting to electrospray biopolymeric solutions. This may be the case of biopolymers with hydrogel-forming capacity, as gelation could occur during processing, hindering the formation of the electrospraying jet (Erencia, Cano, Tornero, Macanás, & Carrillo, 2014). This can be overcome by adjusting the solution properties to prevent gelation. For instance, acidifying the medium was an effective strategy to produce electrosprayed gelatin capsules (Gómez-Mascaraque et al., 2015), preventing gelation during processing.

Another challenge for electrospraying to encapsulate food ingredients is the fact that many of them are lipophilic, so they cannot be readily dissolved in the aqueous biopolymer formulations previously mentioned. One strategy to solve this problem is to prepare oil-in-water emulsions prior to electrospraying. This so-called emulsion-electrospraying technique has been successfully applied to encapsulate a number of lipophilic bioactive ingredients, including fatty acids and carotenoids (Gómez-Mascaraque & López-Rubio, 2016; Pérez-Masiá, Lagaron, & Lopez-Rubio, 2015). Alternatively, the entrapment of lipophilic ingredients within liposome dispersions prior to electrospraying has been recently proposed too (Chang, Stone, & Nickerson, 2018; Gómez-Mascaraque, Sipoli, et al., 2017b). The possibility of using a coaxial configuration in electrospraying as previously discussed also offers a unique opportunity to incorporate lipophilic ingredients within hydrophilic encapsulation matrices, since both solutions flow through independent circuits. This strategy has also been explored for the encapsulation of lycopene within dextran and whey protein concentrate electrosprayed capsules (Pérez-Masiá, Lagaron, & Lopez-Rubio, 2015).

4 Development of edible encapsulation structures by electrospraying

As mentioned earlier, one of the main challenges of the electrospraying technique is the fact that the optimization of the electrospraying conditions must be carried out individually for each encapsulation matrix of interest, given that the solution properties must be very carefully adjusted in order to obtain capsule-like structures (Gómez-Mascaraque, Sanchez, & López-Rubio, 2016). Therefore, research efforts have been made in the last decades to widen the range of edible electrosprayed materials, which could be used for nano- and microencapsulation of food ingredients. Since each compound to be encapsulated has different properties (cf. Section 5), building a wide range of available encapsulation matrices could help future research in selecting the most adequate vehicles for each purpose.

Naturally, the wall materials used for encapsulation of food ingredients must be edible and preferably cheap. Consequently, biopolymers, both proteins and polysaccharides, have been the focus of research in this area, since they have some advantages and disadvantages. Polysaccharides are quite abundant and reasonably inexpensive (Mohan, Rajendran, He, Bazinet, & Udenigwe, 2015). Probably, their most interesting features are their great structural stability as compared with proteins and the resistance of some of them (dietary fibers) to enzymatic degradation during digestion, being potential ideal vehicles for colonic delivery (Fathi, Martin, & McClements, 2014; Gómez-Mascaraque et al., 2018). On the other hand, proteins have a high nutritional value (Ma et al., 2014), and their amphiphilic nature (Malaki Nik, Wright, & Corredig, 2010) provides them with surface activity (Livney, 2010), which facilitates not only the formation and stabilization of emulsions (McClements, 2004) (of interest for lipophilic ingredients) but also the electrospraying process by lowering the surface tension of the formulations. Therefore, electrosprayed nano- and microstructures are being developed using both types of biopolymers.

4.1 Protein-based electrosprayed capsules

The first edible nano- and microstructures incorporating bioactive ingredients that were developed by electrohydrodynamic processing, which begun emerging in the last decade, were based on proteins and had a fibrillar morphology (i.e., were actually obtained by electrospinning). Zein was initially the main protein used to incorporate different bioactive compounds, such as antioxidants (Fernandez, Torres-Giner, & Lagaron, 2009; Li, Lim, & Kakuda, 2009) or later on omega-3 fatty acids (Moomand & Lim, 2014, 2015). The reason why zein attracted so much interest was most probably its solubility in ethanol, as this solvent exhibits lower conductivity, surface tension, and boiling point than water, and thus it is easier to process by electrohydrodynamic processing while being allowed for food purposes. Bearing in mind the convenience of formulating encapsulated ingredients as powdery materials instead of fibrillar mats, electrosprayed zein particles were soon after proposed as delivery vehicles for ω-3 fatty acids (Torres-Giner et al., 2010). Although the obtained ultrathin structures showed to be promising to protect the bioactive ingredients, the morphology of the obtained materials was not thoroughly optimized, resulting in a mixture of microcapsules and nanofibers, which in practice may negatively affect the dispersibility of the capsules within food matrices.

The electrospraying of zein was better optimized in later works, obtaining neat particles free of connecting fibrils both using the uniaxial (Bhushani et al., 2017; Costamagna et al., 2017; Gómez-Estaca et al., 2012) and coaxial (Gómez-Mascaraque et al., 2019) configurations. In the latter case, coaxial electrospraying was found to generally yield higher encapsulation efficiencies and enhanced protective effects of the bioactive ingredients incorporated within them. In their work, Gómez-Estaca et al. (2012) reported that electrosprayed zein nanoparticles/submicroparticles could be obtained from protein concentrations as low as 2.5 wt% and as high as 15 wt%, although the particle sizes were bigger for increasing polymer concentrations and flow rates, and collapsed/shrunken particles rather than spheres were obtained for the highest concentrations.

Although zein is one of the proteins that has been exploited through electrohydrodynamic processing to the greatest extent so far, issues with clogging of the nozzle are usually experienced when using this biopolymer even at low concentrations, due to the volatility of its solvent and the resulting extremely fast drying process (Kanjanapongkul, Wongsasulak, & Yoovidhya, 2010). Very recently, Liu et al. (2018) proposed a strategy to avoid this clogging phenomenon, which consisted of a modified coaxial electrospraying configuration in which ethanol was pumped through the outer orifice as a sheath fluid. Apart from facilitating the electrospraying process, this approach yielded more homogeneous zein particles in terms of size and morphology. The release properties of the capsules were also enhanced when a model drug was incorporated within the electrosprayed structures.

The first electrosprayed capsules obtained from dispersions of proteins in plain water were developed by López-Rubio and Lagaron (2012) and were based on a whey protein concentrate (WPC). As a proof of concept of their ability to encapsulate bioactive ingredients, the authors loaded them with the antioxidant β-carotene. Based on their work, other compounds such as folic acid (Pérez-Masiá, López-Nicolás, et al., 2015), lycopene (Pérez-Masiá, Lagaron, & Lopez-Rubio, 2015), or α-linolenic acid (Gómez-Mascaraque & López-Rubio, 2016) were later on encapsulated using electrosprayed WPC capsules. López-Rubio et al. (2012) also demonstrated the feasibility of producing electrosprayed WPC microcapsules from dispersions of the protein in PBS and skimmed milk, obtaining a better capsule-like morphology in the latter case. The resulting materials were successful in prolonging the viability of probiotics during storage.

WPC is arguably one of the most widely used encapsulation matrices for electrospraying to date, due to both its good electrosprayability and its good performance as carrier for food ingredients. Indeed, several research works have compared WPC with other biopolymers as wall materials for electrospraying, showing that it generally yields higher encapsulation efficiencies and/or protective effects (López-Rubio et al., 2012; Pérez-Masiá, Lagaron, & Lopez-Rubio, 2015; Pérez-Masiá, López-Nicolás, et al., 2015). It has also been shown to be suitable for the production of encapsulation structures through modified electrospraying configurations such as emulsion-electrospraying (Gómez-Mascaraque & López-Rubio, 2016; Gómez-Mascaraque, Perez-Masiá, et al., 2017) or coaxial electrospraying

(Pérez-Masiá, Lagaron, & Lopez-Rubio, 2015). Recently, a novel strategy was developed by Gómez-Mascaraque, Sipoli, et al. (2017a) in which a lipophilic food ingredient (namely, curcumin) was microencapsulated within hybrid liposome/WPC electrosprayed microcapsules. This dual encapsulation approach can be considered an alternative to the emulsion-electrospraying technique in which the ingredient of interest is dispersed in the aqueous phase in the form of liposome vesicles instead of oil droplets. By combining the electrospraying and microfluidic technologies, a semi-continuous process was developed to produce these hybrid liposome/WPC microcapsules (Gómez-Mascaraque, Sipoli, et al., 2017b).

Another protein source that has been used as wall material for encapsulation through electrospraying is soy proteins. However, the electrospraying of native soy proteins has not been reported to be feasible. Pérez-Masiá et al. (2014a) proposed the thermal denaturation of a soy protein isolate (SPI) as a strategy to enhance its electrosprayability. In combination with the addition of surfactants, electrosprayed SPI capsules were successfully produced from aqueous dispersions. This approach was then exploited for the encapsulation of ω-3 fatty acids within SPI capsules by emulsion-electrospraying, showing that this system was effective in delaying the thermal oxidation of the bioactive ingredient (Gómez-Mascaraque & López-Rubio, 2016).

Gelatin was also recently proposed as an encapsulation matrix for food ingredients using electrospraying (Gómez-Mascaraque et al., 2015). This protein has very interesting properties since it can form thermoreversible hydrogels (Peña, de la Caba, Eceiza, Ruseckaite, & Mondragon, 2010), meaning that it can be processed in aqueous solutions while preventing dissolution of the obtained capsules in aqueous foods under certain conditions (Gómez-Mascaraque, Soler, & Lopez-Rubio, 2016). But, in order to avoid gelation during the electrohydrodynamic process when conducted at room temperature from aqueous solutions, gelatin solutions must be acidified (Erencia et al., 2014; Okutan, Terzi, & Altay, 2014). This implies that electrosprayed gelatin capsules are suitable only for certain ingredients resistant to low pH values, such as tea catechins (Gómez-Mascaraque et al., 2015, 2019; Gómez-Mascaraque, Hernández-Rojas, et al., 2017), but not for those sensitive to acidic conditions (Gómez-Mascaraque, Ambrosio-Martín, Perez-Masiá, & Lopez-Rubio, 2017; Gómez-Mascaraque & López-Rubio, 2016). Gómez-Estaca et al. (2015) reported that the production of electrosprayed gelatin microcapsules is also feasible using a ternary mixture of solvents (i.e., water/acetic acid/ethanol 1:1:2 v/v) as a strategy to encapsulate ethanol-soluble ingredients like curcumin. Their work showed that microencapsulation within electrosprayed gelatin particles improved the dispersability of curcumin in a fish product. Gelatin has also been used in combination with other proteins, such as zein, to obtain coaxially electrosprayed capsules (Gómez-Mascaraque et al., 2019). Moreover, a recent work showed that the addition of electrosprayed gelatin capsules to biscuits, that is, a real food product, did not affect their acceptability as perceived by the consumers (Gómez-Mascaraque, Hernández-Rojas, et al., 2017), an aspect that is also of outmost importance for the success of a food product in the market.

Table 2 summarizes different electrospraying studies with proteins as biopolymers for encapsulation of various food bioactive ingredients.

4.2 Polysaccharide-based electrosprayed capsules

The first electrosprayed capsules produced using polysaccharides as wall materials were reported by López-Rubio and Lagaron (2012), who microencapsulated probiotic *bifidobacteria* within pullulan microstructures using skimmed milk as solvent. Soon after, Pérez-Masiá et al. (2014a, 2014b) developed a whole battery of electrosprayed materials based on edible biopolymers, including a number of carbohydrates: maltodextrin, resistant starch, inulin/oligofructose, dextran, and pullulan (Pérez-Masiá et al., 2014a, 2014b). All of them were obtained from solutions of the polysaccharides in water, although strategies such as the addition of gums and/or surfactants to modify the viscosity and/or surface tension of the solutions had to be developed to improve the sprayability of the hydrocolloids. Some of these structures were then used for the encapsulation of a number of different functional ingredients, including *bifidobacteria* (López-Rubio et al., 2012), vitamins (Pérez-Masiá, López-Nicolás, et al., 2015), or carotenoids (Pérez-Masiá, Lagaron, & Lopez-Rubio, 2015), as shown in Table 3.

Pérez-Masiá, Lagaron, & Lopez-Rubio, 2015 also attempted to use chitosan to encapsulate lycopene through electrospraying. However, the highly acidic solvent required to process this particular carbohydrate through electrohydrodynamic processing techniques (Sun & Li, 2011) caused the degradation of the bioactive ingredient. Nevertheless, electrosprayed chitosan proved to be a suitable encapsulation matrix for certain food ingredients, such as green catechins (Gómez-Mascaraque, Sanchez, & López-Rubio, 2016), as it was able to increase the stability of the bioactive compounds at low pH values (Falcó et al., 2017, 2018). It is worth mentioning that processing chitosan by electrospraying is complex due to the rigidity

Table 2 Proteins used to produce nano-/microcapsules by electrospraying

Protein	Solvent	Concentration	Food ingredients encapsulated	Comments	References
Whey protein concentrate (WPC)	Water	40 wt%	β-carotene	Ingredient dissolved in glycerol pH range: 4.5–8.5	López-Rubio and Lagaron (2012)
		30% w/v	Lycopene	Ingredient previously dissolved in soy bean oil and incorporated by emulsion/coaxial electrospraying. Surfactant added (Tween 20)	Pérez-Masiá, Lagaron, and Lopez-Rubio (2015)
			Probiotic bacteria (*L. plantarum*)	Surfactant added (Tween 20)	Gomez-Mascaraque et al. (2016)
		20% w/v	Folic acid	Surfactant added (Span 20)	Pérez-Masiá, López-Nicolás, et al. (2015)
			β-Carotene	Ingredient previously dissolved in soy bean oil and incorporated by emulsion-electrospraying Surfactant added (Tween 20)	Gómez-Mascaraque, Perez-Masiá, et al. (2017)
			α-Linolenic acid	Surfactant added (Tween 20) Ingredient incorporated emulsion-electrospraying	Gómez-Mascaraque and López-Rubio (2016)
			Curcumin	Ingredient incorporated within liposomes prior to electrospraying	Gómez-Mascaraque, Sipoli, et al. (2017a)
	PBS Skimmed milk	30–40 wt%	Probiotic bacteria (*B. animalis* Bb12)	Better capsule-like morphology using milk	López-Rubio et al. (2012)

Zein	85% ethanol	13 wt%	Docosahexaenoic acid	Residual fibrils together with capsules	Torres-Giner et al. (2010)
		12% w/v	Chañar extract	Fiber-free capsules	Costamagna et al. (2017)
			(−)-Epigallo-catechin gallate (EGCG)	Capsules produced using both the uniaxial and the coaxial configurations	Gómez-Mascaraque et al. (2019)
			α-Linolenic acid		
			Green tea extract	–	Gómez-Mascaraque, Hernández-Rojas, et al. (2017)
	80% ethanol		β-carotene	Ingredient previously dissolved in soy bean oil and incorporated by emulsion-electrospraying Surfactant added (Tween 20)	Gómez-Mascaraque, Perez-Masiá, et al. (2017)
		2.5 wt%	Curcumin	Addition of curcumin (up to 10% with respect to zein) did not alter the morphology of the microcapsules	Gómez-Estaca et al. (2012), Gómez-Estaca et al. (2017)
		5–12 wt%	Green tea extract	The most smooth, spherical nanoparticles were obtained at 5%–6% concentration	Bhushani et al. (2017)
Soy protein isolate (SPI)	Water	10% w/v	–	Heat denaturation and/or addition of surfactant (Span 20 5%)	Pérez-Masiá et al. (2014a)
			α-Linolenic acid	Heat denaturation + addition of surfactant (Tween 20) Ingredient incorporated by emulsion-electrospraying	Gómez-Mascaraque and López-Rubio (2016)

Continued

Table 2 Proteins used to produce nano-/microcapsules by electrospraying—cont'd

Protein	Solvent	Concentration	Food ingredients encapsulated	Comments	References
Gelatin	20% v/v acetic acid	8% w/v	EGCG	–	Gómez-Mascaraque et al. (2015)
			Green tea extract	–	Gómez-Mascaraque, Hernández-Rojas, et al. (2017)
			α-Linolenic acid	Ingredient incorporated by emulsion-electrospraying	Gómez-Mascaraque and López-Rubio (2016)
	Water/ethanol/acetic acid 1:2:1 v/v	5% w/v	Curcumin	–	Gómez-Estaca et al. (2015)
Gelatin/zein	20% v/v acetic acid for gelatin 85 wt% aqueous ethanol for zein	5% w/v for gelatin 12% w/v for zein	EGCG, ALA	Produced by coaxial electrospraying (zein in the core, gelatin in the shell)	Gómez-Mascaraque et al. (2019)

Table 3 Polysaccharides used to produce nano-/microcapsules by electrospraying

Polysaccharide	Solvent	Concentration	Food ingredients encapsulated	Comments	References
Pullulan	PBS	15–20 wt%	Probiotic bacteria (*B. animalis* Bb12)	Presence of residual nanofibers	López-Rubio et al. (2012)
	Water	20% w/v	–	Presence of residual nanofibers	Pérez-Masiá et al. (2014a)
Resistant starch	Water	20 wt%	Folic acid	Addition of surfactants was necessary	Pérez-Masiá et al. (2014b), Pérez-Masiá, López-Nicolás, et al. (2015)
Maltodextrin	Water	20 wt%	–	Addition of surfactants was necessary	Pérez-Masiá et al. (2014b)
Dextran	Water	40% w/v	–	–	Pérez-Masiá et al. (2014a)
		25 wt%	Fish oil	The emulsions were not stable, and low oxidative stability of the ingredient was observed	García-Moreno et al. (2017)
		20% w/v	Lycopene	Ingredient previously dissolved in soy bean oil and incorporated by emulsion/coaxial electrospraying. Surfactant added (Tween 20)	Pérez-Masiá, Lagaron, and Lopez-Rubio (2015)
Oligofructose	Water	20 w/v	–	Addition of surfactant (Span 20) was necessary	Pérez-Masiá et al. (2014a)

Continued

Table 3 Polysaccharides used to produce nano-/microcapsules by electrospraying—cont'd

Polysaccharide	Solvent	Concentration	Food ingredients encapsulated	Comments	References
Inulin	Water	20 w/v	–	Addition of surfactant (Span 20) or guar gum was necessary	Pérez-Masiá et al. (2014a)
Chitosan	Acetic acid 90% v/v	2% w/v	Lycopene	Ingredient degraded due to highly acidic conditions	Pérez-Masiá, Lagaron, and Lopez-Rubio (2015)
	Acetic acid 60%–90% v/v	0.5%–5% w/v[a]	(−)-Epigallocatechin gallate (EGCG)	Conditions optimized for chitosan with different molecular weights	Gómez-Mascaraque, Sanchez, and López-Rubio (2016)
Gum arabic	Water	35 wt%	Probiotic bacteria (*L. plantarum*)	Addition of surfactants and thermal treatment (autoclave) facilitated electrospraying	Zaeim et al. (2018)
Alyssum homolocarpum seed gum	Water (+ Tween 20)	0.5 wt%	D-Limonene	Best morphology obtained when emulsion was performed by high speed homogenization. The amount of ingredient incorporated was also decisive on the obtained morphologies	Khoshakhlagh et al. (2017, 2018)

[a]Depending on the molecular weight.

of its structure, its particular behavior in solution, and its polycationic nature (Homayoni, Ravandi, & Valizadeh, 2009). Moreover, the molecular weight of this polysaccharide is usually very high, so that very low concentrations can be usually processed by electrospraying (Neha, Syandan, Nikhil, & S, 2009; Zhang & Kawakami, 2010). Given that its molecular weight is one of its key characteristics that affect its processability, Gómez-Mascaraque, Sanchez, and López-Rubio (2016) thoroughly studied the impact of this parameter on the formation of electrosprayed capsules, showing that there was only a small range of molecular weight-concentration conditions, which yielded neat particles free of nanofibers. Electrosprayed chitosan microparticles have also been produced from solutions of the polysaccharide in aqueous ethanol/acetic acid mixtures (Moreno et al., 2018).

As mentioned before, the emulsifying properties of polysaccharides are generally poorer than those of proteins. As a result, research works comparing the performance of polysaccharides with that of proteins through emulsion-electrospraying found that carbohydrates generally yield lower encapsulation efficiencies due to emulsion instabilities (Pérez-Masiá, Lagaron, & Lopez-Rubio, 2015). García-Moreno et al. (2017) also reported low oxidative stabilization for fish oil encapsulated within electrosprayed dextran capsules due to poor stability of the feed formulations. Nevertheless, polysaccharides may still be used as wall materials for the encapsulation of hydrophobic ingredients through electrospraying, due to the feasibility of using a coaxial configuration in which the bioactive compound and the encapsulation matrix flow through independent circuits. Pérez-Masiá, Lagaron, & Lopez-Rubio, 2015 showed that higher encapsulation efficiencies could be achieved using coaxial electrospraying as compared with emulsion-electrospraying when dextran was used as wall material for the protection of lycopene.

One exception of carbohydrates with surface active properties is some gums, since they usually contain small fractions of polypeptides and proteins. In general, gums are highly branched polysaccharides, which are usually difficult to process by electrohydrodynamic techniques (Stijnman, Bodnar, & Tromp, 2011). Nevertheless, Zaeim et al. (2018) recently reported the successful production of electrosprayed gum arabic capsules, which was aided by the addition of a nonionic surfactant and facilitated by autoclaving the biopolymer solutions. This thermal treatment, which was used to sterilize the materials prior to the encapsulation of probiotic bacteria, positively affected the solution properties (mainly the rheological properties). Khoshakhlagh et al. (2018) also obtained electrosprayed capsules from

Alyssum homolocarpum seed gum emulsions containing a nonionic surfactant and discussed the impact of the emulsifying method on the morphology of the resulting materials as a result of changes in the emulsion properties. The amount of lipophilic ingredient added (i.e., D-limonene) also had a significant effect on the properties of the formulations and the morphology of the obtained nanostructures (Khoshakhlagh et al., 2017), a fact that emphasizes the need of optimizing the electrospraying conditions not only for each individual wall material but also for each wall material-ingredient combination.

Nevertheless, several works comparing the performance of electrosprayed proteins versus polysaccharides as encapsulation vehicles suggested that the encapsulation efficiency and protective effect of proteins are generally greater. This was the case, for instance, of folic acid (Pérez-Masiá, López-Nicolás, et al., 2015), lycopene (Pérez-Masiá, Lagaron, & Lopez-Rubio, 2015), or the probiotic strain *Bifidobacterium animalis* Bb12 (López-Rubio et al., 2012), which were better preserved using WPC than carbohydrates such as resistant starch, dextran and chitosan, and pullulan, respectively.

More details about application of carbohydrates for encapsulation of food bioactive ingredients via electrospraying have been provided in Table 3.

4.3 Electrosprayed capsules based on protein-polysaccharide combinations

The use of biopolymer blends as wall materials for encapsulation is also an interesting strategy since it allows tuning the release properties of the resulting materials by changing the composition of the blends and therefore the structural features of the capsules (Gómez-Mascaraque, Llavata-Cabrero, Martínez-Sanz, Fabra, & López-Rubio, 2018). Accordingly, various protein/polysaccharide combinations have been also exploited as wall materials for nano- and microencapsulation through electrospraying. For instance, Atay et al. (2018) developed electrosprayed chitosan/gelatin microparticles as food-grade delivery vehicles for anthocyanins and confirmed that changes in the matrix composition (protein/carbohydrate ratio) affected the release of these ingredients.

García-Moreno et al. (2018) used combinations of a WPC with carbohydrates, namely, pullulan and dextran or glucose syrup, to encapsulate a fish oil by emulsion-electrospraying, although the protein in their system accounted only for 4%–6% of the total wall material. Several compositions were evaluated in order to obtain fiber-free capsules from mixtures of WPC/pullulan/dextran and WPC/pullulan/glucose syrup, concluding that the capsules containing glucose syrup provided a better oxidative stability to the oil than those containing dextran.

4.4 Electrosprayed lipid nanocapsules

Apart from proteins and carbohydrates, lipids can also be used as nano- or microencapsulation wall materials (Eldem, Speiser, & Hincal, 1991; Lee, Cho, Lee, Jeong, & Yuk, 2003; McClements, Decker, & Weiss, 2007), since their hydrophobic nature makes them ideal as barriers against moisture and facilitates the dispersion of lipophilic food ingredients within them (Akhavan, Assadpour, Katouzian, & Jafari, 2018; Faridi Esfanjani, Assadpour, & Jafari, 2018; Gómez-Mascaraque et al., 2016; Rafiee & Jafari, 2018). Eltayeb et al. (2016) recently reported the production of lipid nanoparticles through electrohydrodynamic processing using stearic acid as wall material and ethanol as solvent. Stearic acid is a natural lipid extracted from animal fat and, thus, suitable for encapsulation of food ingredients. They used ethylvanillin as a model ingredient and confirmed that, due to its lipophilic character, it was located in the core of the nanoparticles surrounded by a shell of stearic acid. Therefore, they demonstrated that core-shell lipid nanoparticles can be produced by electrospraying and studied how the lipid concentration and the lipid/ingredient ratio affected the particle size, encapsulation efficiency, and release profile.

5 Recent advances in the encapsulation of food ingredients through electrospraying

Since each specific food ingredient to be encapsulated has different characteristics and physical-chemical properties, the most adequate wall material for encapsulation needs to be selected individually according to particular needs (Gómez-Mascaraque, Fabra, et al., 2018). This section summarizes the main types of food ingredients, which have been encapsulated using the electrospraying technique so far. Also, a summary of relevant studies has been reported in Table 4.

5.1 Encapsulation of water-soluble polyphenols

Due to their many attributed health benefits, the encapsulation of polyphenols has been a topic of great interest for the development of new functional foods, and a number of successful strategies to encapsulate them by electrospraying have already been reported (Faridi Esfanjani & Jafari, 2016). Among them, tea catechins have been the focus of many research works,

Table 4 Food ingredients encapsulated within electrosprayed capsules

Food ingredient	Edible matrices used	Range of bioactive load	Range of encapsulation efficiencies achieved	References
(−)-Epigallocatechin gallate	Gelatin, chitosan, zein	10 wt%	76%–96%	Gómez-Mascaraque et al. (2015, 2019), Gómez-Mascaraque, Sanchez, and López-Rubio (2016)
Green tea extract	Gelatin, zein	2–20 wt%	87%–97%	Bhushani et al. (2017), Gómez-Mascaraque, Hernández-Rojas, et al. (2017)
Chañar extract	Zein	10 wt%	99%	Costamagna et al. (2017)
Black carrot extract	Gelatin/chitosan	17 wt%	>75%	Atay et al. (2018)
β-Carotene	Whey protein concentrate (WPC), zein	0.1–2.5 wt%	22%–90%	Gómez-Mascaraque, Perez-Masiá, et al. (2017), López-Rubio and Lagaron (2012)
Lycopene	WPC, dextran, chitosan	0.3–1.5 wt%	≤75%	Pérez-Masiá, Lagaron, and Lopez-Rubio (2015)
Docosahexaenoic acid	Zein	33%	Not reported	Torres-Giner et al. (2010)
α-Linolenic acid	Zein, gelatin, WPC, soy protein isolate	10 wt%	23%–89%	Gómez-Mascaraque et al. (2019), Gómez-Mascaraque and López-Rubio (2016)
Fish oil	Dextran	10 wt%	68%–75%	García-Moreno et al. (2017)
Curcumin	Gelatin, zein, WPC	0.1–10 wt%	11%–97%	Gómez-Estaca et al. (2012, 2017, 2015), Gómez-Mascaraque, Sipoli, et al. (2017a)
Folic acid	WPC, resistant starch	1.5 wt%	44%–81%	Pérez-Masiá, López-Nicolás, et al. (2015)
D-Limonene	*Alyssum homolocarpum* seed gum	9–17 wt%	73%–93%	Khoshakhlagh et al. (2017, 2018)
Lactobacillus plantarum	WPC/resistant starch, gum arabic	>8.5 log CFU/g	Viability loss < 1 log CFU	Gomez-Mascaraque et al. (2016), Zaeim et al. (2018)
Bifidobacterium animalis Bb12	WPC, pullulan	~10 log CFU/g	No significant viability loss	López-Rubio et al. (2012)

especially (-)-epigallocatechin gallate (EGCG), which is the most abundant and biologically active form (Barras et al., 2009). Although electrohydrodynamic processing had been previously proposed for its incorporation within protein fibers through electrospinning (Li et al., 2009), it was only recently encapsulated by electrospraying. Different wall materials have been used for this purpose, including gelatin (Gómez-Mascaraque et al., 2015, 2019), chitosan (Gómez-Mascaraque, Sanchez, & López-Rubio, 2016), and zein (Gómez-Mascaraque et al., 2019). High encapsulation efficiencies (76%–96%) were achieved in all cases, due to its good solubility both in aqueous media and ethanol, which allowed its direct dissolution in the biopolymeric formulations. Moreover, the developed encapsulation systems proved to be capable of stabilizing EGCG against degradation in slightly alkaline aqueous solutions in which the free compound is very unstable (Gómez-Mascaraque et al., 2015; Gómez-Mascaraque, Sanchez, & López-Rubio, 2016). Additionally, encapsulation of EGCG through electrospraying was found to have a protective effect on the catechin during the intestinal phase of digestion in vitro (Gómez-Mascaraque et al., 2019).

For practical and economic reasons, enriched functional foods are more likely to be supplemented with a catechin-rich green tea extract rather than with an individual, purified catechin. For this reason, raw green tea extracts have also been encapsulated by electrospraying. Bhushani et al. (2017) used electrosprayed zein to nanoencapsulate a green tea extract and reported that this approach significantly improved the gastrointestinal stability of the catechins in vitro as well as their permeability through Caco-2 cell monolayers. On the other hand, Gómez-Mascaraque, Hernández-Rojas, et al. (2017) used both zein and gelatin electrosprayed capsules to incorporate a green tea extract into a real food matrix, that is, biscuits. Although both carriers greatly improved the thermal stability of the green tea catechins during an in vitro thermal treatment (180°C), their protective ability during the actual baking process to produce the biscuits could not be demonstrated due to the complexity of the systems. Nevertheless, it was found that the addition of electrosprayed protein particles to biscuits did not have a negative impact on the acceptability of the products by the consumers, which suggested the potential of electrosprayed capsules to be incorporated into complex food products without affecting their organoleptic properties.

Other polyphenol-rich extracts that have been encapsulated through electrospraying for food purposes include a chañar (*Geoffroea decorticans*) extract (Costamagna et al., 2017) and an anthocyanin-rich black carrot extract (Atay et al., 2018). Electrosprayed zein capsules similar to those used for the encapsulation of green tea extract were employed to incorporate the chañar extract. Results showed that the biological activity of the extract was preserved upon electrospraying and that encapsulation effectively protected the bioactive polyphenols during an in vitro digestion process (Costamagna et al., 2017). On the other hand, the black carrot extract was encapsulated within electrosprayed capsules based on chitosan/gelatin blends with different carbohydrate/protein ratios, and the impact of the matrix composition on the release of the anthocyanin toward different food simulants was studied (Atay et al., 2018).

5.2 Encapsulation of lipophilic ingredients

5.2.1 ω-3 fatty acids

Due to their many health benefits and high chemical sensitivity, ω-3 fatty acids have been one of the main targets of nano- and microencapsulation in the field of functional foods (Esfahani, Jafari, Jafarpour, & Dehnad, 2019). Docosahexaenoic acid (DHA), a functional long-chain polyunsaturated fatty acid (PUFA), was the first to be selected as a model bioactive compound for its encapsulation within electrosprayed protein capsules. And given its good solubility in ethanol, it was naturally first encapsulated within electrosprayed zein particles (Torres-Giner et al., 2010). Although the encapsulation efficiency of the process was not determined and the morphology of the capsules was not fully optimized, the preliminary results obtained were promising in terms of the protective effect that the protein exerted on the DHA, successfully delaying its oxidation.

α-Linolenic acid, another biologically relevant ω-3 fatty acid, was also encapsulated within electrosprayed zein capsules using both the uniaxial and the coaxial configurations (Gómez-Mascaraque et al., 2019). The encapsulation efficiency of the coaxially electrosprayed capsules was found to be higher than that of the uniaxial system. Moreover, a hybrid zein-gelatin vehicle was also developed by coaxial electrospraying in which the fatty acid was dissolved in the ethanolic zein solution and pumped through the inner orifice of the nozzle while a gelatin solution was pumped through the outer orifice as a protective shell layer. This protein combination leads to an enhanced protection of α-linolenic acid against thermal degradation as compared with capsules prepared from zein only, whereas all the electrosprayed systems successfully improved the stability of the ω-3 fatty acid as compared with the nonencapsulated ingredient (Gómez-Mascaraque et al., 2019).

Emulsion-electrospraying has also been used as an alternative strategy for the encapsulation of lipophilic ingredients such as ω-3 fatty acids within water-soluble or water-dispersible proteins. α-Linolenic acid was encapsulated both by electrospraying and spray-drying after emulsification in WPC, SPI, and gelatin solutions/dispersions (Gómez-Mascaraque & López-Rubio, 2016). While spray-drying resulted in the complete degradation of the fatty acid due to its great thermosensitivity, electrospraying was able to achieve encapsulation efficiencies of up to 67%, proving to be a more suitable alternative. Furthermore, all the electrosprayed protein capsules except those prepared in acidic conditions (gelatin) showed protective effects on the fatty acid against thermal oxidation.

Being a source of ω-3 fatty acids, raw fish oils have also been encapsulated through electrospraying. For instance, García-Moreno et al. (2017) used electrosprayed dextran particles as an encapsulation vehicle for a cod liver oil. But, due to the poor emulsifying properties of the polysaccharide, the proposed encapsulation system provided a low oxidative stability to the ingredient. In a subsequent work, they used combinations of WPC and carbohydrates to microencapsulate the oil. However, the protein only accounted for 4%–6% of the wall materials and they concluded that the oxidative stability of their system needed to be further improved (García-Moreno et al., 2018). Overall and according to the published works to date, the performance of proteins as electrosprayed encapsulation matrices for fatty acids is generally better than that of polysaccharides, mainly attributed to their enhanced surface activity.

5.2.2 Carotenoids

Carotenoids are a group of natural pigments, which may exert a number of health benefits if consumed in sufficient levels (Maiani et al., 2009). But, they are generally insoluble in water and poorly bioavailable (Deng, Chen, Huang, Fu, & Tang, 2014; Rostamabadi, Falsafi, & Jafari, 2019). β-Carotene, one of the most studied carotenoids, was the first to be incorporated as a model bioactive ingredient within electrosprayed protein capsules. In order to overcome its poor solubility in water, β-carotene was first dissolved in glycerol and subsequently dispersed in WPC dispersions prior to electrospraying (López-Rubio & Lagaron, 2012). It was reported that the β-carotene load of these capsules was considerably low (0.1 wt%).

Higher carotenoid loadings, up to 2.5 wt%, were later on achieved through emulsion-electrospraying using zein and WPC as wall materials (Gómez-Mascaraque, Perez-Masiá, et al., 2017). Moreover, this strategy was shown to improve the bioaccessibility of β-carotene, being the gravitational stability of the emulsions prepared prior to electrospraying, one of the key factors affecting the encapsulation efficiency of the process.

Lycopene has also been encapsulated within a range of electrosprayed biopolymers, that is, chitosan, dextran, and WPC chitosan, yielding higher encapsulation efficiencies and achieving a greater protective effect when the protein was used as encapsulation matrix (Pérez-Masiá, Lagaron, & Lopez-Rubio, 2015). The same wall materials were also used by the authors to encapsulate lycopene through spray-drying, obtaining lower encapsulation efficiencies due to its thermosensitivity. Once more, electrospraying proved to be a promising alternative for the encapsulation of labile food ingredients.

5.3 Encapsulation of probiotic bacteria

The first probiotic strain reported to be microencapsulated by electrospraying within edible biopolymers was a bifidobacterium, namely, *B. animalis* subsp. *lactis* Bb12. Two different encapsulation matrices were used, pullulan and WPC, showing the latter to have an enhanced capability to prolong bacterial survival during storage at different relative humidity conditions (López-Rubio et al., 2012). Their pioneering work showed that, despite being subjected to high voltages, the viability of the bacterial cells was not affected by the electrohydrodynamic process, demonstrating the suitability of the electrospraying technique to encapsulate living microorganisms. However, *B. animalis* Bb12 is a well-known commercial strain commonly used in dairy products (Hasnia, Philippe, & Ali, 2012), and more sensitive probiotic strains were later on found to suffer certain viability losses when subjected to electrohydrodynamic processes (Moayyedi et al., 2018).

Gomez-Mascaraque et al. (2016) studied the impact of different process variables on the viability loss of *Lactobacillus plantarum* upon electrospraying within a WPC-based matrix, as well as on the product yield of the encapsulation process, in order to optimize the electrospraying conditions. As a result, low viability losses (below 1 log CFU) could also be achieved for *L. plantarum*, obtaining electrosprayed microcapsules with high cell counts (higher than 8.5 log CFU/g) that proved to better preserve the viability of the bacteria during storage at a high relative humidity than freeze-drying. *L. plantarum* was also encapsulated within electrosprayed gum arabic microparticles with reported viability losses below 0.4 log CFU, obtaining a product with an initial bacterial count of 9.7 log CFU/g (Zaeim et al., 2018), although the protective effect of microencapsulation was not assed.

5.4 Encapsulation of other bioactive ingredients

Other bioactive ingredients such as curcumin, vitamins, or flavoring ingredients have also been encapsulated through electrospraying.

Curcuminoids are a type of phenolic compounds that are highly hydrophobic and, thus, insoluble in aqueous media. Although the solubility of curcumin in water is very poor, it is soluble in ethanol. Accordingly, the first biopolymeric matrix used for its encapsulation through electrospraying was zein, as it is soluble in ethanol too. Gómez-Estaca et al. (2012) used this approach to improve the dispersibility of curcumin in skimmed milk and achieved encapsulation efficiencies as high as 90% due to the good solubility of the ingredient in the protein solution. Later on, Gómez-Estaca et al. (2015) designed a strategy to encapsulate curcumin within electrosprayed gelatin, by using a ternary mixture of solvents (ethanol/water/acetic acid 2:1:1 v/v) to dissolve both the food ingredient and the protein. The resulting capsules could enhance the dispersibility of curcumin in an aqueous food model (a gellified fish product). Gómez-Estaca et al. (2017) also reported an improvement of the antioxidant and antimicrobial properties of curcumin upon encapsulation within electrosprayed gelatin capsules.

Gómez-Mascaraque, Sipoli, et al. (2017a) proposed a new strategy to encapsulate curcumin within electrosprayed biopolymers, which cannot be dissolved in ethanol (i.e., from aqueous formulations). It was consisted of a dual encapsulation approach in which curcumin was first entrapped within liposomes, which were then mixed with WPC prior to electrospraying. The resulting hybrid capsules provided enhanced protection of curcumin as compared with simple entrapment within liposomes and significantly increased the bioaccessibility of the ingredient compared with the free compound. In the light of these promising results, subsequent work was done to develop a semicontinuous process for the production of hybrid liposome/protein electrosprayed microcapsules using microfluidics (Gómez-Mascaraque, Sipoli, et al., 2017b).

Regarding the encapsulation of vitamins through electrospraying, Pérez-Masiá, López-Nicolás, et al. (2015) used resistant starch and WPC to encapsulate folic acid through this technique, obtaining higher encapsulation efficiencies and a greater stabilization effect during storage using proteins than the polysaccharide. The authors attributed this to the ability of WPC to establish intermolecular interactions with the vitamin. They also used the same materials to encapsulate folic acid through spray-drying, obtaining similar results. This highlights that, although electrospraying may be a more appropriate alternative than spray-drying for some ingredients due to their thermosensitivity, it does not necessarily yield better results in all cases.

Flavors have also been the subject of encapsulation through electrospraying. For instance, Khoshakhlagh et al. (2018) encapsulated D-limonene within *A. homolocarpum* seed gum through the emulsion-electrospraying technique, reporting an encapsulation efficiency over 70% despite the volatility of the ingredient and the low concentration of the biopolymeric wall material in the feed formulations used (0.5 wt%). Moreover, the encapsulated ingredient showed a good storage stability, with losses below 10% after 3 months of storage in refrigerated conditions and below 20% at room temperature. In general, and as expected, the encapsulation efficiency was found to decrease with increasing D-limonene loadings (Khoshakhlagh et al., 2017).

6 Future trends and concluding remarks

Electrospraying is a very advantageous technology for the nano- and microencapsulation of bioactive food ingredients, since it features all the benefits of other drying encapsulation techniques while avoiding the need to subject the materials to harsh thermal treatments. Therefore, it represents a promising alternative to the most common encapsulation technologies currently used in the food industry, especially for the protection of thermosensitive ingredients. However, although the process is quite simple to operate once its conditions have been optimized, the complexity of the physics involved in this process results in the need of individually optimizing the formulations and process parameters for every ingredient-encapsulation matrix combination of interest.

Intensive research efforts have been made in the last decade to develop new edible electrosprayed materials based on biopolymers, and a number of proteins and polysaccharides have already been successfully exploited for the nano- and microencapsulation of food ingredients through electrospraying, showing promising results at lab scale. Nevertheless, this battery of electrosprayed wall materials is still somewhat limited and is expected to continue growing in the next decade due to the great potential of this technique and the subsequent interest of the scientific community in better understanding it and further developing new strategies to facilitate its future commercial exploitation.

Although up to date, most of the encapsulation systems have been developed using one single type of biopolymer as wall material for simplicity purposes, recent works have already started exploiting biopolymer blends, since they represent an opportunity for modulating the performance of the nano- and microstructures by adjusting the ratio between both components. Hence, future research is expected to exploit this strategy and provide a new range of edible electrosprayed materials based on biopolymeric blends.

Other important aspect that should be addressed in future research works is the performance of electrosprayed encapsulation vehicles in real food products. To date, the stability, release, and bioactive properties of the encapsulated food ingredients have been mainly assessed in vitro, using buffers or solutions as food simulants and simplified processing or storage conditions in the absence of food matrices. However, food products are complex systems, and their different components may potentially interact with the encapsulation structures, affecting their performance. Therefore, the real benefits of encapsulating food ingredients through electrospraying should also be investigated after incorporating them within real food products, and the impact that these encapsulation structures may have on the organoleptic properties of the final product should also be assessed. Lastly, efforts should be made to scale up the production of encapsulation structures through electrospraying, in order for the technology to be commercially exploited.

References

Akhavan, S., Assadpour, E., Katouzian, I., & Jafari, S. M. (2018). Lipid nano scale cargos for the protection and delivery of food bioactive ingredients and nutraceuticals. *Trends in Food Science & Technology*, *74*, 132–146.

Alehosseini, A., Ghorani, B., Sarabi-Jamab, M., & Tucker, N. (2017). Principles of electrospraying: A new approach in protection of bioactive compounds in foods. *Critical Reviews in Food Science and Nutrition*, 1–18.

Atay, E., Fabra, M. J., Martínez-Sanz, M., Gomez-Mascaraque, L. G., Altan, A., & Lopez-Rubio, A. (2018). Development and characterization of chitosan/gelatin electrosprayed microparticles as food grade delivery vehicles for anthocyanin extracts. *Food Hydrocolloids*, *77*, 699–710.

Barras, A., Mezzetti, A., Richard, A., Lazzaroni, S., Roux, S., Melnyk, P., et al. (2009). Formulation and characterization of polyphenol-loaded lipid nanocapsules. *International Journal of Pharmaceutics*, *379*(2), 270–277.

Bhardwaj, N., & Kundu, S. C. (2010). Electrospinning: A fascinating fiber fabrication technique. *Biotechnology Advances*, *28*(3), 325–347.

Bhushani, J. A., & Anandharamakrishnan, C. (2014). Electrospinning and electrospraying techniques: Potential food based applications. *Trends in Food Science & Technology*, *38* (1), 21–33.

Bhushani, J. A., Kurrey, N. K., & Anandharamakrishnan, C. (2017). Nanoencapsulation of green tea catechins by electrospraying technique and its effect on controlled release and in-vitro permeability. *Journal of Food Engineering, 199*, 82–92.
Bock, N., Dargaville, T. R., & Woodruff, M. A. (2012). Electrospraying of polymers with therapeutic molecules: State of the art. *Progress in Polymer Science, 37*(11), 1510–1551.
Chakraborty, S., Liao, I. C., Adler, A., & Leong, K. W. (2009). Electrohydrodynamics: A facile technique to fabricate drug delivery systems. *Advanced Drug Delivery Reviews, 61* (12), 1043–1054.
Chang, C., Stone, A. K., & Nickerson, M. T. (2018). Microencapsulated food ingredients. In *Reference module in food science*. Elsevier.
Costamagna, M. S., Gómez-Mascaraque, L. G., Zampini, I. C., Alberto, M. R., Pérez, J., López-Rubio, A., et al. (2017). Microencapsulated chañar phenolics: A potential ingredient for functional foods development. *Journal of Functional Foods, 37*, 523–530. Suppl C.
Deng, X. X., Chen, Z., Huang, Q., Fu, X., & Tang, C. H. (2014). Spray-drying microencapsulation of β-carotene by soy protein isolate and/or OSA-modified starch. *Journal of Applied Polymer Science, 131*(12)40399.
Đorđević, V., Balanč, B., Belščak-Cvitanović, A., Lević, S., Trifković, K., Kalušević, A., et al. (2015). Trends in encapsulation technologies for delivery of food bioactive compounds. *Food Engineering Reviews*, 7(4), 452–490.
Doshi, J., & Reneker, D. H. (1995). Electrospinning process and applications of electrospun fibers. *Journal of Electrostatics, 35*(2-3), 151–160.
Eldem, T., Speiser, P., & Hincal, A. (1991). Optimization of spray-dried and-congealed lipid micropellets and characterization of their surface morphology by scanning electron microscopy. *Pharmaceutical Research, 8*(1), 47–54.
Eltayeb, M., Stride, E., Edirisinghe, M., & Harker, A. (2016). Electrosprayed nanoparticle delivery system for controlled release. *Materials Science and Engineering: C, 66*, 138–146.
Erencia, M., Cano, F., Tornero, J. A., Macanás, J., & Carrillo, F. (2014). Resolving the electrospinnability zones and diameter prediction for the electrospinning of the gelatin/water/acetic acid system. *Langmuir, 30*(24), 7198–7205.
Esfahani, R., Jafari, S. M., Jafarpour, A., & Dehnad, D. (2019). Loading of fish oil into nanocarriers prepared through gelatin-gum Arabic complexation. *Food Hydrocolloids, 90*, 291–298.
Falcó, I., Randazzo, W., Gómez-Mascaraque, L., Aznar, R., López-Rubio, A., & Sánchez, G. (2017). Effect of (−)-epigallocatechin gallate at different pH conditions on enteric viruses. *LWT- Food Science and Technology, 81*, 250–257.
Falcó, I., Randazzo, W., Gómez-Mascaraque, L. G., Aznar, R., López-Rubio, A., & Sánchez, G. (2018). Fostering the antiviral activity of green tea extract for sanitizing purposes through controlled storage conditions. *Food Control, 84*, 485–492.
Faridi Esfanjani, A., Assadpour, E., & Jafari, S. M. (2018). Improving the bioavailability of phenolic compounds by loading them within lipid-based nanocarriers. *Trends in Food Science & Technology, 76*, 56–66.
Faridi Esfanjani, A., & Jafari, S. M. (2016). Biopolymer nano-particles and natural nanocarriers for nano-encapsulation of phenolic compounds. *Colloids and Surfaces B: Biointerfaces, 146*, 532–543.
Fathi, M., Martin, A., & McClements, D. J. (2014). Nanoencapsulation of food ingredients using carbohydrate based delivery systems. *Trends in Food Science & Technology, 39*(1), 18–39.
Fernandez, A., Torres-Giner, S., & Lagaron, J. M. (2009). Novel route to stabilization of bioactive antioxidants by encapsulation in electrospun fibers of zein prolamine. *Food Hydrocolloids, 23*(5), 1427–1432.
García-Moreno, P. J., Özdemir, N., Stephansen, K., Mateiu, R. V., Echegoyen, Y., Lagaron, J. M., et al. (2017). Development of carbohydrate-based nano-microstructures

loaded with fish oil by using electrohydrodynamic processing. *Food Hydrocolloids, 69*, 273–285.

García-Moreno, P. J., Pelayo, A., Yu, S., Busolo, M., Lagaron, J. M., Chronakis, I. S., et al. (2018). Physicochemical characterization and oxidative stability of fish oil-loaded electrosprayed capsules: Combined use of whey protein and carbohydrates as wall materials. *Journal of Food Engineering, 231*, 42–53.

Gharsallaoui, A., Roudaut, G., Chambin, O., Voilley, A., & Saurel, R. (2007). Applications of spray-drying in microencapsulation of food ingredients: An overview. *Food Research International, 40*(9), 1107–1121.

Gómez-Estaca, J., Balaguer, M., Gavara, R., & Hernandez-Munoz, P. (2012). Formation of zein nanoparticles by electrohydrodynamic atomization: Effect of the main processing variables and suitability for encapsulating the food coloring and active ingredient curcumin. *Food Hydrocolloids, 28*(1), 82–91.

Gómez-Estaca, J., Balaguer, M. P., López-Carballo, G., Gavara, R., & Hernández-Muñoz, P. (2017). Improving antioxidant and antimicrobial properties of curcumin by means of encapsulation in gelatin through electrohydrodynamic atomization. *Food Hydrocolloids, 70*, 313–320.

Gómez-Estaca, J., Gavara, R., & Hernández-Muñoz, P. (2015). Encapsulation of curcumin in electrosprayed gelatin microspheres enhances its bioaccessibility and widens its uses in food applications. *Innovative Food Science & Emerging Technologies, 29*, 302–307.

Gómez-Mascaraque, L. G., Ambrosio-Martín, J., Fabra, M. J., Pérez-Masiá, R., Lagaron, J. M., & López-Rubio, A. (2016). *Novel nanoencapsulation structures for functional foods and nutraceutical applications*: (pp. 373–395). Nanotechnology in nutraceuticals: CRC Press.

Gómez-Mascaraque, L. G., Ambrosio-Martín, J., Perez-Masiá, R., & Lopez-Rubio, A. (2017). Impact of acetic acid on the survival of L. plantarum upon microencapsulation by coaxial electrospraying. *Journal of Healthcare Engineering, 2017*.

Gómez-Mascaraque, L. G., Fabra, M. J., Castro-Mayorga, J. L., Sánchez, G., Martínez-Sanz, M., & López-Rubio, A. (2018). Nanostructuring biopolymers for improved food quality and safety. In *Biopolymers for food design* (pp. 33–64). Elsevier.

Gómez-Mascaraque, L. G., Hernández-Rojas, M., Tarancón, P., Tenon, M., Feuillère, N., Ruiz, J. F. V., et al. (2017). Impact of microencapsulation within electrosprayed proteins on the formulation of green tea extract-enriched biscuits. *LWT- Food Science and Technology, 81*, 77–86.

Gómez-Mascaraque, L. G., Lagarón, J. M., & López-Rubio, A. (2015). Electrosprayed gelatin submicroparticles as edible carriers for the encapsulation of polyphenols of interest in functional foods. *Food Hydrocolloids, 49*, 42–52. Suppl C.

Gómez-Mascaraque, L. G., Llavata-Cabrero, B., Martínez-Sanz, M., Fabra, M. J., & López-Rubio, A. (2018). Self-assembled gelatin-ι-carrageenan encapsulation structures for intestinal-targeted release applications. *Journal of Colloid and Interface Science, 517*, 113–123.

Gómez-Mascaraque, L. G., & López-Rubio, A. (2016). Protein-based emulsion electrosprayed micro- and submicroparticles for the encapsulation and stabilization of thermosensitive hydrophobic bioactives. *Journal of Colloid and Interface Science, 465*, 259–270. Suppl C.

Gomez-Mascaraque, L. G., Morfin, R. C., Pérez-Masiá, R., Sanchez, G., & Lopez-Rubio, A. (2016). Optimization of electrospraying conditions for the microencapsulation of probiotics and evaluation of their resistance during storage and in-vitro digestion. *LWT- Food Science and Technology, 69*, 438–446.

Gómez-Mascaraque, L. G., Perez-Masiá, R., González-Barrio, R., Periago, M. J., & López-Rubio, A. (2017). Potential of microencapsulation through emulsion-electrospraying to improve the bioaccesibility of β-carotene. *Food Hydrocolloids, 73*, 1–12.

Gómez-Mascaraque, L. G., Sanchez, G., & López-Rubio, A. (2016). Impact of molecular weight on the formation of electrosprayed chitosan microcapsules as delivery vehicles for bioactive compounds. *Carbohydrate Polymers*, *150*, 121–130. Suppl C.

Gómez-Mascaraque, L. G., Sipoli, C. C., de La Torre, L. G., & López-Rubio, A. (2017a). Microencapsulation structures based on protein-coated liposomes obtained through electrospraying for the stabilization and improved bioaccessibility of curcumin. *Food Chemistry*, *233*, 343–350.

Gómez-Mascaraque, L. G., Sipoli, C. C., de La Torre, L. G., & López-Rubio, A. (2017b). A step forward towards the design of a continuous process to produce hybrid liposome/protein microcapsules. *Journal of Food Engineering*, *214*, 175–181.

Gómez-Mascaraque, L. G., Soler, C., & Lopez-Rubio, A. (2016). Stability and bioaccessibility of EGCG within edible micro-hydrogels. Chitosan vs. gelatin, a comparative study. *Food Hydrocolloids*, *61*, 128–138.

Gómez-Mascaraque, L. G., Tordera, F., Fabra, M. J., Martínez-Sanz, M., & Lopez-Rubio, A. (2019). Coaxial electrospraying of biopolymers as a strategy to improve protection of bioactive food ingredients. *Innovative Food Science & Emerging Technologies*, *51*, 2–11.

Hasnia, Z., Philippe, G., & Ali, R. (2012). Calcium alginate-resistant starch mixed gel improved the survival of Bifidobacterium animalis subsp. lactis Bb12 and Lactobacillus rhamnosus LBRE-LSAS in yogurt and simulated gastrointestinal conditions. *International Journal of Food Science & Technology*, *47*(7), 1421–1429.

Homayoni, H., Ravandi, S. A. H., & Valizadeh, M. (2009). Electrospinning of chitosan nanofibers: Processing optimization. *Carbohydrate Polymers*, 77(3), 656–661.

Jafari, S. M., Assadpoor, E., He, Y., & Bhandari, B. (2008). Encapsulation efficiency of food flavours and oils during spray drying. *Drying Technology*, *26*(7), 816–835.

Jafari, S. M., Mahdavi-Khazaei, K., & Hemmati-Kakhki, A. (2016). Microencapsulation of saffron petal anthocyanins with cress seed gum compared with Arabic gum through freeze drying. *Carbohydrate Polymers*, *140*, 20–25.

Jaworek, A., & Sobczyk, A. T. (2008). Electrospraying route to nanotechnology: An overview. *Journal of Electrostatics*, *66*(3–4), 197–219.

Kanjanapongkul, K., Wongsasulak, S., & Yoovidhya, T. (2010). Prediction of clogging time during electrospinning of zein solution: Scaling analysis and experimental verification. *Chemical Engineering Science*, *65*(18), 5217–5225.

Khoshakhlagh, K., Koocheki, A., Mohebbi, M., & Allafchian, A. (2017). Development and characterization of electrosprayed Alyssum homolocarpum seed gum nanoparticles for encapsulation of d-limonene. *Journal of Colloid and Interface Science*, *490*, 562–575.

Khoshakhlagh, K., Mohebbi, M., Koocheki, A., & Allafchian, A. (2018). Encapsulation of D-limonene in Alyssum homolocarpum seed gum nanocapsules by emulsion electrospraying: Morphology characterization and stability assessment. *Bioactive Carbohydrates and Dietary Fibre*, *16*, 43–52.

Kriegel, C., Arrechi, A., Kit, K., McClements, D., & Weiss, J. (2008). Fabrication, functionalization, and application of electrospun biopolymer nanofibers. *Critical Reviews in Food Science and Nutrition*, *48*(8), 775–797.

Kwak, H. -S. (2014). Overview of nano-and microencapsulation for foods. In *Nano-and microencapsulation for foods* (pp. 1–14).

Lee, K., Cho, S., Lee, H., Jeong, S., & Yuk, S. (2003). Microencapsulation of lipid nanoparticles containing lipophilic drug. *Journal of Microencapsulation*, *20*(4), 489–496.

Li, Y., Lim, L. T., & Kakuda, Y. (2009). Electrospun zein fibers as carriers to stabilize (-)-epigallocatechin gallate. *Journal of Food Science*, *74*(3), C233–C240.

Liu, Z.-P., Zhang, Y. -Y., Yu, D. -G., Wu, D., & Li, H. -L. (2018). Fabrication of sustained-release zein nanoparticles via modified coaxial electrospraying. *Chemical Engineering Journal*, *334*, 807–816.

Livney, Y. D. (2010). Milk proteins as vehicles for bioactives. *Current Opinion in Colloid & Interface Science*, *15*(1), 73–83.

López-Rubio, A., & Lagaron, J. M. (2012). Whey protein capsules obtained through electrospraying for the encapsulation of bioactives. *Innovative Food Science & Emerging Technologies*, *13*, 200–206.

López-Rubio, A., Sanchez, E., Wilkanowicz, S., Sanz, Y., & Lagaron, J. M. (2012). Electrospinning as a useful technique for the encapsulation of living bifidobacteria in food hydrocolloids. *Food Hydrocolloids*, *28*(1), 159–167.

Ma, J.-J., Mao, X.-Y., Wang, Q., Yang, S., Zhang, D., Chen, S.-W., et al. (2014). Effect of spray drying and freeze drying on the immunomodulatory activity, bitter taste and hygroscopicity of hydrolysate derived from whey protein concentrate. *LWT- Food Science and Technology*, *56*(2), 296–302.

Mahdavi, S. A., Jafari, S. M., Ghorbani, M., & Assadpoor, E. (2014). Spray-drying microencapsulation of anthocyanins by natural biopolymers: A review. *Drying Technology*, *32* (5), 509–518.

Maiani, G., Periago Castón, M. J., Catasta, G., Toti, E., Cambrodón, I. G., Bysted, A., et al. (2009). Carotenoids: Actual knowledge on food sources, intakes, stability and bioavailability and their protective role in humans. *Molecular Nutrition & Food Research*, *53*(S2), S194–S218.

Malaki Nik, A., Wright, A. J., & Corredig, M. (2010). Interfacial design of protein-stabilized emulsions for optimal delivery of nutrients. *Food & Function*, *1*(2), 141–148.

McClements, D. J. (2004). Protein-stabilized emulsions. *Current Opinion in Colloid & Interface Science*, *9*(5), 305–313.

McClements, D. J., Decker, E. A., & Weiss, J. (2007). Emulsion-based delivery systems for lipophilic bioactive components. *Journal of Food Science*, *72*(8), R109–R124.

Moayyedi, M., Eskandari, M. H., Rad, A. H. E., Ziaee, E., Khodaparast, M. H. H., & Golmakani, M.-T. (2018). Effect of drying methods (electrospraying, freeze drying and spray drying) on survival and viability of microencapsulated Lactobacillus rhamnosus ATCC 7469. *Journal of Functional Foods*, *40*, 391–399.

Mohan, A., Rajendran, S. R., He, Q. S., Bazinet, L., & Udenigwe, C. C. (2015). Encapsulation of food protein hydrolysates and peptides: A review. *RSC Advances*, *5*(97), 79270–79278.

Moomand, K., & Lim, L.-T. (2014). Oxidative stability of encapsulated fish oil in electrospun zein fibres. *Food Research International*, *62*, 523–532.

Moomand, K., & Lim, L.-T. (2015). Properties of encapsulated fish oil in electrospun zein fibres under simulated in vitro conditions. *Food and Bioprocess Technology*, *8*(2), 431–444.

Moreno, J. A. S., Mendes, A. C., Stephansen, K., Engwer, C., Goycoolea, F. M., Boisen, A., et al. (2018). Development of electrosprayed mucoadhesive chitosan microparticles. *Carbohydrate Polymers*, *190*, 240–247.

Nedović, V., Kalušević, A., Manojlović, V., Petrović, T., & Bugarski, B. (2013). Encapsulation systems in the food industry. In *Advances in food process engineering research and applications* (pp. 229–253). Springer.

Neha, A., Syandan, C., Nikhil, D., & S, K. D. (2009). Electrospraying: A facile technique for synthesis of chitosan-based micro/nanospheres for drug delivery applications. *Journal of Biomedical Materials Research Part B: Applied Biomaterials*, *88B*(1), 17–31.

Okutan, N., Terzi, P., & Altay, F. (2014). Affecting parameters on electrospinning process and characterization of electrospun gelatin nanofibers. *Food Hydrocolloids*, *39*, 19–26.

Peña, C., de la Caba, K., Eceiza, A., Ruseckaite, R., & Mondragon, I. (2010). Enhancing water repellence and mechanical properties of gelatin films by tannin addition. *Bioresource Technology*, *101*(17), 6836–6842.

Pérez-Masiá, R., Lagaron, J., & López-Rubio, A. (2014a). Development and optimization of novel encapsulation structures of interest in functional foods through electrospraying. *Food and Bioprocess Technology*, 7(11), 3236–3245.

Pérez-Masiá, R., Lagaron, J. M., & López-Rubio, A. (2014b). Surfactant-aided electrospraying of low molecular weight carbohydrate polymers from aqueous solutions. *Carbohydrate Polymers, 101*, 249–255.

Pérez-Masiá, R., Lagaron, J., & Lopez-Rubio, A. (2015). Morphology and stability of edible lycopene-containing micro- and nanocapsules produced through electrospraying and spray drying. *Food and Bioprocess Technology, 8*(2), 459–470.

Pérez-Masiá, R., López-Nicolás, R., Periago, M. J., Ros, G., Lagaron, J. M., & López-Rubio, A. (2015). Encapsulation of folic acid in food hydrocolloids through nanospray drying and electrospraying for nutraceutical applications. *Food Chemistry, 168*, 124–133.

Persano, L., Camposeo, A., Tekmen, C., & Pisignano, D. (2013). Industrial upscaling of electrospinning and applications of polymer nanofibers: A review. *Macromolecular Materials and Engineering, 298*(5), 504–520.

Rafiee, Z., & Jafari, S. M. (2018). Application of lipid nanocarriers for the food industry. In J. -M. Mérillon & K. G. Ramawat (Eds.), *Bioactive molecules in food* (pp. 1–43). Cham: Springer International Publishing.

Ramakrishna, S. (2005). *An introduction to electrospinning and nanofibers*. World Scientific.

Rosell-Llompart, J., Grifoll, J., & Loscertales, I. G. (2018). Electrosprays in the cone-jet mode: From Taylor cone formation to spray development. *Journal of Aerosol Science, 125*, 2–31.

Rostamabadi, H., Falsafi, S. R., & Jafari, S. M. (2019). Nanoencapsulation of carotenoids within lipid-based nanocarriers. *Journal of Controlled Release, 298*, 38–67.

Silva, D., Favaro-Trindade, C., Rocha, G., & Thomazini, M. (2012). Microencapsulation of lycopene by gelatin–pectin complex coacervation. *Journal of Food Processing and Preservation, 36*(2), 185–190.

Stijnman, A. C., Bodnar, I., & Tromp, R. H. (2011). Electrospinning of food-grade polysaccharides. *Food Hydrocolloids, 25*(5), 1393–1398.

Sun, K., & Li, Z. (2011). Preparations, properties and applications of chitosan based nanofibers fabricated by electrospinning. *Express Polymer Letters, 5*(4).

Tapia-Hernández, J. A., Rodríguez-Félix, F., & Katouzian, I. (2017). 9—Nanocapsule formation by electrospraying. In S. M. Jafari (Ed.), *Nanoencapsulation technologies for the food and nutraceutical industries* (pp. 320–345). Academic Press.

Tapia-Hernández, J. A., Torres-Chávez, P. I., Ramírez-Wong, B., Rascón-Chu, A., Plascencia-Jatomea, M., Barreras-Urbina, C. G., et al. (2015). Micro-and nanoparticles by electrospray: Advances and applications in foods. *Journal of Agricultural and Food Chemistry, 63*(19), 4699–4707.

Torres-Giner, S., Martinez-Abad, A., Ocio, M. J., & Lagaron, J. M. (2010). Stabilization of a nutraceutical omega-3 fatty acid by encapsulation in ultrathin electrosprayed zein prolamine. *Journal of Food Science, 75*(6).

Zaeim, D., Sarabi-Jamab, M., Ghorani, B., Kadkhodaee, R., & Tromp, R. H. (2017). Electrospray assisted fabrication of hydrogel microcapsules by single-and double-stage procedures for encapsulation of probiotics. *Food and Bioproducts Processing, 102*, 250–259.

Zaeim, D., Sarabi-Jamab, M., Ghorani, B., Kadkhodaee, R., & Tromp, R. H. (2018). Electrospray-assisted drying of live probiotics in acacia gum microparticles matrix. *Carbohydrate Polymers, 183*, 183–191.

Zhang, C., Chang, M.-W., Ahmad, Z., Hu, W., Zhao, D., & Li, J. -S. (2015). Stable single device multi-pore electrospraying of polymeric microparticles via controlled electrostatic interactions. *RSC Advances, 5*(107), 87919–87923.

Zhang, L., Huang, J., Si, T., & Xu, R. X. (2012). Coaxial electrospray of microparticles and nanoparticles for biomedical applications. *Expert Review of Medical Devices, 9*(6), 595–612.

Zhang, S., & Kawakami, K. (2010). One-step preparation of chitosan solid nanoparticles by electrospray deposition. *International Journal of Pharmaceutics*, *397*(1), 211–217.
Zhang, C., Yao, Z.-C., Ding, Q., Choi, J. J., Ahmad, Z., Chang, M.-W., et al. (2017). Tri-needle coaxial electrospray engineering of magnetic polymer yolk–shell particles possessing dual-imaging modality, multiagent compartments, and trigger release potential. *ACS Applied Materials & Interfaces*, *9*(25), 21485–21495.
Zuidam, N. J., & Shimoni, E. (2010). Overview of microencapsulates for use in food products or processes and methods to make them. In *Encapsulation technologies for active food ingredients and food processing* (pp. 3–29). Springer.
Zuidam, N. J., & Velikov, K. P. (2018). Choosing the right delivery systems for functional ingredients in foods: An industrial perspective. *Current Opinion in Food Science*, *21*, 15–25.

CHAPTER FOUR

Production of food bioactive-loaded nanoparticles by nano spray drying

Cordin Arpagaus
NTB University of Applied Sciences of Technology Buchs, Institute for Energy Systems, Buchs, Switzerland

Contents

Nanoencapsulation of Food Ingredients by Specialized Equipment
https://doi.org/10.1016/B978-0-12-815671-1.00004-4

1 Introduction

1.1 Bioactive food ingredients market

The global bioactive ingredients market was worth USD 29.8 billion in 2017 according to Research and Markets (2017), and it is projected to grow with about 6.6% to USD 41.1 billion by 2022. The market for encapsulated bioactive and nutraceutical ingredients for functional foods and dietary supplements is largely driven by consumer interest to promote health, well-being, and prevention of disease. People are becoming more aware of the type and source of food and beverages they need for a healthy diet.

After isolation from the original food source, a concentrated extract of the bioactive compound is achieved, and the bioactive ingredients are purified, dried, and then added back into food products to replenish the loss of the component during processing, or they are introduced into foods in which they are typically not present, for example, to create a functional food product or a dietary supplement.

Table 1 lists some major bioactive ingredients of interest to the food and nutraceutical industries including the following (Augustin & Sanguansri, 2017; Garti & McClements, 2012; Gibbs, Kermasha, Alli, & Mulligan, 1999; McClements, 2015):

Table 1 A selection of bioactive food ingredients of interest for the food and nutraceutical industries with potential health effects (Augustin & Sanguansri, 2017; Garti & McClements, 2012; Gibbs et al., 1999; McClements, 2015).

Bioactive food ingredient	Examples	Natural sources	Potential health effects
Omega-3 fatty acids	α-Linolenic acid (ALA)	Flax, perilla, chia	Promoting cardiovascular health
	Eicosapentaenoic acid (EPA) and dodecahexaenoic acid (DHA)	Fish oil, marine algae, krill oil	
Probiotics	Lactobacilli, bifidobacterium	Cultured microorganisms	Improving gut health and immune modulation
Prebiotics	Inulin, oligosaccharides	Chicory root, Jerusalem artichoke, jicama	Promoting gut health and modulation of gut microflora
	β-Glucan	Barley, oats	

Table 1 A selection of bioactive food ingredients of interest for the food and nutraceutical industries with potential health effects (Augustin & Sanguansri, 2017; Garti & McClements, 2012; Gibbs et al., 1999; McClements, 2015).—cont'd

Bioactive food ingredient	Examples	Natural sources	Potential health effects
Carotenoids	β-Carotene	Carrots, sweet potato, palm oil, algae	Reducing risk of diseases and certain cancers
	Lycopene	Tomato, watermelon, red grapefruit	
	Astaxanthin	Green algae	
	Lutein and zeaxanthin	Nastarium (yellow flowers), kale, spinach	
Phenolic compounds	Resveratrol	Japanese knotweed, wine	Reducing risk of cardiovascular disease, cancer, diabetes, and age-related degenerative diseases
	Curcumin	Turmeric	
	Flavonoids (quercetin and rutin)	Onions	
	Flavonoids (hesperidin)	Orange juice	
	Catechins and epicatechins	Cocoa, chocolate, tea	
Essential minerals	Calcium, iron, selenium	Soils, vegetables, water	Maintaining proper human health
Bioactive proteins, peptides, and amino acids	Milk-derived peptides	Milk products	Reducing blood pressure, acting as growth factors, protecting degradation in the gastrointestinal (GI) tracts
Extracts from herbs and spices	Essential oils, various herbal preparations, saffron	Various herbs and spices	A wide range of health benefits

- flavors (e.g., to control taste or aroma of food products)
- antioxidants (e.g., to retard the rate of oxidation in foods)
- antimicrobials (e.g., to inhibit the growth of microorganisms)
- bioactive lipids (e.g., healthy oils, fatty acids, carotenoids, and oil-soluble vitamins)
- bioactive proteins, peptides, and amino acids
- bioactive carbohydrates (e.g., to reduce cholesterol by dietary fibers)
- pre- and probiotics (e.g., to improve gut health)
- essential minerals (e.g., to maintain proper health)
- extracts from herbs and spices

1.2 Colloidal delivery systems for bioactive food ingredients

Due to sensitivity of bioactive ingredients to undesirable environmental influences such as temperature, light, and oxidation, it is important to select an appropriate delivery system and a processing method to maintain their stability and thus their bioactivity. The R&D community is very active in developing new delivery systems for food bioactive ingredients and nutraceuticals in the last couple of years including nanoemulsions, nanostructured lipid carriers, solid lipid nanoparticles, nanosized liposomes, biopolymeric nanoparticles, and micelles made of proteins, polysaccharides, and their complexes or conjugates (Jafari, 2017).

Fig. 1 gives an overview of the variety of colloidal delivery systems available to encapsulate, protect, and control the release of bioactive food ingredients. A challenge is to decide which delivery system to select for a particular food application. In principle, delivery systems for bioactive food ingredients can be formed from different food-grade materials with different processing technologies. The production process has to be economical, reproducible, and robust (Garti & McClements, 2012; McClements, 2015).

The materials used for encapsulated formulations as carrier or matrices typically include a range of proteins (e.g., caseins, whey proteins, soy proteins, wheat proteins), sugars, starches (e.g., maltodextrins), gums (e.g., gum acacia, alginate, pectin, carrageenan), cellulosic materials, chitosan, oils and fats, phospholipids, and food-grade emulsifiers (e.g., Tweens) (Augustin & Sanguansri, 2017).

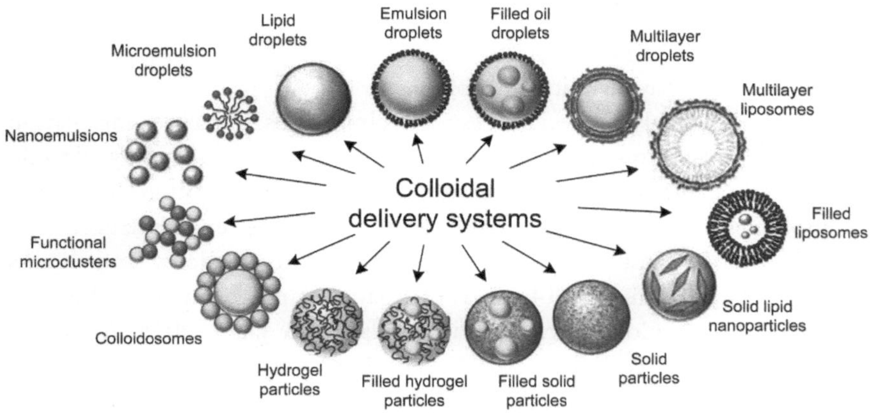

Fig. 1 Examples of different kinds of colloidal delivery systems for encapsulation, protection, and delivery of functional food ingredients.

1.3 Processing technologies, nanotechnology, and encapsulation

Nanotechnology and encapsulation are the two emerging technologies that enable food scientists to realize many innovations in the segment of functional food production. Recent developments in nanoparticle and microparticle delivery systems are revolutionizing colloidal delivery systems in the food industry (Anandharamakrishnan, 2014a; Ezhilarasi, Karthik, Chhanwal, & Anandharamakrishnan, 2013; Garti & McClements, 2012; Jafari, 2017; Jafari, Fathi, & Mandala, 2015; Jafari & McClements, 2017; Livney, 2017; McClements, 2015; Quintanilla-Carvajal et al., 2010).

The nanoencapsulation technologies to fabricate such delivery systems are typically classified into top-down approaches that consist in decreasing the size of macrostructures down to the nanosize scale (e.g., homogenization, dispersions, grinding, injection, and spraying), bottom-up techniques in which arrangements of atoms, molecules, or single particles are induced (e.g., assembly processes) or combined approaches.

The main aims of encapsulating bioactive food ingredients are as follows (Fang & Bhandari, 2012, chap. 12):

- protection from chemical, physical or biological degradation processes
- masking of flavors
- targeted and controlled release of the bioactive food ingredients at a specific location

- extension of shelf life
- improvement of handling and usage
- increase of bioavailability and solubility

The most commonly applied processing technologies for nanoencapsulation and delivery systems of bioactive food ingredients are spray drying, spray cooling/chilling, freeze-drying, fluidized bed coating, extrusion technologies, emulsification, coacervation, liposome encapsulation, and cyclodextrin encapsulation, as outlined in many reviews and books (Anandharamakrishnan, 2014a, 2014b; Anandharamakrishnan & Ishwarya, 2015a; Arpagaus, Collenberg, Rütti, Assadpour, & Jafari, 2018; Arpagaus, John, Collenberg, & Rütti, 2017, chap. 10; Celli, Ghanem, & Brooks, 2015; De Vos, Faas, Spasojevic, & Sikkema, 2010; Ezhilarasi et al., 2013; Fang & Bhandari, 2012, chap. 12, 2017; Garti & McClements, 2012; Grumezescu, 2016; Jafari, 2017; Jafari et al., 2015; Jafari, Paximada, Mandala, Assadpour, & Mehrnia, 2017, chap. 2; Livney, 2017; Ray, Raychaudhuri, & Chakraborty, 2016).

Despite the very active R&D community working on colloidal delivery systems, the list of encapsulation technologies with industrial real usage is surprisingly short, especially when considering natural flavors that are widely used in nutrition. Based on a market analysis by Porzio (2007, 2008) and subjective experiences by Garti and McClements (2012), about 78% of the market of the most commonly employed flavor delivery systems was covered by spray drying, with the remaining portions mostly dominated by melt extrusion, emulsification, fluidized bed coating, plating, coacervation, and others (Fig. 2).

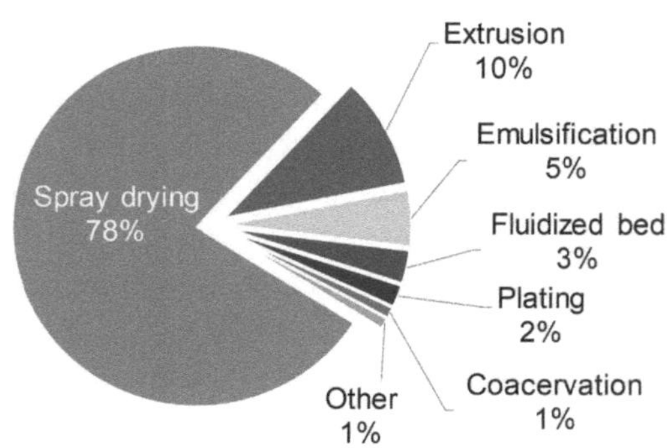

Fig. 2 Estimated worldwide production of flavor delivery systems structured by processing technologies. Spray drying is the most important commercial encapsulation processing technology for flavors. *(Data adapted from Garti, N., McClements, D.J. (2012).* Encapsulation technologies food ingredients and nutraceuticals. *Cambridge, UK: Woodhead Publishing).*

1.4 Benefits of spray drying for the encapsulation of bioactive food ingredients

Spray drying is a simple, fast, continuous, scalable, and cost-effective drying technology that is very well established in the food industries. Spray drying has been employed for decades to effectively encapsulate a wide range of food bioactive ingredients such as vitamins, minerals, salts, colorants (Akhavan Mahdavi, Jafari, Assadpoor, & Dehnad, 2016; Akhavan Mahdavi, Jafari, Assadpour, & Ghorbani, 2016; Mahdavee Khazaei, Jafari, Ghorbani, & Hemmati Kakhki, 2014), flavors (Esfanjani, Jafari, Assadpoor, & Mohammadi, 2015; Gharsallaoui, Roudaut, Chambin, Voilley, & Saurel, 2007; Rajabi, Ghorbani, Jafari, Mahoonak, & Rajabzadeh, 2015), spices (Jafari, He, & Bhandari, 2007a), fish oils (Mehrad, Shabanpour, Jafari, & Pourashouri, 2015; Pourashouri et al., 2014a, 2014b), lipids, polyphenols, carotenoids (Rostamabadi, Falsafi, & Jafari, 2019), antioxidants (Jafari, Ghalegi Ghalenoei, & Dehnad, 2017; Murugesan & Orsat, 2012), probiotic living cells (Celli et al., 2015; De Vos et al., 2010), proteins, peptides, and enzymes (Abdel-Mageed et al., 2019; Bürki, Jeon, Arpagaus, & Betz, 2011; Dahili & Feczkó, 2015; Sarabandi, Sadeghi Mahoonak, Hamishekar, Ghorbani, & Jafari, 2018). Spray drying technology is still gaining increasing interest due to its wide range of applications. Numerous reviews, research reports, and books have been published on the application of spray drying in the food industries. Table 2 summarizes the main benefits of spray drying technology for encapsulating bioactive food ingredients.

The spray-dried powder form offers high stability; protection of the bioactive food ingredients from oxidation, light, and temperature; easier handling and storage; and redispersibility in aqueous solutions. Spray drying can handle both water- and oil-soluble delivery systems equally well. The produced powders have typically a low moisture content, which increase the food shelf life during storage (Ray et al., 2016). The rapid evaporation of the solvent (e.g., water) during spray drying creates a cooling effect and keeps the droplet temperature relatively low (i.e., below the outlet drying gas temperature); thus, it is possible to produce encapsulated products with heat-sensitive bioactive cores (Murugesan & Orsat, 2012). Spray drying equipment is readily available by many suppliers in the laboratory, pilot, and production scale. Compared with other drying technologies like freeze-drying, the production cost of spray drying is about 30 to 50 times smaller (Arpagaus et al., 2017, chap. 10; Celli et al., 2015; Gharsallaoui et al., 2007).

Table 2 The main benefits of spray drying for the encapsulation of food bioactive ingredients, summarized from various literature (Anandharamakrishnan, 2014b; Anandharamakrishnan & Ishwarya, 2015a, 2015b, chap. 8; Arpagaus, 2007; Arpagaus et al., 2018, 2017, 2013, chap. 18; Arpagaus & Schwartzbach, 2008; Assadpour & Jafari, 2017, 2019; Assadpour, Jafari, & Maghsoudlou, 2017; Augustin & Sanguansri, 2017; Celli et al., 2015; Esfanjani et al., 2015; Fang & Bhandari, 2017; Faridi Esfanjani & Jafari, 2016; Gharsallaoui et al., 2007; Jafari, Assadpoor, Bhandari, & He, 2008, Jafari, Assadpoor, He, & Bhandari, 2008, Jafari et al., 2007a; Li et al., 2015; Mahdavi et al., 2014; Masters, 1991; Murugesan & Orsat, 2012; Nandiyanto & Okuyama, 2011; Porzio, 2007; Rajabi et al., 2015; Sosnik & Seremeta, 2015; Suna et al., 2014; Thybo, Hovgaard, Lindeløv, Brask, & Andersen, 2008; Wong & John, 2015).

Main benefits of spray drying technology for the encapsulation of food bioactive ingredients
• High flexibility to control of particle size, shape, and morphology (e.g., amorphous/crystalline form and porosity) • One-step process to directly transform various liquid feeds (e.g., solutions, emulsions, suspensions, and slurries) into dry powders • Process simplicity and ease of operation • Possible encapsulation of highly sticky feeds with drying aids and surfactants • Low operating costs, energy-efficient technology, and a fast process • Scale-up capability (it is one of the oldest processes for the encapsulation of food ingredients, technology is well established, and equipment is readily available) • Open- or closed-cycle design for aqueous or organic solvents • Suitable for drying of heat-sensitive bioactives with a low risk of degradation • Possibility of working with highly viscous feeds through preheating • Design of particles with controlled release properties • Applicable to both hydrophilic and hydrophobic food ingredients • High encapsulation efficiency and extended shelf life for the obtained powders • Versatile technique with numerous applications and applicability of various design of experiment (DOE) techniques for optimizing the process

Although widely used, spray drying also has some disadvantages. Some authors argue that spray drying is more of an immobilization than a true encapsulation technology, as some of the bioactive agents may be exposed at the surface (Celli et al., 2015). This can be a problem for probiotics, for example, as they can leak from the capsule into the product and affect their viability. Some loss of volatile oils and flavors during spray drying encapsulation seems to be inevitable. Reducing the infeed emulsion size to the nanoscale (about 100 to 300 nm) and thereby increasing the difference between emulsion size and spray-dried powder size are a key in achieving encapsulated spray-dried powders with high retention of volatile oil, low unencapsulated oil at the surface, and maximum encapsulation efficiency

(Jafari et al., 2007a; Jafari, Assadpoor, Bhandari, & He, 2008; Jafari, Assadpoor, He, & Bhandari, 2008; Jafari, He, & Bhandari, 2007b). Another drawback can be too high temperatures used for spray drying. The heat can cause cracks on the surface of the particles, which can adversely affect the stability of the encapsulated bioactive ingredients (Celli et al., 2015). Nevertheless, evaporation of the water from the droplets is very fast, at most in the range of a few seconds, and the particle surface exposed to the heat remains below the outlet drying gas temperature.

1.5 Wall materials, particle morphology and particle size range

The selection of a suitable wall material for nanoencapsulation is based on similar factors as for microencapsulation (Gharsallaoui et al., 2007). Relevant criteria include compatibility with the encapsulated bioactive component, suitable release properties, high encapsulation efficiency, mechanical strength, storage stability, easy emulsifiability, water solubility, and edibility. The ideal wall material should have (Jafari, Assadpoor, He, & Bhandari, 2008)

- emulsifying properties,
- be a good film former and oxygen barrier,
- low viscosity a high solid levels,
- exhibit low hygroscopicity,
- controllable release properties,
- be low in cost,
- bland in taste.

Different types of food-grade carriers and encapsulating wall materials are used for spray drying of bioactive food ingredients. Table 3 presents a brief summary of wall materials along with their encapsulation-related properties. The major wall materials used are carbohydrates (e.g., starches, maltodextrins, gum arabic, and cyclodextrins), proteins (e.g., whey proteins, caseinates, and gelatin), and other biopolymers (Fang & Bhandari, 2012, chap. 12, 2017). Maltodextrin and gum arabic are the classical wall materials used for the encapsulation of essential oils by spray drying. A typically adopted core to wall material ratio is 1:4 (20% core material) (Arpagaus, 2007; Büchi Labortechnik AG, 2002; Jafari, Assadpoor, He, & Bhandari, 2008; Jafari et al., 2007a; Ray et al., 2016).

As shown in Fig. 3, various morphologies of encapsulated powders can be obtained by spray drying such as single-core, irregular, multiwall, multicore, composites, and pure bioactives. The particle surface can be either smooth and spherical, collapsed, dimpled, wrinkled, raisin-like, highly

Table 3 Commonly used wall materials and their properties for spray drying encapsulation.

Wall material	Example	Encapsulation-related properties
Carbohydrates	Hydrolyzed starches (corn syrup solids, maltodextrins, etc.)	• Very good oxygen barrier • Low viscosity at high solids • No/limited emulsion stabilization • Low cost
	Modified starches (acetylated starch, monostarch phosphate, etc.)	• Sometimes varying quality • Constricted usage due to regulatory situation • Low cost
	Gums: agar, gum arabic, sodium alginate, etc.	• Good emulsion stabilization • Good retention of volatiles • Varying quality • Constricted usage due to regulatory situation • Sometimes impurities • Low cost (price depends on availability)
	Cyclodextrins: α-, β-, γ-cyclodextrins	• Good inclusion of volatiles • Excellent oxygen barrier • Relatively expensive
Proteins	Milk proteins: whey proteins, caseinates, skim milk powders Other proteins: soy protein, egg protein, gelatin	• Good emulsions • Properties depend on other factors such as pH and ionic strength • Allergenic potential • Relatively expensive
Other biopolymers	Soluble soy polysaccharides, chitosan, Maillard reaction products, modified celluloses	• Varied properties • May provide additional benefit to the stability of bioactives

From Fang, Z., Bhandari, B. (2017). Spray drying of bioactives. In Y. H. Roos & Y. D. Livney (Eds.), *Engineering foods for bioactives stability and delivery, Food engineering series* (pp. 261–284). New York, NY: Springer New York; Fang, Z., Bhandari, B. (2012). Encapsulation techniques for food ingredient systems. In B. Bhandari & Y. H. Roos (Ed.), *Food materials science and engineering* (pp. 320–348). Blackwell Publishing Ltd. with permission.

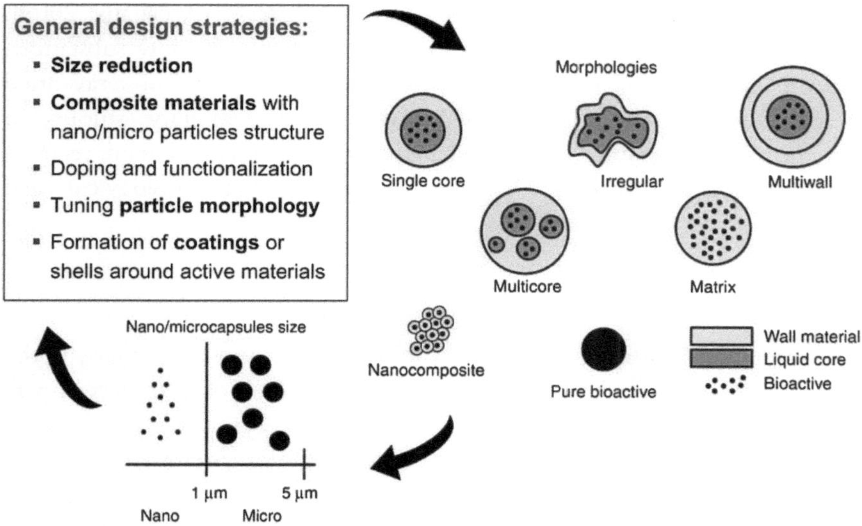

Fig. 3 Different particle morphologies and size ranges (Arpagaus, 2018a; Arpagaus et al., 2017).

crumpled, or folded (Arpagaus, 2018a; Arpagaus et al., 2017; Esfanjani et al., 2015; Mahdavi, Jafari, Ghorbani, & Assadpoor, 2014; Nandiyanto & Okuyama, 2011).

The sizes of the particles formed through encapsulation can be classified into macro (>5 mm), micro (1 μm to 5 mm), and nano (<1 μm) (Fang & Bhandari, 2012). Nanoparticles are commonly defined as solid colloidal particles with sizes below 1 μm (Kaialy & Al Shafiee, 2016; Lee, Heng, Ng, Chan, & Tan, 2011; Li, Anton, Arpagaus, Belleteix, & Vandamme, 2010; Wong, 2015; Wong & John, 2015). Capsules smaller than 1 μm in diameter are known as nanocapsules (Arpagaus et al., 2017; Couvreur, Dubernet, & Puisieux, 1995; Jafari, Assadpoor, Bhandari, & He, 2008; Jafari, Assadpoor, He, & Bhandari, 2008; Jafari et al., 2007a). Nanocapsules, due to their smaller size and larger specific surface, offer higher absorption rates and bioavailability of the encapsulated bioactive food ingredient compared with microcapsules (Lee et al., 2011).

According to the studies of Friends of the Earth (2008, 2014), the number of nanofood products has increased more than 10-fold in the last 6 years. A list of some 200 foods and beverages known to contain nanomaterials could be compiled. Besides the many benefits of nanomaterials, the formulation of nanoparticles involves certain risks, such as potential toxicological hazards. Therefore, it is crucial to be sure about product safety before releasing a new nanofood ingredient onto the market (Arpagaus, 2009). Nanomaterials must be tested as if they were completely new substances, and it is necessary to look at each individual case in detail. For spray-dried nanoproducts, it is recommended to conduct feasibility studies in laboratory-scale first with small powder amounts, and if an ingredient receives the green light for launch to consumers, the spray drying technology can then be scaled-up on an industrial basis.

1.6 Nano spray drying

The recent rise in nanotechnological applications in foods particularly in the area of food functionality like enriched and fortified food products in particulate dosage form (Grumezescu, 2016; Oprea & Grumezescu, 2017) has increased the pressure on existing spray drying systems to produce nanoparticles with a high yield and a narrow size distribution.

A remarkable growth of the research activity in the field of nano spray drying can be recognized in the past 20 years. A literature search in the online databases of Web of Science and SCOPUS with the topic "nano spray drying" resulted 658 and 720 hits, respectively, from the years 1997 to 2017 (Fig. 4). For instance, the number of publications increased to approximately 100 in 2017. Considerable research activities on nano spray drying are taking place worldwide, particularly in the fields of pharmaceuticals, materials technology, and bioactive food ingredients. Obviously the potential of nano spray drying has not yet been fully exploited. Compared with traditional spray drying in the micrometer scale, the current level of nano spray drying is still in its infancy (Assadpour & Jafari, 2019), but it is generally expected that more and more products will be developed in the coming years.

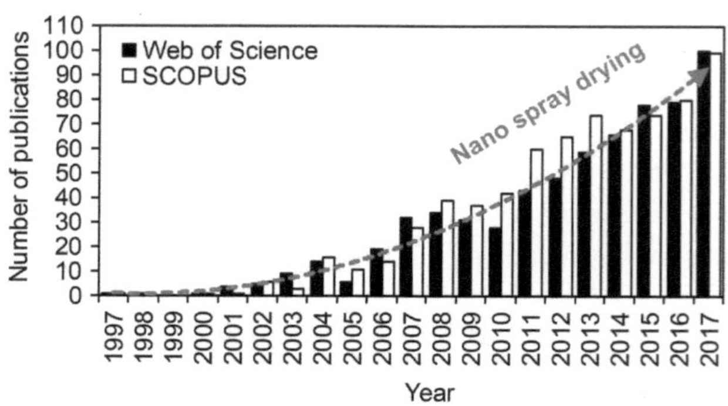

Fig. 4 Number of papers published in the online databases of Web of Science (http://www.webofscience.com) and SCOPUS (http://www.scopus.com) with the topic "nano spray drying" since 1997 (accessed on December 31, 2018).

Nanoscale powders by spray drying represent a new platform for many applications in food technology (Arpagaus, 2009).

Several excellent reviews on spray drying technology for nanoparticle formation have already been published (Nandiyanto & Okuyama, 2011; Okuyama, Abdullah, Lenggoro, & Iskandar, 2006; Okuyama & Lenggoro, 2003; Sosnik & Seremeta, 2015; Wang, Zhang, Zhang, Mujumdar, & Huang, 2005) and in particular on nano spray drying technology (Arpagaus, 2012, 2018a, 2018b, 2018c; Arpagaus et al., 2018, 2017, chap. 10; Assadpour & Jafari, 2019; Heng, Lee, Ng, & Tan, 2011; Lee et al., 2011; Suna, Sinir, & Çopur, 2014).

1.7 Objectives of this chapter

This chapter explains the specialized technology of nano spray drying for stabilizing bioactive food ingredients. First, the laboratory product Nano Spray Dryer B-90 developed by the Swiss company Büchi Labortechnik AG is presented with focus on explaining the essential technological and practical aspects of the nano spray drying process.

The second section discusses the influence of process parameters on the respective powder properties for successfully nano spray drying of bioactive food ingredients. In particular, the chapter highlights the way of reducing the size of spray-dried particles by an order of magnitude to reach submicron sizes.

The final part gives an overview of recent food and nutraceutical applications performed on a laboratory scale. The encapsulation of bioactive food ingredients in several polymeric wall materials with a high encapsulation efficiency is presented. In particular, the potential of nanoemulsions and solid lipid nanoparticles as colloidal delivery systems for nano spray-dried formulations is demonstrated.

Overall, this chapter attempts to summarize the most important findings of the last 15 years on nano spray drying of bioactive food ingredients.

2 Nano spray drying technology

2.1 Process steps

The process steps of a nano spray dryer are essentially the same as in a traditional spray dryer, which includes

(1) heating of the drying gas,
(2) droplet formation by atomization of the fluid supply,
(3) drying of the droplets in the drying gas and formation of dry particles,

(4) particle separation and collection of the dry particles from the drying gas. However, compared with a traditional spray dryer, some technological modifications on the experimental setup are necessary to produce and collect nanoscale particles, that is,

(1) the nozzle system should be able to produce smaller droplets;

(2) the drying gas flow needs to be gentle, laminar, and cocurrent with the sprayed droplets; and

(3) the particle collector must be highly efficient in separating submicron particles.

In 2009, the Swiss company Büchi Labortechnik AG launched the laboratory product Nano Spray Dryer B-90 for producing small quantities of submicron powders in the gram scale with very narrow distributions and high yields (Arpagaus, 2009; Arpagaus, Friess, & Schmid, 2009). Fig. 5 shows the setup and explains the implemented technologies for spray generation, heating of the drying gas, and particle collection.

2.2 Droplet generation by vibrating mesh technology

The droplet generation is based on vibration mesh technology, which has been adapted from nebulizers for inhalation therapy to the nano spray dryer application (Dhand, 2002; Knoch & Keller, 2005; Lass, Sant, & Knoch, 2006; Smart et al., 2002; Vecellio, 2006). Fig. 6A shows an exemplary picture of the generated spray of droplets and (B) to (E) some microscopic images of the holes in a spray mesh. The spray head is connected by plastic tubes to the feed reservoir, and a peristaltic pump with variable speed circulates the fluid uniformly through the spray head into the spray mesh and back to the reservoir (see Fig. 5). This circulation mode enables efficient and continuous atomization.

An electronically driven piezoelectric actuator vibrates the thin spray mesh that contains an array of laser drilled micron-sized holes in the center (approximately 1500 holes in a 5.5-μm spray mesh). The holes have a conical shape with about 50-μm diameter toward the fluid supply side and a narrow diameter (i.e., 5.5 μm) facing the front of the droplet release. Spray meshes are available with 4.0-, 5.5-, and 7.0-μm holes (first-generation spray head). The second generation of vibrating mesh technology in the Nano Spray Dryer B-90 HP launched in 2017 offers optimized productivity and better handling during installation and replacement. These nebulizers are available in small, medium, and large size.

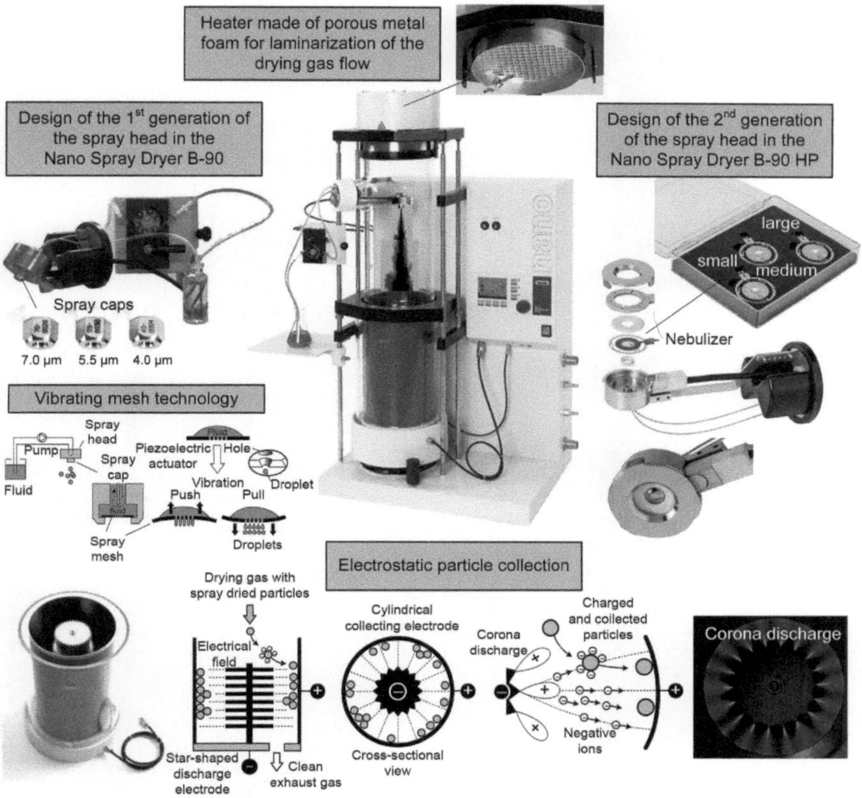

Fig. 5 Nano Spray Dryer B-90 (short setup) with the first and second generation of vibrating mesh technology for spray generation, the electrostatic particle collector, and the heater made of porous metal foam for gas laminarization. *(Pictures courtesy of Büchi Labortechnik AG).*

The vibration at ultrasonic frequency, which is adjustable from 80 to 140 kHz, deforms the mesh into the fluid side, loads the holes with fluid, and pushes the fluid through the mesh into the drying chamber. The output is fine low-velocity droplets with very narrow size distribution optimized for nano spray drying. At a vibration frequency of 100 kHz and 1000 active holes in the spray mesh, approximately 100 million precisely dimensioned droplets are produced per second. The uniformity of the droplet size is mainly defined by the uniformity of the holes. The droplet size depends on the mesh size and the physicochemical properties of the fluid, such as viscosity and surface tension. The mean droplet size of water is approximately 4.8, 6.2, and 7.2 μm (SPAN $(d_{90} - d_{10})/d_{50}$ from 1.1 to 1.3, reproducibility

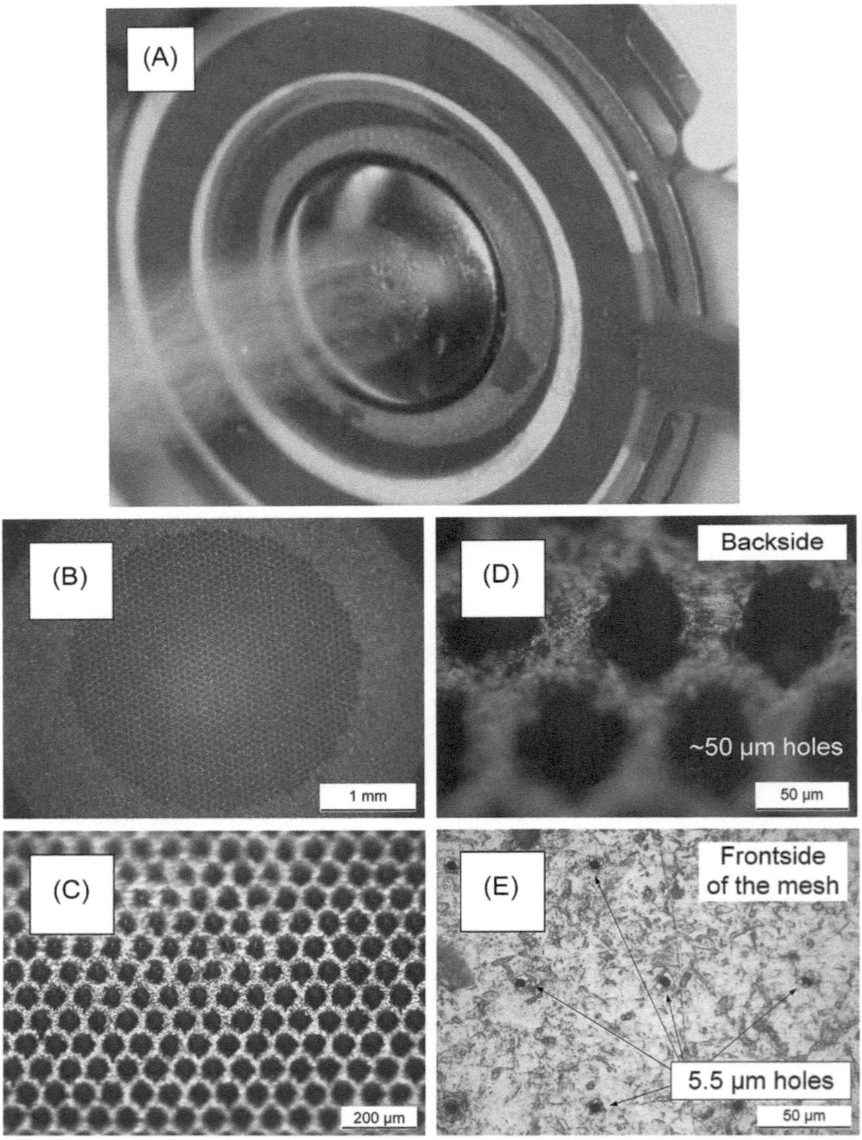

Fig. 6 Vibrating mesh atomizer with perforated metallic membrane. (A) Picture from TouchSpray (eFlow) illustrating the generated spray of droplets (Geller, 2008). A low-velocity mist is generated through the holes. (B) Microscopic backside view of a 5.5-μ m spray mesh (in contact with the liquid feed reservoir) with approximately 1500 precise tapered laser drilled holes (Nano Spray Dryer B-90). (C) and (D) Higher resolutions of the backside of the spray mesh. (E) Frontside view of the spray mesh (toward the drying chamber) with 5.5-μm holes.

Table 4 Droplet size of water sprayed with spray meshes of 4.0, 5.5, and 7.0 μm hole size (determined by laser diffraction).

Spray mesh (μm)	d_{10} (μm)	d_{50} (μm)	d_{90} (μm)	SPAN $(d_{90}-d_{10})/d_{50}$
4.0	3.3	4.8	8.5	1.11
5.5	4.4	6.2	11.5	1.15
7.0	5.2	7.2	14.7	1.32

Data from Arpagaus, C., John, P., Collenberg, A., Rütti, D. (2017). Nanocapsules formation by nano spray drying. In S. M. Jafari (Ed.), *Nanoencapsulation technologies for the food and nutraceutical industries* (pp. 346–401). Elsevier Inc.; Schmid, K., 2011. *Spray drying of protein precipitates and evaluation of the nano spray dryer B-90* (Ph.D. thesis). Munich: Ludwig-Maximilians-University; Schmid, K., Arpagaus, C., Friess, W. (2011). Evaluation of the nano spray dryer B-90 for pharmaceutical applications. *Pharmaceutical Development and Technology 16*, 287–294.

±0.5 μm) using spray mesh 4.0, 5.5, and 7.0 μm (see Table 4) (Arpagaus et al., 2017, chap. 10; Schmid, 2011; Schmid, Arpagaus, & Friess, 2011).

A limitation of the vibrating mesh technology is the risk to clog some of the tiny holes in the spray mesh if the product is too concentrated or if the primary particle sizes in the fluid are bigger than about 1/10 of the nebulizer hole diameter (i.e., particles > about 0.4 to 0.7 μm). Some solutions may be too viscous to pass through the mesh, for example, sodium carboxymethyl cellulose, which is a strong viscosifier (Oliveira, Guimarães, Cerize, Tunussi, & Poço, 2013). This may result in intermittent droplet generation or even cessation of droplet generation. The maximum liquid viscosity is about 5 to 10 mPa s (Arpagaus, Schafroth, & Meuri, 2010a). Obviously, regular cleaning of the spray mesh and nebulizer is necessary to maintain the efficient function of the mesh. It is recommended to clean the spray mesh and nebulizer in an ultrasonic bath for 1 to 2 min. Moreover, it is considerable using a lower sample concentration, filtering of the feed solution, or changing to a larger nebulizer. As stated by Schmid (2011), virtually any substance can be nano spray-dried successfully as long as the correct test arrangement is used (i.e., solvent, solid concentration, feed composition, and drying conditions).

2.3 Laminar drying conditions

The drying gas is heated up to the set inlet temperature (T_{in}) in a compact heater at the top of the nano spray dryer. The heating unit consists of a porous metal foam with an embedded electrical heating coil (Fig. 7). This coil ensures efficient heat transfer to the metal foam and uniform heat distribution in the entire heating volume. The porous metal surface generates a laminarization of the gas flow for gentle drying of the sprayed droplets. This

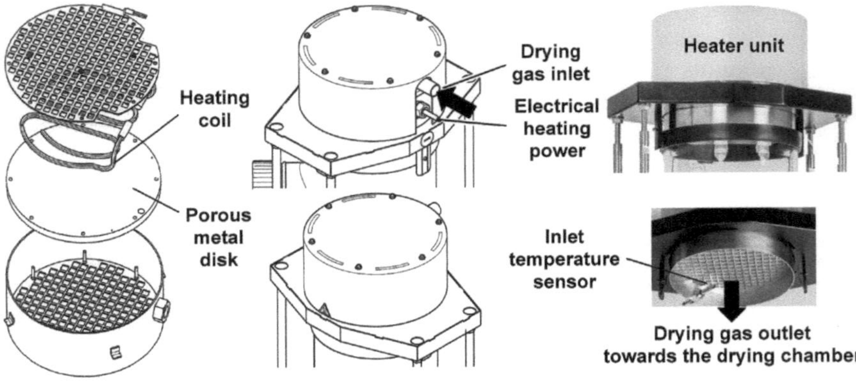

Fig. 7 Compact heater unit at the top of the Nano Spray Dryer B-90 consisting of an electrical heating coil pressed into a porous metal foam for laminarization of the drying gas. *(Pictures courtesy of Büchi Labortechnik AG, Patent EP 2056037 A1, Schön, M., & Baumgartner, R. (2009).* Apparatus with spray drier and electrostatic separator as well as method for separating particles. *Patent EP 2056037 (A1), priority date: 2007-10-30).*

is crucial because turbulence would lead to uncontrolled spray formation and particle depositions on the sidewalls of the drying chamber. Thus, nano spray drying is a very gentle drying technology suitable for heat-sensitive products with a low risk of degradation or loss of activity.

The modular glass assembly of the drying cylinders allows simple modification of the length of the drying chamber, it is easy to clean, and considerable process time can be saved in daily lab work. The Nano Spray Dryer B-90 can be operated with a short or long version of the drying chamber, which corresponds to approx. 0.32-m and 0.77-m drying length from the spray nebulizer to the particle collector. The drying gas flow rate can be adjusted within a range of 80 to 160 L/min, which is equivalent to gas velocities of 0.05 to 0.1 m/s in the drying chamber (tube inner diameter = 0.18 m). The Reynolds number is between 330 and 660 at 120°C inlet temperature and 40% humid air. The average gas residence time in the drying chamber is about 3–6 s in the short setup and 7 to 15 s in the tall setup, respectively.

2.4 Electrostatic particle collection

A highly efficient electrostatic particle collector separates the dried particles from the gas stream. The electrostatic particle collector consists of a stainless steel cylinder (anode = particle collection electrode) and a star-shaped counter electrode (cathode) inside the cylinder (see Fig. 5). During the nano spray drying process, a high voltage of approx. 15 kV is applied between the electrodes, and the dried particles are electrically charged deflected to the inner wall of the cylinder electrode. The electrostatic particle collector is able to capture submicron particles ($<1\,\mu m$) at a separation efficiency >99% for small batches of 10 mg to 2.7 g powder (Arpagaus & Meuri, 2010; Arpagaus, Schafroth, & Meuri, 2010b; Bürki et al., 2011; Lee et al., 2011; Li et al., 2010; Schmid, 2011; Schmid, Arpagaus, & Friess, 2009; Schmid et al., 2011). It can even collect thin-walled particles without breaking (Feng et al., 2011; Sun, Song, Wang, & Yu, 2011).

A special feature of the electrostatic particle collector is a segregation effect over the cylinder length. Larger particles are captured earlier than smaller ones because bigger particles have a larger surface charge and a larger electrostatic force acting on the particles. As a result, the particles collected on the lower part of the collecting electrode are slightly smaller than those collected on the upper part (Brinkmann-Trettenes, Barnert, & Bauer-Brandl, 2014; Li et al., 2010; Suryaprakash, Lohmann, Wagner, Abel, & Varga, 2014). Moreover, the particle deposition is axially symmetrical in the collection cylinder (Brinkmann-Trettenes et al., 2014).

After completion of the nano spray drying process, the fine dried powder particles are gently collected from the inner surface of the collection electrode cylinder using the particle scraper and the particle collection paper included in the scope of delivery of the laboratory instrument (Fig. 8). Finally, the particles are filled into airtight glass vials and stored in a controlled and dry atmosphere (e.g., in a desiccator over silica gel at room temperature) until further usage and examination to prevent crystallization and moisture absorption (Schmid, Arpagaus, & Friess, 2011). Yield fluctuations and minimal losses can occur when collecting powder manually with a scraper.

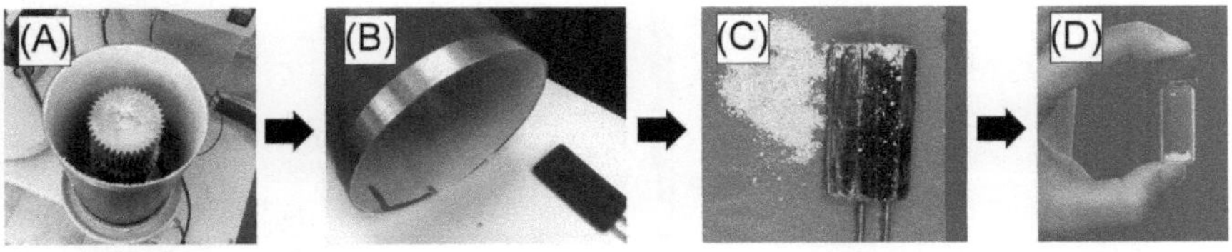

Fig. 8 Particle collection process in the laboratory-scale Nano Spray Dryer B-90. (A) Dried particles deposited on the collection cylinder (minor losses on the star electrode in the center), (B) collection cylinder with particle scraper, (C) manually recovered powder on collecting paper, (D) powder filled in air-tight glass vial.

2.5 Influence of process parameters and formulation variables on powder properties

For nano spray drying of bioactive food ingredients, there are several process parameters that can be varied to optimize the yield, bioactive loading, encapsulation efficiency, particle size, release profile, stability, and morphology (Arpagaus, 2018a, 2018b, 2018c; Arpagaus et al., 2018, 2017; Arpagaus, Rütti, & Meuri, 2013). Fig. 9 illustrates the adjustable process parameters and formulation variables for a nano spray dryer. Table 5 provides an overview of the main process parameters and their influence on the final product properties. The thickness of each arrow illustrates the strength of the relationship.

Depending on the application, an optimized set of process parameters can be found. Design of experiment (DOE) studies are suitable to optimize the nano spray drying conditions, as shown by many researchers (Abdel-Mageed et al., 2019; Arpagaus et al., 2017, chap. 10; Bürki et al., 2011; Draheim, de Crécy, Hansen, Collnot, & Lehr, 2015; Durli et al., 2014; Gu, Linehan, & Tseng, 2015; Harsha et al., 2017; Lee et al., 2011;

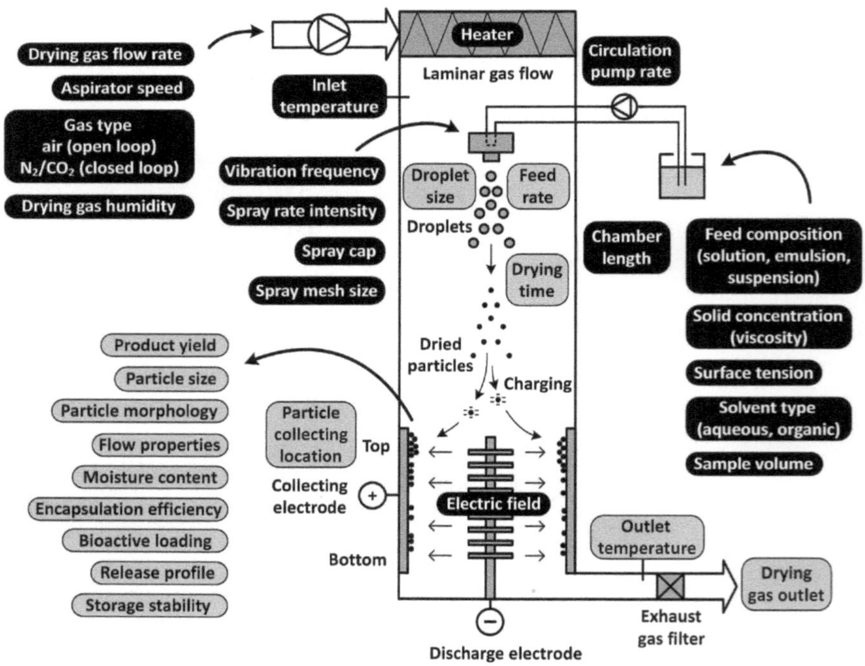

Fig. 9 Process parameters and formulation variables for the production of bioactive food-loaded nanoparticles by nano spray drying (Arpagaus et al., 2018, 2017).

Table 5 Influence of the main process parameters in nano spray drying (**↑**/**↓** strong increasing/decreasing influence, ↑/↓ weak increasing/decreasing influence, – minimal or no influence).

Process parameter	Outlet temperature	Droplet size	Particle size	Feed rate	Moisture content	Yield	Stability
Drying gas flow rate ↑	**↑**	–	–	–	**↓**	–	–
Drying gas humidity ↑	↑	–	–	–	**↑**	↓	–
Inlet temperature ↑	**↑**	–	↑	–	↓	↑	**↓**
Spray mesh size ↑	↓	**↑**	**↑**	**↑**	–	–	↑
Spray frequency ↑	**↓**	↑	↑	**↑**	↑	–	**↓**
Circulation pump rate ↑	–	↑	↑	↑	–	–	↑
Solid concentration (viscosity) ↑	↑	–	**↑**	**↓**	↓	↑	–
Surfactant/stabilizer in feed ↑	–	↓	↓	↑	–	↑	**↑**
Organic solvent instead of water	**↑**	↓	↓	↑	**↓**	↑	–

Based on Arpagaus, C., John, P., Collenberg, A., Rütti, D., 2017. Nanocapsules formation by nano spray drying. In S. M. Jafari (Ed.), *Nanoencapsulation technologies for the food and nutraceutical industries* (pp. 346–401). Elsevier Inc.

Li et al., 2010; Littringer et al., 2013; Schafroth, Arpagaus, Jadhav, Makne, & Douroumis, 2012; Schoubben, Giovagnoli, Tiralti, Blasi, & Ricci, 2014).

Assuming a one-to-one transformation of droplets to dried particles, the final particle size (d_P) is directly related to the solid concentration in the feed solution (C_S) and the droplet size (d_D) (Maa, Nguyen, Sit, & Hsu, 1998). The relationship can be approximated by Eq. (1):

$$d_P = d_D \cdot \left(\frac{C_S}{\rho}\right)^{\frac{1}{3}} \tag{1}$$

where ρ is the particle density. The formula assumes spherical droplets and solid nonporous particles, negligible volatile content in the particles, and that all droplets contain the same concentration of solids. C_S is typically in the order of 0.1 to 0.01 g/mL, and ρ is mainly around 0.9 to 1.5 g/cm^3, respectively. Obviously, a smaller spray mesh leads to smaller droplets and consequently to smaller dried particles. The submicron particle size is typically reached when using a 4.0-μm spray cap and diluted solutions of about

0.1% to 1% (w/v), as demonstrated in many studies (Baba & Nishida, 2013; Beck-Broichsitter, Paulus, Greiner, Kissel, 2015; Beck-Broichsitter et al., 2012; Bürki et al., 2011; De Cicco, Porta, Sansone, Aquino, & Del Gaudio, 2014; Lee et al., 2011; Li et al., 2010; Littringer et al., 2013; Nandiyanto & Okuyama, 2011; Ngan et al., 2014; Pérez-Masiá et al., 2015; Schmid, 2011; Schmid et al., 2009, 2011). Table 6 presents some particle size data of food-grade materials, which have been nano spray-dried with different spray mesh sizes and solid concentrations. The SEM images in Fig. 10 show the effect of varying spray mesh size on nano spray-dried particle size and morphology.

Table 6 Influence of spray mesh size and solid concentration on nano spray-dried particle size (– = data not available).

Substance	Solvent	Solid concentration (%, w/v)	Particle size (in nm) obtained with spray mesh			Reference
			4.0 μm	5.5 μm	7.0 μm	
Chitosan (low-density)	0.5% acetic acid	0.025	95	215	265	Ngan et al. (2014)
		0.05	180	–	–	
		0.075	236	–	–	
		0.1	358	–	–	
Ethambutol (bacteriostatic and antituberculosis agent)	Water	1	222	–	–	Ahmad, Ungphaiboon, and Srichana (2014)
Gum arabic	Water	0.1	355	–	–	Li et al. (2010)
		1	580	–	–	
Whey protein	Water	0.1	420	–	–	Li et al. (2010)
		1	595	–	–	
Sodium chloride	Water	0.1	515	–	–	Li et al. (2010)
		1	995	–	–	
Disodium phosphate	Water	0.1	500	–	–	Schmid et al. (2011) and Schmid (2011)
Trehalose	Water	0.1	800	–	–	Schmid et al. (2011) and Schmid (2011)
Bovine serum albumin (with surfactant)	Water	0.1	460	–	–	Lee et al. (2011)
		1	700	1700	2600	
Sodium alginate	Water	0.13	760	–	–	Blasi, Schoubben, Giovagnoli, Rossi, and Ricci (2010)
Curcumin in albumin	Water	0.5	715	–	–	Jain (2014)

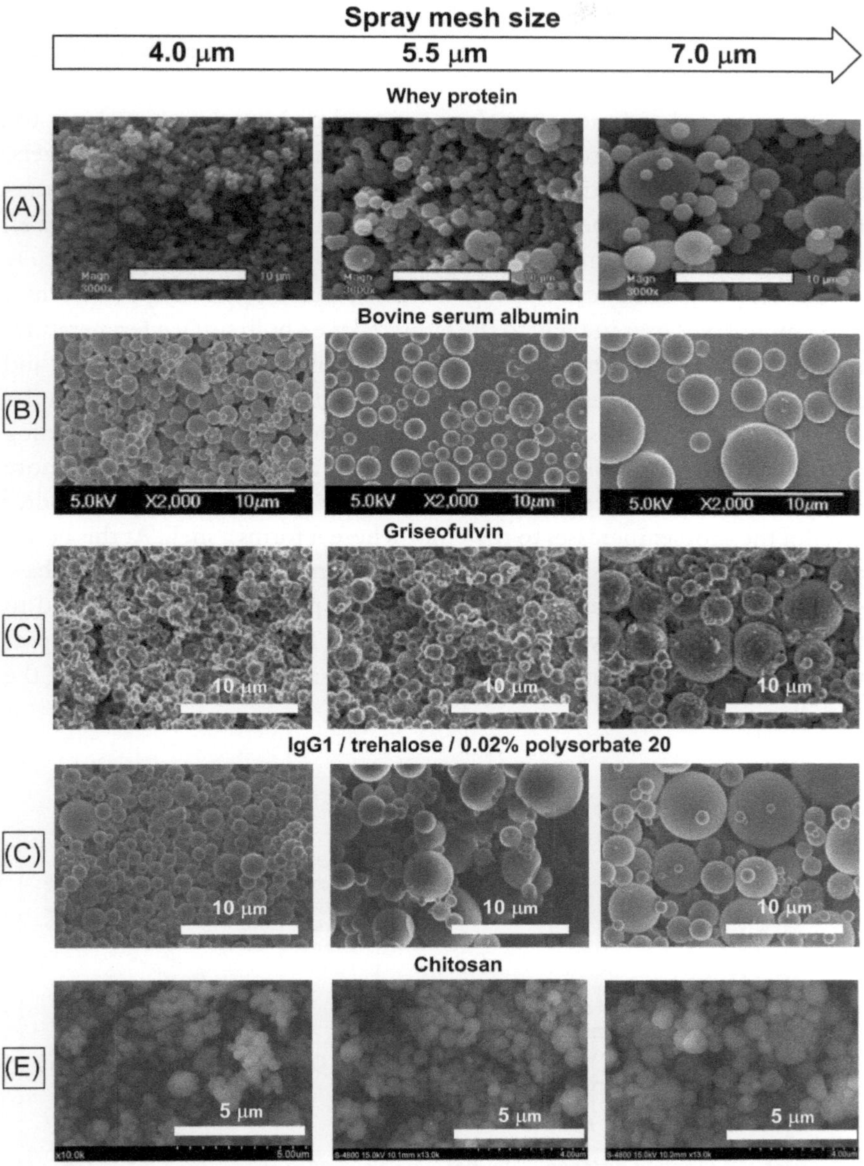

Fig. 10 Effect of the spray mesh size 4.0, 5.5, and 7.0 μm on the dried particle size. (A) Whey protein at 0.1%, 1%, and 10% (w/v) solid concentration (Li et al., 2010), (B) bovine serum albumin at 1%, particle size: 0.7, 1.7, and 2.6 μm (Lee et al., 2011), (C) griseofulvin at 0.5% in methanol/acetone, particle size: 3.4, 4.9, and 6.5 μm (Schmid, 2011), (D) monoclonal IgG_1 antibody in trehalose (2.5%, ratio 70:30) with 0.02% polysorbate 20 (Schmid, 2011), (E) chitosan (276 cP) at 0.025%, particle size: 95, 215, and 265 nm (Ngan et al., 2014).

Other parameters to consider when nano spray drying are the inlet and outlet temperatures of the drying gas, the spray rate of the vibrating mesh atomizer, the drying gas type, the drying gas flow rate, and its humidity. The feed-related parameters, such as the feed composition (e.g., core-to-wall materials ratio, bioactive content, surfactants, emulsifiers, stabilizers, and polymer glass transition temperature), the viscosity, the solid concentration, and the solvent type (e.g., boiling point, aqueous, organic, or mixture), will mainly affect the final powder properties.

As shown schematically in Fig. 11 (left), the droplet temperature in a nano spray dryer rises initially to the saturated wet-bulb surface temperature, remains constant during evaporation (constant-rate drying phase), and approaches the temperature of the surrounding gas at dryer outlet (falling-rate drying phase). The solvent evaporation cools the surrounding drying gas, and the droplet temperature remains constant. As more and more of the solvent evaporates from the droplet, the solid content of the outer layer of the droplet increases to the point where it forms a shell. At this point, a particle with a solid shell but wet core is formed, and the drying phase switches to the falling-rate phase. During this phase, heat is transferred to the particle from the drying gas as sensible heat. The temperature of the particle is raised to fully evaporate the remaining solvent from the core of the particle.

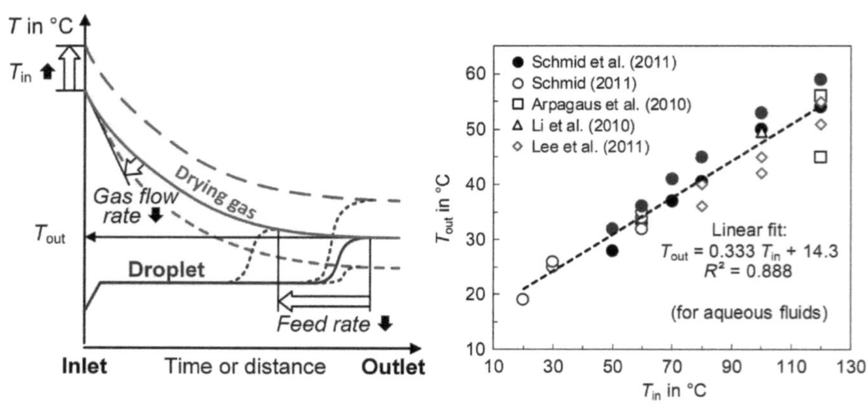

Fig. 11 Typical temperature curve of the drying gas and the droplets during nano spray drying. Left: influence of drying gas temperature, gas flow rate, and feed rate on the droplet temperature. Right: linear fit between the inlet (T_{in}) and outlet (T_{out}) temperature in the Nano Spray Dryer B-90 for aqueous fluids, that is, $T_{out} = 0.333 \cdot T_{in} + 14.3°C$ (Arpagaus, 2018a; Arpagaus et al., 2017).

The outlet temperature of the drying gas is directly related to the set inlet temperature, the flow rate, the sample concentration, and the drying gas flow rate. Changes in any of these parameters will either increase or decrease the outlet temperature. A lower drying gas flow rate leads to a lower outlet temperature; a smaller feed rate increases the outlet temperature. A higher inlet temperature reduces the relative humidity of the drying gas, resulting in a drier and less sticky powder. Residual moisture contents of lower than 0.5% in mannitol, 2% to 5% in trehalose, and 7% for β-cyclodextrin and hydroxytyrosol powders have been observed (Malapert et al., 2019; Schmid, 2011). For aqueous applications, the outlet temperatures are between 28°C and 59°C and follow a linear relationship of $T_{out}=0.333 \cdot T_{in}+14.3°C$ as a first approximation (Fig. 11, right) (Arpagaus, 2018a; Arpagaus et al., 2017), which makes nano spray drying a suitable process for heat-sensitive substances (Amsalem et al., 2017; Aquino et al., 2014; Heng et al., 2011).

The estimated drying time of water droplets in a Nano Spray Dryer B-90 is in the order of 10 ms, assuming droplets of 7-μm diameter, 75°C drying temperature, and 100 L/min drying air flow rate (Feng et al., 2011). Because the total residence time of the particles in the nano spray dryer is much longer than the required drying time, the particles are dry when arriving at the electrostatic particle collector.

A compromise needs to be found between feed rate (productivity in mL/min), solid concentration, and particle size. The feed rate increases primarily with the spray mesh size and the setting of the relative spray frequency, and it depends on the feed formulation (i.e., solid concentration, solvent type, and the addition of surfactant) (see experimental data in Table 7).

For pure water, the specified feed rates are in the ranges of 10–20, 25–50, and 80–150 mL/h for a 4.0-, 5.5-, and 7.0-μm spray mesh, respectively (Büchi Labortechnik AG, 2010). A higher solid concentration results in a lower feed rate. The addition of organic solvents or a small amount of surfactant into the feed tends to increase the feed rates, which can be attributed to the lower surface tension.

The yield of nano spray-dried particles can be calculated from the total weight of the recovered particles and the original weight of the bioactive and polymer. Table 8 shows some examples of optimized product yields that typically range from about 43% to 95% for small sample quantities of less than 1 g. Particle losses can be attributed to deposits on the walls of the drying chamber and losses during manual removal of particles from the surface of the electrostatic particle collector.

Table 7 Influence of spray mesh size and solid concentration on the feed rate for aqueous substances at 100% spray rate in the Nano Spray Dryer B-90.

Substance	Solid concentration (%, w/v)	Solvent	Spray mesh size (μm)	Feed rate (mL/h)	Reference
Water	–	Water	4.0	10–20	Büchi
			5.5	25–50	Labortechnik
			7.0	80–150	AG (2010)
Gum arabic, maltodextrin, polyvinyl alcohol, modified starch	1	Water	4.0	3–25	Li et al. (2010)
Trehalose	0.1	Water	4.0	11	Schmid (2011)
	1			8	
	10			5	

Table 8 Examples of optimized yields achieved by nano spray drying for small sample amounts (– = not available).

Product yield (%)	Solid sample amount (mg)	Substance	Reference
43–95	30–300	Gum arabic, maltodextrin, polyvinyl alcohol, modified starch, and whey protein	Li et al. (2010)
75–94	500	α-Amylase enzyme in sucrose and Tween 80	Abdel-Mageed et al. (2019)
60–94	500	β-Galactosidase enzyme in trehalose	Bürki et al. (2011)
75–91	800	Hypromellose (enteric film coating and emulsifier)	Gu et al. (2015)
>90	n.a.	Sodium alginate (polysaccharide)	Blasi et al. (2010)
81–85	50–500	Sodium chloride NaCl	Li et al. (2010)
80	300	Resveratrol in poly(ε-caprolactone), sodium deoxycholate, and trehalose	Dimer, Ortiz, Pohlmann, and Guterres (2015)
50–78	10–50	Trehalose, mannitol or disodium phosphate surfactant, polysorbate	Schmid (2011) and Schmid et al. (2009, 2011)
68–76	–	Bovine serum albumin (protein)	Lee et al. (2011)
60–63	–	Chitosan (antibacterial activity)	Ngan et al. (2014)
53	–	Hydroxytyrosol-β-cyclodextrin	Malapert et al. (2019)

In general, the slow and gentle drying in a nano spray dryer yields spherical and compact particles. Small amounts of surface-active compounds (e.g., polysorbate) in the formulations are typically used to optimize the smoothness and sphericity of the particles, as shown by several researchers (Bürki et al., 2011; Li et al., 2010; Schmid, 2011; Schmid et al., 2009, 2011). Surfactants balance the surface-to-viscous forces inside the drying droplet and enable the formation of a smooth spherical surface on the dry particle (Moghbeli, Jafari, Maghsoudlou, & Dehnad, 2019). However, hollow, porous, and encapsulated structures with wrinkled, shriveled, or even doughnut-like shapes are also possible (Arpagaus et al., 2017).

After nano spray drying, the activity of the encapsulated product is preserved if the powder is stored under controlled conditions and if a stabilizer is added into the feed formulation. Most nano spray-dried powders are amorphous due to the short drying time, which is too short to form crystalline structures. To prevent recrystallization of amorphous drugs, the powders are stored under dry conditions. For example, Pérez-Masiá et al. (2015) observed bioactive stability of folic acid (vitamin B_9) in whey protein after 60 days under dry storage conditions and in darkness. Merchant et al. (2014) found no evidence of change in the crystallinity of chitosan powder after storage for 60 days at room temperature and ambient humidity. In addition, the nano spray drying process itself has a negligible or only a marginal impact on the product degradation or aggregation. Schmid (2011) detected a slight reduction in L-lactic dehydrogenase enzyme activity during pump circulation. Bürki et al. (2011) showed that nano spray-dried β-galactosidase enzyme in trehalose lost only 3% of its activity during 3 weeks of storage at 70°C. Optimized 600-nm-sized powder of α-amylase enzyme in sucrose retained 72% of its activity after 60 days (Abdel-Mageed et al., 2019). An option to minimize the potential of activity loss during the nano spray drying process is to cool the feed vessel in ice (Amsalem et al., 2017; Schoubben et al., 2013; Torge, Grützmacher, Mücklich, & Schneider, 2017) or to use a larger spray mesh to reduce the mechanical shear of atomization (Abdel-Mageed et al., 2019; Bürki et al., 2011).

Table 9 provides some data on encapsulation efficiency and loading of different bioactive food ingredients by nano spray drying. For example, Kyriakoudi and Tsimidou (2018) encapsulated aqueous saffron extracts in maltodextrin by nano spray drying. The encapsulation efficiency of the crocins was found to range between 54% and 81%. It was also found that the dried particles had a low moisture content of 3.3 to 3.9%.

Table 9 Encapsulation efficiencies and bioactive compound loadings achieved by nano spray drying (– = not available).

Bioactive compound	Wall material	Bioactive loading	Encapsulation efficiency	Reference
Folic acid (vitamin B_9)	Resistant starch	–	53%	Pérez-Masiá et al. (2015)
Folic acid (vitamin B_9)	Whey protein	–	84%	Pérez-Masiá et al. (2015)
Hydroxytyrosol (from olive oil)	β-Cyclodextrin	–	84 ± 3%	Malapert et al. (2019)
Resveratrol	Poly(ε-caprolactone, sodium deoxycholate, and trehalose	31%	–	Dimer, Ortiz, et al. (2015)
Saffron extracts (i.e., hydrophilic apocarotenoids crocins and picrocrocin)	Maltodextrin	1:5, 1:10, 1:20 core-wall ratio (w/w)	54%–81%	Kyriakoudi and Tsimidou (2018)

The size, shape, surface charge, wall composition, molecular weight, etc. play a major role on the release and bioavailability of bioactive food ingredients. The mathematical models for controlled release of substances are, for example, summarized by Estevinho, Rocha, Santos, and Alves (2013). The release of a bioactive compound in an ideal system may follow zero (constant release rate, pure material, and no encapsulation)-, half (matrix particles)-, or first (core is a solution)-order kinetics. The equations of Higuchi or Korsmeyer-Peppas are generally used to characterize the kinetic mechanism of controlled release of substances.

Commonly applied analytical methods to characterize the nano spray-dried powders are scanning electron microscopy (SEM) to determine particle size and morphology, laser diffraction to measure particle size (dynamic light scattering, DLS), and X-ray diffraction (XRD) or differential scanning calorimetry (DSC) to identify the amorphous/crystalline state.

3 Food and nutraceutical applications

Since its market launch in 2009, the Nano Spray Dryer B-90 has been used in the laboratory primarily in pharmaceutical research (Arpagaus, 2011,

2012, 2018b, 2018c). The use of nano spray drying for the nanoencapsulation of bioactive food ingredients is still at an early stage but is constantly evolving (Anandharamakrishnan & Ishwarya, 2015a; Assadpour & Jafari, 2019). This section presents published works on bioactive food applications performed with a nano spray dryer, along with a brief overview of the results obtained.

3.1 Polymeric wall materials

A number of research studies have demonstrated the ability of nano spray drying to produce submicron powders from polymeric wall materials. Table 10 shows experimental conditions that can be used as a first guide for the application of a Nano Spray Dryer B-90 with similar substances. Fig. 12 shows several SEM images of representative nano spray-dried polymeric wall materials.

Table 10 Process conditions for nano spray drying of polymeric wall materials for food applications.

Polymeric wall material	*T* in (°C)	*T* out (°C)	Drying gas (L/min)	Particle size (μm)	Reference
Gum arabic, whey protein, maltodextrin (DE 12), polyvinyl alcohol, and modified starch	100	38–60	100	0.2–1.1	Li et al. (2010)
Gum arabic (1%)	120	50–60	130	0.8–3.7	Oliveira et al. (2013)
Gum arabic (1%)	80–100	54–65	133	0.4–2.5	Büchi Labortechnik AG (2009)
Sodium alginate	110	50	100	0.4–1.2	Blasi et al. (2010)
Sodium alginate (0.1%)	120	50–60	130	0.8–5.5	Oliveira et al. (2013)
Sodium alginate/pectin	90	45	100	0.3–1.0	De Cicco et al. (2014)
Maltodextrin (DE 19) (1%)	100	33–39	120	0.5–2.1	Büchi Labortechnik AG (2017a)
Gelatine (1%)	80–90	44–50	133	0.3–2.0	Büchi Labortechnik AG (2009)
Trehalose (0.1% and 1%) with 0.05% polysorbate 20	60–100	30–45	115	0.3–5.0	Büchi Labortechnik AG (2009)

Continued

Table 10 Process conditions for nano spray drying of polymeric wall materials for food applications.—cont'd

Polymeric wall material	*T* in (°C)	*T* out (°C)	Drying gas (L/min)	Particle size (μm)	Reference
Trehalose, mannitol	60–100	30–45	115	0.3–3.0	Schmid (2011) and Schmid et al. (2009, 2011)
Leucine and trehalose	75	45	100	2.1–5.4	Feng et al. (2011)
Carboxymethyl cellulose (0.1%)	120	50–60	130	1.0–3.7	Oliveira et al. (2013)
Mannitol (1%)	80–90	40–50	133	0.4–2.0	Büchi Labortechnik AG (2009)
Nanosuspension of mannitol and poly(lactic-*co*-glycolic acid) (PLGA)	80	32–39	140	1.1–7.2	Torge et al. (2017)
Chitosan (30,000 Mv) in 1% acetic acid	120	55	130	0.6–1.6	Gautier et al. (2010)
Chitosan (low-density, 267–1200 cP) in 0.5% acetic acid	120	80	130	0.1–0.3	Ngan et al. (2014)
Chitosan/Tween 20 in 1% acetic acid	80–120	n.a.	100–150	0.3	O'Toole et al. (2012)
Bovine serum albumin with surfactant, polyoxyethylene, sorbitan monooleate	80–120	36–55	150 (N_2)	0.5–2.6	Lee et al. (2011)
Bovine serum albumin, 0.1% with 0.05% Tween 80	100	51–61	150	0.1–2.0	Büchi Labortechnik AG (2017b)

The comparison of the morphologies reveals that the primary particles are almost spherical in shape and that the most part is submicron in size. The wall materials used for the encapsulation of bioactive food ingredients by nano spray drying are mainly water-soluble polymers such as sodium alginate (Blasi et al., 2010; De Cicco et al., 2014; Oliveira et al., 2013), gum arabic (Büchi Labortechnik AG, 2017a; Li et al., 2010; Li, Anton, & Vandamme, 2015; Oliveira et al., 2013), gelatine (Büchi Labortechnik AG, 2009), whey protein (Li et al., 2010, 2015), polyvinyl alcohol (Li et al., 2010), modified starch (Li et al., 2010), and maltodextrin (Büchi Labortechnik AG, 2009, 2017a; Li et al., 2010).

Maltodextrins of various dextrose equivalents (DE) are typically used as carriers for flavors, fragrances and oils. Low-viscosity sodium alginate derived from marine algae is widely used as an emulsifier and immobilizer in food formulations. Gelatine is a tasteless animal protein and is often used as

Fig. 12 SEM pictures of nano spray-dried particles obtained with the Nano Spray Dryer B-90. (A) Maltodextrin (DE 12), 1% (w/v) aqueous solution (Li et al., 2010), (B) maltodextrin (DE 19), 1% (Büchi Labortechnik AG, 2009), (C) sodium alginate, 0.1% (Oliveira et al., 2013), (D) gum arabic, 1% (Oliveira et al., 2013), (E) gum arabic, 1% (Li et al., 2010), (F) cashew nut gum, 1% (Oliveira et al., 2013), (G) modified starch, 1% (Li et al., 2010), (H) gelatine, 1% (Büchi Labortechnik AG, 2009), (I) trehalose, 1% with 0.05% polysorbate 20 (Tween 20) (Büchi Labortechnik AG, 2009), (J) chitosan, 0.025% in 1% acetic acid (O'Toole et al., 2012), (K) chitosan, 0.1% in 1% acetic acid (Gautier et al., 2010), (L) chitosan (276 cP), 0.1% in 0.5% acetic acid (Ngan et al., 2014), (M) carboxymethyl cellulose, 1% (Oliveira et al., 2013), (N) bovine serum albumin, 0.1% with 0.05% Tween 80 (Lee et al., 2011), (O) bovine serum albumin, 0.1% with 0.05% Tween 80 (Büchi Labortechnik AG, 2017b). *(All images reprinted with permission.)*

a gelling agent and for encapsulation in food. Mannitol (Torge et al., 2017), chitosan (Gautier et al., 2010; Ngan et al., 2014; O'Toole et al., 2012), leucine (Feng et al., 2011), and trehalose (Schmid, 2011; Schmid et al., 2009, 2011; Vereman, Thysens, Derdelinckx, Van Impe, & Van de Voorde, 2019) are also used as food encapsulating agents due to their high aqueous solubility and low toxicity. Trehalose is applied as a protein stabilizer increasing shelf life (Schmid et al., 2009; Vereman et al., 2019). Leucine is an amino acid and a useful dispersing agent (Feng et al., 2011). Chitosan offers advantages for mucosal delivery, such as low toxicity, good biodegradability, and antibacterial activity (Cerchiara et al., 2015; Dimer, de Souza Carvalho-Wodarz, Haupenthal, Hartmann, & Lehr, 2015; Merchant et al., 2014; Ngan et al., 2014; Nguyen, Nguyen, Wang, Vo, & Nguyen, 2017; Rampino, Borgogna, Blasi, Bellich, & Cesàro, 2013). Moreover, nano spray-dried chitosan particles strongly inhibited bacterial growth and have much potential as fat blocker (Gautier et al., 2010). Bovine serum albumin (BSA) is a protein derived from cows and has many applications in life science. It is often used as a model protein in numerous biochemical applications and in spray drying to evaluate the process as a heat-sensitive substance. Submicron particles are achieved with the 4.0-μm spray mesh at a BSA concentration of 0.1% (w/v) and a surfactant concentration of 0.05% (w/v) (Tween 80) (Arpagaus, 2012; Büchi Labortechnik AG, 2017b; Lee et al., 2011).

In general, a dilution of the solution leads to an end product with a smaller particle size. Typically, the solid concentrations are 0.1%–1% (w/v). Since a single encapsulating agent may not have all the ideal wall material properties, recent research has focused on mixtures of carbohydrates, gums, and proteins. There are many other common wall materials for encapsulation by spray drying (see Table 3), which shows in particular the potential for further applications in nano spray drying.

3.2 Water-soluble vitamins, polyphenols, and extracts

Vitamins are vital amines needed for the normal growth and function of the human body. However, vitamins are sensitive and unstable to exposure to high temperatures, oxygen, light, or moisture, which can lead to the loss of their functions (Katouzian & Jafari, 2016). Spray drying is the most common method for encapsulating vitamins. Vitamins are usually classified as fat soluble (vitamins A, D, E, and K) and water soluble (vitamins B and C). So far, there are a few studies on the encapsulation of water-soluble vitamins with nano spray drying (Oliveira et al., 2013; Pérez-Masiá et al., 2015). Table 11 lists the identified optimal experimental process parameters.

Table 11 Process conditions for nano spray drying of water-soluble vitamins, polyphenols, and extracts (– = not available).

Bioactive ingredient	Encapsulating wall material	Solvent, solid concentration	*T* in (°C)	*T* out (°C)	Drying gas (L/min)	Particle size (μm)	Reference
Vitamin B_{12}	Gum arabic, cashew nut gum, sodium alginate, carboxymethyl cellulose, Eudragit RS100	0.1%–1% (w/v) in water	120	50–60	130	0.2–5.5	Oliveira et al. (2013)
Folic acid (synthetic vitamin B_9)	Guar gum, whey protein, resistant starch	Water, 0.4% (w/v)	90	45	140	0.2–4.5	Pérez-Masiá et al. (2015)
Curcumin	Chitosan/Tween 20	Water with 1% acetic acid	80–120	–	100–150	0.3	O'Toole et al. (2012)
Curcumin	Human serum albumin	Water	120	–	150	0.2–0.7	Jain (2014)
Curcumin in nanogels	Egg yolk low-density lipoprotein (LDL), pectin, or carboxy-methyl cellulose (CMC)	Water	70–120	50–60	120	0.5–1.5	Zhou et al. (2018, 2016)
Curcumin	Structural change without wall material	Ethanol	65	–	120	0.4–1.3	Büchi Labortechnik AG (2017c)
Hydroxytyrosol (biophenol of olive oil)	β-Cyclodextrin (β-CD)	Water	100	–	100	0.4–3.2	Malapert et al. (2019)
Resveratrol	Poly(ε-caprolactone), sodium deoxycholate, trehalose	Water/ acetone (50:50)	55	–	110 (N_2/CO_2)	1–5	Dimer, Ortiz, et al. (2015)
Guava leaf extracts	Aqueous extract	Water	105	38	–	–	Camarena-Tello et al. (2018)
Saffron extracts (apocarotenoids crocins and picrocrocin)	Maltodextrin	Water, 2.5% (w/w)	100	45–50	100	1.5 (mean)	Kyriakoudi and Tsimidou (2018)

In particular, Oliveira et al. (2013) encapsulated hydrophilic vitamin B_{12} in gum arabic, cashew nut gum, sodium alginate, and carboxymethyl cellulose. A deficiency of vitamin B_{12} can be associated with coronary artery disease. The influence of the different polymer properties was evaluated in terms of controlled release performance. Aqueous solutions with a viscosity of 6.3–9.2 mPa s were successfully atomized with the vibrating mesh technology of the Nano Spray Dryer B-90. SEM micrographs displayed spherical submicron particles. The reduction of the particles size and the increase of the specific surface area contributed to speed up the release rate kinetics.

More recently, Pérez-Masiá et al. (2015) encapsulated water-soluble folic acid (synthetic vitamin B_9) in whey protein and resistant starch through nano spray drying. Folic acid has important roles in cell metabolism and can be obtained in its natural form (folate) by consuming green vegetables or dietary supplements (Oprea & Grumezescu, 2017). Specific dosages of folic acid are needed in the daily diet during phases of cell division and growth such as pregnancy and infancy. By nano spray drying, spherical submicron and micron capsules were obtained. Whey protein–based capsules achieved an encapsulation efficiency of about 84%, compared with about 53% with resistant starch, which was attributed to the different interactions between the bioactive and the protein matrix. It was further observed that whey protein conserved the folic acid better against degradation during storage in an aqueous solution as well as dry storage. The capsules made of whey protein were able to keep the bioactive stability at almost 100% in darkness and under dry storage conditions after 60 days.

A number of research studies have shown that nano spray drying can be used as an appropriate method for the nanoencapsulation of phenolic compounds and antioxidants (Büchi Labortechnik AG, 2017c; Camarena-Tello et al., 2018; Dimer, Ortiz, et al., 2015; Jain, 2014; Malapert et al., 2019; O'Toole et al., 2012; Zhou et al., 2018; Zhou, Wang, Hu, & Luo, 2016). The optimized process conditions and the obtained particle sizes are summarized in Table 11. Polyphenols are found naturally in fruits and vegetables and primarily consist of flavonoids and nonflavonoid polyphenolics (Faridi Esfanjani, Assadpour, & Jafari, 2018). The antioxidant properties prevent degenerative diseases such as cancer, inflammation, and neurodegenerative diseases (Faridi Esfanjani & Jafari, 2016).

Curcumin is a polyphenolic compound derived from the rhizome of the turmeric plant *Curcuma longa* L. Its therapeutic potential as an antioxidant, antiinflammatory, antimutagenic, and anticancer agent is limited by poor uptake in the body due to its insolubility in water, rapid metabolism by the intestinal mucosa and liver, and quick excretion. Specific delivery vehicles are used to enhance the bioavailability of curcumin (Rafiee, Nejatian, Daeihamed, & Jafari, 2019).

O'Toole et al. (2012) encapsulated curcumin in in chitosan/Tween 20 via nano spray drying. Chitosan offers improved mucous adhesiveness, nontoxicity, biocompatibility, and biodegradability. Spherical particles with 285 ± 30-nm diameter were produced. Release studies in 1% acetic acid and buffer solution revealed nearly 40% of curcumin release in the first 5 min of the experiment. All detectable curcumin release was complete within 2 h. Jain (2014) encapsulated curcumin in cross-linked human serum albumin. The particles were smooth, spherical, and submicron-sized in a range of 0.2–0.7 μm. The cross-linked albumin implied a slower release of curcumin and the data followed a first-order release profile.

In an application note of Büchi Labortechnik AG (2017c), curcumin particles of 367 to 1290 nm could be obtained directly from a 0.1% (w/w) curcumin solution prepared in ethanol. The structural change of the curcumin powder was established by operating a Nano Spray Dryer B-90 HP in closed loop configuration using an Inert Loop B-295 and inert gas N_2/CO_2. The gas flow rate was set up between 120 and 150 L/min, the inlet temperature of 65°C, and the spray rate at 80%. SEM images of nano spray-dried curcumin powder showed a spherical morphology, while the original curcumin had a needle-like morphology in the 100-μm scale (see Fig. 13).

Dimer, Ortiz, et al. (2015) encapsulated resveratrol, a polyphenol derived, for example, from grapes, with antioxidant properties, in poly(ε-caprolactone), sodium deoxycholate, and trehalose. Most particles were in the size range of 1–5 μm and showed a great potential for the treatment

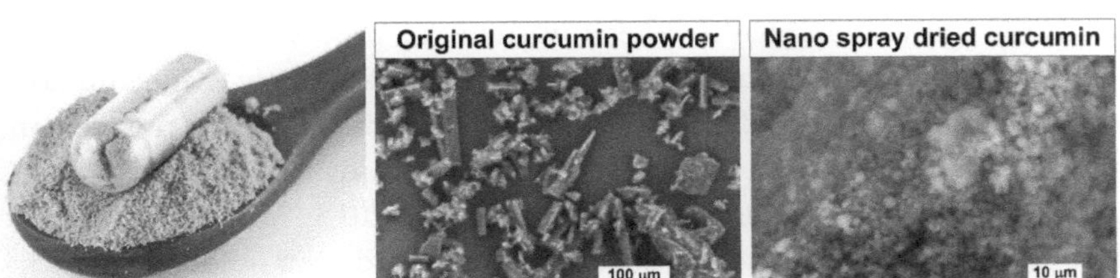

Fig. 13 (Left & middle) Original curcumin powder and (right) after nano spray drying as 0.1% (w/w) ethanol solution, illustrating the structural nanoionization of a pure bioactive substance (Büchi Labortechnik AG, 2017c).

of pulmonary arterial hypertension. The nano spray drying process provided high yields of approximately 80% with low moisture content <2%. The in vitro release of resveratrol was sustained for up to 12h. After 30min, 25% of the resveratrol was released.

As an example for the encapsulation of plant extracts, Kyriakoudi and Tsimidou (2018) encapsulated aqueous saffron extracts (i.e., hydrophilic apocarotenoids crocins and picrocrocin) in maltodextrin by nano spray drying. Spherical particles with smooth surfaces (mean particle size 1.5 μm) were obtained using a 4-μm spray mesh. The product yield was found to be 71% to 87%, the encapsulation efficiency of the crocins in a range between about 54% and 81%. The particles were found to have low moisture content of 3.3% to 3.9%. The nanoencapsulation resulted in increased stability of crocins and picrocrocin under thermal and gastrointestinal (GI) conditions. As another example, Camarena-Tello et al. (2018) nano spray-dried the aqueous extract of guava leaves and quantified the total polyphenols and flavonoids contents and the in vitro ability to scavenge hydroxyl radicals.

Malapert et al. (2019) encapsulated hydroxytyrosol (HT) into β-cyclodextrin (β-CD) by nano spray drying. HT is found in olive oil and leaves and is used as a functional food additive or ingredient to protect blood fats from oxidative damage. Cyclodextrins (e.g., natural cyclic oligosaccharides from the degradation of starch) offer a hydrophobic cavity and can accommodate guest compounds to form inclusion complexes, while their hydrophilic outer surface provides water solubility. An equimolar aqueous solution of HT and β-CD (final concentration of 5 mM each) was prepared and filtered before spray drying. Spherical particles from 0.4 to 3.4 μm were formed by nano spray drying while freeze-drying only provided irregular shapes. An encapsulation efficiency of 84% and a final moisture content of 7% were obtained.

Nano spray drying has also been utilized as a process to dry formulated nanoparticles containing bioactive food compounds, namely, nanoemulsions (Büchi Labortechnik AG, 2017a; Hu, Gerhard, Upadhyaya, Venkitanarayanan, & Luo, 2016; Li et al., 2010, 2015; Prasad Reddy, Padma Ishwarya, & Anandharamakrishnan, 2019; Veneranda et al., 2018; Wang, Soyama, & Luo, 2016), nanogels (Zhou et al., 2018, 2016), and solid lipid nanoparticles (SLNs) (Wang, Bae, Lee, & Luo, 2018; Wang, Hu, Zhou, Xia, et al., 2016; Wang, Hu, Zhou, Xue, & Luo, 2016; Wang, Ma, Lei, and Luo, 2016; Wang et al., 2017; Xue, Wang, Hu, Zhou, & Luo, 2017, 2018). In most cases, the formulated nanoparticles are mixed with additional excipients and dispersing agents prior to nano spray drying to fabricate fine powders. Further details on the applications and process conditions are given in the following sections.

3.3 Nanoemulsions with lipophilic bioactive compounds

Nanoemulsions are particularly suitable for solubilization of lipophilic bioactive compounds, such as essential oils in aqueous phase. Essential oils are derived from a variety of natural sources like vegetables, herbs, and fruits and contain bioactive compounds with well-known antioxidant and antimicrobial activities. The main challenges are their poor water solubility and high volatility (Veneranda et al., 2018). Using nanoemulsions in food applications offers the potential to increase the absorption rate, improve bioavailability, enable target delivery, and so forth. Thereby, surfactants and emulsifiers play a decisive role in the production of nanoscale oil droplets by lowering the interfacial tension between the aqueous and oil phases.

In the food industry, nanoemulsions from essential oils are typically prepared by high-energy methods such as high-pressure homogenization, ultrasonication, and microfluidization (Büchi Labortechnik AG, 2017a; Hu et al., 2016; Jafari et al., 2015; Quintanilla-Carvajal et al., 2010; Wang, Soyama, et al., 2016) or low-energy methods like self-emulsification (Jafari et al., 2015; Jafari, Paximada, et al., 2017; Li et al., 2010, 2015; Veneranda et al., 2018). Fig. 14 schematically illustrates the manufacturing processes of various oil nanoemulsions prior to nano spray drying.

Table 12 summarizes the experimental process parameters of published studies on nano spray drying for the encapsulation of nanoemulsions (Büchi Labortechnik AG, 2017a; Hu et al., 2016; Li et al., 2010, 2015; Prasad Reddy et al., 2019; Veneranda et al., 2018; Wang, Soyama, et al., 2016). Fig. 15 shows some SEM images of the resulting nano spray dried powders containing nanoemulsions.

In particular, Li et al. (2010, 2015) encapsulated oil-in-water nanoemulsions of vitamin E acetate (oil droplets below 100 nm) in gum arabic, modified starch, maltodextrin, and whey protein (1 w% wall material) (Fig. 14A). Vitamin E acetate was chosen as a model for the oily phase. Vitamin E is an antioxidant that can scavenge free radicals and is also involved in immune function (Oprea & Grumezescu, 2017). The nanoemulsions were obtained spontaneously by low-energy emulsification after mixing the two phases with 1% (w/v) aqueous wall material solutions in a weight ratio of 1:4. A surfactant immediately stabilized the formed nanodrops (Li et al., 2015). Optimal drying conditions of 100°C inlet

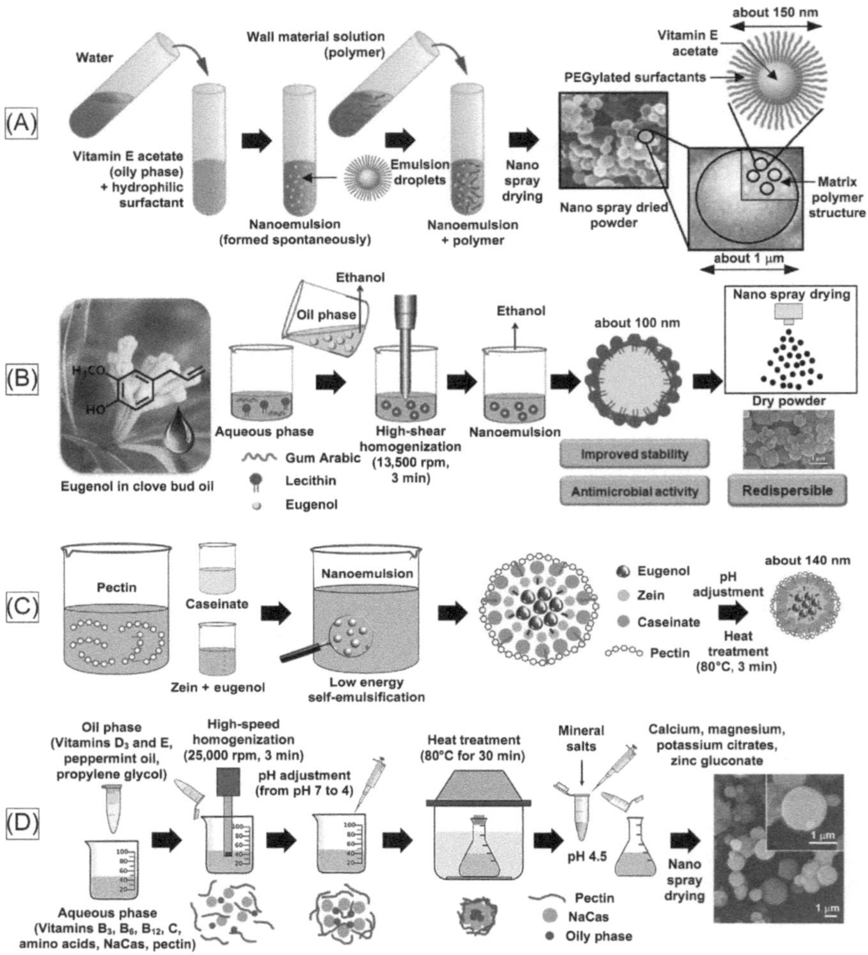

Fig. 14 Preparation procedures of (A) spontaneously formed vitamin E oil-in-water nanoemulsion <100 nm in whey protein (Li et al., 2010, 2015), (B) antimicrobial eugenol oil nanoemulsion with gum arabic and lecithin (Hu et al., 2016), (C) eugenol-loaded pectin/NaCas/zein nanoparticles by a pH- and heat-induced complexation process (Veneranda et al., 2018), and (D) functional drink containing fat- and water-soluble nutrients (Wang, Soyama, et al., 2016).

temperature and 40–60°C outlet temperature ensured no damage of the food ingredients during nano spray drying. Submicron solid particles (460–730-nm distributions) were produced with almost 100% encapsulation efficiency (Fig. 15A and B). Redissolving the spray-dried nanoparticles in water led to nanoemulsions without any significant degradation or size

Table 12 Process parameters of published studies on nano spray-dried nanoemulsions (–=not available).

Bioactive substance, application	Encapsulating wall material	Solvent	*T* in (°C)	*T* out (°C)	Drying gas (L/min)	Dried particle size (μm)	Reference
Vitamin E acetate oil-in-water (vitamin supplement)	Whey protein, gum arabic, modified starch, maltodextrin	Water	100	38–60	100	0.4–1.1	Li et al. (2010, 2015)
Eugenol oil (antimicrobial)	Gum arabic, lecithin	Water, ethanol	100	–	100–110	0.2–0.5	Hu et al. (2016)
Eugenol oil (antimicrobial)	Zein, sodium caseinate (NaCas), pectin	Water	100	–	120	1.0–1.5	Veneranda et al. (2018)
Functional drink as the nanoemulsion of antimicrobial peppermint oil, vitamins (D3, E, B3, B6, B12, and C), minerals, amino acids	NaCas, pectin, propylene glycol	Water	100	–	120	1–2	Wang, Soyama, et al. (2016)
Sunflower oil	Maltodextrin (DE 19), Tween 20	Water	120	45	100	1.0–2.5	Büchi Labortechnik AG (2017a)
Omega-3 fatty acids	Lactoferrin	Water	80	40–46	–	0.8–3 (agglomerated)	Nunes et al. (2018)
Coffee bean oil	Whey protein, Tween 80	Water	90	45	90–110	0.2–0.4	Prasad Reddy et al. (2019)

Fig. 15 Nano spray-dried powders obtained from (A) vitamin E oil-in-water nanoemulsion in gum arabic (Li et al., 2010) and (B) in whey protein (Li et al., 2010); (C) eugenol oil nanoemulsion in gum arabic and lecithin (Hu et al., 2016); (D) eugenol oil nanoemulsion in pectin/NaCas/zein complex (Veneranda et al., 2018); (E) oil- and water-soluble vitamins, amino acids, and mineral salts as functional drink (Wang, Soyama, et al., 2016); (F) sunflower oil emulsion in maltodextrin (1%, w/v) (Büchi Labortechnik AG, 2017a).

increase (Li et al., 2010, 2015). Overall, the nano spray drying technology was able to encapsulate nanoemulsions smaller than 100 nm (an order of magnitude smaller than the sprayed droplets) into separated solid particles and to preserve the integrity of the nanoemulsion system.

Hu et al. (2016) and Veneranda et al. (2018) prepared eugenol oil-loaded nanoemulsions with enhanced antimicrobial activity (i.e., inhibiting bacterial growth), good physical stability, and excellent redispersibility after nano spray drying (Fig. 14B and C). Eugenol is a naturally occurring essential oil extracted from cloves and is known for its strong bactericidal and antioxidant properties, which is seen as promising to be applied in the food industry as a food preservative. Gum arabic and lecithin were used as food-grade natural emulsifiers (ratio of 4:1 w/w), and nanoemulsions of around 100 nm were obtained by high-speed homogenization at 13,500 rpm for 3 min. The addition of ethanol as a cosurfactant to the eugenoil oil improved the morphology and homogeneity of the powders during nano spray drying. Spherical powders in the range of 200 to 500 nm with smooth surface could be produced (Fig. 15C), which exhibited excellent redispersibility in water. The eugenol oil droplets were well surrounded and trapped by lecithin and gum arabic. The powders were about 20 to 50 times smaller than those dried by conventional spray dryers (i.e., usually 10 to 25 μm).

Veneranda et al. (2018) prepared about 140-nm-sized nanoemulsions by a pH- and heat-induced self-emulsification method (80°C for 60 min) using food polymers zein, sodium caseinate (NaCas), and pectin. Zein is derived from corn and is a suitable storage protein for liquid-liquid dispersions due to its unique solubility. Caseinate and pectin are known for their pH- and heating-induced electrostatic interactions. Stable eugenol-loaded biopolymeric complex nanoparticles could be prepared by premixing eugenol with zein and then complexing among the biopolymers caseinate and pectin. The colloidal eugenol-loaded complex nanoparticles were nano spray-dried and transformed to fine 1 to 1.5-μm-sized powders (Fig. 15D) enabling a long-term storage in powder form with exceptional microbial stability (i.e., no growth of bacteria up to 2 months), a fast redispersion capability, and thus a greater potential to the applications in food industry.

Wang, Soyama, et al. (2016) developed a novel nanoemulsion-based functional drink from all natural food ingredients using casein/pectin nanocomplex particles as a delivery system (Fig. 14D). Nano spray drying was applied to form instant powders to improve the shelf life and transportation. This property also gains a ready-to-drink character by dissolving them in water, milk, or coffee. The functional drink consisted of both lipophilic (vitamin D_3 and E) and hydrophilic vitamins (B_3, B_6, B_{12}, and C), mineral salts (i.e., calcium citrate, magnesium citrate, potassium citrate, and zinc gluconate), and amino acids, as well as antimicrobial peppermint oil dissolved in propylene glycol. All water-soluble vitamins and the amino acid were completely dissolved in pectin solution. Pectin from citrus peel is a naturally occurring polysaccharide with excellent gelling properties and can stabilize casein micelles. The oil phase was first homogenized at high speed (25,000 rpm for 3 min) into aqueous phase with NaCas from bovine milk as natural emulsifier to obtain a nanoemulsion. Then, the pH value of the nanoemulsion was reduced by the addition of citric acids drops, followed by a thermal treatment at 80°C for 30 min to form complex nanoparticles (Luo, Pan, & Zhong, 2015). The final samples were either freeze-dried or nano spray-dried to produce dry powders of the prepared functional drink. The freeze-dried powders exhibited a flake-like morphology, while the nano spray-dried powders were spherical particles of about 1 to 2 μm size (Fig. 15E). A higher concentration of biopolymeric NaCas/pectin matrix was beneficial to obtain a better morphology of the dry powders.

Transparent sunflower oil emulsions of about 100 nm were also encapsulated in maltodextrin (Büchi Labortechnik AG, 2017b). Eight grams of maltodextrin (DE 19) was dissolved in 80-mL deionized water, and

12-mL sunflower oil was added and high shear homogenized with 8-mL Tween 20 surfactant. Particles with 1.0–2.56 μm size (Fig. 15F) with a recovery yield of approx. Ninety percent of 45-mg sample quantities were obtained using a Nano Spray Dryer B-90 HP. The optimized process parameters were 5.0 μm spray mesh size, 120 L/min drying air flow, and 100°C inlet temperature.

Nunes, Bourbon, Martins, Pastrana, and Vicente (2018) developed nanoemulsions of omega-3 (ω-3) fatty acids with water-soluble lactoferrin as a natural emulsifier and dried the nanoemulsions in a Nano Spray Dryer B-90 HP with a small nebulizer. Omega-3 (ω-3) fatty acids offer health benefits, such as the reduction of cardiovascular disease. The inlet temperature of the spray drier was kept constant at 80°C, and the outlet temperature varied between 40°C and 46°C. The nanoemulsions were produced by high-pressure homogenization (five cycles, 20,000 psi) with 2% (w/w) lactoferrin and 5% (w/w) ω-3 fatty acids. Transmission electron microscopy (TEM) images showed an average nanoemulsion size of about 170 nm. Nano spray drying produced particle aggregates of about 0.8 to 3-μm-sized primary particles. The ω-3 fatty acids were successfully encapsulated. The slow release of the ω-3 fatty acids under different simulated gastrointestinal conditions (e.g., 20% ethanol) is promising for potential incorporation in foods with lipophilic character such as ice cream and mayonnaise.

Recently, Prasad Reddy et al. (2019) reported on the nanoencapsulation of roasted coffee bean oil in whey protein by nano spray drying. The powders showed a smooth and spherical particle morphology with a mean particle size of about 305 ± 100 nm and a homogeneous size distribution. The coffee oil emulsions were formed with a rotor-stator homogenizer operated at 20,000 rpm for 5 min. First, an aqueous wall material solution (0.4%, w/v) was prepared; then, roasted coffee oil and Tween 80 (emulsifier) were added at 3% (w/w) and 5% (w/w), respectively. The viscosity was about 0.9 mPa s. The results were considered promising for the food ingredient and flavor industries to increase the oxidative stability of coffee oil and facilitate controlled release of hydrophobic flavor compounds.

3.4 Nanogels made of egg yolk low-density lipoprotein

Nanogels made of egg yolk low-density lipoprotein (LDL) and polysaccharides are also investigated as nutrient delivery systems with promising encapsulation and release potentials. Zhou et al. (2016) reported a pH- and

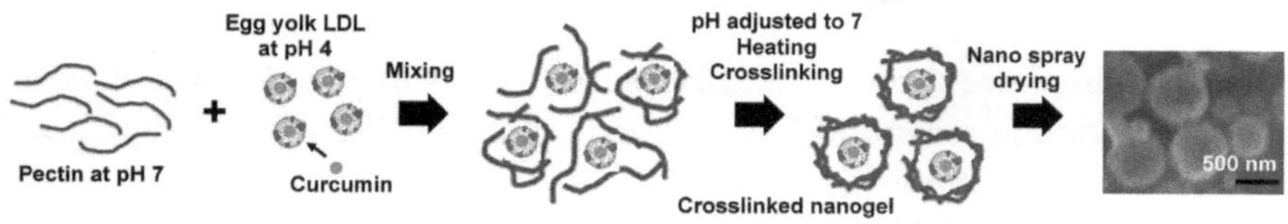

Fig. 16 Schematic illustration of the formulation of egg yolk low-density lipoprotein (LDL)/pectin nanogels. *(Based on Zhou, M., Khen, K., Wang, T., Hu, Q., Xue, J., Luo, Y. (2018). Chemical crosslinking improves the gastrointestinal stability and enhances nutrient delivery potentials of egg yolk LDL/polysaccharide nanogels.* Food Chemistry, 239, *840–847; Zhou, M., Wang, T., Hu, Q., Luo, Y. (2016). Low density lipoprotein/pectin complex nanogels as potential oral delivery vehicles for curcumin.* Food Hydrocolloids, 57, *20–29).*

heat-induced method for the fabrication of such nanogels with a diameter of only 58 nm. The nano spray-dried powders produced from the nanogels exhibited an average particle size of about 1 μm and could be redispersed in water while maintaining their size in the nanoscale. Encapsulation of curcumin, as a model nutrient, into the nanogels was effectively achieved. Fig. 16 illustrates the encapsulation procedure of curcumin in nanogels. Curcumin was first dissolved in ethanol at 2 mg/mL and then added to LDL solutions to reach a curcumin/LDL ratio of 1%. The encapsulated curcumin in the nanogels showed a pH-dependent controlled release profile in simulated acidic gastrointestinal (GI) conditions (pH 2, 4, and 7.5) compared with free curcumin, demonstrating as potential oral delivery vehicles for lipophilic nutrients. The extreme acidic conditions in the human stomach are still a major challenge for further application of egg yolk LDL–based nanogels.

To improve the GI stability, Zhou et al. (2018) chemically cross-linked protein–pectin and protein–carboxymethyl cellulose nanogels by 1-ethyl-3-(3-dimethylaminopropyl) and carbodiimide/*N*-hydroxysuccinimide. Compared with original un-cross-linked nanogels, cross-linking significantly enhanced the resistance against biodegradation by digestive enzymes and aggregation under acidic pH conditions. Moreover, curcumin exhibited a sustained release profile in simulated GI fluids compared with noncross-linked nanogels. The encapsulation efficiency was about 80%, and the particle size and morphology of the nanogels did not change due to cross-linking.

3.5 Solid lipid nanoparticles

Solid lipid nanoparticles (SLNs) are regarded as promising nanoscale delivery systems for the encapsulation of lipophilic bioactives to enhance their stability and improve their bioavailability in foods. SLNs offer a high loading capacity in their lipid core compared with other colloidal delivery systems. Several studies have shown that SLNs can be successfully encapsulated using nano spray drying technology (Wang et al., 2018, 2017; Wang, Hu, Zhou, Xia, et al., 2016; Wang, Hu, Zhou, Xue, et al., 2016; Wang, Ma, et al., 2016; Xue et al., 2017, 2018). Table 13 summarizes possible material combinations for nano spray drying of SLNs.

Wang et al. (2018, 2017) recently developed a new oral delivery platform for lipophilic bioactives from all-natural biomaterials, that is, solid lipid–polymer hybrid nanoparticles (SLPNs), consisting of SLNs as the inner core, bovine serum albumin (BSA)–dextran (BD) Maillard conjugate as natural emulsifier, and pectin as secondary polymeric coating. Fig. 17 shows a schematic diagram of the fabrication process.

Table 13 Process conditions for nano spray drying of solid lipid nanoparticles (SLNs), 100°C/50°C[a] inlet/outlet drying temperature, 120 L/min drying gas flow rate, and water as solvent.

Bioactive substance	Encapsulation wall material	Particle size (μm)	Reference
Curcumin-loaded solid lipid nanoparticles (SLNs)	Stearic acid, sodium caseinate (NaCas) (emulsifier), pectin (stabilizer)	0.5–1.0	Xue et al. (2017, 2018)
Curcumin-loaded SLNs (layer by layer coated)	Compritol ATO 888 (glyceryl behenate), NaCas, pectin	0.5–1.0 3–5	Wang, Hu, Zhou, Xia, et al. (2016) and Wang, Ma, et al. (2016)
Polysaccharide-coated SLNs	Pectin, gum arabic, alginate, CMC	1–10	Wang, Hu, Zhou, Xue, et al. (2016)
Curcumin loaded in caseinate-zein-polysaccharide complex nanoparticles	Zein, caseinate, pectin	1–4	Chang et al. (2017)
Curcumin in solid lipid–polymer hybrid nanoparticles (SLPNs) (SLNs as core and BSA-dextran Maillard conjugate/pectin as shell)	Bovine serum albumin (BSA), dextran, pectin (secondary polymeric coating)	1–1.5	Wang et al. (2018)

[a]Extrapolated from Fig. 11.

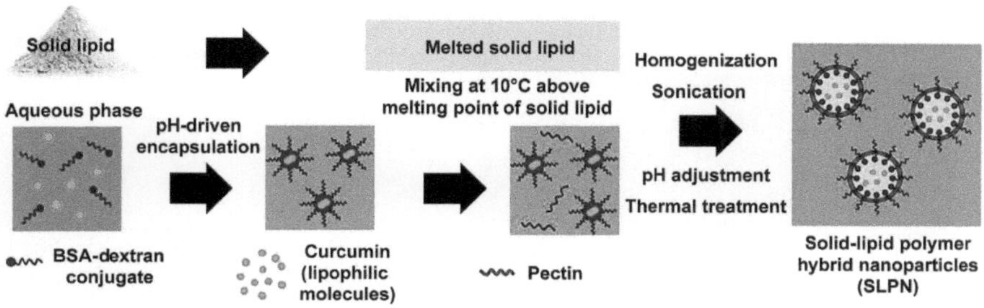

Fig. 17 Schematic diagram of fabrication of pectin-coated and curcumin-encapsulated solid lipid–polymer hybrid nanoparticles (SLPNs). *(Based on Wang, T., Bae, M., Lee, J., Luo, Y. (2018). Solid lipid-polymer hybrid nanoparticles prepared with natural biomaterials: a new platform for oral delivery of lipophilic bioactives.* Food Hydrocolloids, 84, *581–592).*

Two glyceride lipids (i.e., Compritol 888 ATO and Precirol ATO 5) and four saturated fatty acids (i.e., behenic acid, stearic acid, palmitic acid, and myristic acid) were tested together with curcumin as a model lipophilic bioactive. Briefly, the SLPNs were prepared by emulsifying various solid lipids using homogenization and sonication techniques. Pectin solution was mixed with BD solution to form an aqueous phase. Then, the melted solid lipid was added into the aqueous phase under homogenization followed by sonication and cooling down in an ice bath to recrystallize the solid lipid core. The pH of the BD solutions was first adjusted to pH=12 to encapsulate curcumin powder, followed by stirring, and neutralizing to pH=7. The colloidal SLPN samples were then spray-dried in a Nano Spray Dryer B-90 under following operating conditions: 100°C inlet temperature, 120 L/min drying gas flow rate, and 5.5 μm spray mesh size (Wang et al., 2018, 2017). The produced nano spray-dried powders were in the range of 1 to 1.5 μm and showed excellent stability and redispersibility in water, as shown in Fig. 18.

Although the original size of the SLPN particles of about 250 nm was slightly increased in the spray-dried powders, they were able to reassemble their colloidal nanostructures when redispersed in water. The encapsulation efficiency of curcumin in the various solid lipids was in the range of 39% and 82%. SLPNs prepared with glyceryl lipids (ATO 888 and ATO 5) exhibited a higher encapsulation efficiency than fatty acids groups, which was attributed to the higher hydrophobicity of the longer hydrocarbon chains leading to higher accommodation of lipophilic payload. The in vitro release profiles of curcumin from SLPNs under GI environment (pH=2 and pH=7) after 6 h incubation were about 40% to 56% compared with the free curcumin control.

Fig. 18 Characterization of nano spray-dried and water-redispersed solid lipid–polymer hybrid nanoparticles (SLPNs) made of dissolved Compritol 888 ATO, BSA-dextran conjugate, and pectin. The particle size distribution was measured by dynamic light scattering. *(Data from Wang, T., Xue, J., Hu, Q., Zhou, M., Chang, C., Luo, Y. (2017). Synthetic surfactant- and cross-linker-free preparation of highly stable lipid-polymer hybrid nanoparticles as potential oral delivery vehicles.* Scientific Reports, *7, 2750).*

In previous studies, the feasibility of nano spray drying to transform caseinate/lecithin- and polysaccharide-coated colloidal SLNs and nanostructured lipid carriers (NLCs) to fine powders was demonstrated by Wang, Hu, Zhou, Xia, et al. (2016) and Wang, Hu, Zhou, Xue, et al. (2016) (Fig. 19).

Wang, Hu, Zhou, Xia, et al. (2016) prepared "all natural" layer-by-layer redispersible SLNs by nano spray drying. The SLNs were made of Compritol 888 ATO (glyceryl behenate) and NaCas and coated with pectin. Wang, Hu, Zhou, Xue, et al. (2016) coated SLNs made from Compritol 888 ATO with various natural polysaccharides (i.e., pectin, gum arabic, alginate, carboxymethyl cellulose (CMC), and carrageenan) together with lecithin by nano spray drying. Among the five polysaccharides, carrageenan exhibited the best effect on producing small, spherical, and well-separated powders, followed by pectin, alginate, CMC, and gum arabic.

Wang, Ma, et al. (2016) further tested the feasibility of cross-linking the polymeric double layer made of NaCas and pectin for oral delivery of curcumin. The cross-linking significantly improved the physicochemical properties of the SLNs, resulting in a higher encapsulation efficiency and loading capacity, better stability, and slower release profile in simulated GI conditions. The cross-linking further helped forming homogenous nano spray-dried powders of 0.5 to 1 μm size. Overall, the biopolymer coatings played a critical role in producing spherical and uniform dry particles that were redispersible in water (Wang, Hu, Zhou, Xia, et al., 2016; Wang, Hu, Zhou, Xue, et al., 2016; Wang, Ma, et al., 2016). Optimal pH conditions (pH = 5) and thermal treatment (80°C for 30 min) were identified as

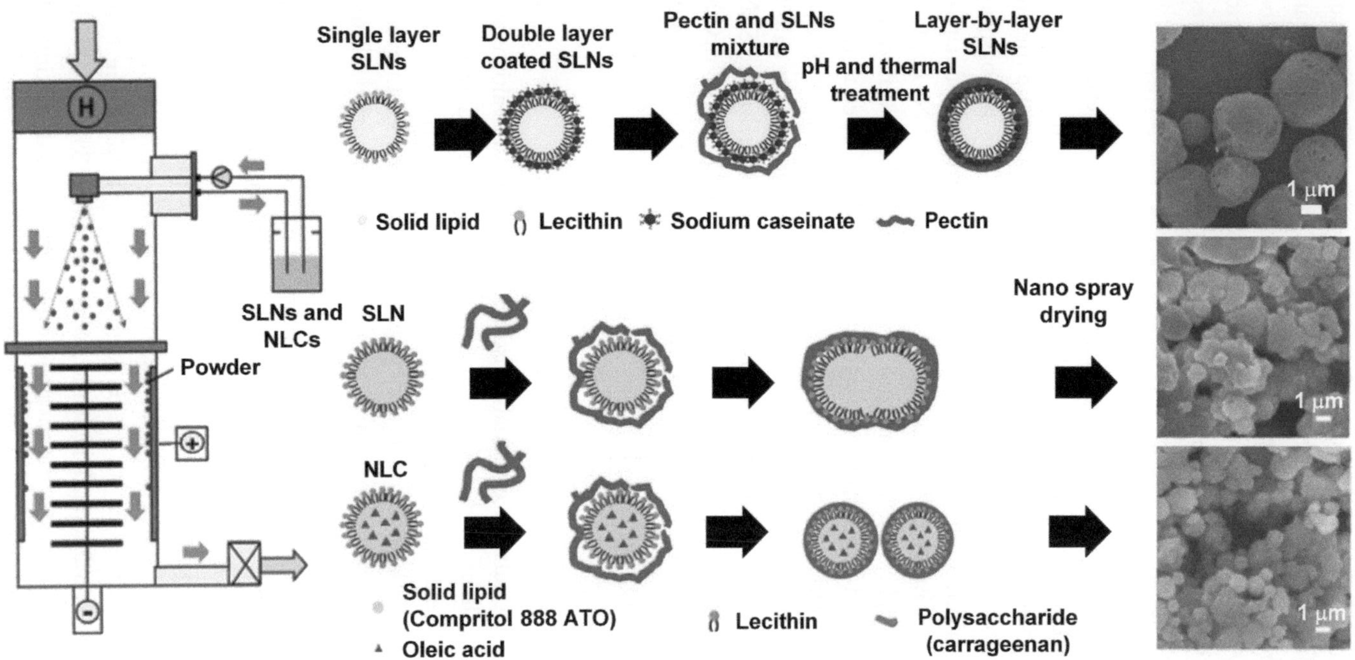

Fig. 19 Schematic diagrams of the fabrication of layer-by-layer coated solid lipid nanoparticles (SLNs) and nanostructured lipid carriers (NLCs) with nano spray drying. *(Based on Wang, T., Hu, Q., Zhou, M., Xia, Y., Nieh, M.-P., Luo, Y. (2016). Development of "all natural" layer-by-layer redispersible solid lipid nanoparticles by nano spray drying technology.* European Journal of Pharmaceutics and Biopharmaceutics, 107, *273–285; Wang, T., Hu, Q., Zhou, M., Xue, J., Luo, Y. (2016). Preparation of ultra-fine powders from polysaccharide-coated solid lipid nanoparticles and nanostructured lipid carriers by innovative nano spray drying technology.* International Journal of Pharmaceutics, 511, *219–222).*

essential to obtain well-separated uniform and spherical particles after nano spray drying (Wang, Hu, Zhou, Xia, et al., 2016). Chang, Wang, Hu, and Luo (2017) proposed zein-caseinate-pectin complex nanoparticles as potential oral delivery vehicles for curcumin, owing to their high encapsulation efficiency, slow release, enhanced antioxidant activity, and exceptional redispersibility after drying. The chemically cross-linked protein–polysaccharide nanoparticles revealed an excellent GI stability.

Xue et al. (2017) fabricated organic solvent-free SLNs by emulsifying melted stearic acid as solid lipid matrix via sonication directly in an aqueous phase containing NaCas and pectin, followed by pH adjustment and thermal treatment to reinforce the polymeric coating network. The nano spray-dried powders exhibited spherical morphology, smooth surface, and narrow size distribution in a range of 500 to 800 nm. The colloidal SLNs could be effectively redispersed in water after nano spray drying without variation of dimension, shape, and morphology. In a follow-up study, Xue et al. (2018) encapsulated curcumin in biopolymer-emulsified SLNs and produced nano spray-dried powders of 0.5 to 1 μm size with uniform distribution and no visible aggregation. To sum up, it can be said that the bioactive-loaded SLNs can be successfully processed into stable powders by nano spray drying, which holds a promising potential for future applications of lipophilic nutrients.

3.6 Salts

Mineral nutrients and salts are essential constituents for human health as they play an important role in the development of bones and teeth and strengthen human muscles. For example, sodium (Na) regulates osmotic pressure, nerve function, acid-base balance, and blood pressure (Oprea & Grumezescu, 2017). Table 14 lists some application examples of nano spray dried salt substances that have a potential use in food applications.

Li et al. (2010) demonstrated the basic ability to nanostructure salt substances by nano spray drying using the example of sodium chloride (NaCl) solutions. Particle sizes from 517 to 993 nm were produced at salt concentrations of 0.1% to 1% (w/w) using a 4.0-μm spray mesh (Fig. 20A). Moncada et al. (2015) used higher concentrations, that is, 3%, 5%, 10%, and 20%, to achieve a higher salt yield. Approximately 80% of the NaCl particles were between 500 and 1900 nm, when drying a 3% (w/v) solution. Fig. 20B and C illustrates the morphology of the fine salt particles. In particular, Moncada et al. (2015) investigated the effect of sensory and

Table 14 Process conditions for nano spray drying of salt solutions (– = not available).

Substance	Food application	Solid concentration (w/v)	*T* in/*T* out (°C)	Drying gas (L/min)	Particle size (μm)	Reference
Sodium chloride (NaCl)	Surface-salted cheese crackers	3%	95/-	125	0.5–1.9	Moncada et al. (2015)
Sodium chloride (NaCl)	Model of water-soluble salt	0.1%–1.0%	100/-	100	0.5–1.0	Li et al. (2010)
Disodium phosphate (Na_2HPO_4)	Anticaking agent, pH adjustment	0.1%	120/-	120	0.5	Schmid et al. (2011)
Tartaric acid ($C_4H_6O_6$)	Acidifier	10%	80/34–46	133	1–4	Büchi Labortechnik AG (2009)
Sodium sulfite (Na_2SO_3)	Food preservative	1%	80–110/43–51	133	0.5–2	Büchi Labortechnik AG (2009)

Fig. 20 SEM pictures of nano spray-dried salt particles obtained with the Nano Spray Dryer B-90. (A) Sodium chloride (NaCl), 0.1% (w/v) aqueous solution (Li et al., 2010); (B) NaCl salt, 3% (Moncada et al., 2015); (C) NaCl salt, 5% (Moncada et al., 2015); (D) disodium phosphate (Na_2HPO_4), 0.1% (Schmid et al., 2011); (E) tartaric acid ($C_4H_6O_6$), 10% (Büchi Labortechnik AG, 2009); (F) sodium sulfite (Na_2SO_3), 1% (Büchi Labortechnik AG, 2009).

microbiological characteristics of surface-salted cheese crackers. Nano spray-dried salts resulted in a significantly higher level of preferred saltiness than the control. Interestingly, the small particles produced by nano spray drying would reduce sodium intake by 25% to 50% if used in surface-salted cheese biscuits. By reducing the particle size, the dissolution rate and the solubility will increase.

Schmid et al. (2011) examined the possibility of producing submicron particles of disodium phosphate (Na_2HPO_4), which is used mostly in foods to adjust pH. Its presence also prevents coagulation in the preparation of condensed milk. Similarly, it is used as an anticaking additive in powdered products. Fig. 20D shows a SEM photograph of nano spray-dried Na_2HPO_4 with a mean particle size of about 500 nm using a highly diluted solution of 0.1% (w/v).

Büchi Labortechnik AG (2009) provides supplementary application notes for sodium sulfite (Na_2SO_3) or tartaric acid ($C_4H_6O_6$), which demonstrates the versatility of nano spray drying. Sodium sulfite, for example, is used in foods as a preservative and antioxidant to prevent the discoloration of dried fruits. Tartaric acid is found in many fruits, especially grapes. Its salt, potassium bitartrate, develops naturally during wine production. It is most likely recognized by wine drinkers as the "wine diamonds," the small crystals that form spontaneously on the cork or bottom of the wine bottle. In the food industry, it is used to acidify many kinds of products like drinks, wine gums, and ice cream. As Fig. 20E and F shows, with these substances at 1% solid concentration, particle sizes are in the range of 0.5 to 4 μm.

4 Conclusions and final remarks

The food industry is increasingly turning to encapsulation technologies to add value to certain bioactive ingredients. Spray drying plays an important role as an encapsulation technology and has evolved considerably in recent years. The main objectives of a food scientist are the production of bioactive-loaded powders with a high yield, maximum encapsulation efficiency and loading, extended storage stability, and controlled release under gastrointestinal conditions.

Innovative spray drying technologies are becoming even more important for food engineers. While the focus was on microencapsulation, the potential of nanoencapsulation is increasingly being investigated. A technological innovation was the introduction of the laboratory-scale Nano Spray Dryer B-90 from the Swiss company Büchi Labortechnik AG in 2009, which enables to produce submicron particles of various bioactive food ingredients with a controllable particle size and adjustable release profile. The implemented core technologies are vibration mesh technology for the production of small droplets of a few microns, electrostatic particle separation for highly efficient collection of submicron particles, and laminar drying gas flow to maintain the stability and activity of heat-sensitive bioactive food ingredients.

Depending on the application, an optimized set of process parameters can be found by trial and error or design of experimental studies. The most important process parameters are the inlet and outlet temperatures, the drying gas flow rate, the spray mesh size, and the solid concentration of the used excipients, stabilizers, and surfactants. This gives food engineers and scientists a whole range of formulation playgrounds. Each application demands a defined set of encapsulation materials that meet the requirements of a particular bioactive food ingredient. The main requirements for the encapsulation wall material are food safety, cost-effectiveness, and water solubility.

The number of publications on nano spray drying of bioactive-loaded nanoparticles for food applications has increased rapidly in the last 10 years. More than 150 contributions have been identified, reviewed, and summarized in this chapter. Nevertheless, the current state of research in nano spray drying for the nanocapsulation of bioactive food ingredients is still at an early stage compared with pharmaceuticals but is constantly evolving. So far, nano spray drying has been successfully applied in the following nanoscale food applications:

- Production of submicron powders from different encapsulating food-grade wall materials. The submicron particle size range is typically reached with diluted solutions of 0.1 to 1% (w/v) solid concentration. Due to the short drying times and the gentle drying conditions, the particles usually have a spherical shape with a smooth or structured surface.
- Encapsulation of water-soluble vitamins (e.g., vitamin B_{12}, B_9), polyphenols (e.g., curcumin, tyrosol, and resveratrol), and extracts (e.g., from guava leaves and saffron) in different wall materials like sodium alginate, gum arabic, whey protein, albumin, pectin, modified starch, chitosan, and maltodextrin.
- Encapsulation of various nanoemulsions (e.g., eugenol oil, peppermint oil, sunflower oil, coffee bean oil, omega-3 fatty acids, and vitamin E acetate) in polymeric wall materials (e.g., gum arabic, lecithin, zein, sodium caseinate, pectin, whey protein, and maltodextrin) and drying to submicron solid particles.
- Drying of nanogels made of egg yolk low-density lipoprotein and polysaccharides with the encapsulation of curcumin.
- Drying of colloidal solid lipid nanoparticles (SLNs) with encapsulated lipophilic bioactive substances (e.g., curcumin) to enhance their stability and bioavailability.
- Nanostructuring of water-soluble salts (e.g., sodium chloride). Tiny salt particles with a size of a few hundred nanometres are formed, which mainly depend on the droplet size (i.e., the size of the spray mesh) and the solid concentration of the feed.

It is foreseeable that the encapsulation of food ingredients by nano spray drying will continue to grow as instruments become more widespread. In particular, this new technology is easy to use and very versatile and allows parameter studies to be carried out on a laboratory scale. Obviously, more applications are to be expected, as nanoscale food applications are an emerging field. The reduction of the spray-dried particle size to the nanoscale offers many new perspectives for the use in functional foods such as carotenoids, polyphenols, minerals, colorants, flavors, antioxidants, and vitamins. Future work will extend the application spectrum, especially with regard to the encapsulation of nanoemulsions containing lipophilic bioactive food ingredients.

In order to further explore the potential of nano spray drying, it is desirable to commercialize this technology on an industrial scale. There is an increasing need to scale up the process to larger powder quantities, for example, by multiplying the number of vibrating mesh nebulizers. Further technological innovations are expected in atomizer technology to further reduce the droplet size.

References

Abdel-Mageed, H. M., Fouad, S. A., Teaima, M. H., Abdel-Aty, A. M., Fahmy, A. S., Shaker, D. S., et al. (2019). Optimization of nano spray drying parameters for production of α-amylase nanopowder for biotheraputic applications using factorial design. *Drying Technology*, *0*, 1–9.

Ahmad, M. I., Ungphaiboon, S., & Srichana, T. (2014). The development of dimple-shaped chitosan carrier for ethambutol dihydrochloride dry powder inhaler. *Drug Development and Industrial Pharmacy*, *41*, 791–800.

Akhavan Mahdavi, S., Jafari, S. M., Assadpoor, E., & Dehnad, D. (2016). Microencapsulation optimization of natural anthocyanins with maltodextrin, gum Arabic and gelatin. *International Journal of Biological Macromolecules*, *85*, 379–385.

Akhavan Mahdavi, S., Jafari, S. M., Assadpour, E., & Ghorbani, M. (2016). Storage stability of encapsulated barberry's anthocyanin and its application in jelly formulation. *Journal of Food Engineering*, *181*, 59–66.

Amsalem, O., Nassar, T., Benhamron, S., Lazarovici, P., Benita, S., & Yavin, E. (2017). Solid nano-in-nanoparticles for potential delivery of siRNA. *Journal of Controlled Release*, *257*, 144–155.

Anandharamakrishnan, C. (2014). Nanoencapsulation of food bioactive compounds. In *Techniques for nanoencapsulation of food ingredients, SpringerBriefs in food, health, and nutrition,* New York: Springer, pp. 1–6 (Chapter 1).

Anandharamakrishnan, C. (2014b). Drying techniques for nanoencapsulation. *In SpringerBriefs in food, health, and nutrition Techniques for nanoencapsulation of food ingredients,* (pp. 51–60). New York: Springer (Chapter 6).

Anandharamakrishnan, C., & Ishwarya, P. S. (2015a). *Spray drying techniques for food ingredient encapsulation*. John Wiley & Sons, Ltd.

Anandharamakrishnan, C., & Ishwarya, P. S. (2015b). Spray drying for nanoencapsulation of food components. In *Spray drying techniques for food ingredient encapsulation* (pp. 180–197). Chichester, UK: John Wiley & Sons, Ltd.

Aquino, R. P., Stigliani, M., Del Gaudio, P., Mencherini, T., Sansone, F., & Russo, P. (2014). Nanospray drying as a novel technique for the manufacturing of inhalable NSAID powders. *The Scientific World Journal*, 1–7. https://doi.org/10.1155/2014/838410, Article ID 838410.

Arpagaus, C. (2007). Part B, Spray drying. In *The laboratory assistant* (pp. 67–87). Flawil: Büchi Labortechnik AG.

Arpagaus, C. (2009). Food technology at the nano scale: Advances in spray drying. *Food Engineering & Ingredients*, *34*, 26–27.

Arpagaus, C. (2011). *Nano spray dryer B-90: Literature review and applications*. best@buchi Information Bulletin, Number 63/2011.

Arpagaus, C. (2012). A novel laboratory-scale spray dryer to produce nanoparticles. *Drying Technology*, *30*, 1113–1121.

Arpagaus, C. (2018a). Nano spray drying of pharmaceuticals. In *IDS'2018—21st international drying symposium, Valencia, Spain, 11–14 September 2018*, (pp. 1–8).

Arpagaus, C. (2018b). Nano spray drying of pharmaceuticals. *American Journal of Nanotechnology and Nanomedicine, 1*, 64–68.
Arpagaus, C. (2018c). A short review on nano spray drying of pharmaceuticals. *Nanomedicine and Nanoscience Research*, 1–5.
Arpagaus, C., Collenberg, A., Rütti, D., Assadpour, E., & Jafari, S. M. (2018). Nano spray drying for encapsulation of pharmaceuticals. *International Journal of Pharmaceutics, 546*, 194–214.
Arpagaus, C., Friess, W., & Schmid, K. (2009). Nano spray drying in laboratory scale. *G.I.T. Lab Journal, 11–12*, 30–31.
Arpagaus, C., John, P., Collenberg, A., & Rütti, D. (2017). Nanocapsules formation by nano spray drying. In S. M. Jafari (Ed.), *Nanoencapsulation technologies for the food and nutraceutical industries* (pp. 346–401). Elsevier Inc.
Arpagaus, C., & Meuri, M. (2010). Laboratory scale spray drying of inhalable particles: A review. *Respiratory Drug Delivery*, 469–476.
Arpagaus, C., Rütti, D., & Meuri, M. (2013). Enhanced solubility of poorly soluble drugs via spray drying. In D. Douroumis & A. Fahr (Eds.), *Drug delivery strategies for poorly water-soluble drugs* (pp. 551–585). John Wiley & Sons, Ltd.
Arpagaus, C., Schafroth, N., & Meuri, M. (2010a). *Laboratory scale spray drying of inhalable drugs: A review.* best@buchi Information Bulletin, Number 59/2010.
Arpagaus, C., Schafroth, N., & Meuri, M. (2010b). *Laboratory scale spray drying of lactose: A review.* best@buchi Information Bulletin, Number 57/2010.
Arpagaus, C., & Schwartzbach, H. (2008). *Scale-up from the Büchi mini spray dryer B-290 to the Niro MOBILE MINOR.* best@buchi Information Bulletin, Number 52/2008.
Assadpour, E., & Jafari, S. M. (2017). Spray drying of folic acid within nano-emulsions: Optimization by Taguchi approach. *Drying Technology, 35*, 1152–1160.
Assadpour, E., & Jafari, S. M. (2019). Advances in spray-drying encapsulation of food bioactive ingredients: From microcapsules to nanocapsules. *Annual Review of Food Science and Technology, 10*, 8.1–8.29.
Assadpour, E., Jafari, S. M., & Maghsoudlou, Y. (2017). Evaluation of folic acid release from spray dried powder particles of pectin-whey protein nano-capsules. *International Journal of Biological Macromolecules, 95*, 238–247.
Augustin, M. A., & Sanguansri, L. (2017). Microencapsulation technologies. In *Engineering foods for bioactives stability and delivery,* New York: Springer Science + Business Media, pp. 119–142, (Chapter 6).
Baba, K., & Nishida, K. (2013). Steroid nanocrystals prepared using the nano spray dryer B-90. *Pharmaceutics, 5*, 107–114.
Beck-Broichsitter, M., Paulus, I. E., Greiner, A., & Kissel, T. (2015). Modified vibrating-mesh nozzles for advanced spray-drying applications. *European Journal of Pharmaceutics and Biopharmaceutics, 92*, 96–101.
Beck-Broichsitter, M., Schweiger, C., Schmehl, T., Gessler, T., Seeger, W., & Kissel, T. (2012). Characterization of novel spray-dried polymeric particles for controlled pulmonary drug delivery. *Journal of Controlled Release, 158*, 329–335.
Blasi, P., Schoubben, A., Giovagnoli, S., Rossi, C., & Ricci, M. (2010). Alginate micro- and nanoparticle production by spray drying. In *Proceedings of meeting on lactose as a carrier for inhalation products, Parma, Italy, September 26–28, 2010* (pp. 137–138).
Brinkmann-Trettenes, U., Barnert, S., & Bauer-Brandl, A. (2014). Single step bottom-up process to generate solid phospholipid nano-particles. *Pharmaceutical Development and Technology, 19*, 326–332.
Büchi Labortechnik AG (2002). *Training papers spray drying*: (pp. 1–19). Order code 97758.
Büchi Labortechnik AG (2009). *Nano spray dryer B-90 application booklet.*
Büchi Labortechnik AG (2010). *Nano spray dryer B-90 Brochure, Flawil, Switzerland. 11592236 en 1012, available http://www.buchi.com.*

Büchi Labortechnik AG (2017a). *Nano spray drying maltodextrin from solution and emulsion.* Application note no. 275/2017.

Büchi Labortechnik AG (2017b). *Sub-micron bovine serum albumin particles*. Application note no. 272/2017.

Büchi Labortechnik AG (2017c). *Curcumin sub-micrometer particles by nano spray drying*. Application note no. 276/2017.

Bürki, K., Jeon, I., Arpagaus, C., & Betz, G. (2011). New insights into respirable protein powder preparation using a nano spray dryer. *International Journal of Pharmaceutics, 408*, 248–256.

Camarena-Tello, J., Martínez-Flores, H., Garnica-Romo, M., Padilla-Ramírez, J., Saavedra-Molina, A., Alvarez-Cortes, O., et al. (2018). Quantification of phenolic compounds and in vitro radical scavenging abilities with leaf extracts from two varieties of Psidium guajava L. *Antioxidants*, 7, 34.

Celli, G. B., Ghanem, A., & Brooks, M. S.-L. (2015). Bioactive encapsulated powders for functional foods—a review of methods and current limitations. *Food and Bioprocess Technology, 8*, 1825–1837.

Cerchiara, T., Abruzzo, A., di Cagno, M., Bigucci, F., Bauer-Brandl, A., Parolin, C., et al. (2015). Chitosan based micro- and nanoparticles for colon-targeted delivery of vancomycin prepared by alternative processing methods. *European Journal of Pharmaceutics and Biopharmaceutics, 92*, 112–119.

Chang, C., Wang, T., Hu, Q., & Luo, Y. (2017). Caseinate-zein-polysaccharide complex nanoparticles as potential oral delivery vehicles for curcumin: Effect of polysaccharide type and chemical cross-linking. *Food Hydrocolloids, 72*, 254–262.

Couvreur, P., Dubernet, C., & Puisieux, F. (1995). Controlled drug delivery with nanoparticles—current possibilities and future trends. *European Journal of Pharmaceutics and Biopharmaceutics, 41*, 2–13.

Dahili, L. A., & Feczkó, T. (2015). Cross-linking of horseradish peroxidase enzyme to fine particles generated by nano spray dryer B-90. *Periodica Polytechnica, Chemical Engineering, 59*, 209–214.

De Cicco, F., Porta, A., Sansone, F., Aquino, R. P., & Del Gaudio, P. (2014). Nanospray technology for an in situ gelling nanoparticulate powder as a wound dressing. *International Journal of Pharmaceutics, 473*, 30–37.

De Vos, P., Faas, M. M., Spasojevic, M., & Sikkema, J. (2010). Encapsulation for preservation of functionality and targeted delivery of bioactive food components. *International Dairy Journal, 20*, 292–302.

Dhand, R. (2002). Nebulizers that use a vibrating mesh or plate with multiple apertures to generate aerosol. *Respiratory Care, 47*, 406–416.

Dimer, F., de Souza Carvalho-Wodarz, C., Haupenthal, J., Hartmann, R., & Lehr, C. -M. (2015). Inhalable clarithromycin microparticles for treatment of respiratory infections. *Pharmaceutical Research, 32*, 3850–3861.

Dimer, F., Ortiz, M., Pohlmann, A. R., & Guterres, S. S. (2015). Inhalable resveratrol microparticles produced by vibrational atomization spray drying for treating pulmonary arterial hypertension. *Journal of Drug Delivery Science and Technology, 29*, 152–158.

Draheim, C., de Crécy, F., Hansen, S., Collnot, E. -M., & Lehr, C. -M. (2015). A design of experiment study of nanoprecipitation and nano spray drying as processes to prepare PLGA nano- and microparticles with defined sizes and size distributions. *Pharmaceutical Research, 32*, 2609–2624.

Durli, T. L., Dimer, F. A., Fontana, M. C., Pohlmann, a. R., Beck, R. C. R., & Guterres, S. S. (2014). Innovative approach to produce submicron drug particles by vibrational atomization spray drying: Influence of the type of solvent and surfactant. *Drug Development and Industrial Pharmacy, 40*, 1011–1020.

Esfanjani, A. F., Jafari, S. M., Assadpoor, E., & Mohammadi, A. (2015). Nano-encapsulation of saffron extract through double-layered multiple emulsions of pectin and whey protein concentrate. *Journal of Food Engineering*, *165*, 149–155.

Estevinho, B. N., Rocha, F., Santos, L., & Alves, A. (2013). Microencapsulation with chitosan by spray drying for industry applications—a review. *Trends in Food Science and Technology*, *31*, 138–155.

Ezhilarasi, P. N., Karthik, P., Chhanwal, N., & Anandharamakrishnan, C. (2013). Nanoencapsulation techniques for food bioactive components: A review. *Food and Bioprocess Technology*, *6*, 628–647.

Fang, Z., & Bhandari, B. (2012). Encapsulation techniques for food ingredient systems. In B. Bhandari & Y. H. Roos (Eds.), *Food Materials Science and Engineering* (pp. 320–348). Blackwell Publishing Ltd.

Fang, Z., & Bhandari, B. (2017). Spray drying of bioactives. Y. H. Roos & Y. D. Livney (Eds.), *Engineering foods for bioactives stability and delivery* (pp. 261–284). *Food engineering series*New York, NY: Springer New York.

Faridi Esfanjani, A., Assadpour, E., & Jafari, S. M. (2018). Improving the bioavailability of phenolic compounds by loading them within lipid-based nanocarriers. *Trends in Food Science & Technology*, *76*, 56–66.

Faridi Esfanjani, A., & Jafari, S. M. (2016). Biopolymer nano-particles and natural nano-carriers for nano-encapsulation of phenolic compounds. *Colloids and Surfaces. B, Biointerfaces*, *146*, 532–543.

Feng, A. L., Boraey, M. A., Gwin, M. A., Finlay, P. R., Kuehl, P. J., & Vehring, R. (2011). Mechanistic models facilitate efficient development of leucine containing microparticles for pulmonary drug delivery. *International Journal of Pharmaceutics*, *409*, 156–163.

Friends of the Earth (2008). *Out of the laboratory and on to our plates—Nanotechnology in food & agriculture*.

Friends of the Earth (2014). *Tiny ingredients big risks—Nanomaterials rapidly entering food and farming*.

Garti, N., & McClements, D. J. (2012). *Encapsulation technologies food ingredients and nutraceuticals*. Cambridge, UK: Woodhead Publishing.

Gautier, S., Arpagaus, C., Schafroth, N., Meuri, M., Deschamps, A., & Maquet, V. (2010). Very fine chitosan microparticles with narrow & controlled size distribution using spray-drying technologies. *Drug Delivery and Technology*, *10*, 30–37.

Geller, D. E. (2008). The science of aerosol delivery in cystic fibrosis. *Pediatric Pulmonology*, *43*, 5–17.

Gharsallaoui, A., Roudaut, G., Chambin, O., Voilley, A., & Saurel, R. (2007). Applications of spray-drying in microencapsulation of food ingredients: An overview. *Food Research International*, *40*, 1107–1121.

Gibbs, B. F., Kermasha, S., Alli, I., & Mulligan, C. N. (1999). Encapsulation in the food industry: A review. *International Journal of Food Sciences and Nutrition*, *50*, 213–224.

Grumezescu, A. M. (2016). *Encapsulations—Nanotechnology in the agri-food industry. Vol. 2*. Amsterdam: Elsevier.

Gu, B., Linehan, B., & Tseng, Y.-C. (2015). Optimization of the Büchi B-90 spray drying process using central composite design for preparation of solid dispersions. *International Journal of Pharmaceutics*, *491*, 208–217.

Harsha, S., Al-Dhubiab, B. E., Nair, A. B., Attimarad, M., Venugopala, K. N., & Sa, K. (2017). Pharmacokinetics and tissue distribution of microspheres prepared by spray drying technique: Targeted drug delivery. *Biomedical Research*, *28*, 3387–3396.

Heng, D., Lee, S. H., Ng, W. K., & Tan, R. B. H. (2011). The nano spray dryer B-90. *Expert Opinion on Drug Delivery*, *8*, 965–972.

Hu, Q., Gerhard, H., Upadhyaya, I., Venkitanarayanan, K., & Luo, Y. (2016). Antimicrobial eugenol nanoemulsion prepared by gum arabic and lecithin and evaluation of drying technologies. *International Journal of Biological Macromolecules*, *87*, 130–140.

Jafari, M. (2017). *Nanoencapsulation of food bioactive ingredients—Principles and applications*. Elsevier.
Jafari, S. M., Assadpoor, E., Bhandari, B., & He, Y. (2008). Nano-particle encapsulation of fish oil by spray drying. *Food Research International*, *41*, 172–183.
Jafari, S. M., Assadpoor, E., He, Y., & Bhandari, B. (2008). Encapsulation efficiency of food flavours and oils during spray drying. *Drying Technology*, *26*, 816–835.
Jafari, S. M., Fathi, M., & Mandala, I. (2015). Emerging product formation. In *Food waste recovery* (pp. 293–317). Elsevier.
Jafari, S. M., Ghalegi Ghalenoei, M., & Dehnad, D. (2017). Influence of spray drying on water solubility index, apparent density, and anthocyanin content of pomegranate juice powder. *Powder Technology*, *311*, 59–65.
Jafari, S. M., He, Y., & Bhandari, B. (2007a). Encapsulation of nanoparticles of d-limonene by spray drying: Role of emulsifiers and emulsifying techniques. *Drying Technology*, *25*, 1069–1079.
Jafari, S. M., He, Y., & Bhandari, B. (2007b). Role of powder particle size on the encapsulation efficiency of oils during spray drying. *Drying Technology*, *26*, 1081–1089.
Jafari, S. M., & McClements, D. J. (2017). Nanotechnology approaches for increasing nutrient bioavailability. In *Advances in food and nutrition research* (pp. 1–30). Academic Press.
Jafari, S. M., Paximada, P., Mandala, I., Assadpour, E., & Mehrnia, M. A. (2017). Encapsulation by nanoemulsions. In *Nanoencapsulation technologies for the food and nutraceutical industries* (pp. 36–73). London, UK: Elsevier.
Jain, I. (2014). *Crosslinking albumin for drug release from spray dried particles* [Master thesis, electronic thesis and dissertations, paper 674]. USA: University of Louisville.
Kaialy, W., & Al Shafiee, M. (2016). Recent advances in the engineering of nanosized active pharmaceutical ingredients: Promises and challenges. *Advances in Colloid and Interface Science*, *228*, 71–91.
Katouzian, I., & Jafari, S. M. (2016). Nano-encapsulation as a promising approach for targeted delivery and controlled release of vitamins. *Trends in Food Science & Technology*, *53*, 34–48.
Knoch, M., & Keller, M. (2005). The customised electronic nebuliser: A new category of liquid aerosol drug delivery systems. *Expert Opinion on Drug Delivery*, *2*, 377–390.
Kyriakoudi, A., & Tsimidou, M. Z. (2018). Properties of encapsulated saffron extracts in maltodextrin using the Büchi B-90 nano spray-dryer. *Food Chemistry*, *266*, 458–465.
Lass, J. S., Sant, A., & Knoch, M. (2006). New advances in aerosolised drug delivery: Vibrating membrane nebuliser technology. *Expert Opinion on Drug Delivery*, *3*, 693–702.
Lee, S. H., Heng, D., Ng, W. K., Chan, H.-K., & Tan, R. B. H. (2011). Nano spray drying: A novel method for preparing protein nanoparticles for protein therapy. *International Journal of Pharmaceutics*, *403*, 192–200.
Li, X., Anton, N., Arpagaus, C., Belleteix, F., & Vandamme, T. F. (2010). Nanoparticles by spray drying using innovative new technology: The Büchi nano spray dryer B-90. *Journal of Controlled Release*, *147*, 304–310.
Li, X., Anton, N., & Vandamme, T. F. (2015). Spray-drying of nano- and microcapsules of nutraceuticals. In J. Shi (Ed.), *Functional food ingredients and nutraceuticals—Processing technologies* (2nd ed., pp. 455–480). *Boca Raton: CRC Press, Taylor & Francis Group*.
Littringer, E. M., Zellnitz, S., Hammernik, K., Adamer, V., Friedl, H., & Urbanetz, N. a. (2013). Spray drying of aqueous salbutamol sulfate solutions using the nano spray dryer B-90—the impact of process parameters on particle size. *Drying Technology*, *31*, 1346–1353.
Livney, Y. D. (2017). Nanoencapsulation technologies. In *Foods for bioactives stability and delivery* (pp. 143–169). New York: Springer Science + Business Media.
Luo, Y., Pan, K., & Zhong, Q. (2015). Casein/pectin nanocomplexes as potential oral delivery vehicles. *International Journal of Pharmaceutics*, *486*, 59–68.

Maa, Y. F., Nguyen, P. A., Sit, K., & Hsu, C. C. (1998). Spray-drying performance of a bench-top spray dryer for protein aerosol powder preparation. *Biotechnology and Bioengineering*, *60*, 301–309.
Mahdavee Khazaei, K., Jafari, S. M., Ghorbani, M., & Hemmati Kakhki, A. (2014). Application of maltodextrin and gum Arabic in microencapsulation of saffron petal's anthocyanins and evaluating their storage stability and color. *Carbohydrate Polymers*, *105*(1), 57–62.
Mahdavi, S. A., Jafari, S. M., Ghorbani, M., & Assadpoor, E. (2014). Spray-drying microencapsulation of anthocyanins by natural biopolymers: A review. *Drying Technology*, *32*, 509–518.
Malapert, A., Reboul, E., Tourbin, M., Dangles, O., Thiéry, A., Ziarelli, F., et al. (2019). Characterization of hydroxytyrosol-β-cyclodextrin complexes in solution and in the solid state, a potential bioactive ingredient. *LWT*, *102*, 317–323.
Masters, K. (1991). *Spray drying handbook* (5th ed.). USA: Longman Scientific & Technical, Harlow.
McClements, D. J. (2015). *Nanoparticle- and microparticle-based delivery systems: Encapsulation, protection and release of active compounds*. Boca Raton, FL: CRC Press.
Mehrad, B., Shabanpour, B., Jafari, S., & Pourashouri, P. (2015). Characterization of dried fish oil from Menhaden encapsulated by spray drying. *AACL Bioflux*, *8*(1), 57–69.
Merchant, Z., Taylor, K. M. G., Stapleton, P., Razak, S. A., Kunda, N., Alfagih, I., et al. (2014). Engineering hydrophobically modified chitosan for enhancing the dispersion of respirable microparticles of levofloxacin. *European Journal of Pharmaceutics and Biopharmaceutics*, *88*, 816–829.
Moghbeli, S., Jafari, S. M., Maghsoudlou, Y., & Dehnad, D. (2019). Influence of pectin-whey protein complexes and surfactant on the yield and microstructural properties of date powder produced by spray drying. *Journal of Food Engineering*, *242*, 124–132.
Moncada, M., Astete, C., Sabliov, C., Olson, D., Boeneke, C., & Aryana, K. J. (2015). Nano spray-dried sodium chloride and its effects on the microbiological and sensory characteristics of surface-salted cheese crackers. *Journal of Dairy Science*, *98*, 5946–5954.
Murugesan, R., & Orsat, V. (2012). Spray drying for the production of nutraceutical ingredients—a review. *Food and Bioprocess Technology*, *5*, 3–14.
Nandiyanto, A. B. D., & Okuyama, K. (2011). Progress in developing spray-drying methods for the production of controlled morphology particles: From the nanometer to submicrometer size ranges. *Advanced Powder Technology*, *22*, 1–19.
Ngan, L. T. K., Wang, S. -L., Hiep, Đ. M., Luong, P. M., Vui, N. T., Đinh, T. M., et al. (2014). Preparation of chitosan nanoparticles by spray drying, and their antibacterial activity. *Research on Chemical Intermediates*, *40*, 2165–2175.
Nguyen, T. V., Nguyen, T. T. H., Wang, S. L., Vo, T. P. K., & Nguyen, A. D. (2017). Preparation of chitosan nanoparticles by TPP ionic gelation combined with spray drying, and the antibacterial activity of chitosan nanoparticles and a chitosan nanoparticle–amoxicillin complex. *Research on Chemical Intermediates*, *43*, 3527–3537.
Nunes, R., Bourbon, A. I., Martins, J. T., Pastrana, L., & Vicente, A. A. (2018). Structural characterization and release profile of omega-3 fatty-acids encapsulated in nanoemulsions. In *FSD 2018 3rd food structure & design conference, University of Debrecen, Hungary, 20–22 September 2018*, (pp. 48–49).
O'Toole, M. G., Henderson, R. M., Soucy, P. A., Fasciotto, B. H., Hoblitzell, P. J., Keynton, R. S., et al. (2012). Curcumin encapsulation in submicrometer spray-dried chitosan/Tween 20 particles. *Biomacromolecules*, *13*, 2309–2314.
Okuyama, K., Abdullah, M., Lenggoro, I. W., & Iskandar, F. (2006). Preparation of functional nanostructured particles by spray drying. *Advanced Powder Technology*, *17*, 587–611.
Okuyama, K., & Lenggoro, I. W. (2003). Preparation of nanoparticles via spray route. *Chemical Engineering Science*, *58*, 537–547.

Oliveira, A. M., Guimarães, K. L., Cerize, N. N., Tunussi, A. S., & Poço, J. G. (2013). Nano spray drying as an innovative technology for encapsulating hydrophilic active pharmaceutical ingredients (API). *Journal of Nanoscience and Nanotechnology*, *4*, 1–6.

Oprea, A. E., & Grumezescu, A. M. (2017). *Nanotechnology applications in food—Flavor, stability, nutrition and safety*. Elsevier.

Pérez-Masiá, R., López-Nicolás, R., Periago, M. J., Ros, G., Lagaron, J. M., & López-Rubio, A. (2015). Encapsulation of folic acid in food hydrocolloids through nanospray drying and electrospraying for nutraceutical applications. *Food Chemistry*, *168*, 124–133.

Porzio, M. (2007). Spray Drying—an in-depth look at the steps in spray drying and the different options available to flavorists. *Perfumer and Flavorist*, *32*, 34–37.

Porzio, M. (2008). Melt extrusion and melt injection an in-depth look at the strengths, limitations and applications of these two processes. *Perfumer and Flavorist*, *33*, 48–53.

Pourashouri, P., Shabanpour, B., Razavi, S. H., Jafari, S. M., Shabani, A., & Aubourg, S. P. (2014a). Oxidative stability of spray-dried microencapsulated fish oils with different wall materials. *Journal of Aquatic Food Product Technology*, *23*(6), 567–578.

Pourashouri, P., Shabanpour, B., Razavi, S. H., Jafari, S. M., Shabani, A., & Aubourg, S. P. (2014b). Impact of wall materials on physicochemical properties of microencapsulated fish oil by spray drying. *Food and Bioprocess Technology*, 7(8), 2354–2365.

Prasad Reddy, M. N., Padma Ishwarya, S., & Anandharamakrishnan, C. (2019). Nanoencapsulation of roasted coffee bean oil in whey protein wall system through nanospray drying. *Journal of Food Processing & Preservation*,e13893.

Quintanilla-Carvajal, M. X., Camacho-Díaz, B. H., Meraz-Torres, L. S., Chanona-Pérez, J. J., Alamilla-Beltrán, L., Jimenéz-Aparicio, A., et al. (2010). Nanoencapsulation: A new trend in food engineering processing. *Food Engineering Reviews*, *2*, 39–50.

Rafiee, Z., Nejatian, M., Daeihamed, M., & Jafari, S. M. (2019). Application of different nanocarriers for encapsulation of curcumin. *Critical Reviews in Food Science and Nutrition*, 1–77.

Rajabi, H., Ghorbani, M., Jafari, S. M., Mahoonak, S. A., & Rajabzadeh, G. (2015). Retention of saffron bioactive components by spray drying encapsulation using maltodextrin, gum Arabic and gelatin as wall materials. *Food Hydrocolloids*, *51*, 327–337.

Rampino, A., Borgogna, M., Blasi, P., Bellich, B., & Cesàro, A. (2013). Chitosan nanoparticles: Preparation, size evolution and stability. *International Journal of Pharmaceutics*, *455*, 219–228.

Ray, S., Raychaudhuri, U., & Chakraborty, R. (2016). An overview of encapsulation of active compounds used in food products by drying technology. *Food Bioscience*, *13*, 76–83.

Research and Markets (2017). *Global bioactive ingredients market—Forecasts from 2017 to 2022*.

Rostamabadi, H., Falsafi, S. R., & Jafari, S. M. (2019). Nanoencapsulation of carotenoids within lipid-based nanocarriers. *Journal of Controlled Release*, *298*, 38–67.

Sarabandi, K., Sadeghi Mahoonak, A., Hamishekar, H., Ghorbani, M., & Jafari, S. M. (2018). Microencapsulation of casein hydrolysates: physicochemical, antioxidant and microstructure properties. *Journal of Food Engineering*, *237*, 86–95.

Schafroth, N., Arpagaus, C., Jadhav, U. Y., Makne, S., & Douroumis, D. (2012). Nano and microparticle engineering of water insoluble drugs using a novel spray-drying process. *Colloids and Surfaces.s B, Biointerfaces*, *90*, 8–15.

Schmid, K. (2011). *Spray drying of protein precipitates and evaluation of the nano spray dryer B-90*. (Ph.D. thesis). Munich: Ludwig-Maximilians-University.

Schmid, K., Arpagaus, C., & Friess, W. (2009). Evaluation of a vibrating mesh spray dryer for preparation of submicron particles. *Respiratory and Drug Delivery*, 323–326.

Schmid, K., Arpagaus, C., & Friess, W. (2011). Evaluation of the nano spray dryer B-90 for pharmaceutical applications. *Pharmaceutical Development and Technology*, *16*, 287–294.

Schoubben, A., Blasi, P., Marenzoni, M. L., Barberini, L., Giovagnoli, S., Cirotto, C., et al. (2013). Capreomycin supergenerics for pulmonary tuberculosis treatment: Preparation, in vitro, and in vivo characterization. *European Journal of Pharmaceutics and Biopharmaceutics*, *83*, 388–395.

Schoubben, A., Giovagnoli, S., Tiralti, M. C., Blasi, P., & Ricci, M. (2014). Capreomycin inhalable powders prepared with an innovative spray-drying technique. *International Journal of Pharmaceutics*, *469*, 132–139.

Smart, J., Berg, E., Nerbrink, O., Zuban, R., Blakey, D., New, M., et al. (2002). TouchSpray technology: comparison of the droplet size with cascade impaction and laser diffraction. *Respiratory and Drug Delivery VIII*, *2*, 525–528.

Sosnik, A., & Seremeta, K. P. (2015). Advantages and challenges of the spray-drying technology for the production of pure drug particles and drug-loaded polymeric carriers. *Advances in Colloid and Interface Science*, *223*, 40–54.

Sun, Y., Song, X., Wang, J., & Yu, J. (2011). Preparation of lithium carbonate hollow spheres by spray pyrolysis. *Crystal Research and Technology*, *46*, 173–177.

Suna, S., Sinir, G. Ö., & Çopur, Ö. U. (2014). Nano-spray drying applications in food industry. *Bulgarian Chemical Communications*, *46*, 137–141.

Suryaprakash, R. C., Lohmann, F. P., Wagner, M., Abel, B., & Varga, A. (2014). Spray drying as a novel and scalable fabrication method for nanostructured CsH2PO4, Pt-thin-film composite electrodes for solid acid fuel cells. *RSC Advances*, *4*, 60429–60436.

Thybo, P., Hovgaard, L., Lindeløv, J. S., Brask, A., & Andersen, S. K. (2008). Scaling up the spray drying process from pilot to production scale using an atomized droplet size criterion. *Pharmaceutical Research*, *25*, 1610–1620.

Torge, A., Grützmacher, P., Mücklich, F., & Schneider, M. (2017). The influence of mannitol on morphology and disintegration of spray-dried nano-embedded microparticles. *European Journal of Pharmaceutical Sciences*, *104*, 171–179.

Vecellio, L. (2006). The mesh nebulizer: A recent technical innovation for aerosol delivery. *Breathe*, *2*, 252–260.

Veneranda, M., Hu, Q., Wang, T., Luo, Y., Castro, K., & Madariaga, J. M. (2018). Formation and characterization of zein-caseinate-pectin complex nanoparticles for encapsulation of eugenol. *LWT*, *89*, 596–603.

Vereman, J., Thysens, T., Derdelinckx, G., Van Impe, J., & Van de Voorde, I. (2019). Extraction and spray drying of class II hydrophobin HFBI produced by Trichoderma reesei. *Process Biochemistry*, *77*, 159–163.

Wang, T., Bae, M., Lee, J., & Luo, Y. (2018). Solid lipid-polymer hybrid nanoparticles prepared with natural biomaterials: A new platform for oral delivery of lipophilic bioactives. *Food Hydrocolloids*, *84*, 581–592.

Wang, T., Hu, Q., Zhou, M., Xia, Y., Nieh, M.-P., & Luo, Y. (2016). Development of "all natural" layer-by-layer redispersible solid lipid nanoparticles by nano spray drying technology. *European Journal of Pharmaceutics and Biopharmaceutics*, *107*, 273–285.

Wang, T., Hu, Q., Zhou, M., Xue, J., & Luo, Y. (2016). Preparation of ultra-fine powders from polysaccharide-coated solid lipid nanoparticles and nanostructured lipid carriers by innovative nano spray drying technology. *International Journal of Pharmaceutics*, *511*, 219–222.

Wang, T., Ma, X., Lei, Y., & Luo, Y. (2016). Solid lipid nanoparticles coated with cross-linked polymeric double layer for oral delivery of curcumin. *Colloids and Surfaces B, Biointerfaces*, *148*, 1–11.

Wang, T., Soyama, S., & Luo, Y. (2016). Development of a novel functional drink from all natural ingredients using nanotechnology. *LWT*, *73*, 458–466.

Wang, T., Xue, J., Hu, Q., Zhou, M., Chang, C., & Luo, Y. (2017). Synthetic surfactant- and cross-linker-free preparation of highly stable lipid-polymer hybrid nanoparticles as potential oral delivery vehicles. *Scientific Reports*, *7*, 2750.

Wang, B., Zhang, W., Zhang, W., Mujumdar, A. S., & Huang, L. (2005). Progress in drying technology for nanomaterials. *Drying Technology*, *23*, 7–32.

Wong, T. W. (2015). Nanospray drying technology: Existing limitations and future challenges. *Recent Patents on Drug Delivery & Formulation*, *9*, 185–186.

Wong, T. W., & John, P. (2015). Advances in spray drying technology for nanoparticle formation. In M. Aliofkhazraei (Ed.), *Handbook of nanoparticles* (pp. 1–16). Springer International Publishing.

Xue, J., Wang, T., Hu, Q., Zhou, M., & Luo, Y. (2017). A novel and organic solvent-free preparation of solid lipid nanoparticles using natural biopolymers as emulsifier and stabilizer. *International Journal of Pharmaceutics*, *531*, 59–66.

Xue, J., Wang, T., Hu, Q., Zhou, M., & Luo, Y. (2018). Insight into natural biopolymer-emulsified solid lipid nanoparticles for encapsulation of curcumin: Effect of loading methods. *Food Hydrocolloids*, *79*, 110–116.

Zhou, M., Khen, K., Wang, T., Hu, Q., Xue, J., & Luo, Y. (2018). Chemical crosslinking improves the gastrointestinal stability and enhances nutrient delivery potentials of egg yolk LDL/polysaccharide nanogels. *Food Chemistry*, *239*, 840–847.

Zhou, M., Wang, T., Hu, Q., & Luo, Y. (2016). Low density lipoprotein/pectin complex nanogels as potential oral delivery vehicles for curcumin. *Food Hydrocolloids*, *57*, 20–29.

Further reading

Schön, M., & Baumgartner, R., (2009). *Apparatus with spray drier and electrostatic separator as well as method for separating particles*. Patent EP 2056037 (A1), priority date: 2007-10-30.

CHAPTER FIVE

Production of food bioactive-loaded nanostructures by micro-/nanofluidics

Talita Aline Comunian[a] and Seid Mahdi Jafari[b]
[a]Department of Food Engineering, School of Food Engineering, University of Campinas, Campinas, Brazil
[b]Department of Food Materials and Process Design Engineering, Faculty of Food Science and Technology, Gorgan University of Agricultural Sciences and Natural Resources, Gorgan, Iran

Contents

Nanoencapsulation of Food Ingredients by Specialized Equipment
https://doi.org/10.1016/B978-0-12-815671-1.00005-6

1 Introduction

Bioactive compounds are naturally found in plant and animal foods, which have beneficial properties for the consumer health. Functional foods—foods rich in bioactive compounds—are, therefore, foods that reduce the risk of various diseases in addition to their basic nutritional functions. The main classes of bioactive compounds are vitamins, carotenoids, natural pigments, essential fatty acids, phenolic compounds, probiotics, and prebiotics (Jafari, 2017, chap. 1).

However, the manipulation and application of these compounds in food products are hampered due to their instability when in contact with adverse conditions such as high temperature, high concentration of oxygen, moisture, and extreme values of pH (Akhavan, Assadpour, Katouzian, & Jafari, 2018; Assadpour & Jafari, 2018). For this reason, bioactive compounds and different ways for their protection have been of great interest to researchers in the food, pharmaceutical, and cosmetic fields.

The nano-/microencapsulation process is an alternative to minimize the problems obtained in relation to the instability of these compounds; besides promoting protection, it also allows greater control of bioactive release at the time/place of interest, masking unpleasant tastes and odors and preventing loss of compounds by evaporation (volatile compounds) (Faridi Esfanjani & Jafari, 2016). The concept of encapsulation arose from the cellular model in which the membrane surrounds and protects the cytoplasm, in addition to controlling the entry and exit of compounds from the cell. The first reported studies are from the 1930s to 1940s in which one of the major precursors of the area encapsulated oil in gelatin (Vandegaer, 1973).

There are several techniques of nano- and microencapsulation very widespread in the area of Food Science, such as spray-drying (Carmo et al., 2018; Encina et al., 2018; Francisco et al., 2018), spray chilling (Oriani et al., 2018; Silva et al., 2018), complex coacervation (Comunian, Nogueira, et al., 2018; Duhoranimana et al., 2018), ionic gelation (Moura, Berling, Germer, Alvim, & Hubinger, 2018; Vasile, Judis, & Mazzobre, 2018), liposome entrapment (Azzi, Jraij, Auezova, Fourmentin, & Greige-Geres, 2018; Ota et al., 2018), and high-pressure homogenization (Akbas, Soyler, & Oztop, 2018; Juttulapa, Piriyaprasarth, Takeuchi, & Sriamornsak, 2017). However, one technique that has aroused the interest of scientists in the field of encapsulation with limited applications to food ingredients is nano-/microfluidic (Comunian, Abbaspourrad, Favaro-Trindade, & Weitz, 2014; Comunian,

Nogueira, et al., 2018; Comunian et al., 2017; Michelon, Oliveira, Furtado, Torre, & Cunha, 2017) to obtain food bioactive-loaded nano-/ microstructures.

The technique of nano-/microfluidic is a promising method based on flows of multiple fluids, being possible the elaboration of simple water-in-oil (W/O) or oil-in-water (O/W) emulsions, water-in-oil-in-water (W/O/W) or oil-in-water-in-oil (O/W/O) double emulsion, or multiple compartment emulsions (W/O/W/O or O/W/O/W). From these emulsions, depending on the type of material used in the middle phase (in the case of double or multiple emulsions), it is possible to combine other techniques and obtain solid particles, as will be discussed in this chapter. All of these systems are very common in the food industry. Systems such as mayonnaise and milk are examples of simple O/W emulsions. Margarines and butters are examples of W/O emulsions. In the case of "light" products, W/O/W double emulsions are used as water replaces part of the oil phase. There are also products that are bases of emulsions composed of air in one of its phases, such as ice cream and chocolate, in which air-in-water or air-in-oil systems are essential for the sensorial characteristics of the final product (Muijlwijk, Berton-Carabin, & Schroën, 2016). The basic principle in nano-/ microfluidics is the interfacial tension between the different fluids, which allows the formation of spherical droplets. Moreover, the particle size and the release of the encapsulated bioactive can be controlled very accurately according to the materials, type of device, and flows used. Another important factor in this method is the production of highly monodisperse drops and, consequently, nano-/microcapsules that can be manipulated with an impressive degree of accuracy (Datta et al., 2014). The fact that this technique produces monodisperse drops is an important feature for the stability of the system since it is not necessary to use a large amount of emulsifier, besides resulting in a lower rate of creaming, flocculation, and sedimentation (Muijlwijk et al., 2016).

Application of the nano-/microfluidic technique is very broad, covering research in the field of chemistry, medicine, and biology. Its application in the area of Food Science is still recent and is focused on the development of systems that promote the protection and controlled release of bioactive compounds. It is still a laboratory scale technique but with great potential to scale up in the future (Maan, Nazir, Khan, Boom, & Schroën, 2015).

In accordance with the earlier explanations, the main elements of a microfluidic/nanofluidic device, the types of devices, combined methods, and the materials used in this technique will be discussed in the next sections.

2 Nano versus microencapsulation by nano-/microfluidics

As mentioned throughout the chapter, several bioactive compounds are encapsulated in order to obtain their greater protection, controlled release, and application in several products. The encapsulation process involves the formation of structures within the nano- and microscale, and this classification is based basically on the size of these structures. Besides the size, other important factor is the structure, obtaining these types of systems: capsules, emulsions and nanoemulsions, and particles and nanoparticles. It is known that the smaller the particle size of the encapsulated systems, the better their application in food products due to the nonperception in the mouth by the consumer and occurrence of no changes in the texture and flavor of the final product. However, even in the microscale, it is possible to apply these structures when they have a size of up to 100 μm (Comunian & Favaro-Trindade, 2016).

In addition, another major difference among these structures is where the bioactive compound is located. In the case of capsules (micro- or nanocapsules), the active (or encapsulated) material is surrounded by the encapsulating material(s) that is also known as a core-shell structure. This type of structure is known as a "true capsule," since all material is protected. In the case of emulsions or nanoemulsions, in addition to being considered one of the stages of capsule formation in some cases, they are systems in which at least two immiscible phases are in equilibrium due to the addition of an emulsifying agent, as already mentioned. In the case of nano- or microspheres, the encapsulated material is dispersed in the encapsulating matrix. In this case, the encapsulating material is called carrier material, since the active compound is not surrounded by a protective layer but dispersed throughout the system, being found both inside and on the surface of the particle.

So, it can be stated that nano-/microfluidics encapsulation offers potential for the application of these materials in foods due to the great control obtained by this technique, as presented in the previous sections. In the case of obtaining bioactive-loaded nano- and microcapsules by different methods such as complex coacervation or combined methods (complex coacervation with spray-drying and freeze drying), the size and the morphology of the microcapsules are not controlled and sometimes present a low encapsulation efficiency. On the other hands, nano- and microcapsules obtained by microfluidic devices can be highly controlled due to the use of different flow rates, obtaining structures with desired particle sizes, monodisperse, and with high values of encapsulation efficiency. Moreover, it is possible to add one or more compounds in the inner phase in the nano- and microcapsules obtained by microfluidic devices, encapsulating hydrophilic and hydrophobic materials in the same system (Sanjay et al., 2017). So, based on explanations in the previous sections and according to Sanjay et al. (2017), microfluidics is a powerful tool to obtain nano- and microcapsules of desired characteristics (composition, size, and particle size distribution), demonstrating a great potential for bioactive compound-loaded carrier synthesis and controlled release. There are different researches that studied the encapsulation and controlled release of bioactive compounds encapsulated within nano- and microcapsules by microfluidic devices, such as encapsulation of rhodamine B and insulin (Zhang et al., 2013), bovine serum albumin (Liu et al., 2013; Pessi, Santos, Miroshnyk, Yliruusi, & Weitz, 2014; Yeh, Chen, & Lin, 2011), ascorbic acid (Comunian et al., 2014), echium oil (Comunian et al., 2017), and natural pigments (Comunian, Ravanfar, Alcaine, & Abbapsourrad, 2018).

In the case of emulsions—colloidal dispersion multiphase systems—nanoemulsions present small droplets in the range of 50–500 nm, and emulsions can reach up to 100 μm (McClements & Jafari, 2018a, chap. 1, 2018b; Sanjay et al., 2017). Nanoemulsions exhibit a higher stability due to their smaller droplet size, resulting in lower agglomeration and coalescence rates. These types of systems are suitable for the encapsulation and controlled release of several bioactive compounds since they present a high surface area, tunable rheology, easy application in different types of food products, and efficient penetration of the encapsulated compounds (Jafari, Paximada, Mandala, Assadpour, & Mehrnia, 2017, chap. 2). Moreover, there are a lot of food-grade ingredients for use in nanoemulsion formulations (Durukan, Kahraman, Parlevliet, Geistbeck, & Seyhan, 2016; Patel & Joshi, 2012; Sanjay et al., 2017; Sun, Dai, Liu, & Gao, 2016; Uluata, Decker, & McClements, 2016; Yildiz, Demirkesen, & Mert, 2016).

There are various technologies for the preparation of single and double emulsions such as conventional homogenizers (rotor-stator systems), ultrasonication, and high-pressure homogenization (Jafari, Assadpoor, He, & Bhandari, 2008b); however, as already mentioned in previous sections, the microfluidic technique can be considered as one of the most controlled methods to obtain this type of system. Several researches have been carried out with the aim of encapsulating important bioactive compounds by nano-/microfluidic devices. An interesting research was done by Khan et al. (2015) when they studied a method for synthesis of poly(acrylamide) microparticles, and, for that, the production of nanoemulsion was necessary. According to these authors, for the production of these nanoemulsions, it was necessary an elongational flow micromixer linked to a microfluidic device.

And finally, the particles and nanoparticles, which have aroused great interest of researchers when obtained by nano-/microfluidics devices, have been studied in different researches. Mostly this class of structure is formed by the use of lipids or polymers due to their biocompatibility with food products, in addition to their low toxicity (Sanjay et al., 2017). There are several researches focused on obtaining particles by microfluidic devices that can be applied in the area of Food Science in the future (Leung et al., 2012; Majedi et al., 2013; Shamsi, Zahedi, Ghourchian, & Minaeian, 2017; Zhigaltsev et al., 2012).

All structures mentioned herein (in nano- and microscale) are produced with different encapsulating materials (lipid of high and low melting point and biopolymers) and analyzed by different characterization techniques (optical, confocal, and scanning electron microscopy; particle size; encapsulation efficiency; stability; zeta potential; and controlled release). In this way, due to their particularities and importance, the following sections will address these materials and analyze in more detail.

3 Elements of a micro-/nanofluidic device

Nano-/microfluidic devices consist of systems in which a channel must be in the range of micrometer; the disperse and continuous phases of emulsions flow inside these channels. The devices are set of channels molded or built on a base, which can be made of glass or PDMS. Channels are interconnected to allow the possible flow of materials in different directions in order to achieve a specific objective. Liquids or gases are injected and removed from the nano-/microfluidic devices using tubing, syringes, and pumps or hydrostatic pressure.

Different materials are used in the manufacture of micro- and nanofluidic devices. Jena, Yue, and Lam (2012) used cyclic olefin copolymer (COC) in the manufacture of microfluidic devices in order to study various process parameters to develop a low-cost alternative system. These authors used a not embossing technique to construct the devices and evaluated the embossing temperature, load, the holding time, and the demolding temperature and demonstrated a new type of system, with the optimal parameters:

embossing temperature for COC, 10°C above its T_g (glass transition temperature); optimal embossing load, 2.94kN; and holding time, 180s. In another interesting research, Eusner, Hale, and Hardt (2010) studied the use of polymethylmethacrylate (PMMA) in the fabrication of microfluidic devices. These authors also used the hot embossing technology to develop the system and showed that it is a repeatable process, capable of producing quality parts at different operating conditions and for features on the order of 50 μm.

Nie, Xu, Seo, Lewis, and Kumacheva (2005) studied the production of polymer particles with various shapes and morphologies in microfluidic devices fabricated in polyurethane (PU) elastomer by using standard soft lithography. These authors reported the production of polymer spherical capsules and particles with nonspherical shapes using silicon oil and monomer in the oil and aqueous phases, respectively. Kawakatsu, Kikuchi, and Nakajima (1997) studied the production of regular-sized cell in microchannel produced with silicon designed and prepared by using semiconductor technology. These authors showed a method for both O/W and W/O emulsion cell production, showing the possibility for creating artificial biological cells.

The production of monodisperse double emulsions in microfluidic devices was already studied in 2004 by Okushima, Nisisako, Torii, and Higuchi. These authors evaluated two chip configurations: (a) one which the hydrophobic junction and the hydrophilic junction are on the same plate and (b) one which these junctions are on separate chips. They fabricated devices on Pyrex glass by repeated isotropic etching and showed that the size of the droplets (nano- or microscale) and the number of internal droplets (nano- or microscale) can be influenced by flow conditions. They also showed that emulsions with internal droplets of two colors were produced using this type of material and devices.

Michelon et al. (2017) studied the production of liposomes using hydrodynamic flow-focusing microfluidic devices. In this case, these authors encapsulated β-carotene into liposomes with a high encapsulation efficiency (approximately 60%) using a microfluidic device obtained by the conventional soft lithography technique. Moreover, the device was casted with a polydimethylsiloxane (PDMS) prepolymer and curing agent mixture and degassed at 6Pa for 1h. In another research, Nisisako, Torii, and Higuchi (2004) studied the formation of acrylic monomer droplets. The devices were prepared with quartz glass in T-shaped and sheath-flow channels, obtaining particles with the same size.

Based on these studies, it is observed that there is a wide range of materials that can be used in the construction of nano- and microfluidic devices. However, the two more common materials used in devices for food bioactive encapsulation and controlled release are glass and polydimethylsiloxane, which we are going to detail them in the next sections.

4 Glass nano-/microfluidic devices

Glass nano- and microfluidic devices are a class of device composed of cylindrical capillaries inserted into a square capillary, whose inner dimension is slightly larger than the outer diameter of the cylindrical capillaries (Datta et al., 2014). The construction of these devices depends on the type of emulsion that will be produced: single, double, or multiple emulsions.

4.1 Single emulsions

In the case of single W/O or O/W emulsions, only one internal capillary within the square capillary is enough for droplet formation. In this way, the dispersed phase (internal phase) and the continuous phase (external phase), in the liquid format, flow through channels—the dispersed phase into the continuous phase—resulting in monodisperse droplets (Fig. 1) at the moment in which the two flows meet each other at the pore opening; the formation of the droplets is done one by one.

Obtaining the single emulsions as well as double emulsions and multiple compartments, as will be seen in the succeeding text, is dependent on a number of factors, such as the size of nozzle and diameters of the cylindrical and square capillaries, flow rates of dispersed and continuous phases, composition of the emulsion (lipids, proteins, carbohydrates, and surfactants), phase viscosities, and wettability of the devices (Maan et al., 2015).

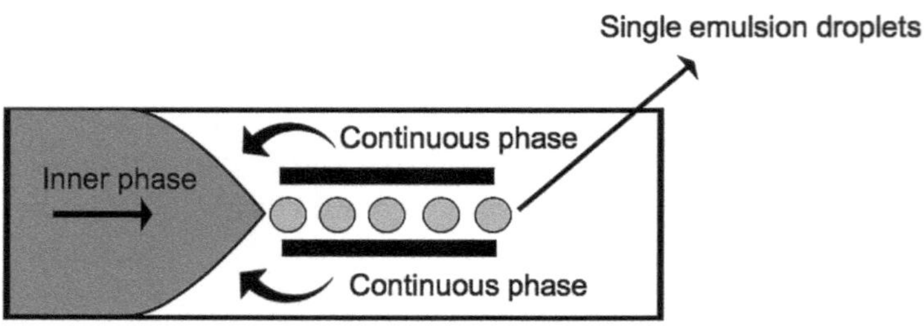

Fig. 1 Scheme of the microfluidic device for the production of single emulsions (flow focusing system).

The larger the diameter and nozzle of the capillaries, the larger the droplet size of the emulsions, since the volume of internal phase and outer phase will be larger. In the case of flows, the droplet size is extremely dependent on the relationship between the dispersed phase and the continuous phase flows. In general, the larger the flow of the continuous phase, the smaller the droplet size; in contrast, the higher the dispersed phase flow, the higher the amount of encapsulated bioactive compound. However, it must be taken into account that the higher this flow, the more difficult to maintain the stability of the system. For this reason, the ideal to obtain the right flows is to start the studies with the dispersed phase flowing at a low speed and the continuous phase at medium speed and increase the velocities of flows of both phases until obtaining a stable system and ideal particle size for protection of bioactive compounds and future applications. Since the main purpose is to obtain small droplets with large amount of the encapsulated material (a high encapsulation efficiency), it is allowed to increase the flow of the internal phase as long as its stability is not lost.

In the case of phase viscosity, it is important to analyze the material to be used prior to the injection of the solutions into the capillaries. High viscosity for both phases (dispersed and continuous) results in larger droplet sizes. However, even for obtaining larger size particles, continuous phase high viscosity prevents rapid collapse resulting in monodisperse droplets (Maan et al., 2015; Nie et al., 2008). Regarding to wettability, according to Hughes et al. (2013), the surface should be wetted by the continuous phase to keep the formation of droplets stable since the contact with the wall is very small; otherwise the system will not work properly. It is important to note that the effect of wettability is directly associated with the size of capillaries.

According to Favaro-Trindade, Pinho, and Rocha (2008), the ideal particle size for food application should be below 100 μm. In this way, it can be affirmed that single emulsions obtained by glass microfluidic devices present great potential for food applications since it is much easier to control only the two-phase flow (dispersed phase and continuous phase) in comparison with a larger number of phases from other types of emulsions. In other words, with greater flow control, it is possible to control the particle size of systems loaded with bioactive compounds, at both nano- and microscale.

4.2 Double emulsions

For the formation of W/O/W or O/W/O double emulsions, the device presents a different configuration from that used for the formation of single emulsions. In this case, two internal capillaries are required, and these tapered cylindrical capillaries are inserted into the opposite ends of a square capillary. Again, the square capillary inner dimension is slightly larger than the outer diameter of the cylindrical capillaries (Fig. 2). Particular attention is required for the construction of the double emulsion devices in relation to the alignment, which the cylindrical capillaries are positioned coaxially. If the alignment of the two internal capillaries is not exactly in the same direction, there will be deviation of the flows, and it will not be possible to form stable systems.

For the formation of double emulsions, the inner phase (oily in the case of O/W/O emulsions or aqueous in the case of W/O/W emulsions) is injected through the left inner cylindrical capillary. In order for the drop to be inserted into the intermediate phase, this phase (aqueous in the case of O/W/O emulsions or oily in the case of W/O/W emulsions) is injected in the same direction of inner phase, however, between the internal capillary and the square. The continuous phase is injected in the opposite direction to these two flows through the space formed between the square and the capillary (Fig. 2). The three flows meet at the tip of tapered capillary and form the drops when used at the ideal flow rates. Monodisperse droplets are formed since a system called dripping regime is used.

An important factor is to remember the density difference between the inner and intermediate phases. Double emulsions formed by nano- and microfluidic devices produce monodisperse droplets, however with varying thickness, depending on the flow rates. When the inner droplet is lighter than the intermediate one, the first rises; in case the inner droplet is denser than the intermediate droplet, the opposite occurs. In these cases, the thickness of the double emulsion is of extreme importance to maintain the stability of the system. If it is very thin, the system can be broken due to its fragility; however, if it presents a considered thickness, it can present a great

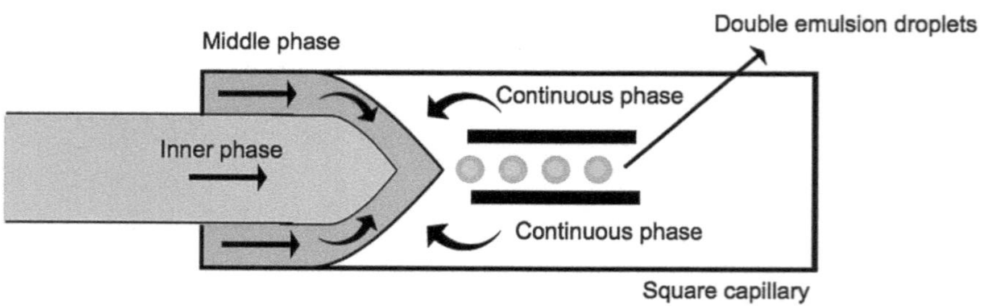

Fig. 2 Scheme of the microfluidic device for the production of double emulsions.

difference of density comparing with the external droplet, thus breaking it. For this reason, the thickness needs to be studied in depth, acquiring its ideal size for each system and materials used.

Depending on the material used in the intermediate phase of the double emulsions, it is possible to obtain solid particles from polymerization reactions (when there is an aqueous phase) or solidification of fatty acids (when there is an oil phase). In the first case, it is necessary to have a monomer and a cross-linker or photoinitiator for the formation of a polymer. In the second case, high melting point lipids are used. In order to allow the injection of this phase, the material is heated during the whole passage through the capillary, remaining in the liquid state; the droplets are collected at a low temperature, for example, in an ice bath, causing immediate solidification due to variation of temperature.

4.3 Multiple emulsions

Multiple emulsions are not as common as single and double emulsions as mentioned earlier. They are composed of four phases, which can be either water-in-oil-in-water-in-oil (W/O/W/O) or oil-in-water-in-oil-in-water (O/W/O/W). Nano-/microfluidics is the essential way to produce this type of emulsion in one step. In this case, instead of injecting one inner phase in the capillary, two or more inner phases can be injected together in different capillaries. The multiple system is formed in the same way that single and double emulsions are formed: at the tip of the tapered capillary. The great advantage of this system is the encapsulation of different compounds—hydrophilic and hydrophobic—within the same emulsion. However, it is more difficult to maintain its stability for a long period of time since droplets of different compositions and densities are within a single system, which facilitates their rupture, leading to the coalescence of the droplets and release of the materials that would be protected.

4.4 Coflow and flow-focusing nano-/microfluidic devices

As shown in Figs. 1 and 2, the formation of single, double, or multiple emulsions can occur with different flow types: coflow or flow focusing. In the coflow system, all the phases of an emulsion flow in the same direction into the capillary(s) and between the internal capillary and the square (Fig. 3A and B). As mentioned in the previous sections, the faster the flow, the easier the internal droplet phase to break in droplets due to the formation of a thin film (Utada, Fernandez-Nieves, Stone, & Weitz, 2007), influencing

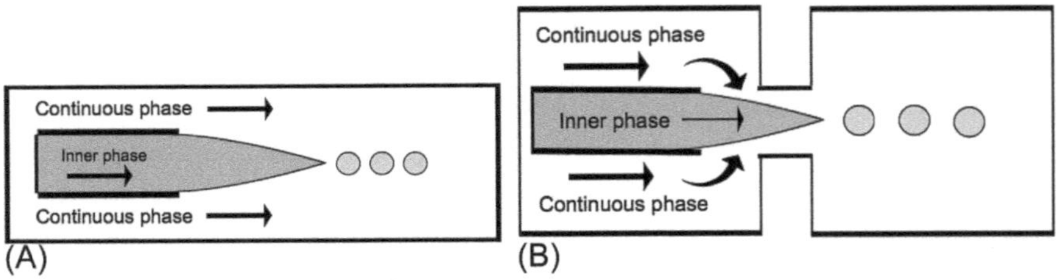

Fig. 3 Schemes for the production of single emulsions by microfluidics in coflowing junction (A) and flow focusing junction (B).

the particle size of the systems. In other words, the faster the flow of the continuous phase, the smaller the droplet size. On the other hand, the faster the dispersed phase flow, the larger the droplet size.

The system presented in flow-focusing devices differs by the direction in which the phases of an emulsion are injected. In the same way as in the coflow system, the phases of the emulsion, whether the system is single, double, or multiple, are forced through a small orifice (Garstecki, Gañán-Calvo, & Whitesides, 2005). However, in flow-focusing devices, the phases are injected in opposite directions (Figs. 1 and 2). That is, the dispersed phase flows through the internal capillary, and the continuous phase flows between the inner capillary and the square in the opposite direction to the dispersed phase (in the case of single emulsions). For double systems such as W/O/W or O/W/O emulsions, the internal and intermediate phases flow in the same direction, while the continuous phase flows in the opposite direction (Fig. 2). According to Nie et al. (2005), due to the shear stress and pressure exerted by the continuous phase, the internal phase ruptures, forming the droplets of the emulsions. The main advantage of the flow-focusing system is the possibility of producing emulsions with smaller droplet sizes than the capillary orifice, avoiding the clogging of the system, and expanding the application areas of this technique (Vladisavljevic et al., 2013).

5 Polydimethylsiloxane (PDMS) nano-/microfluidic devices

In addition to the glass devices presented earlier, another type of commonly used material, as already mentioned, is the PDMS. The manufacture of this type of device involves more steps than the manufacturing process of the glass devices, besides being necessary knowledge in design programs such as Autocade. However, even with more detailed and robust manufacturing, this type of device has attracted the attention of researchers for producing droplets, emulsions, and stable systems and for food bioactive encapsulation.

According to Vladisavljevic et al. (2013), the manufacturing technique used in the production of the PDMS devices is soft lithography, which produced a master with positive relief features. This master can be used to produce many PDMS replicas. Initially, the master is produced using a spin-coating negative photoresist onto a substrate and is exposed to UV light. Then, catalyst and curing agent are added in the master, and the mixture is submitted to high temperature and, after that, peeled off the master. To finalize the process, the mold can be sealed in a glass slide or another PDMS block (Vladisavljevic et al., 2013).

Different types of PDMS devices can also be obtained. In this case, the cylindrical, planar, and three-dimensional channels deserve prominence. In the first example, there is a material in which T-shaped and S-shaped channels can be applied, and there is no limitation on substrate size or channel length that can be generated (Vladisavljevic et al., 2013; Wilson et al., 2011). In the second example, nonspherical particles are produced due to droplet contact with the channel wall and presents less problems related to fragile interface and wettability (Kumacheva & Garstecki, 2011; Takeuchi, Garstecki, Weibel, & Whitesides, 2005; Vladisavljevic et al., 2013). The last one has the same mechanism as glass microfluidic devices, which a dispersed phase is surrounded by the continuous phase. Moreover, in the three-dimensional PDMS devices, it is possible to insert glass capillaries (Takeuchi et al., 2005; Vladisavljevic et al., 2013).

The formation of droplets in PDMS nano-/microfluidic devices is influenced by the geometry of the devices: T-, X-, and Y-junction. The simplest structure is T-junction, which the device is in the form of the letter T (Fig. 4A), and the dispersed phase is injected through the orthogonal channel and flows in the same direction of the continuous phase. Moreover, T-junction can also be applied when the continuous phase is injected through the orthogonal channel and the dispersed phase injected into the main channel, the two streams flowing in the same direction (Fig. 4B). Still with the T-junction geometry, there is a so-called head-on geometry, which the dispersed phase and the continuous phase flow in opposite directions (Fig. 4C) (Shui, Mugele, Van den Berg, & Eijkel, 2008; Van der Graaf, Steegmans, Van der Sman, Schroën, & Boom, 2005; Vladisavljevic et al., 2013; Xu, Li, Tan, Wang, & Luo, 2006). The formation of the droplet occurs basically due to the shear stress formed in the continuous phase flow in combination with the dispersed phase pressure, resulting in its distortion and disruption, forming the emulsions. An advantage

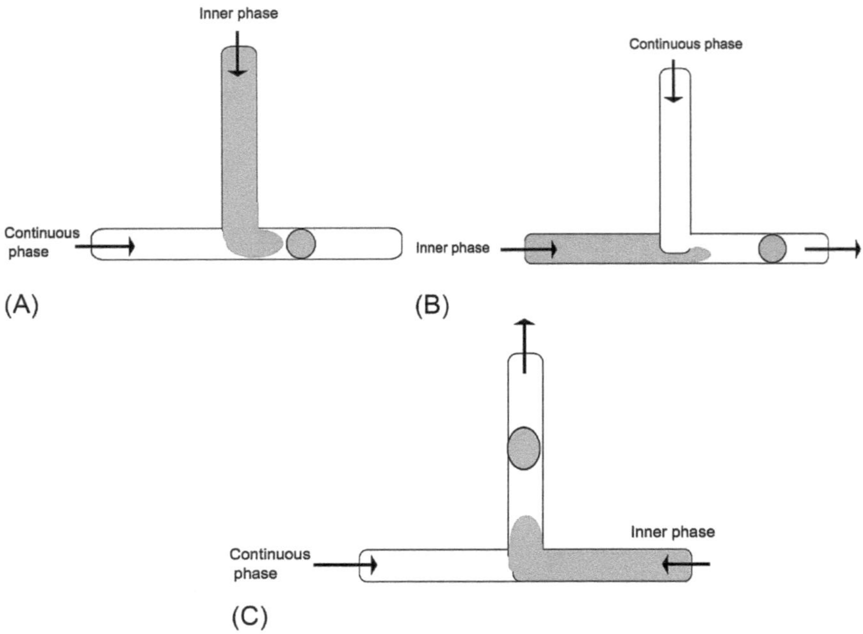

Fig. 4 Schemes for the production of single emulsions with T-junction, where (A) the dispersed phase is injected through the orthogonal channel and flows in the same direction of the continuous phase, (B) continuous phase is injected through the orthogonal channel and the dispersed phase injected into the main channel, and (C) dispersed phase and the continuous phase flow in opposite directions.

of the devices with T-junction geometry is that O/W and W/O emulsions can be formed, depending on the materials used for their stabilization and wettability of the channel walls (Vladisavljevic et al., 2013). In addition to the emulsions obtained by glass nano-/microfluidic devices, the droplet size obtained in that geometry can be controlled from the flow rates of each phase of the emulsions. However, instead of syringe pumps, microvalves can be used, and in this case, the flows are controlled by opening and closing these valves, varying the opening time and cycle time of the valve (Lin & Su, 2008; Vladisavljevic et al., 2013).

In the case of X- and Y-junction (Fig. 5A and B), the systems change a little bit. In the first case (cross junction, Fig. 5A), the disperse phase flows against two streams of the continuous phase and then flows in the same direction. In the second case the device is in the format of the letter "Y" (Fig. 5B), where the dispersed and the continuous phase are injected into separate channels and meet each other, forming O/W or W/O emulsion droplets. Unlike the T-junction system, the droplet size formed in the Y-junction system does

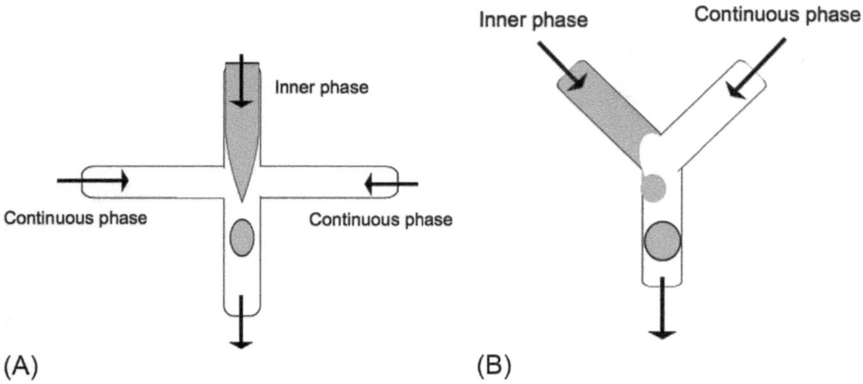

Fig. 5 Schemes for the production of single emulsions with (A) X- and (B) Y-junction.

not depend on the composition of each phase and the flow rates (Steegmans, Schroën, & Boom, 2009; Vladisavljevic et al., 2013).

6 Combined encapsulation methods by micro-/nanofluidics

In addition to the production of emulsions as mentioned earlier, the combination of techniques is being explored with the aim of producing more rigid and/or more stable systems, resulting in greater protection of encapsulated bioactive compounds.

The use of nano-/microparticles to stabilize an emulsion system is called Pickering emulsion. This type of emulsion can be applied in nano-/microfluidic through addition of nano-/microparticle solutions in one of the phases. Ravanfar, Comunian, Dando, and Abbaspourrad (2018) studied the optimization of microcapsule shell structure using whey protein isolate (WPI) microgels as aqueous phase in a fish O/W emulsion by a glass microfluidic device. Besides the application of Pickering emulsions, these authors also combined these methods to electrostatic interaction, since the droplets were collected in the solution of pectin of different charge compared with the protein so that electrostatic interaction between the polymers occurred. Finally, in order to obtain an even more stable system, the laccase enzyme was added for cross-linking and formation of a more protected system. Schröder, Sprakel, Schroën, Spaen, and Berton-Carabin (2018) studied approaches to obtain food-grade Pickering emulsions by a cross-flow microfluidic device and observed spherical morphology and unimodal size distribution with particle sizes around 170 nm, characteristics that are already expected for samples obtained by nano-/microfluidic.

In another interesting research, Balbino et al. (2013) produced liposomes using microfluidic devices that were constructed using soft lithography in PDMS. The production of cationic liposomes was done by the single and double hydrodynamic focusing methods. These authors obtained liposomes with particle sizes in the range of 100–130 nm, using a continuous process with reduced number of steps, showing the advantage of methods combination. The combination of liposomes and microfluidics was also studied by Correia, Briuglia, Niosi, and Lamprou (2017). These authors evaluated the production of phospholipid nanoparticles by microfluidics, showing how this method can be simple and fast. They obtained stable liposomes, with a high encapsulation efficiency and particle sizes in nanoscale.

Still with the combination of microfluidic devices and formation of liposomes, Leung, Morales, Britton, Kutter, and Chan (2018) studied the encapsulation of phages into liposomes using a microfluidic flow-focusing setup. They evaluated two types of *Pseudomonas* phages (*Podovirus* and *Myovirus*), which are completely new in the nano-/microfluidic area. They showed that microfluidic technique was able to encapsulate phages with a reasonable encapsulation efficiency (50%–59%).

In addition to the types of samples presented in the previous sections, there are also the Janus droplets and particles, which are extremely important for the simultaneous encapsulation of bioactive compounds of different and/or similar solubility properties; as in the same way as double or multiple emulsions, Janus droplets and particles have multicompartments, however, with two physically and chemically distinct exposed segments (Hu, Zhou, Sun, Fang, & Wu, 2012; Liang, Zhang, & Yang, 2014; Lone & Cheong, 2014; Nisisako, 2016; Pang, Wan, Wang, & Lin, 2014). In general, the formation of this system is based on two infused dispersed phases in a parallel stream emulsified by a coflowing continuous phase (Nisisako, 2016). There are different types of devices for the production of Janus droplets and particles, such as a planar microfluidic in a Y-shape channel made of different materials (polymers, silicone or glass) (Nisisako, 2016; Nisisako & Torii, 2007; Nisisako, Torii, Takahashi, & Takizawa, 2006), a 3D microcapillary with side-by-side geometry (Nisisako, 2016; Yang et al., 2015), a 2D lithography process and higher replication with respect to parallelization (Xu et al., 2015), or a centrifuge-based droplet formation system with a multibarreled capillary (Maeda, Onoe, Takinoue, & Takeuchi, 2012; Zhao et al., 2013).

It is important to highlight some extremely important factors in the formation of Janus droplets and particles when the two dispersed phases are miscible or not. First of all, the two dispersed phases should have similar viscosities at the same flow rates. According to Nisisako (2016), if it does not happen, there will be no balance between the two parallel dispersed phases, and then this difference can disturb the interface of the two miscible solutions. Moreover, to keep the system stable and without mixing the two dispersed phases, it is necessary to combine another process as soon as after the formation of the droplets in order to have them "locked," clearly displaying the interface between these two dispersed phases (Nisisako, 2016; Nisisako et al., 2006). In the case of two dispersed immiscible phases, it is necessary that the balance of the three interfacial tensions (two dispersed phase and one continuous phase) be taken into account (Nisisako & Hatsuzawa, 2016; Torza & Mason, 1969; Zhang, Zhao, Xu, & Luo, 2015).

The formation of Janus droplets and particles is also possible when using only one dispersed phase and two continuous phase (Nisisako, 2016; Subramaniam, Abkarian, & Stone, 2005). This idea was reported by Subramaniam et al. (2005) who evaluated the colloidal crystal growth on curved interfaces, which were injected in a microfluidic device. However, there are not many works in the literature that report this type of system for application in the area of food and encapsulation of bioactive compounds, which could be an innovative idea for future works.

The obtaining of Janus droplets and particles follows different protocols. The formation of Janus droplets can occur from core-shell droplets comprising three immiscible phases (Abate, Thiele, & Weitz, 2011; Nisisako, 2016; Nisisako, Okushima, & Torii, 2005; Okushima, Nisisako, Torii, & Higuchi, 2004; Utada et al., 2005) or dewetting from W/O/W emulsion droplets (Bayley et al., 2008; Deshpande, Caspi, Meijering, & Dekker, 2016; Hayward, Utada, Dan, & Weitz, 2006; Shum, Zhao, Kim, & Weitz, 2011). In the case of the formation of Janus particles, the process can occur mainly from Janus emulsion droplets (Seiffert, Romanowsky, & Weitz, 2010; Shepherd et al., 2006; Yu et al., 2016; Zhao et al., 2011) or phase separation in single emulsion droplets (Hu et al., 2013; Hu, Wang, Abbaspourrad, & Ardekani, 2015; Kim, Hwangbo, Lee, & Cho, 2014; Shang et al., 2013; Wang, Feng, Ma, Yao, & Ge, 2016; Yamagami, Kitayama, & Okubo, 2014).

So, it could be concluded that all these literature researches on Janus droplets and particles were not done for Food Science area, which means that this type of droplets and particles obtained by nano- and microfluidic devices is still in the beginning for the encapsulation of bioactive compounds.

7 Encapsulating materials for nano-/microfluidics

As noted throughout this chapter, various materials may be used for the formation of nano- and microstructures for the encapsulation of bioactive compounds by nano-/microfluidic devices, such as lipids of high and low melting point, polysaccharides, and proteins, depending on the main objective and the physicochemical properties of the bioactive compound.

7.1 Lipids of high melt point

Lipids of high melt point are widely used for the production of solid lipid capsules or particles since they solidify at low temperatures or at room temperature. Examples of high melt point lipids are palm oil, palmitic acid, and stearic acid. It is also possible to mix different oils to achieve the optimum melt temperature for the desired process by mixing a high melt point oil with a low melting point oil, causing the decrease of melt temperature, but it is still sufficient for the lipid phase to solidify under certain conditions.

In order to obtain solid lipid capsules or particles by micro-/nanofluidics, the formation of W/O/W double emulsion is the first step in the process. Comunian et al. (2014) studied the encapsulation of vitamin C using a microfluidic technique. In this case, the authors produced a W/O/W double emulsion with an aqueous solution containing vitamin C in the inner phase, palm oil in the intermediate phase, and an aqueous solution of polyvinyl alcohol (PVA) with a concentration of 10% (w/w) in the continuous phase. The palm oil phase was heated during all the process to keep this phase in the liquid state, but the emulsion droplets were collected in a solution in ice bath in order to solidify the intermediate phase and then obtain solid lipid microcapsules containing ascorbic acid.

In another interesting research, Sun, Shum, Holtze, and Weitz (2010) also studied the formation of solid lipid microcapsules by W/O/W double emulsions followed by solidification of oil phase. However, in this case, these authors used water, glycerol, and PVA as the continuous phase; oils including Suppocire AIM oil, paraffin, and eicosane as intermediate phase; and fluorescein isothiocyanate-dextran, fluorescein sodium salt, toluidine blue, and rhodamine B as model bioactives in the inner phase. The most interesting part of this research is the simultaneous encapsulation of multiple bioactives, which can be applied for encapsulating incompatible or reactive ingredients while keeping them separate (Sun et al., 2010).

The same idea of producing solid lipid particles was adopted by Comunian, Ravanfar, Alcaine, et al. (2018) when they encapsulated natural pigments (elderberry extract and norbixin) within W/O/W double emulsions by glass microfluidic devices, followed by solidification of oil phase (palm oil). In brief, the use of high melt point lipids in the nano- and microfluidic technology has attracted the attention of researchers to obtain solid lipid capsules and particles since they promote controlled release of the bioactive compound under desired conditions (controlled temperature) and also promote greater absorption by the body.

7.2 Lipids of low melt point

On the other hand, the lipids of low melting point do not solidify at high temperatures or ambient temperature, remaining in the liquid state throughout the process and application. Examples of lipids of low melting point are canola oil, soybean oil, corn oil, and sunflower oil. In this case, no capsules or particles are obtained, but single W/O or O/W emulsions or double emulsions are produced. For this goal, the addition of emulsifiers is necessary to obtain a stable system, which is not necessary in the case of obtaining capsules or solid lipid particles.

Lao, Wang, and Lee (2009) evaluated a microfluidic platform for the formation of single and double emulsion droplets. These authors used deionized water solution as inner aqueous phase and corn oil as oil phase in the single W/O emulsions, using 1% of dodecyl glycerol itaconate as emulsifier. Then, still with this single emulsion, a double emulsion was prepared using deionized water solution with 5% Triton X100 (emulsifier) as aqueous continuous phase. So, it can be said that lipids of low melting point are also important in the production of different formulations by nano- and microfluidic devices, protecting and promoting the controlled release of bioactive compounds.

7.3 Alginate

In addition to the use of lipids as encapsulating materials, biopolymers are also widely used in nano-/microfluidic systems. One of the most commonly used materials is alginate. It is extracted from algae and widely used in the food industry as an emulsifier and also to increase the viscosity of aqueous solutions. A small amount of this biopolymer results in a high-viscosity solution, which makes it to be used in low concentrations (maximum 2%), which is an advantage for the industry since little amount of this material is sufficient for the desired applications. Another important property of alginate is its ability to make gels in the presence of ions. This property makes it very popular in the area of encapsulation, mainly by the technique of nano- and microfluidic.

Comunian et al. (2017) studied the coencapsulation of echium oil (an oil rich in omega-3 and omega-6 fatty acids) and phenolic compounds (quercetin and sinapic acid) by O/W/O emulsions obtained in a glass microfluidic device. These authors combined two methods: microfluidics and internal ionic gelation using a mixture of sodium alginate and calcium-EDTA complex (1:1) in the middle aqueous phase and collecting the emulsion droplets in corn oil with lecithin and glacial acetic acid. This procedure promoted the gelation of alginate due to the reduction of pH (presence of acetic acid), which stimulated the release of calcium ions from the complex. Utech et al. (2015) encapsulated cell by the same method, using alginate and the complex of Ca-EDTA in the middle phase of an O/W/O emulsion obtained by microfluidic devices, showing then the importance of alginate in this type of procedure. In another interesting research, Ren, Ju, Xie, and Chu (2010) also studied the formation of monodisperse microcapsules obtained by microfluidic devices from an O/W/O emulsion. However, these authors designed a two-stage microcapillary microfluidic device. In this case, the alginate aqueous solution was injected in the middle phase and the calcium chloride aqueous phase solution into the downstream.

8 Encapsulated food bioactives by micro-/nanofluidics

8.1 Oils rich in polyunsaturated fatty acids

Fatty acids are the main components of lipids. Unsaturated fatty acids have double bonds in their structure, and those with more than one double bond are called polyunsaturated fatty acids (PUFAs). The main constituents of this group are those belonging to the families of omega-3 and omega-6, which have double bonds in the third and sixth carbon from the methyl group, respectively (Esfahani, Jafari, Jafarpour, & Dehnad, 2019). They are molecules not synthesized by the body, so they need to be ingested and are found in vegetable oils or marine fish oil (Comunian & Favaro-Trindade, 2016).

Omega-3 and omega-6 fatty acids are widely studied in the food, pharmaceutical, and cosmetic areas due to their health benefits, mainly in reducing the risk of cardiovascular diseases. However, these compounds are susceptible to oxidation when in contact with light, heating, oxygen and moisture, and also in the case of fish oil, it has a strong characteristic odor, making it difficult to apply these oils to food and pharmaceutical products (Ghorbanzade, Jafari, Akhavan, & Hadavi, 2017). For this reason, as mentioned, the encapsulation process has been an alternative to minimize these problems, and glass microfluidic devices have been one of the techniques used to protect these compounds, in addition to masking their unpleasant odor.

Comunian, Ravanfar, Selig, and Abbaspourrad (2018) studied the influence of protein type on the stability of fish oil-in-water emulsions obtained by glass microfluidic device. These authors studied formulations of fish oil in gelatin, casein, and hydrolyzed whey protein in single O/W emulsions and observed that the combination of 0.5% (w/w) gelatin and casein (ratio 1:1) promoted the best protection to the fish oil against oxidation, and the microfluidic method presented by them offered a new alternative to protect the PUFAs in fish oil, enabling the future application of this material in food products.

In another work, Ravanfar et al. (2018) compared the protection of fish oil in O/W emulsions obtained by two methods: glass microfluidic devices and conventional homogenization using whey protein microgel and beet pectin as wall materials and laccase and $CaCl_2$ as cross-linking agents. These authors obtained big emulsion droplets by microfluidics (about 10 times larger than the droplets produced by homogenization), however more physicochemical stable systems than those produced via homogenizer, showing that the microfluidic encapsulation technique is a promising method for the protection of bioactive compounds.

Echium seed oil is a vegetable oil rich in stearidonic acid, an omega-3 fatty acid and intermediate in the biosynthesis of EPA and DHA, also important omega-3 fatty acids. Comunian et al. (2017) studied the coencapsulation of echium seed oil and phenolic compounds by glass microfluidic devices and O/W/O double emulsion. The double emulsion was composed of echium oil in the internal oil phase, sodium alginate aqueous solution in the middle phase (wall material), and corn oil and soy lecithin in the continuous phase. Still, quercetin or sinapic acid (phenolic compounds) were added to the inner or middle phase, respectively. In this case, to form the microcapsules, the internal ionic gelation technique was used by adding a calcium-EDTA complex in the sodium alginate solution; the samples were collected in a beaker containing corn oil, 2% (w/w) soy lecithin, and 5% (v/v) glacial acetic acid, which promoted the gelation of sodium alginate due to the acid diffusion into the droplets, causing the pH reduction and the dissociation of calcium ions from Ca-EDTA complex. In this way, it was possible to show that the use of the microfluidic technique combined with ionic gelation was feasible, since these authors obtained high values of encapsulation yield and protection of the echium oil stored at 40°C.

8.2 Vitamins

Vitamins are essential micronutrients for the body. They are classified as hydrophilic and hydrophobic, depending on their solubility in water or oil. The fat-soluble vitamins are A, D, E, and K, while water-soluble vitamins are B complex and vitamin C, also known as ascorbic acid (Katouzian & Jafari, 2016).

Vitamin C is also widely used as an antioxidant; however, it oxidizes easily when in contact with high temperature, oxygen, light, and moisture. For this reason, the objective of Comunian et al. (2014) was to encapsulate ascorbic acid using a microfluidic technique to protect this bioactive compound against environmental conditions. These authors produced a double emulsion (W/O/W) using ascorbic acid aqueous solution (3% and 20%, w/w) in the inner phase, palm oil in the oil phase, and an aqueous solution of PVA with a concentration of 10% (w/w) in the continuous phase. Solid lipid microcapsules were obtained since a thermal jacket was used to heat the palm oil to 66°C within the glass syringe prior to injection and melt it down for injection into the capillary device and then collected in an ice bath to solidify the palm oil and convert it to capsules with a consolidated shell. Moreover, these authors also analyzed the incorporation of Na_2CO_3 in the internal aqueous phase along with vitamin C and $CaCl_2$ in the solution in which the microcapsules were collected and stored. This was done in order to decrease the release of ascorbic acid through the pores since these salts should diffuse through the lipid phase and form a precipitate, clogging the pores of the lipid phase. In this way, Comunian et al. (2014) demonstrated that encapsulation by glass microfluidic devices can also be applied to hydrophilic bioactive compounds, promoting their protection and enabling future applications in the food and pharmaceutical industry.

8.3 Pigments

The color of a product is one of the most valued sensory attributes by consumers when deciding to purchase a product. Natural and synthetic pigments are the compounds responsible for the coloring of food products. Due to the growing concern of consumers in the search for foods from natural sources, natural pigments have gained more attention from researchers and food scientists due to their safety and beneficial health properties (Mahdavi, Jafari, Ghorbani, & Assadpoor, 2014). However, various natural pigments are unstable under specific conditions of pH and temperature and also oxidize when in contact with heat, light, and oxygen (Akhavan Mahdavi, Jafari, Assadpoor, & Dehnad, 2016; Mahdavee Khazaei, Jafari, Ghorbani, & Hemmati Kakhki, 2014).

The microfluidic technique has been an alternative for the protection of natural pigments. Comunian, Ravanfar, Alcaine, et al. (2018) studied the encapsulation of anthocyanin and norbixin by glass microfluidic devices. These authors prepared a W/O/W double emulsion composed of elderberry extract aqueous solution 10% (w/w) (source of anthocyanins) or norbixin aqueous solution 1% (w/w) in the internal aqueous phase, palm oil in the middle oil phase, and aqueous solution with 5% (w/w) of Pluronic F-108 in the continuous phase. To obtain the solid lipid microcapsules, the oil phase was injected using a syringe coupled to a heating with the temperature maintained at 60°C and collected in ice bath to ensure the solidification of palm oil. Comunian, Ravanfar, Alcaine, et al. (2018) observed that this method was feasible to promote the protection of pigments under conditions that they would have been unstable; moreover, the use of palm oil was important for future application of these materials in products that are submitted to pasteurization process, keeping the colors stable and allowing the controlled release at high temperatures.

8.4 Proteins/peptides

Amino acids are compounds composed of carbon, hydrogen, oxygen, and nitrogen, and their structure is formed by an amine group and a carboxyl group. They can be divided into polar, nonpolar, and neutral. In the case of peptides, they are molecules formed by two or more amino acids from peptide bonds. Proteins are molecules formed by one or more chains of amino acids and perform essential functions to the body related to the transport of molecules, response to stimuli, participate in reactions, and act as enzymes, in addition to structural functions (actin and myosin). However, proteins and peptides are sensitive under various conditions, causing their denaturation, aggregation, precipitation, or degradation. For this reason, micro/nanoencapsulation has a fundamental role in protecting these compounds, allowing their application in foods (Sarabandi, Sadeghi Mahoonak, Hamishekar, Ghorbani, & Jafari, 2018).

Table 1 Examples of encapsulated bioactive compounds by microfluidic devices.

Bioactive compound	System	Device	Reference
Fish oil	O/W emulsion	Glass microfluidic devices	Comunian, Ravanfar, Selig, et al. (2018)
Fish oil	O/W emulsion	Glass microfluidic devices and conventional homogenization	Ravanfar et al. (2018)
Echium seed oil	O/W/O emulsion	Glass microfluidic devices	Comunian et al. (2017)
Vitamin C	W/OW emulsion	Glass microfluidic devices	Comunian et al. (2014)
Anthocyanin and norbixin	W/O/W emulsion	Glass microfluidic devices	Comunian, Ravanfar, Alcaine, et al. (2018)
Bovine serum albumin	W/O/W emulsion	Glass microfluidic devices	Pessi et al. (2014)

The microencapsulation of bovine serum albumin (BSA) by glass microfluidic devices was studied by Pessi et al. (2014). These authors evaluated the microencapsulation of BSA in a double emulsion (W/O/W), which was composed of 1% (w/w) BSA in the inner phase with PVA, polycaprolactone, and polyethylene glycol, and obtained spherical microcapsules, intact and monodisperse in size, which was already expected due to the method of encapsulation. Furthermore, they observed a high encapsulation efficiency (84%), indicating that this method was a feasible way to develop a formulation template for protection and also controlled release of proteins.

Some examples of these encapsulated bioactive compounds, type of emulsions, and devices are summarized in Table 1.

9 Characterization of encapsulated ingredients produced by nano-/microfluidics

9.1 Optical, confocal, scanning, and transmission electron microscopy

The morphological study of systems obtained by nano- and microfluidics is essential for evaluation of the process, materials used, and types of devices. This study can be performed in different ways, such as optical, confocal, and/or scanning electron microscopy, depending on the specific objective. Optical microscopy aims to observe the morphology in general, being the sample single, double, or multiple emulsion, particles, or microcapsules, and can be performed with the sample wet or dry. This type of micrograph offers a generalized view of the sample; however, for the analysis of more specific data such as particle size, surface study, or evaluation of the presence and number of pores, more robust analyzes are necessary.

With confocal or fluorescence microscopy, in addition to visualizing the morphology in general, as well as optical microscopy, it is possible to observe the specific location of some compounds or aqueous/oil phase of a given emulsion (Jafari & Esfanjani, 2017, chap. 14). This requires the use of specific fluorophores (dyes) to color oil, water, or protein, for example. Samples may be liquid or dry. In the case of scanning electron microscopy (SEM), it is possible to study in depth the surface of the materials (smooth or rough) and also to analyze the presence/absence of pores. In this case, the materials (particles, capsules, or emulsions) need to be dry (Jafari, Esfanjani, Katouzian, & Assadpour, 2017, chap. 10). Samples of sizes within the nanoscale are best analyzed by transmission electron microscopy (TEM) since this technique presents better conditions for more detailed analyzes, as well as higher resolutions. It is also possible to study all the layers of the samples; in the case of capsules or emulsions with several layers, it is possible to analyze the structure both internally and on the surface, a characteristic not possible by other microscopy techniques, and the samples can be wet.

9.2 Particle size and particle size distribution

Particle size and particle size distribution analysis is another factor that is directly related to the morphology and stability of the different systems obtained by nano- and microfluidics. In this case, the results for emulsions, capsules, or particles obtained by these devices are in most cases unimodal, which shows the control achieved, as already mentioned, by this encapsulation technique. The average particle size may vary from nano- to microscale, depending on the type and material used in the device construction and final research objective, as can be seen throughout this chapter.

The size and particle size distribution are analyzed with the wet samples and can be evaluated by laser diffraction (Mastersize or Zetasizer), depending on the type and size range of the samples.

9.3 Encapsulation efficiency

The encapsulation efficiency quantifies how much of the bioactive compound was actually encapsulated (Jafari, Assadpoor, He, & Bhandari, 2008a). In other words, how much of the compound is within the emulsion (protected), capsule, or particle in relation to the amount effectively present in one of the flows injected into the capillaries of the device. The encapsulation efficiency for samples obtained by nano- and microfluidics is usually around 100%, which, again, proves the efficacy and control during the process within these types of devices. The high encapsulation efficiency is directly related to the unimodal behavior observed in the particle size distribution and can be measured by different techniques, depending on the type of encapsulated bioactive compound, such as UV-Vis spectroscopy, high-performance liquid chromatography (HPLC), or gas chromatography (GC).

9.4 Stability of the system and encapsulated bioactive compound

The stability of the system (emulsion, capsule, or particle) and the encapsulated bioactive compounds is the most important analysis since it is the main objective of encapsulation process and the technique presented in this chapter. Stable single, double, and multiple emulsions are obtained by nano- and microfluidics due to the excellent droplet formation and control achieved by this technique. This control is possible through the study of different flow rates, as mentioned in the previous sections of this chapter. Thus, with the formation of a stable system, a high encapsulation efficiency, and unimodal particle size behavior, it is possible to obtain encapsulated bioactive compounds stable for a long time when compared with the same nonencapsulated compounds, factor of extremely important for nano-/microfluidics.

In the same way as for encapsulation efficiency analysis, the quantification of bioactive compounds over time can also be measured by different techniques depending on the type of encapsulated compound, such as UV-Vis spectroscopy, HPLC, or GC. In addition, the stability study can be performed under accelerated conditions when samples are submitted to stress conditions, such as extreme temperatures (freezing or heating), different concentrations of salt (ionic strength change), and different pH values.

9.5 Zeta potential

The optimization of emulsion formulations and, consequently, the prediction of their stability can be controlled from analyzes of zeta potential measurements. The zeta potential is considered the electric potential at the surface of dispersed droplets/particles and depends on several factors, such as the capsule or particle surface, the materials used in each phase of the emulsion, the pH of the medium, the conductivity, and the concentration of the compounds. The results obtained for zeta potential may have positive or negative values depending on the nature and properties of the analyzed materials and are measured in mV. In the case of proteins, the zeta potential is positive below their isoelectric point (IP) and negative above their IP.

In the cases of systems as single, double, or multiple emulsions and capsules/particles obtained by both the nano- and microfluidic technique and by other encapsulation techniques, they exhibit stable behavior when the zeta potential values are greater than −30 mV or +30 mV (McClements & Jafari, 2018a, chap. 1, 2018b). In addition, when such systems are in suspension, as is the case of emulsions, a layer is formed around the system. It is known that emulsions present Brownian motion—this layer formed around the system also moves according to this movement, and the zeta potential is the potential of this layer.

9.6 Controlled release

Last but not least, controlled release studies are increasingly being the focus of researchers working with different encapsulation techniques. One of the main goals of the encapsulation process, in addition to protecting bioactive compounds, is to promote controlled release at the desired time and place (Jafari, Katouzian, Rajabi, & Ganje, 2017, chap. 13). Since the bioactive compound is protected within the internal phase of emulsions or capsules/particles and the system applied in a food product, the next goal is that the compound is released only into the intestinal fluid for nutraceutical delivery purposes. Depending on the structure of the system (number of layers and presence/absence of pores) and properties of the materials (lipids, proteins, and polysaccharides), the bioactive compounds can be released at different times during eating and ingestion, from chewing to the end of the digestive process. For this reason, the controlled release study of encapsulated compounds by nano- and microfluidic techniques needs to be studied in order to obtain a viable whole process.

10 Limitations of encapsulation by micro-/nanofluidic devices

As shown in this chapter, it can be seen that nano-/microfluidics are ideal techniques for controlled production of emulsions, particles, and capsules of the desired size for each type of application. However, some limitations of these techniques need to be mentioned, since they restrict the use of these methods in the food, pharmaceutical, and cosmetic industry. The first limitation is the scaling up for the food industry. The flow rates used for the devices, being glass or PDMS, do not exceed the average of 15–20 mL/h, which makes it difficult to obtain a considerable sample volume for an industrial scale (Celli & Abbaspourrad, 2018; Nisisako & Torii, 2008). In order to produce a larger volume, an idea is the production of parallel devices, so that the volume of produced sample is multiplied according to the number of such devices. However, even so, the volume is considered small for application in food products (Celli & Abbaspourrad, 2018; Tetradis-Meris et al., 2009).

There are few researches in the literature that study the increase in the scale of the production of systems by nano-/microfluidics and much less in the area of food, which makes the idea totally innovative for future works. A first example is the study done by Nisisako and Torii (2008). These authors were able to obtain a production of 320 mL/h when using microfluidic large-scale integration on a chip. In this case, they used the glass microfluidic chip using a module having 128 cross junctions. These authors obtained monodisperse droplets with mean diameter of 96.4 μm. In the same research, these authors also produced a system that could produce Janus droplets in a flow rate of 128 mL/h, obtaining droplets with a mean diameter of 143 μm. According to these authors, these systems could be applied to different areas, so it would be an idea for encapsulation of bioactive compounds.

In another interesting research, Tetradis-Meris et al. (2009) studied a novel parallel integration of microfluidic devices, and they were possible to scale up to 180 microfluidic devices with a low coefficient of variation (5%) and obtained monodisperse water-in-oil emulsions. According to these authors, there are three factors that are important to have this system. First of all, they analyzed how fabrication tolerances could affect the droplet formation. For this step, a model simulation was used. Then, they improved the performance of the system by producing separate draining manifolds. After that, the droplets were formed in a large opening short channel downstream.

It is also important to note that, according to Kobayashi, Wada, Uemura, and Nakajima (2010), to obtain the scale-up for single emulsions by nano-/microfluidic devices, the possible reduction of system efficiency should be taken into account when particle size reduction is desired and mass parallelized systems are used. According to Abrahamse, van Lierop, van der Sman, van der Padt, and Boom (2002) and Kobayashi et al. (2010), this can be explained due to the lower Laplace pressure of a droplet relative to the Laplace pressure of an interface on the terrace. However, Van Dijke, Schrën, van der Padt, and Boom (2010) showed an efficient parallelized system for food-grade ingredients. In this case, these authors used the edge-based droplet generation (EDGE) emulsification method to produce the emulsions and developed a stable and robust technology. They obtained single (O/W) and double (W/O/W) emulsions based on whey protein concentrate, skim milk, and sunflower oil in an upscaled microdevice, with no changes in time, wettability, and fouling. These authors obtained EDGE lengths to produce 1 m^3/h of product with an effective volume of the system around 250–300 L for emulsions with a thickness of 1 mm for the EDGE microchips.

Also, Nisisako, Ando, and Hatsuzawa (2012) studied the scale-up production of single emulsions in a microfluidic parallelization coupled with coaxial annular chip. In this case, the authors evaluated a system of two elements. One is a planar square chip with many microfluidic droplet generators (MFDG) arranged and, the second one, a cubic module with coaxial annular channels. They parallelized 144 MFDGs, produced emulsions with droplets of 90 μm and rate of 180 mL/h. They also produced Janus droplets, double emulsions, and triple emulsions with 31–128 MFDGs, showing the possibility of scale-up in this case. Romanowsky, Abate, Rotem, Holtze, and Weitz (2012) presented a system for increasing the scale of the production of double emulsions. In this interesting work, these authors succeeded in increasing the production at rates over 1 kg/day and with diameter variation less than 6%, with the use of a design with 15 dropmaker units. So, according to Romanowsky et al. (2012), it is possible to encapsulate bioactive compounds within double emulsions on a larger scale than those shown in other works using nano-/microfluidics. Following the same idea, it is also possible to produce nano-/microcapsules on this larger scale since the production of double emulsions is one of the steps for obtaining such systems by nano-/microfluidics. However, as noted in this chapter, additional control is required for the production of double emulsion, which makes its scale-up a more difficult process than for single emulsions (Maan et al., 2015).

Therefore, according to these explanations, still with the increase of scale presented by these authors, this technique remains impracticable for the application in food products since it is possible to obtain scales much larger by other techniques already known.

Another limitation as important as scale-up is the type of material to be used for the construction of devices that are mostly nonfood grade. Glass microfluidic device is considered the most suitable option for the food area, since it does not promote interaction with the material used for the formation of emulsions, droplets, capsules, and/or particles. PDMS is another widely used material, as discussed in this chapter; however it needs to be used carefully so that there is no interaction between it and the biopolymers or lipids of the emulsions. In addition, glass microfluidic devices may be used at high temperatures, which cannot be done with PDMS devices.

In this way, it can be said that the scale-up for these types of systems is still a challenge, but it is a start. It was possible to increase the scale according to some authors; however, the particle size obtained by them is still big for application in foods, and the scale remains low. However, even with these limitations, nano- and microfluidic devices have many advantages as discussed earlier, and they need to be taken into account as they are critical for achieving controlled and stable systems, which promote the controlled release of encapsulated bioactive compounds. Due to this importance, the scale-up of these systems should be explored, so the application in the food, pharmaceutical, and cosmetic industries could be possible.

11 Conclusions

In general, it can be concluded that the production of food nanostructures by micro-/nanofluidics is considered a new and emerging technology for the protection and controlled release of important bioactive compounds for the food area and food industry. It is a process of great capacity for the production of stable single, double, or multiple emulsions and of desired, controlled, and homogeneous droplet size. In addition, various types of food-grade materials may be used in the intermediate and continuous phase of the emulsions, such as lipids, proteins, and polysaccharides, which makes this technique closer to the desired formulations in the food industry. It is important to emphasize that it is a technique that does not use solvents, high energy, and high temperatures, which are highly desired by companies since these factors can degrade bioactive compounds and alter the composition of foods. Emulsions/capsules obtained by micro-/

nanofluidics present controlled particle size according to the flow rates of each phase of the emulsions, the viscosity of the solutions, and the type (glass or PDMS) and geometry (T-, X-, and Y-junctions) of devices. For this reason, its application in food products is possible without the interference on the texture of the final product and consequently in the acceptance of the consumer. The scale-up process is still in a slow process of adaptation, since it requires operating systems that demand more attention and study. For this to be possible, building a system with parallel devices would be necessary, since it would multiply the volume of sample produced per hour of process. In other words, nano-/microfluidics can be applied in the food, cosmetic, and pharmaceutical areas in a near future, improving the protection and controlled release of bioactive compounds, resulting in food products with a high quality and safety for consumers.

References

Abate, A. R., Thiele, J., & Weitz, D. A. (2011). One-step formation of multiple emulsions in microfluidics. *Lab on a Chip*, *11*, 253–258.

Abrahamse, A. J., van Lierop, R., van der Sman, R. G. M., van der Padt, A., & Boom, R. M. (2002). Analysis of droplet formation and interactions during cross-flow membrane emulsification. *Journal of Membrane Science*, *204*, 125–137.

Akbas, E., Soyler, B., & Oztop, M. H. (2018). Formation of capsaicin loaded nanoemulsions with high pressure homogenization and ultrasonication. *LWT- Food Science and Technology*, 266–273.

Akhavan, S., Assadpour, E., Katouzian, I., & Jafari, S. M. (2018). Lipid nano scale cargos for the protection and delivery of food bioactive ingredients and nutraceuticals. *Trends in Food Science & Technology*, *74*, 132–146.

Akhavan Mahdavi, S., Jafari, S. M., Assadpoor, E., & Dehnad, D. (2016). Microencapsulation optimization of natural anthocyanins with maltodextrin, gum Arabic and gelatin. *International Journal of Biological Macromolecules*, *85*, 379–385.

Assadpour, E., & Jafari, S. M. (2018). A systematic review on nanoencapsulation of food bioactive ingredients and nutraceuticals by various nanocarriers. *Critical Reviews in Food Science and Nutrition*, 1–47.

Azzi, J., Jraij, A., Auezova, L., Fourmentin, S., & Greige-Geres, H. (2018). Novel findings for quercetin encapsulation and preservation with cyclodextrins, liposomes, and drug-in-cyclodextrin-in-liposomes. *Food Hydrocolloids*, *81*, 328–340.

Balbino, T. A., Aoki, N. T., Gasperini, A. A. M., Oliveira, C. L. P., Azzoni, A. R., Cavalcanti, L. P., et al. (2013). Continuous flow production of cationic liposomes at high lipid concentration in microfluidic devices for gene delivery application. *Chemical Engineering Journal*, *226*, 423–433.

Bayley, H., Croning, B., Heron, A., Hoden, M. A., Hwang, W. L., Syeda, R., et al. (2008). Droplet interface bilayers. *Molecular BioSystems*, *4*, 1191–1208.

Carmo, E. L., Teodoro, R. A. R., Felix, P. H. C., Fernandez, R. V. B., Oliveira, E. R., Veiga, T. R. A., et al. (2018). Stability of spray-dried beetroot extract using oligosaccharides and whey proteins. *Food Chemistry*, *249*, 51–59.

Celli, G. B., & Abbaspourrad, A. (2018). Tailoring delivery system functionality using microfluidics. *Annual Review of Food Science and Technology*, *9*, 481–501.

Comunian, T. A., Abbaspourrad, A., Favaro-Trindade, C. S., & Weitz, D. A. (2014). Fabrication of solid lipid microcapsules containing ascorbic acid using a microfluidic technique. *Food Chemistry*, *152*, 271–275.
Comunian, T. A., & Favaro-Trindade, C. S. (2016). Microencapsulation using bio- polymers as an alternative to produce food enhanced with phytosterols and ômega-3 fatty acids: A review. *Food Hydrocolloids*, *61*, 442–457.
Comunian, T. A., Nogueira, M., Scolaro, B., Thomazini, M., Ferro-Furtado, R., Castro, I. A., et al. (2018). Enhancing stability of echium seed oil and beta-sitosterol by their coencapsulation by complex coacervation using different combinations of wall materials and crosslinkers. *Food Chemistry*, *252*, 277–284.
Comunian, T. A., Ravanfar, R., Alcaine, S. D., & Abbapsourrad, A. (2018). Water-in-oil-in-water emulsion obtained by glass microfluidic device for protection and heat-triggered release of natural pigments. *Food Research International*, *106*, 945–951.
Comunian, T. A., Ravanfar, R., Castro, I. A., Dando, R., Favaro-Trindade, C. S., & Abbaspourrad, A. (2017). Improving oxidative stability of echium oil emulsions fabricated by microfluidics: Effect of ionic gelation and phenolic compounds. *Food Chemistry*, *233*, 125–134.
Comunian, T. A., Ravanfar, R., Selig, M. J., & Abbaspourrad, A. (2018). Influence of the protein type on the stability of fish oil in water emulsion obtained by glass microfluidic device. *Food Hydrocolloids*, *77*, 96–106.
Correia, M. G. S., Briuglia, M. L., Niosi, F., & Lamprou, D. A. (2017). Microfluidic manufacturing of phospholipid nanoparticles: Stability, encapsulation efficacy, and drug release. *International Journal of Pharmaceutics*, *516*, 91–99.
Datta, S. S., Abbaspourrad, A., Amstad, E., Fan, J., Kim, S.-H., Romanowsky, M., et al. (2014). 25th Anniversary article: Double emulsion templated solid microcapsules: Mechanics and controlled release. *Advanced Materials*, 1–14. https://doi.org/10.1002/adma.201305119.
Deshpande, S., Caspi, Y., Meijering, A. E. C., & Dekker, C. (2016). Octanol-assisted liposome assembly on chip. *Nature Communications*, *7*, 10447.
Duhoranimana, E., Yu, J., Mukeshimana, O., Habinshuti, I., Karangwa, E., Xu, X., et al. (2018). Thermodynamic characterization of gelatin-sodium carboxymethyl cellulose complex coacervation encapsulating conjugated linoleic acid (CLA). *Food Hydrocolloids*, *80*, 149–159.
Durukan, O., Kahraman, I., Parlevliet, P., Geistbeck, M., & Seyhan, A. T. (2016). Microfluidization, time-effective and solvent free processing of nanoparticle containing thermosetting matrix resin suspensions for producing composites with enhanced thermal properties. *European Polymer Journal*, *85*, 575–587.
Encina, C., Marquez-Ruiz, G., Holgado, F., Giménez, B., Vergara, C., & Robert, P. (2018). Effect of spray-drying with organic solvents on the encapsulation, release and stability of fish oil. *Food Chemistry*, *263*, 283–291.
Esfahani, R., Jafari, S. M., Jafarpour, A., & Dehnad, D. (2019). Loading of fish oil into nanocarriers prepared through gelatin-gum Arabic complexation. *Food Hydrocolloids*, *90*, 291–298.
Eusner, T., Hale, M., & Hardt, D. E. (2010). Process robustness of hot embossing microfluidic devices. *Journal of Manufacturing Science and Engineering*, *132*, 030920-1–030920-8.
Faridi Esfanjani, A., & Jafari, S. M. (2016). Biopolymer nano-particles and natural nanocarriers for nano-encapsulation of phenolic compounds. *Colloids and Surfaces B: Biointerfaces*, *146*, 532–543.
Fávaro-Trindade, C. S., Pinho, S. C., & Rocha, G. A. (2008). Microencapsulação de ingredientes alimentícios. *Brazilian Journal of Food Technology*, *11*(2), 103–112.

Francisco, C. R. L., Heleno, S. A., Fernandes, I. P. M., Barreira, J. C. M., Calhelha, R. C., Barros, L., et al. (2018). Functionalization of yogurts with *Agaricus bisporus* extracts encapsulated in spray-dried maltodextrin crosslinked with citric acid. *Food Chemistry, 245*, 845–853.

Garstecki, P., Gañán-Calvo, A., & Whitesides, G. (2005). Formation of bubbles and droplets in microfluidic systems. *Bulletin of the Polish Academy of Sciences Technical Sciences, 53*, 361–372.

Ghorbanzade, T., Jafari, S. M., Akhavan, S., & Hadavi, R. (2017). Nano-encapsulation of fish oil in nano-liposomes and its application in fortification of yogurt. *Food Chemistry, 216*, 146–152.

Hayward, R. C., Utada, A. S., Dan, N., & Weitz, D. A. (2006). Dewetting instability during the formation of polymersomes from block-copolymer-stabilized double emulsions. *Langmuir, 22*, 4457–4461.

Hu, Y., Wang, S., Abbaspourrad, A., & Ardekani, A. M. (2015). Fabrication of shape controllable Janus alginate/pNIPAAm microgels via microfluidics technique and off-chip ionic crosslinking. *Langmuir, 31*, 1885–1891.

Hu, Y., Wang, J., Li, C., Wang, Q., Wang, H., Zhu, J., et al. (2013). Janus photonic crystal microspheres: Centrifugation-assisted generation and reversible optical property. *Langmuir, 29*, 15529–15534.

Hu, J., Zhou, S., Sun, Y., Fang, X., & Wu, L. (2012). Fabrication, properties and applications of Janus particles. *Chemical Society Reviews, 41*, 4356–4378.

Hughes, E., Maan, A. A., Acquistapace, S., Burbidge, A., Johns, M. L., Gunes, D. Z., et al. (2013). Microfluidic preparation and self-diffusion PFG-NMR analysis of monodisperse water-in-oil-in-water double emulsions. *Journal of Colloid and Interface Science, 389*, 147–156.

Jafari, S. M. (2017). An overview of nanoencapsulation techniques and their classification. In *Nanoencapsulation technologies for the food and nutraceutical industries* (pp. 1–34). Academic Press.

Jafari, S. M., Assadpoor, E., He, Y., & Bhandari, B. (2008a). Encapsulation efficiency of food flavours and oils during spray drying. *Drying Technology, 26*(7), 816–835.

Jafari, S. M., Assadpoor, E., He, Y., & Bhandari, B. (2008b). Re-coalescence of emulsion droplets during high-energy emulsification. *Food Hydrocolloids, 22*(7), 1191–1202.

Jafari, S. M., & Esfanjani, A. F. (2017). Instrumental analysis and characterization of nanocapsules. In *Nanoencapsulation technologies for the food and nutraceutical industries* (pp. 524–544). Academic Press.

Jafari, S. M., Esfanjani, A. F., Katouzian, I., & Assadpour, E. (2017). Release, characterization, and safety of nanoencapsulated food ingredients. In *Nanoencapsulation of food bioactive ingredients* (pp. 401–453). Academic Press.

Jafari, S. M., Katouzian, I., Rajabi, H., & Ganje, M. (2017). Bioavailability and release of bioactive components from nanocapsules. In *Nanoencapsulation technologies for the food and nutraceutical industries* (pp. 494–523). Academic Press.

Jafari, S. M., Paximada, P., Mandala, I., Assadpour, E., & Mehrnia, M. A. (2017). Encapsulation by nanoemulsions. In *Nanoencapsulation technologies for the food and nutraceutical industries* (pp. 36–73). Academic Press.

Jena, R. K., Yue, C. Y., & Lam, Y. C. (2012). Micro fabrication of cyclic olefin copolymer (COC) based microfluidic devices. *Microsystem Technologies, 18*, 159–166.

Juttulapa, M., Piriyaprasarth, S., Takeuchi, H., & Sriamornsak, P. (2017). Effect of high-pressure homogenization on stability of emulsions containing zein and pectin. *Asian Journal of Pharmaceutical Sciences, 12*, 21–27.

Katouzian, I., & Jafari, S. M. (2016). Nano-encapsulation as a promising approach for targeted delivery and controlled release of vitamins. *Trends in Food Science & Technology, 53*, 34–48.

Kawakatsu, T., Kikuchi, Y., & Nakajima, M. (1997). Regular-sized cell creation in microchannel emulsification by visual microprocessing method. *Journal of the American Oil Chemists' Society*, *74*(3), 317–321.

Khan, I. U., Serra, C. A., Anton, N., Er-Rafik, M., Blanck, C., Schmutz, M., et al. (2015). Microfluidic conceived Trojan microcarriers for oral delivery of nanoparticles. *International Journal of Pharmaceutics*, *493*, 7–15.

Kim, S. W., Hwangbo, K. H., Lee, J. K., & Cho, K. Y. (2014). Microfluidic fabrication of microparticles with multiple structures from a biodegradable polymer blend. *RSC Advances*, *4*, 46536–46540.

Kobayashi, I., Wada, Y., Uemura, K., & Nakajima, M. (2010). Microchannel emulsification for mass production of uniform fine droplets: Integration of microchannel arrays on a chip. *Microfluidics and Nanofluidics*, *8*, 255–262.

Kumacheva, E., & Garstecki, P. (2011). Microfluidic synthesis of polymer particles with non-conventional shapes. In E. Kumacheva & P. Garstecki (Eds.), *Microfluidic reactors for polymer particles* (pp. 192–214). Chichester, UK: John Wiley & Sons.

Lao, K. -L., Wang, J. -H., & Lee, G. -B. (2009). A microfluidic platform for formation of double-emulsion droplets. *Microfluidics and Nanofluidics*, 7, 709–719.

Leung, A. K. K., Hafez, I. M., Baoukina, S., Belliveau, N. M., Zhigaltsev, I. V., Afshinmanesh, E., et al. (2012). Lipid nanoparticles containing siRNA synthesized by microfluidic mixing exhibit an electrondense nanostructured core. *The Journal of Physical Chemistry C*, *116*, 18440–18450.

Leung, S. S. Y., Morales, S., Britton, W., Kutter, E., & Chan, H. -K. (2018). Microfluidic-assisted bacteriophage encapsulation into liposomes. *International Journal of Pharmaceutics*, *545*, 176–182.

Liang, F., Zhang, C., & Yang, Z. (2014). Rational design and synthesis of Janus composites. *Advanced Materials*, *26*, 6944–6949.

Lin, B. C., & Su, Y. C. (2008). On-demand liquid-in-liquid droplet metering and fusion utilizing pneumatically actuated membrane valves. *Journal of Microelectromechanical Systems*, *18*, 115005.

Liu, L., Wu, F., Ju, X. J., Xie, R., Wang, W., Niu, C. H., et al. (2013). Preparation of monodisperse calcium alginate microcapsules via internal gelation in microfluidic-generated double emulsions. *Journal of Colloid and Interface Science*, *404*, 85–90.

Lone, S., & Cheong, I. W. (2014). Fabrication of polymeric Janus particles by droplet microfluidics. *RSC Advances*, *4*, 13322–13333.

Maan, A. A., Nazir, A., Khan, M. K. I., Boom, R., & Schroën, K. (2015). Microfluidic emulsification in food processing. *Journal of Food Engineering*, *147*, 1–7.

Maeda, K., Onoe, H., Takinoue, M., & Takeuchi, S. (2012). Controlled synthesis of 3D multicompartmental particles with centrifuge-based microdroplet formation from a multi-barreled capillary. *Advanced Materials*, *24*, 1340–1346.

Mahdavee Khazaei, K., Jafari, S. M., Ghorbani, M., & Hemmati Kakhki, A. (2014). Application of maltodextrin and gum Arabic in microencapsulation of saffron petal's anthocyanins and evaluating their storage stability and color. *Carbohydrate Polymers*, *105*(1), 57–62.

Mahdavi, S. A., Jafari, S. M., Ghorbani, M., & Assadpoor, E. (2014). Spray-drying microencapsulation of anthocyanins by natural biopolymers: A review. *Drying Technology*, *32* (5), 509–518.

Majedi, F. S., Hasani-Sadrabadi, M. M., Emami, S. H., Shokrgozar, M. A., VanDersarl, J. J., Dashtimoghadam, E., et al. (2013). Microfluidic assisted self-assembly of chitosan based nanoparticles as drug delivery agents. *Lab on a Chip*, *13*, 204–207.

McClements, D. J., & Jafari, S. M. (2018a). General aspects of nanoemulsions and their formulation. In *Nanoemulsions* (pp. 3–20). Academic Press.

McClements, D. J., & Jafari, S. M. (2018b). Improving emulsion formation, stability and performance using mixed emulsifiers: A review. *Advances in Colloid and Interface Science*, *251*, 55–79.

Michelon, M., Oliveira, D. R. B., Furtado, G. F., Torre, L. G., & Cunha, R. L. (2017). High-throughput continuous production of liposomes using hydrodynamic flow-focusing microfluidic devices. *Colloids and Surfaces B: Biointerfaces*, *156*, 349–357.

Moura, S. C. S. R., Berling, C. L., Germer, S. P. M., Alvim, I. D., & Hubinger, M. D. (2018). Encapsulating anthocyanins from Hibiscus sabdariffa L. calyces by ionic gelation: Pigment stability during storage of microparticles. *Food Chemistry*, *241*, 317–327.

Muijlwijk, K., Berton-Carabin, C., & Schroën, K. (2016). Cross-flow microfluidic emulsification from a food perspective. *Trends in Food Science & Technology*, *49*, 51–63.

Nie, Z., Seo, M., Xu, S., Lewis, P., Mok, M., Kumacheva, E., et al. (2008). Emulsification in a microfluidic flow-focusing device: Effect of the viscosities of the liquids. *Microfluidics and Nanofluidics*, *5*(5), 585–594.

Nie, Z., Xu, S., Seo, M., Lewis, P. C., & Kumacheva, E. (2005). Polymer particles with various shapes and morphologies produced in continuous microfluidic reactors. *Journal of the American Chemical Society*, *127*, 8058–8063.

Nisisako, T. (2016). Recent advances in microfluidic production of Janus droplets and particles. *Current Opinion in Colloid & Interface Science*, *25*, 1–12.

Nisisako, T., Ando, T., & Hatsuzawa, T. (2012). High-volume production of single and compound emulsions in a microfluidic parallelization arrangement coupled with coaxial annular world-to-chip interfaces. *Lab on a Chip: Miniaturisation for Chemistry, Biology & Bioengineering*, *12*, 3426–3435.

Nisisako, T., & Hatsuzawa, T. (2016). Microfluidic fabrication of oil-filled polymeric microcapsules with independently controllable size and shell thickness via Janus to core-shell evolution of biphasic droplets. *Sensors and Actuators B: Chemical*, *223*, 209–216.

Nisisako, T., Okushima, S., & Torii, T. (2005). Controlled formulation of monodisperse double emulsions in a multiple-phase microfluidic system. *Soft Matter*, *1*, 23–27.

Nisisako, T., & Torii, T. (2007). Formation of biphasic Janus droplets in a microfabricated channel for the synthesis of shape-controlled polymer microparticles. *Advanced Materials*, *19*, 1489–1493.

Nisisako, T., & Torii, T. (2008). Microfluidic large-scale integration on a chip for mass production of monodisperse droplets and particles. *Lab on a Chip*, *8*, 287–293.

Nisisako, T., Torii, T., & Higuchi, T. (2004). Novel microreactors for functional polymer beads. *Chemical Engineering Journal*, *101*, 23–29.

Nisisako, T., Torii, T., Takahashi, T., & Takizawa, Y. (2006). Synthesis of monodisperse bicolored Janus particles with electrical anisotropy using a microfluidic co-flow system. *Advanced Materials*, *18*, 1152–1156.

Okushima, S., Nisisako, T., Torii, T., & Higuchi, T. (2004). Controlled production of monodisperse double emulsions by two-step droplet breakup in microfluidic devices. *Langmuir*, *20*, 9905–9908.

Oriani, V. B., Alvim, I. D., Paulino, B. N., Procópio, F. R., Pastore, G. M., & Hubinger, M. D. (2018). The influence of the storage temperature on the stability of lipid microparticles containing oleoresin. *Food Research International*, *109*, 472–480.

Ota, A., Istenic, K., Skrt, M., Segatin, N., Znidarsic, N., Kogej, K., et al. (2018). Encapsulation of panthothenic acid into liposomes and into alginate or alginate-pectin microparticles loaded with liposomes. *Journal of Food Engineering*, *229*, 21–31.

Pang, X., Wan, C., Wang, M., & Lin, Z. (2014). Strictly biphasic soft and hard Janus structures: Synthesis, properties, and applications. *Angewandte Chemie International Edition*, *53*, 5524–5538.

Patel, R. P., & Joshi, J. R. (2012). An overview on nanoemulsion: A novel approach. *International Journal of Pharmaceutical Sciences and Research*, *3*, 4640.

Pessi, J., Santos, H. A., Miroshnyk, I., Yliruusi, J., & Weitz, D. A. (2014). Microfluidics-assisted engineering of polymeric microcapsules with high encapsulation efficiency for protein drug delivery. *International Journal of Pharmaceutics*, *472*, 82–87.

Ravanfar, R., Comunian, T. A., Dando, R., & Abbaspourrad, A. (2018). Optimization of microcapsules shell structure to preserve labile compounds: A comparison between microfluidics and conventional homogenization method. *Food Chemistry, 241*, 460–467.
Ren, P.-W., Ju, X.-J., Xie, R., & Chu, L.-Y. (2010). Monodisperse alginate microcapsules with oil core generated from a microfluidic device. *Journal of Colloid and Interface Science, 343*, 392–395.
Romanowsky, M. B., Abate, A. R., Rotem, A., Holtze, C., & Weitz, D. A. (2012). High throughput production of single core double emulsions in a parallelized microfluidic device. *Lab on a Chip, 12*, 802–807.
Sanjay, S. T., Zhou, W., Dou, M., Tavakoli, H., Ma, L., Xu, F., et al. (2017). Recent advances of controlled drug delivery using microfluidic platforms. *Advanced Drug Delivery Reviews, 128*, 3–28.
Sarabandi, K., Sadeghi Mahoonak, A., Hamishekar, H., Ghorbani, M., & Jafari, S. M. (2018). Microencapsulation of casein hydrolysates: Physicochemical, antioxidant and microstructure properties. *Journal of Food Engineering, 237*, 86–95.
Schröder, A., Sprakel, J., Schroën, K., Spaen, J. N., & Berton-Carabin, C. C. (2018). Coalescence stability of pickering emulsions produced with lipid particles: A microfluidic study. *Journal of Food Engineering, 234*, 63–72.
Seiffert, S., Romanowsky, M. B., & Weitz, D. A. (2010). Janus microgels produced from functional precursor polymers. *Langmuir, 26*, 14842–14847.
Shamsi, M., Zahedi, P., Ghourchian, H., & Minaeian, S. (2017). Microfluidic-aided fabrication of nanoparticles blend based on chitosan for a transdermal multidrug delivery application. *International Journal of Biological Macromolecules, 99*, 433–442.
Shang, L., Shangguan, F., Cheng, Y., Lu, J., Xie, Z., Zhao, Y., et al. (2013). Microfluidic generation of magnetoresponsive Janus photonic crystal particles. *Nanoscale, 5*, 9553–9557.
Shepherd, R. F., Conrad, J. C., Rhodes, S. K., Link, D. R., Marquez, M., Weitz, D. A., et al. (2006). Microfluidic assembly of homogeneous and Janus colloid-filled hydrogel granules. *Langmuir, 22*, 8618–8622.
Shui, L., Mugele, F., Van den Berg, A., & Eijkel, J. C. T. (2008). Geometry-controlled droplet generation in head-on microfluidic devices. *Applied Physics Letters, 93*, 153113-1–153113-3.
Shum, H. C., Zhao, Y., Kim, S. H., & Weitz, D. A. (2011). Multicompartment polymersomes from double emulsions. *Angewandte Chemie International Edition, 50*, 1648–1651.
Silva, M. P., Tulini, F. L., Matos-Jr., F. E., Oliveira, M. G., Thomazini, M., & Favaro-Trindade, C. S. (2018). Application of spray chilling and electrostatic interaction to produce lipid microparticles loaded with probiotics as an alternative to improve resistance under stress conditions. *Food Hydrocolloids, 83*, 109–117.
Steegmans, M. L. J., Schroën, K. G. P. H., & Boom, R. M. (2009). Characterization of emulsification at flat microchannel Y junctions. *Langmuir, 25*, 3396–3401.
Subramaniam, A. B., Abkarian, M., & Stone, H. A. (2005). Controlled assembly of jammed colloidal shells on fluid droplets. *Nature Materials, 4*, 553–556.
Sun, C. X., Dai, L., Liu, F. G., & Gao, Y. X. (2016). Dynamic high pressure microfluidization treatment of zein in aqueous ethanol solution. *Food Chemistry, 210*, 388–395.
Sun, B. J., Shum, H. C., Holtze, C., & Weitz, D. A. (2010). Microfluidic melt emulsification for encapsulation and release of actives. *Applied Materials & Interfaces, 2*(12), 3411–3416.
Takeuchi, S., Garstecki, P., Weibel, D. B., & Whitesides, G. M. (2005). An axisymmetric flow-focusing microfluidic device. *Advanced Materials, 17*, 1067–1072.

Tetradis-Meris, G., Rossetti, D., de Torres, C. P., Cao, R., Lian, G., & Janes, R. (2009). Novel parallel integration of microfluidic device network for emulsion formation. *Industrial & Engineering Chemistry Research, 48*, 8881–8889.

Torza, S., & Mason, S. G. (1969). Coalescence of two immiscible liquid drops. *Science, 163*, 813–814.

Uluata, S., Decker, E. A., & McClements, D. J. (2016). Optimization of nanoemulsion fabrication using microfluidization: Role of surfactant concentration on formation and stability. *Food Biophysics, 11*, 52–59.

Utada, A. S., Fernandez-Nieves, A., Stone, H. A., & Weitz, D. A. (2007). Dripping to jetting transitions in coflowing liquid streams. *Physical Review Letters, 99*, 094502.

Utada, A. S., Lorenceau, E., Link, D. R., Kaplan, P. D., Stone, H. A., & Weitz, D. A. (2005). Monodisperse double emulsions generated from a microcapillary device. *Science, 308*, 537–541.

Utech, S., Prodanovic, R., Mao, A. S., Ostafe, R., Mooney, D. J., & Weitz, D. A. (2015). Microfluidic generation of monodisperse, structurally homogeneous alginate microgels for cell encapsulation and 3D cell culture. *Advanced Healthcare Materials, 4*, 1628–1633.

Van der Graaf, S., Steegmans, M. L. J., Van der Sman, R. G. M., Schroën, C. G. P. H., & Boom, R. M. (2005). Droplet formation in a T-shaped microchannel junction: A model system for membrane emulsification. *Colloids and Surfaces, A: Physicochemical and Engineering Aspects, 266*, 106–116.

Van Dijke, K. C., Schrën, K., van der Padt, A., & Boom, R. (2010). EDGE emulsification for food-grade dispersions. *Journal of Food Engineering, 97*, 348–354.

Vandegaer, J. E. (1973). *Microencapsulation*. New York: Plenum Press.

Vasile, F. E., Judis, M. A., & Mazzobre, M. F. (2018). Impact of Prosopis alba exudate gum on sorption properties and physical stability of fish oil alginate beads prepared by ionic gelation. *Food Chemistry, 250*, 75–82.

Vladisavljevic, G. T., Khalid, N., Neves, M. A., Kuroiwa, T., Nakajima, M., Uemura, K., et al. (2013). Industrial lab-on-a-chip: Design, application ans scale-up for drug discovery and delivery. *Advanced Drug Delivery Reviews, 65*, 1626–1663.

Wang, X., Feng, X., Ma, G., Yao, L., & Ge, M. (2016). Amphiphilic Janus particles generated via a combination of diffusion-induced phase separation and magnetically driven dewetting and their synergistic self-assembly. *Advanced Materials, 28*, 3131–3137.

Wilson, M. E., Kota, N., Kim, Y., Wang, Y., Stolz, D. B., LeDuc, P. R., et al. (2011). Fabrication of circular microfluidic channels by combining mechanical micromilling and soft lithography. *Lab on a Chip, 11*, 1550–1555.

Xu, K., Ge, X. H., Huang, J. P., Dang, Z. X., Xu, J. H., & Luo, G. S. (2015). A region-selective modified capillary microfluidic device for fabricating water–oil Janus droplets and hydrophilic-hydrophobic anisotropic microparticles. *RSC Advances, 5*, 46981–46988.

Xu, J. H., Li, S. W., Tan, J., Wang, Y. J., & Luo, G. S. (2006). Preparation of highly monodisperse droplet in a T-junction microfluidic device. *AIChE Journal, 52*, 3005–3010.

Yamagami, T., Kitayama, Y., & Okubo, M. (2014). Preparation of stimuli-responsive "mushroom-like" Janus polymer particles as particulate surfactant by site-selective surface-initiated AGET ATRP in aqueous dispersed systems. *Langmuir, 30*, 7823–7832.

Yang, Y. T., Wei, J., Li, X., Wu, L. J., Chang, Z. Q., & Serra, C. A. (2015). A side-by-side capillaries-based microfluidic system for synthesizing size- and morphology-controlled magnetic an-isotropy janus beads. *Advanced Powder Technology, 26*, 156–162.

Yeh, C. H., Chen, Y. C., & Lin, Y. C. (2011). Generation of droplets with different concentrations using gradient-microfluidic droplet generator. *Microfluidics and Nanofluidics, 11*, 245–253.

Yildiz, E., Demirkesen, I., & Mert, B. (2016). High pressure microfluidization of agro by-product to functionalized dietary fiber and evaluation as a novel bakery ingredient. *Journal of Food Quality*, *39*, 599–610.

Yu, X., Zhang, C., You, S., Liu, H., Zhang, L., Liu, W., et al. (2016). Microfluidic synthesis of multiferroic Janus particles with disk-like compartments. *Applied Physics Letters*, *108*, 073504.

Zhang, M. J., Wang, W., Xie, R., Ju, X. J., Liu, L., Gu, Y. Y., et al. (2013). Microfluidic fabrication of monodisperse microcapsules for glucose-response at physiological temperature. *Soft Matter*, *9*, 4150–4159.

Zhang, M. Y., Zhao, H., Xu, J. H., & Luo, G. S. (2015). Controlled coalescence of two immiscible droplets for Janus emulsions in a microfluidic device. *RSC Advances*, *5*, 32768–32774.

Zhao, Y., Gu, H., Xie, Z., Shum, H. C., Wang, B., & Gu, Z. (2013). Bioinspired multifunctional Janus particles for droplet manipulation. *Journal of the American Chemical Society*, *135*, 54–57.

Zhao, Y., Shum, H. C., Chen, H., Adams, L. L. A., Gu, Z., & Weitz, D. A. (2011). Microfluidic generation of multifunctional quantum dot barcode particles. *Journal of the American Chemical Society*, *133*, 8790–8793.

Zhigaltsev, I. V., Belliveau, N., Hafez, I., Leung, A. K., Huft, J., Hansen, C., et al. (2012). Bottom-up design and synthesis of limit size lipid nanoparticle systems with aqueous and triglyceride cores using millisecond microfluidic mixing. *Langmuir*, *28*, 3633–3640.

CHAPTER SIX

Production of food bioactive-loaded nanostructures by high-pressure homogenization

C. Fernandez-Avila[a], E. Hebishy[b], F. Donsì[c], E. Arranz[d] and A.J. Trujillo[a]

[a]Center for Innovation, Research and Knowledge Transfer in Food Technology (CIRTTA), XaRTA, TECNIO, Department of Animal and Food Science, Faculty of Veterinary Medicine, Universitat Autònoma de Barcelona, Bellaterra, Spain

[b]National Center for Food Manufacturing, College of Sciences, University of Lincoln, Holbeach Campus, Spalding, United Kingdom

[c]Department of Industrial Engineering, University of Salerno, Fisciano, Italy

[d]Teagasc Food Research Centre, Moorepark, Fermoy Co. Cork, Ireland

Contents

1 Introduction

Biologically bioactive compounds (lipids, vitamins, peptides, fatty acids, antioxidants, minerals, and living cells such as probiotics) are available

Nanoencapsulation of Food Ingredients by Specialized Equipment
https://doi.org/10.1016/B978-0-12-815671-1.00006-8

on the market in different forms. Bioactive compounds have health benefits that can contribute to the prevention and cure of a range of diseases. Natural bioactive compounds, when added into food products in a free/unprotected form, may be sensitive to environmental conditions, which results in their instability under food processing conditions and low bioavailability in the human body (Chen, Remondetto, & Subirade, 2006; de Vos, Faas, Spasojevic, & Sikkema, 2010). These compounds, when exposed to light, oxygen, moisture, or heat, could be chemically unstable and be easily oxidized. Oxidation may deteriorate bioactive compounds, leading to the generation of free radicals (Chen, Ren, Zhou, Wei, & Dai, 2016) and unpleasant tastes and off-odors, and subsequently poor shelf stability, sensory characteristics, and overall low consumer acceptability (Ariyarathna & Karunaratne, 2015).

The bioavailability of food bioactive compounds in food products can be improved through their incorporation into nanosized structures by nanoencapsulation, which reduces the damage and undesirable changes through processing stages as well as digestive conditions (Assadpour & Jafari, 2018; Jafari & McClements, 2017). Nanostructures could also improve water solubility of many poor water-soluble nutraceuticals (i.e., flavonoids, carotenoids, and some colors such as curcumin) and newly developed drug candidates (Huang, Yu, & Ru, 2010).

Nanostructures can be produced by top-down, also known as high-energy, and bottom-up, or low-energy, approaches. While bottom-up approaches typically require large amounts of surfactants and/or organic solvents (Sanguansri & Augustin, 2006), top-down approaches enable the use of formulations more suitable for food applications, despite the larger energy expenditure. Among the different available top-down fabrication methods, high-pressure homogenization (HPH) and microfluidization (see Chapter 7) stand out for their high throughput, repeatability, and industrial scalability. HPH or dynamic HPH enables pressures 10–15 times higher than traditional homogenizers and covers the range of 100–400 MPa, within which the pressure range of 200–400 MPa has been referred to as ultrahigh-pressure homogenization (UHPH). Pressure levels of up to 400 MPa can be reached and maintained by these UHPH systems, although the maximum pressure achieved depends on the engineering design. Moreover, UHPH has been documented to play a key role in the delivery of bioactive compounds (Dumay et al., 2013). In the food industry, low water-soluble compounds have gained interest to be incorporated into beverage products. In recent years, emergent UHPH-produced emulsions have been studied to boost the bioavailability of lipophilic bioactives, such as curcumin and conjugated linoleic acid (Fernandez-Avila, Arranz, Guri, Trujillo, & Corredig, 2016; Fernandez-Avila & Trujillo, 2017; Yucel, Quagliariello, Iaffaioli, Ferrari, & Donsì, 2015; Zou et al., 2016). There are other alternatives within HPH equipment such as the Microfluidizer® processor, which finely controls the level of shearing and offers an ideal solution for the production of liposomes and for achieving ambition particle size reduction targets by using less energy (Henry, Fryer, Frith, & Norton, 2010; Jafari, He, & Bhandari, 2007). It has been discussed throughly in Chapter 7.

HPH or UHPH are technologies that have been applied in the food, pharmaceutical, and cosmetic areas. Particularly, UHPH technology has been applied to different food systems such as milk and vegetable beverages to improve their physical stability and microbial quality (Donsì, Ferrari, & Maresca, 2009). UHPH has been suggested as an alternative to thermal pasteurization, due to its effect on inactivating pathogenic and other food spoilage microorganisms (Georget et al., 2014; Georget, Miller, Callanan, Heinz, & Mathys, 2014). UHPH technology has also been widely used to create stable nanoemulsion systems (Fernandez-Avila & Trujillo, 2016; Hebishy, Buffa, Guamis, Blasco-Moreno, & Trujillo, 2015; Hebishy, Buffa, Juan, Blasco-Moreno, & Trujillo, 2017; Hebishy, Zamora, Buffa, Blasco-Moreno, & Trujillo, 2017). The coarse solution containing shell and bioactive materials is exposed to a high shear force under high pressure in the range of 10–150 MPa for HPH and up to 400 MPa for UHPH. HPH and microfluidization techniques have also been used to improve functionalities of different biopolymers such as gums (Gulrez, Al-Assaf, Fang, Phillips, & Gunning, 2012; Harte & Venegas, 2010; Porto, Augusto, Terekhov, Hamaker, & Cristianini, 2015; Villay et al., 2012) and proteins (Dissanayake & Vasiljevic, 2009; Donsì, Senatore, Huang, & Ferrari, 2010; Grácia-Juliá et al., 2008; Hu, Zhao, Sun, Zhao, & Ren, 2011; Iordache & Jelen, 2003; Keerati-U-Rai & Corredig, 2009; Shen & Tang, 2012).

HPH or UHPH, as one of the top-down approaches, is recommended technique to formulate food delivery systems containing biological compounds to improve their physicochemical stability and to increase their delivery efficiency in the target organs and tissues. HPH technologies have been used to produce nanostructures to protect bioactive and labile ingredients prone to environmental stresses such as polyunsaturated fatty acids

(PUFAs and CLAs) (Fernandez-Avila et al., 2016), polyphenols such as anthocyanins (Arranz et al., 2017; Syamaladevi, Insan, Dhawan, Andrews, & Sablani, 2012) and (−)-epigallocatechin-3-gallate (EGCG) (Park, Garcia, Shin, & Kim, 2015; Ru, Yu, & Huang, 2010; Sabouri, Arranz, Guri, & Corredig, 2018), vitamins (Chevalier-Lucia, Blayo, Gràcia-Julià, Picart-Palmade, & Dumay, 2011; Haham et al., 2012), curcumin (Zou et al., 2016), carotenoids (Anarjan, Mirhosseini, Baharin, & Tan, 2011; Khalid et al., 2017; Liang, Huang, Ma, Shoemaker, & Zhong, 2013; Mao et al., 2009; Silva et al., 2011; Sotomayor-Gerding et al., 2016; Tan & Nakajima, 2005; Wang, Tian, & Chen, 2011; Yang, Zhou, & Chen, 2014), folic acid (Ding & Yao, 2013), flavor compounds such as D-limonene (Jafari et al., 2007) and orange oil (Kwan & Davidov-Pardo, 2018), antimicrobials such as essential oils (Donsì, Annunziata, Sessa, & Ferrari, 2011; Donsì, Annunziata, Vincensi, & Ferrari, 2012; Donsì, Sessa, & Ferrari, 2012; Donsì, Annunziata, & Ferrari, 2013; Liang et al., 2012; Zahi, Wan, Liang, & Yuan, 2014), and minerals such as iron (Prichapan, McClements, & Klinkesorn, 2018) and calcium (Márquez, Wagner, & Palazolo, 2018).

This chapter discusses the current (and potential) applications of HPH/UHPH to produce bioactive-loaded nanostructures for encapsulation of food bioactive compounds (lipids, vitamins, phenolic antioxidants, minerals, flavors, or colors) and outlines recent research on nanostructured materials aimed at producing scalable and nontoxic products.

2 Homogenization techniques

Mechanical homogenization refers to the capability of producing a homogeneous size distribution of particles suspended in a liquid, by forcing the liquid under the effect of high pressure through a disruption valve. The first invention for "intimately mixing milk" and other liquids was patented in 1899 by Auguste Gaulin using pressures up to 30 MPa (US Patent no. 756953). Since then, the pressure range of conventional homogenization was extended to 60 MPa, by forcing the fluid through a gap of approximatively 100–300 μm. Recent developments in HPH technology and the design of new homogenization valves capable of withstanding pressures up to 350–400 MPa have opened new opportunities for homogenization processing in food industries. HPH technology should not be confused with high hydrostatic pressure (Trujillo, Capellas, Saldo, Gervilla, & Guamis, 2002), as HPH is a continuous process that combines both homogenization and pressure leading to shear forces, turbulence, and cavitation, produced due to the pass of the fluid throughout the HPH valve. Currently, there are numerous pilot- or lab-scale equipment that use the HPH processing, although the maximum pressure achieved depends on the engineering design. In Table 1, high-pressure processors that combine superior (flow rate and pressure) and scaleable results for pilot and commercial scale environments have been explained.

Microfluidization is a HPH technique (high-pressure impinging jet device) with a proprietary homogenization valve design, which consists of an interaction chamber with microchannels of fixed geometry, where the streams interact producing intense disruptive forces, which results in collapsing of the coarse droplets into fine nanoscale droplets (Jafari et al., 2007). When the liquid exits the chamber, elongational flow applies to the fluid, as in HPH system, facilitating adsorption of the emulsifier at the newly created interface (Henry et al., 2010). Chapter 7 presents more details focusing on microfluidization.

Table 1 High-pressure homogenizers currently available on the market and their operating conditions (flow rate and pressure) suitable for pilot and production environments.

Corporation	Model	Maximum operating pressure (MPa)	Flow rate (L/h)
Microfluidics (the United States)	Microfluidizer M-7250	275	816
BEE International (the United States)	DeBEE 2000P-250/45	310	1500
Avestin (Canada)	EmulsiFlex-C1000	207	1000
Gea Niro Soavi (Italy)	Ariete NS5355	150	60,000
APV (the United Kingdom)	Rannie 75	150	800
Stansted Fluid Power (the United Kingdom)	SPF-nG11300	400	240
Ypsicon (Spain)	UHPH-B1	350	150
	UHPH-C1	350	1200
	UHPH-D1	300	3000
	UHPH-E1	300	10,000

2.1 Ultrahigh-pressure homogenization (UHPH)

The continuous UHPH system is a revolutionary product on the market, as it enables the simultaneous stabilization and sanitization (pasteurization or sterilization depending on the pressure and the inlet temperature of the product applied) of liquid products. The operation can be divided into three basic steps: (1) preheating the product to reach the required inlet temperature of the product (T_{in}), (2) application of the continuous ultrahigh-pressure homogenization, and (3) instant cooling.

Fig. 1 represents the complete UHPH system, which is made up of a product tank (1), filters (2), heat exchangers (3), a UHPH system including pumps (4), a UHPH valve gap of few μm in width (5), heat exchangers using cold water (6), and an aseptic tank (7). The UHPH system can be sterilized prior to processing, using steam or water heated at 140°C for 45 min in a closed circuit, allowing the stabilization of sterilized products, according to the UHPH conditions applied. This UHPH system, which is capable of sterilizing pumpable liquids, has been patented by a group of researchers from Universitat Autònoma de Barcelona (UAB) (Guamis et al., 2015; Lopez et al., 2015).

The production starts by filling the tank with the liquid product (1). The product is then immediately pumped into the heat exchanger (2), where depending on its characteristics, it can be kept at cold/room temperature or preheated up to a specific temperature to reduce its viscosity and/or guarantee the aimed temperature peak reached at the UHPH valve (3) according to the objective of the treatment applied. Once the right temperature has been reached, the product is pumped (4) to the homogenizer, where the

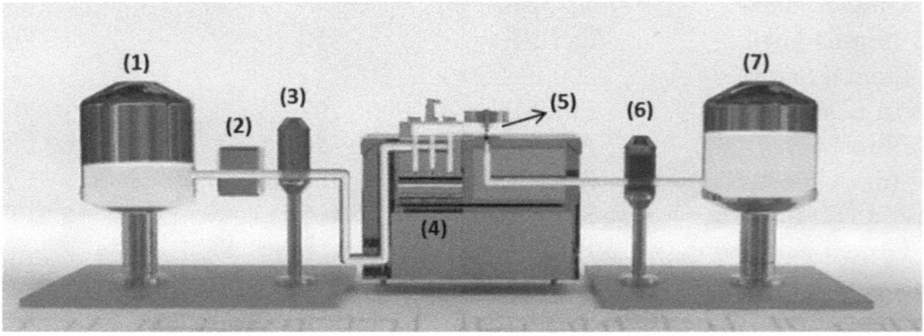

Fig. 1 Graphic scheme representing the complete ultrahigh-pressure homogenization (UHPH) system patented by the Autonomous University of Barcelona (UAB). The system is made up of a product tank (1), filters (2), heat exchangers (3), a UHPH system including pumps (4), a UHPH valve gap of few μm in width (5), heat exchangers using cooled water (6), and an aseptic tank (7).

fluid is forced to pass through the UHPH valve (5) at a specific pressure (up to 400 MPa maximum). At the UHPH valve, such an increase in pressure modifies the temperature of the product, about 2–2.5°C per 10 MPa, in less than a fraction of a second. This short time exposure to temperature might cause insignificant damages to the product. For this reason, and although the temperature reached in the UHPH valve seems to have a crucial effect on sanitization conditions of the product, the UHPH technology could be considered as a nonthermal treatment for the processing of pumpable fluids (Diels & Michiels, 2006; Zamora & Guamis, 2015). Right after the UHPH valve, the homogenized product enters to a cooling system (6) to guarantee its instant cooling, in order to reduce thermal damage. When sterilization conditions are achieved, the fluid can be sent to an aseptic tank (7) for subsequent aseptic packaging, such as in cardboard, polyethylene, or glass.

The UHPH process is an emerging technique, and the innovative design of the homogenization valve, originally developed by Stansted Fluid Power Ltd., was modified regarding the geometry of the valve and the use of ceramic materials, which is known to withstand ultrahigh-pressure levels (Fig. 2B). This type of equipment can attain pressures up to 150–200 MPa, considered as HPH, and up to 300–400 MPa as UHPH (Zamora & Guamis, 2015).

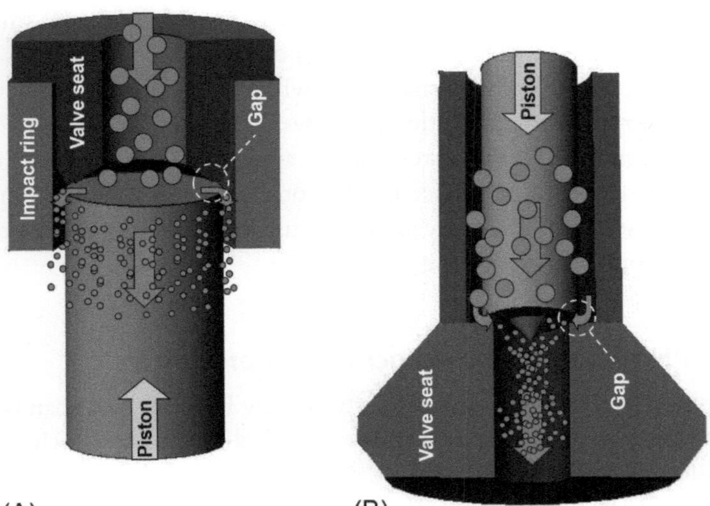

Fig. 2 Homogenizing valves geometries from Manton-Gaulin APV (A) and stansted fluid power (B) (Donsì et al., 2009).

In the UHPH valve, as it is shown in Fig. 2B, the fluid is first fed axially along the mobile part of the valve and then accelerates radially through the narrow gap between the valve and valve seat. On contrary, in the classic valve design from APV-Gaulin, the flow direction of fluid is fed in the opposite direction (Fig. 2A). The geometry of the valve is a critical design element with reference to the quality of the finished product, as well as the physical processes that take place (Donsì et al., 2013; Donsì, Annunziata, et al., 2012; Donsì, Sessa, et al., 2012). Different physical processes have been proposed as the cause of disruption of droplets and cells in UHPH, and each mechanism shows up at a different extent as a result of various factors such as the viscosity of the product and the operating parameters (pressure, temperature, and flow rate): (1) shear stress, which is defined as the viscous forces within particles through a material cross section. It is the only mechanism that contributes to droplet disruption in a laminar flow (Floury, Legrand, & Desrumaux, 2004); however, both laminar and turbulence regimes coexist in this type of homogenizer (Fig. 2B); (2) turbulence, which is a mechanism caused by the inertial forces generated in the highly turbulent region, located just at the exit of the valve (Donsì et al., 2009); (3) cavitation, defined as a phenomenon when liquid passes from low pressure, in which cavities/air bubbles are formed, to high-pressure region, when they will be collapsing between each other. Intensities of cavitation strongly increase with homogenizing pressure (Floury et al., 2004) and decrease with higher fluid viscosity (Diels, Callewaert, Wuytack, Masschalck, & Michiels, 2005); (4) impingement, which is the impact of the droplets or cells against the walls of the UHPH valve; and (5) extensional stress, which stretches the droplets/cells which are disrupted by elongational stresses (Floury et al., 2004).

The continuous UHPH system can be applied in a wide variety of sectors, such as the food, pharmaceutical, biotechnology, chemical, and cosmetic industries. Generally, it can also be used for any pumpable product, chemically compatible with the system.

2.2 UHPH effects on physicochemical properties

The physical properties of suspensions treated by HPH/UHPH can be characterized by a combination of various analytical techniques, such as visual observation, rheological methods, ultrasound profiling, electroacoustic spectroscopy (zeta potential), surface protein concentration, microscopic analysis, nuclear magnetic resonance, and optical methods (dynamic light scattering [DLS], diffusing wave spectroscopy, and Turbiscan™). In practice, DLS has become of vital importance to characterize particle sizing of emulsions, along with the other available techniques. To date a number of papers have been published on the effect of homogenization pressure on the stability of food emulsions (Dumay et al., 2013; Georget, Miller, Aganovic, et al., 2014; Georget, Miller, Callanan, et al., 2014). Many studies have confirmed that homogenization pressure, especially above 100 MPa, significantly influences emulsion stability (Hebishy et al., 2015; Poliseli-Scopel, Hernández-Herrero, Guamis, & Ferragut, 2012). Appropriate pressure is believed to be beneficial to the stabilization and emulsifying properties of protein-stabilized emulsions, and an excessive pressure may cause detrimental effect on particle size distribution and flocculation (Floury, Desrumaux, & Legrand, 2002).

2.2.1 Particle size and viscosity

The particle size of suspensions or dispersions is an important factor in defining structural, oxidative, and sensory characteristics of the products. Numerous studies have reported the effect on reducing particle size by increasing the pressure of the UHPH treatment of food dispersions, such as emulsions (Cortés-Muñoz, Chevalier-Lucia, & Dumay, 2009; Dumay et al., 2013; Floury et al., 2004). The particle size is reported to have a strong influence on the bulk or macroproperties, such as rheology, which gives an indication of emulsion colloidal state and interactions between the different constituents. In general, at a constant emulsifier concentration, viscosity increases with reducing particle size by means of UHPH (Floury et al., 2002; Hebishy et al., 2015). As an example, UHPH-treated milk achieved the narrowest particle size distribution with increasing pressures up to 200 MPa (Pereda, Ferragut, Quevedo, Guamis, & Trujillo, 2007; Roach & Harte, 2008). However, in those studies, when 300 MPa was applied, an increase in the particle size was observed, probably due to the fact that there is insufficient protein for the larger surface area created, and as a result, flocculation occurs. By contrast, when there is enough protein at a constant volume oil fraction, viscosity increases with reducing the polydispersity of size distribution profiles (Cortés-Muñoz et al., 2009). Moreover, at a constant number of particles, the flow behavior will generally change from Newtonian to shear thinning (pseudoplastic) with increasing the oil volume fraction (Chevalier-Lucia, Cortés-Muñoz, Picart-Palmade, & Dumay, 2009). Nevertheless, other factors (e.g., protein denaturation) influence on the viscosity of the emulsions, and they should be carefully taken into account.

2.2.2 Protein denaturation

Proteins respond differently to heat and pressure treatments (Floury et al., 2002). The addition of protein ligands improves physical protein stability, delaying the formation of higher molecular weight aggregates when heat or pressure treatments are applied (Chi, Krishnan, Randolph, & Carpenter, 2003). In this sense, some authors employed heat (40–93°C) and high hydrostatic pressure treatment (50–800 MPa for 30 min) on β-lactoglobulin (β-LG) B to determine the extent of unfolding and aggregation in the presence of different ligands such as conjugated linoleic acid and myristic acid (Considine, Patel, Singh, & Creamer, 2007). It was found that when pressurizing β-LG B dispersions at higher pressures (from 50 to 800 MPa), only conjugated linoleic acid had the ability to inhibit the protein transition from (1) mainly nonnative monomers and dimers, to (2) larger polymers. In contrast, UHPH technology possess different physical processes along with pressure and temperature to contribute to protein denaturation. UHPH processing can be used as a tool to modify the functional aspects of proteins in (1) liquid products, such as dairy-based foods, and (2) the development of new ingredients to change rheological/textural properties of different food products.

In liquid products, UHPH treatment at 300 MPa has shown to cause the same extent of denaturation in soymilk proteins (Cruz et al., 2007) and an increase in the surface hydrophobicity of almond UHPH-beverages (200 and 300 MPa) (Valencia-Flores, Hernández-Herrero, Guamis, & Ferragut, 2013), in comparison with UHT treatment. Surprisingly, cow milk UHPH treated (200 and 300 MPa), regardless of the increase of temperature experienced in the UHPH valve (<85°C and 100°C, respectively), suffered less whey protein denaturation compared with a commercial pasteurized milk (90°C, 15 s), which leads to a better nutritional value of the milk (Pereda, Ferragut, Quevedo, Guamis, & Trujillo, 2009). In a pilot-scale study, researchers showed that the amount of denatured β-LG was greater (17%) for UHPH-treated milk at 200 MPa, which reached approximately 75°C for a very short time, than for light-pasteurized milk at 72°C for 15 s (Zamora, Ferragut, Jaramillo, Guamis, & Trujillo, 2007). Such results corroborated the idea that not only heat but also physical processes and pressure during UHPH treatment induce the denaturation of whey protein.

Moreover, UHPH has been used as a tool to modify the functional aspects of proteins in skim milk (Roach & Harte, 2008), especially casein micelles, which have been used as potential and efficient carrier systems of hydrophobic drugs (Corzo-Martínez, Mohan, Dunlap, & Harte, 2015) and α-tocopherol acetate (Chevalier-Lucia et al., 2011). Conjugates of casein and polysaccharides, such as casein-hydroxypropyl cellulose, have been modified using UHPH processing at 300 MPa with improved stability of the aqueous system (Ye & Harte, 2014). Also the partial denaturation of whey proteins from cow milk treated by UHPH (200 and 300 MPa; single stage) improved the cheese-making properties of milk, in comparison with conventional homogenized–pasteurized milk (Zamora et al., 2007).

2.2.3 Physical stability

Among optical methods, Turbiscan™ overcomes the major drawback of light scattering techniques that is the need of sample dilution, as it modifies the structure and composition of the real emulsion (Huang et al., 2010). In Table 2, an overview of the current UHPH food emulsions have been studied, and their stabilization phenomena have been summarized. The best physical stability of those food emulsions reviewed depends not only on smaller particle size but also on the absence of large particle aggregates (Poliseli-Scopel et al., 2014), the possibly denaturation of milk proteins by the intense temperatures reached at the exit of the UHPH valve (Addo & Ferragut, 2015), and particularly on the increase of protein content and viscosity (Hebishy et al., 2015), which will determine the acceptability by consumers.

2.2.4 UHPH effects on microbial quality

UHPH has been suggested as an alternative to thermal pasteurization to preserve liquid food products (Pereda et al., 2007; Saldo, Suárez-Jacobo, Gervilla, Guamis, & Roig-Sagués, 2009; Tahiri, Makhlouf, Paquin, & Fliss, 2006). The best efficiency of UHPH against most microorganisms depends on the pressure applied and number of homogenization cycles (Diels & Michiels, 2006; Tahiri et al., 2006). All the physical processes explained earlier, along with higher T_{in} of samples, will increase the temperature at the exit of UHPH valve, provoking a very short time thermal effect, which could also induce lethality of microorganisms (Roig-Sagués et al., 2009). In fact, this ultrashort time thermal effect, along with the pressure applied, might be the reason of achieving sterilization conditions (Georget, Miller, Aganovic, et al., 2014). In Table 3 the effect of UHPH inactivation of naturally present bacterial spores in various food emulsions have been reported. UHPH treatment at 200–300 MPa has been shown to be as effective

Table 2 An overview of the stabilization phenomena contributing to the physical stability of different food emulsions prepared by ultrahigh-pressure homogenization (UHPH).

Food emulsion	Pressure (MPa)	T_{in} (°C)	Best physical stability	Stabilization phenomena	Reference
UHPH O/W emulsion (1%, 2%, and 4% whey protein isolate, WPI; 15% sunflower oil; and 5% olive oil)	15 100 and 200	60 25	Emulsions with 4% WPI treated at 100 and 200 MPa against creaming	↑Protein concentration ↑Aqueous phase viscosity ↓Droplet diameter	Hebishy et al., 2015
UHPH soymilk	200 Control (pasteurized: 90°C, 30 s)	55 75	Both UHPH samples against sedimentation	↓Droplet diameter	Poliseli-Scopel, Hernández-Herrero, Guamis, & Ferragut, 2013
UHPH soymilk	300 Control (UHT: 142°C, 6 s)	80	Soymilk treated at 300 MPa against sedimentation	↓Droplet diameter, and same $d_{4,3}$ values during storage indicated no increase in particle aggregates	Poliseli-Scopel, Hernández-Herrero, Guamis, & Ferragut, 2014
Tiger Nut beverage (*orxata de xufla*)	200 300 Control (homogenized-pasteurized: 18 + 4 MPa and 80°C, 15 s)	40 40	Tiger nut beverage treated at 300 MPa against creaming	↓Droplet diameter	Codina-Torrella, Guamis, Ferragut, & Trujillo, 2017

Table 2 An overview of the stabilization phenomena contributing to the physical stability of different food emulsions prepared by ultrahigh-pressure homogenization (UHPH).—cont'd

Food emulsion	Pressure (MPa)	T_{in} (°C)	Best physical stability	Stabilization phenomena	Reference
Donkey milk	100	40	Donkey milk treated at 100 MPa against creaming and sedimentation	↓Droplet diameter ↓T (°C) at the exit of the UHPH valve (high T causes denaturation and aggregation)	Addo & Ferragut, 2015
	200	40			
	300	40			
	Control (pasteurized: 70 and 85°C, 60 s)				
Cow milk	200	30	All UHPH samples against creaming	↓Droplet diameter	Pereda et al., 2007
	200	40			
	300	30			
	300	40			
	Control (high pasteurized: 90°C, 15 s)				

Table 3 An overview of the literature dealing with successful pasteurization and sterilization of beverage emulsions by ultrahigh-pressure homogenization (UHPH) for inactivation of naturally present bacterial spores.

Food emulsion	Pressure (MPa)	T_{in} (°C)	Max *T* (°C) UHPH valve	Initial count (log cfu/mL)	Maximal microbial stability	Reference
Almond beverage	200 300	55 65 75	129.3 ± 12.6	4.00	ND[a] after 20 days at 30°C (300 MPa/65–75°C)	Valencia-Flores et al., 2013
Soymilk	200 300	55 65 75	135.7 ± 1.5	1.54	ND after 20 days at 30°C (300 MPa/75°C)	Poliseli-Scopel et al., 2012
Milk 3.5% fat	200 300	55 65 75 85	139.0 ± 2.7	1.00	ND after 15 days at 30°C and 7 days at 45°C (300 MPa/75–85°C)	Amador-Espejo, Suàrez-Berencia, Juan, Bárcenas, & Trujillo, 2014
Soymilk	300	80	144	2.18	ND after 20 days at 30°C and 10 days at 55°C	Poliseli-Scopel et al., 2014
Almond and soymilk beverage	200 300	55 65 75	135.7 ± 1.5	3.18 (soymilk) 4.21 (almond)	ND (200 MPa/75°C and 300 MPa/65–75°C)	Ferragut et al., 2014

[a]*ND*, not detected.

as thermal pasteurization (Poliseli-Scopel et al., 2012; Valencia-Flores et al., 2013) and sterilization (Amador-Espejo et al., 2014; Poliseli-Scopel et al., 2014) to ensure the microbiological quality. The literature reviewed showed that the pressure and T_{in} combination had a significant impact on the lethal effect of UHPH treatment (Table 3). In milk and soymilk, sterilization has been achieved when a maximum temperature of the fluid passing through the UHPH valve was closer to 140°C (Amador-Espejo et al., 2014; Poliseli-Scopel et al., 2014). In fact the first attempt on the validation of the UHPH for bacterial spores inactivation has been conducted with thermostable strains such as the wet heat sterilization indicator *Geobacillus stearothermophilus* ATCC7953 spores in a model buffer system, and up to 2 $\log_{10}$ were inactivated at 300 MPa and >145°C at the exit of the UHPH valve (Georget, Miller, Callanan, et al., 2014). Moreover, it is important to note that characteristics of the food matrix, such as fat content and viscosity, had also been reported in the literature to influence microbial inactivation (Diels et al., 2005; Roig-Sagués et al., 2009). In this sense, higher viscosity and fat content can contribute to the development of UHPH emulsions for the purpose of bacterial inactivation.

2.2.5 UHPH effects on oxidative stability

Lipid oxidation in emulsions has become a renewed concern, and it is important to develop new technologies to replace the use of synthetic antioxidants (Waraho, Cardenia, Decker, & McClements, 2010). When oil is oxidized, it produces a series of breakdown products in stages, starting with primary oxidation products (peroxides, dienes, and free fatty acids); then secondary products (carbonyls, aldehydes, and trienes), particularly aldehydes, responsible for undesirable odors and flavors; and finally tertiary products (Gunstone & Martini, 2010). Many factors influence the rate of oxidation such as temperature, light exposure, the presence of metals, oxygen, oil composition, and higher specific surface area of emulsion droplets (Waraho et al., 2010).

So far, however, there has been little discussion about the effect of UHPH on oxidative stability of oil-in-water (O/W) emulsions. In particular, Hebishy et al. (2015) examined the effect of UHPH (100 and 200 MPa) on oxidative stability of whey protein-stabilized emulsions with 20% sunflower oil. It was found that UHPH emulsions were more oxidatively stable than their counterparts prepared by conventional homogenization at 15 MPa.

Nonetheless, there is a reliable evidence that UHPH could act as an antioxidant technology on emulsion beverages as the following studies have shown, especially at 200 MPa. Milk UHPH treated at 300 MPa showed a lower aldehyde content compared with commercial milk samples (Pereda, Ferragut, Buffa, Guamis, & Trujillo, 2008), where, in the later samples, heat treatment induced the formation of volatile compounds related to heat and oxidized off-flavors, such as aldehydes and methyl ketones. Similarly, other researchers have successfully applied the UHPH technology (200 and 300 MPa) on "tiger nut beverage" (*orxata de xufla*) as an alternative method of production and preservation, compared with conventionalz homogenized-pasteurized beverages and their combination with preservatives (Codina-Torrella, 2014). After the production, all UHPH-tiger nut beverages resulted in less aldehyde content compared with conventional treatments. Within UHPH-tiger nut beverages, which were stored under refrigeration conditions, the treatment performed at 300 MPa drastically increased the total aldehyde content at day 14 and then kept constant. Comparing with the treatment at 200 MPa, the authors attributed this increase to the temperature reached at the exit of the UHPH valve in the treatment at 300 MPa (~116°C). Similarly, milk samples UHPH treated at 300 MPa presented a higher malondialdehyde and hexanal content compared with those treated at 200 MPa, possibly due to the fact that higher temperature values (up to 100°C) were reached after the exit of the UHPH valve during <1 s when 300 MPa was chosen (Pereda, Ferragut, Quevedo, Guamis, & Trujillo, 2008).

In another study, UHPH soymilk treated at 300 MPa, and reaching 144° C at the exit of the UHPH valve (Poliseli-Scopel et al., 2014), showed decreased values of hydroperoxide index and hexanal levels in UHPH soymilk comparing with the counterpart soymilk treated by ultrahigh temperature (UHT). Interestingly, some authors observed that UHPH (0–350 MPa) did not significantly affect the fatty acid content and the possible loss by oxidation/isomerization of conjugated linoleic acid isomer profile in cow, ewe, and goat milk (Rodríguez-Alcalá, Harte, & Fontecha, 2009). In contrast, UHT sterilization resulted in the isomerization of *cis*9, *trans*11 to *trans*9, *trans*11 isomers, and this extent was greater in those milks that were naturally enriched in polyunsaturated fatty acids (PUFAs). In this sense, there is a huge potential market for HPH to reduce lipid oxidation in the preparation of nanostructured carriers, as long as they comply with the desired shelf life targeted.

2.2.6 UHPH effect on sensory attributes

Sensory evaluation is a discipline that analyzes and measures human responses to the composition of food and drink, for example, appearance, touch, odor, texture, temperature, and taste. Sensory evaluation panels (a group of trained testers) are increasingly being employed to help make vital decisions about food product improvement, quality control, new product development, production process change, and competitor benchmarking. In the literature reviewed, nanoemulsion liquid beverages were found to be evaluated regarding sensory attributes. In a thesis dissertation, UHPH-treated milk (100–300 MPa) and conventionally pasteurized milk were studied (Pereda, 2009) by a trained sensory panel. The panel considered milk UHPH treated at 200 MPa as "rancid," which could have been related to their increase in free fatty acids, as that treatment could not inactivate the native milk lipoprotein lipase (Pereda, Ferragut, Quevedo, et al., 2008). However, when 300 MPa was performed, almost no defects were noticed as no significant differences were observed in proteolysis and lipolysis between UHPH at 300 MPa and pasteurized milk (Pereda et al., 2008; Pereda, Ferragut, Quevedo, et al., 2008). Nevertheless, UHPH milk treated at 300 MPa showed a higher aldehyde content compared with pasteurized milk (Pereda et al., 2008), albeit just 25% of people from the sensory panel characterized the UHPH milk treated at 300 MPa as "oxidized" (Pereda, 2009). Nonetheless, UHPH milk at 300 MPa was preferred to conventionally pasteurized milk by the sensory panel (Pereda, 2009). A possible explanation for this might be the retention of hexanal with increasing homogenization pressure, as it has been previously observed in HPH emulsions (0, 18, 100, and 150 MPa) of sodium caseinate-whey protein (2% + 2%) in which interactions between hexanal and caseinate were suggested (Innocente, Marchesini, & Biasutti, 2014). In a similar way, Codina-Torrella (2014) concluded that UHPH-tiger nut beverages (200 and 300 MPa) showed similar sensory characteristics and acceptability by the sensory panel during their shelf life, even though the treatment at 300 MPa showed a higher aldehyde content. Similar results were obtained with soymilk, where Poliseli-Scopel et al. (2014) concluded that chemical changes that occurred during storage did not impact on the organoleptic characteristics of soymilk treated at 300 MPa when comparing with UHT (sterile beverages). On the contrary, when nonsterile soymilk (treated at 200 MPa) was compared with a pasteurized sample, better acceptance of UHPH-treated soymilk was achieved (67%) (Poliseli-Scopel et al., 2012).

3 Bioactive-loaded nanostructures

Various techniques to produce bioactive-loaded nanostructures/nanocapsules have been proposed depending on the coating material used as a carrier during nanoencapsulation (lipids, carbohydrates, or proteins). Lipid coating materials are typically used in nanoemulsions, nanostructured lipid carriers (NLCs), nanosuspensions, solid-lipid nanoparticles (SLNs), and nanosized liposomes (Akhavan, Assadpour, Katouzian, & Jafari, 2018; Fathi, Mozafari, & Mohebbi, 2012; Rafiee & Jafari, 2018). On the other hand, carbohydrate and protein coatings, or their complexes or conjugates, are used to formulate biopolymer nanoparticles and micelles (Fathi, Martín, & McClements, 2014). These techniques yield nanoscale droplet sizes (10–1000 nm).

Properties of food bioactive-loaded nanostructures (e.g., shelf life, physical and oxidative stability, appearance, texture, and flavor) are determined by their particle size (McClements, 2005). Consequently, it is important for food scientists to be able to reliably predict, control, measure, and report the particle size in a wide range of current food nanostructures, from the nanoscale (nm) to the microscale (μm).

As mentioned before, nanostructures are normally produced by two different approaches; top-down and bottom-up approaches. The top-down approach uses dry milling, HPH or microfluidization, and ultrasound emulsification to reduce the size to nanoscale (Degant & Schwechten, 2002) and require a significant amount of energy for the production of new surfaces. In contrast the bottom-up approach involves self-assembling of individual atoms to build the nanomaterials in a natural and self-regulating manner (Sanguansri & Augustin, 2006), with the energy consumption mainly associated to mixing. In real food systems the use of nanostructures formed by the bottom-up approach is limited, because of the high concentrations of emulsifiers and/or solvents required, and also, for some techniques, environmental modification (dilution, pH, or temperature variation) might bring to their disassembly.

Conventional milling techniques are not highly efficient to reduce the size of product to the nanoscale, and thus products present a high polydispersity that requires subsequent removal of microparticles (Labrune & Palmino, 2004) and also toxical excipients or materials from the production equipment. HPH or UHPH is a highly recommended technique for the production of nanostructures due to its efficiency to reduce particle size, processing highly concentrated suspensions with a high reproducibility, lack of organic solvents usage, less time consumption, and industrial scalability compared with other methods. HPH can be performed under hot (Silva et al., 2011) or cold (Wise, 2000) conditions. The cold mode is more applied for thermal labile compounds, while lipid nanoparticles are usually obtained by using the hot mode to achieve a more uniform morphology.

When producing nanostructures, two things are generally needed: encapsulating materials and encapsulation technique. For instance, some compounds such as curcumin can be well encapsulated within different carriers by using various encapsulation methods; curcumin can be encapsulated in lipids by emulsification (Aditya et al., 2015), in gelatin by electrospinning (Deng, Kang, Liu, Feng, & Zhang, 2017) and in zein by antisolvent precipitation (Dai et al., 2018). However, a delivery system that is successful in protecting the payload compounds (in product behavior) might not be similarly effective in delivering them through the biological barriers for a high bioavailability (in body behavior). As an example, the use of O/W emulsions promoted the synergistic interactions between various antioxidants (ascorbic acid, gallic acid, (−)-epicatechin, and EGCG), but its effectiveness upon intestinal absorption has not been successfully verified (Durand, Zhao, Coupland, & Elias, 2015).

Typically, nanocarrier systems used for encapsulating and protecting food bioactives are carbohydrate, protein, or lipid based. This section aims at summarizing the different nanostructures used for delivery of bioactive compounds based on these three components.

3.1 Lipid-based nanostructures

Lipids have an impact on the nutritional, organoleptic, and physicochemical properties of foods in a variety of ways. Lipids are essential nutrients and considered as a major source of energy in our diets; however, overconsumption of certain types of lipids (saturated fat and *trans*fatty acids) and underconsumption of polyunsaturated lipids has been linked to various health concerns, such as obesity, cardiovascular disease, diabetes, and cancer (PREDIMED Study Investigators, 2015). Subsequently, in the past decade, there has been a trend in the society and the food industry towards consumption of low-fat food, which resulted in a decrease in the intake of lipophilic bioactives (Gao et al., 2014).

One of the most recent nutritional recommendations has been to increase the amounts of polyunsaturated fats incorporated in food (EFSA

Panel on Dietetic Products 2010). In this sense a challenge for food scientists is to manufacture a product with similar characteristics to the original but enriched in polyunsaturated lipids and/or lipophilic bioactives (McClements, 2005). The lipid phase can act as a solvent of these lipophilic bioactives, such as oil-soluble vitamins (A, D, E, and K), or essential oils, to formulate functional foods.

The development of new engineered food systems is a growing area to improve encapsulation, controlled release, and bioefficacy of bioactives in the human body (McClements, 2009). Efficient delivery systems could protect bioactive molecules through their food matrix, maintain their bioactivity during digestion, and trigger their release in the appropriate site of the organism to exert their bioactivity (Faridi Esfanjani, Assadpour, & Jafari, 2018). Typically, lipophilic bioactives are trapped within lipid-based colloidal particles in food matrices, and their release depends on the digestion of triacylglycerols and the formation of mixed micelles for their solubilization (McClements, 2013). When using viscous or gel-like matrices and delivery systems, the extent of lipid digestion is reduced (Li, Hu, Du, Xiao, & McClements, 2011; Li & McClements, 2011), with the concomitant reduction of micelles available to solubilize lipophilic bioactives.

The functional properties of the lipid-based nanostructures, starting from droplet size and surface charge, are significantly affected by the type and concentration of oil used (Komaiko, Sastrosubroto, & McClements, 2015; Sun & Gunasekaran, 2009) and the homogenization technique. In particular, HPH represents an easily scalable, reproducible, high-throughput, and robust mechanical disruption process for the production of submicrometric lipid-based nanostructures (Donsì et al., 2013; Vincekovic et al., 2017). Remarkably, both type/concentration of oil and particle size strongly impact on the rheology and flavor of food lipid-based nanostructures. Generally, viscosity increases when increasing droplet size (Sun & Gunasekaran, 2009). However, particle size plays an important role also on the bioaccessibility of bioactive compounds (Donsì, Cuomo, Marchese, & Ferrari, 2014; Sessa et al., 2014; Yucel et al., 2015).

Lipid-based nanostructures are generally classified into nanosized liposomes, nanoemulsions, solid-lipid nanoparticles (SLNs), and nanostructured lipid carriers (NLCs) (Fig. 3).

3.1.1 Nanoemulsions

An emulsion is a mixture of two or more nonmiscible liquids, usually oil and water; one of them is dispersed as droplets in the other. Addition of

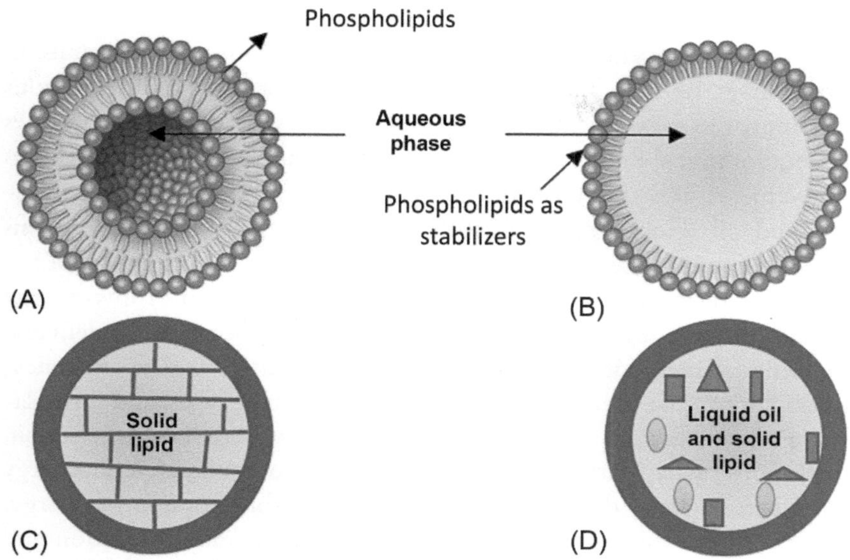

Fig. 3 Classification of lipid-based nanostructures: (A) nanosized liposomes, (B) nanoemulsions, (C) solid-lipid nanoparticles (SLNs), and (D) nanostructured lipid carriers (NLCs).

emulsifiers stabilizes the dispersion and prevents droplets to come together. Emulsions are categorized depending on their droplet size into three different types: macro-, nano-, or microemulsions (Windhab, Dressler, Feigl, Fischer, & Megias-Alguacil, 2005). Macroemulsions have a droplet size in the range of 0.5–100 μm and are normally formulated by classic homogenization methods such as colloid mills. Nanoemulsions are milky to transparent formulated using high-energy methods (HPH and sonication) and low-energy methods, named spontaneous emulsification (Jafari, Assadpoor, He, & Bhandari, 2008). The droplet size of nanoemulsions falls in the range of 100–1000 nm. Finally, microemulsions are transparent in appearance with a droplet size of 10–100 nm, and they are thermodynamically stable systems, which form spontaneously (i.e., self-emulsification).

Nanoemulsions have certain inherent advantages that attracted the food industry. Owing to a small droplet size, nanoemulsions have a high stability against creaming, due to low gravity forces such as the Brownian motion and low attractive forces acting between oil droplets (McClements, 2005). One of the key advantages of nanoemulsions is that they can be prepared in a transparent form for incorporation in clear beverages (Qian & McClements, 2011).

For nanoemulsion preparation, low-energy emulsification (referred to also as spontaneous emulsification) allows microemulsions or nanoemulsion formation (<100nm) and produces bioactive containing nanoemulsions with improved solubility and bioavailability. It is a simple method that does not require much energy and production costs; however, this approach is limited to use only nonionic small molecule surfactants at high concentrations and oil (Huang et al., 2010; Lee et al., 2011). Furthermore, this approach needs the use of solvents, and therefore the total removal of the solvents used becomes very difficult. Thus the application of low-energy methods in food applications is limited and mainly used in drug delivery. High-energy methods have several advantages such as higher efficiency, availability to scale up, using various oil and emulsifier types, and formulation of emulsions with desired characteristics (Salvia-Trujillo, Rojas-Graü, Soliva-Fortuny, & Martín-Belloso, 2013). On the other hand, high-energy methods have some drawbacks such as the high consumption of energy, chemical degradation, and particle coalescence after homogenization and their limits in particle size that can be produced (Jafari et al., 2017).

In the food industry, HPH at pressures up to 350MPa is the most common equipment used to produce fine emulsions (Santana, Perrechil, & Cunha, 2013). When used to treat a preexisting emulsion, if sufficient emulsifier is present to cover the newly created oil-water interface, HPH is effective in reducing the size of the droplets to the nanorange. Increasing the number of passes through the homogenizer could also result in lower droplet size (McClements, 2011). HPH is available for both lab and industrial scale (Setya, Talegaonkar, & Razdan, 2014). The three main mechanisms of the size reduction of droplets into smaller ones are turbulence, shear, and cavitation forces (Schultz, Wagner, Urban, & Ulrich, 2004).

Several recent papers are looking into the issues of emulsion formulation using HPH processing, to develop a better understanding of the mechanisms of nanoemulsion fabrication and control of their properties. For example, the type and concentration of surfactants; hydrophilic-lipophilic balance; lipid/aqueous phase ratio; surfactant/oil ratio; payload composition in nanoemulsion; processing conditions including pressure, temperature, and number of cycle; and homogenization chamber geometry are of the main factors that affect emulsion stability. They all affect the droplet size, ζ-potential, color, and interfacial tension of the nanoemulsion (Donsì, Sessa, Mediouni, Mgaidi, & Ferrari, 2011; Donsì, Annunziata, et al., 2012; Donsì, Sessa, et al., 2012; Lee & Norton, 2013). As an example, Hebishy et al. (2015) reported that UHPH at moderate pressures (below 200MPa) using whey proteins as an emulsifier produced physically and chemically stable emulsions. However, overprocessing at pressures higher than 200 MPa resulted in an increase of droplet/globule size. Appropriate emulsifier(s) and its concentration are critical parameters for the successful design of a nanoemulsion as well; it adsorbs on dispersed-phase surface to help droplet disruption and avoid recoalescence (Ziani, Fang, & McClements, 2012). In this sense, small molecule surfactants can easily form small particles, in comparison with biopolymers such as protein and polysaccharides. Although biopolymers are more efficient stabilizing agents due to their ability to form compact and viscoelastic layers at the interface (Lee, Lefèvre, Subirade, & Paquin, 2009).

Nanoemulsions produced by HPH display a high stability against creaming over conventional systems due to their small droplet size (Lee et al., 2009). Nanoemulsions with a rigid interfacial membrane could be produced by HPH or UHPH, which may help improving the stability of the emulsion and to retard degradation and oxidation (Fernandez-Avila & Trujillo, 2016; Hebishy et al., 2015; Hebishy, Buffa, et al., 2017; Hebishy, Zamora, et al., 2017; Phoon, Paul, Burgner, San Martin-Gonzalez, & Narsimhan, 2014).

Sonication is another technique to form nanoemulsions in which the mechanism is attributed to bubble cavitation (Gharibzahedi & Jafari, 2018; Mason, 1992). Sonication can produce nanoemulsions like those prepared using a microfluidizer at 100 MPa. However, when working at an industrial scale, high-pressure homogenizer or microfluidization is often preferred (Samer & Schork, 1999).

Among the high-energy encapsulation techniques used for producing nanoemulsions at nanosize range, microfluidization was reported to be more effective than valve homogenization (Pinnamaneni, Das, & Das, 2003) and sonication (Jafari, He, & Bhandari, 2006), with smaller size and longer storage stability.

O/W nanoemulsions have successfully been used as delivery systems for the encapsulation and targeted delivery of lipophilic bioactives like PUFAs such as fish oil; essential fatty acids like conjugated linoleic acid; carotenoids such as β-carotene; essential oils; and fat-soluble vitamins such as vitamin A, D, and E (Pathakoti, Manubolu, & Hwang, 2017). Using novel technologies like HPH offers a higher protection and release of bioactive compounds. Water-in-oil (W/O) nanoemulsions, on the other hand, are considered as good vehicles for encapsulating hydrophilic compounds such as polyphenols, water-soluble vitamins, and minerals (Donsì, Sessa, et al., 2011;

Donsì, Wang, & Huang, 2011). Recently, double nanoemulsions, also called emulsion in emulsion or primary emulsion in a secondary emulsion (Dickinson, 2011a), as more complex systems have received increased interest (Assadpour, Maghsoudlou, Jafari, Ghorbani, & Aalami, 2016a, 2016b; Esfanjani, Jafari, & Assadpour, 2017; Mehrnia, Jafari, Makhmal-Zadeh, & Maghsoudlou, 2017). Water-in-oil-in-water (W/O/W) emulsions are the most common form of multiple emulsions, which have been used for encapsulation of hydrophilic compounds (Assadpour et al., 2016a, 2016b; Esfanjani, Jafari, Assadpoor, & Mohammadi, 2015; Hemar, Cheng, Oliver, Sanguansri, & Augustin, 2010). Oil-in-water-in-oil (O/W/O) emulsions can also be used in the encapsulation of lipophilic compounds (Esfanjani et al., 2015). One single double emulsion can be used for encapsulation of both hydrophilic and lipophilic compounds (Aditya et al., 2015).

Up to date, double emulsion formulation is restricted to microscale due to the instability of the double structure, when the size reaches scales lower than 100 nm (Florence & Whitehill, 1985), probably due to the great increase in the Laplace pressure in the inner aqueous droplets. In a recent study (Ding et al., 2017), HPH combined with a low-energy emulsification method was used in a two-step approach for the formulation of double nanoemulsions with a final size below 200 nm. A primary W/O nanoemulsion was formulated by HPH, followed by the reemulsification of this primary emulsion by a low-energy method, assuming that low-energy emulsification would be a smooth method that preserves the inner droplet structure.

Bou, Cofrades, and Jiménez-Colmenero (2014) used HPH to prepare W/O/W double nanoemulsions with four different lipid sources to encapsulate vitamin B_2 with the goal of using this process as functional healthier fat food ingredients. With the aid of dynamic high-pressure microfluidization, double emulsions were found to be an effective approach to encapsulate both water soluble such as vitamins and phenolic compounds (Li, Chen, et al., 2015; Li, Yang, et al., 2015; Bou et al., 2014; Mohammadi, Jafari, Assadpour, & Esfanjani, 2016) and insoluble bioactive compounds (Esfanjani et al., 2015). In a study by Li, Chen, et al. (2015) and Li, Yang, et al. (2015), complex nanoliposomes have been fabricated using double emulsion technique by employing microfluidization to encapsulate vitamin C and medium-chained fatty acids with a high encapsulation efficiency (64%) and stability for 90 days of storage at 4°C.

3.1.2 Solid-lipid nanoparticles (SLNs)

SLNs are colloidal delivery systems that comprise a solid core. They typically contain biocompatible lipids, which are selected specifically for protecting the active compounds from degradation, exhibiting controlled release properties, and increasing bioavailability (Wen, Chen, & Chen, 2018). SLNs are produced via the hot homogenization technique, by preheating the ingredients at temperatures above the melting point of the lipids and processing them by HPH to produce a hot O/W nanoemulsion, which is then rapidly cooled to room temperature to induce fat crystallization (Donsì et al., 2013).

SLNs offer the advantages of combining slower release profile than nanoemulsions and the possibility of targeted drug delivery with an excellent physical stability (Naseri, Valizadeh, & Zakeri-Milani, 2015). For instance, slow release nisin-loaded SLNs were produced by hot homogenization technique to avoid the limited bioactivity due to the interaction with food components, to exert its antimicrobial activity (Prombutara, Kulwatthanasal, Supaka, Sramala, & Chareonpornwattana, 2012).

The crystalline structure of SLNs generally limits loading capacity of bioactive compounds, due to their exclusion after polymorphic transition during storage. The crystalline morphology leads to less space for bioactive compounds to be accommodated (Katouzian, Faridi Esfanjani, Jafari, & Akhavan, 2017). Then the polymorphic transition generates a food active compound–enriched shell, resulting in a burst release of food active compounds (Wen et al., 2018).

SLNs are not widely used in food applications since they require the use of solid lipids, mostly saturated lipids, which consumers, in recent years, tend to avoid (Sagalowicz & Leser, 2010). Therefore, recent studies focused on the incorporation of cocoa butter in the core of SLNs, which is suitable for food applications (Shtay, Tan, & Schwarz, 2018). The formulation with vegetable lipids, however, does not overcome the limitations of SLNs on loading capability.

3.1.3 Nanostructured lipid carriers (NLCs)

NLCs comprise lipid droplets that are partially crystallized and have a less ordered crystalline structure or an amorphous solid structure, which were developed to overcome the limitations of emulsion delivery systems and SLNs (Wen et al., 2018). Therefore, NLCs are considered the next generation of SLNs, because the modified core structure might improve the particles stability and their loading capacity (Naseri et al., 2015).

In general, NLCs are stabilized only by surfactants rather than biopolymers, since surfactant can rapidly adsorb onto the surface of droplets and reduce the interfacial tension during HPH processing (McClements, 2011). Recently, hot HPH was exploited to produce small and stable NLCs, which were developed for a successful controlled release of vitamin D_3 for food applications (Park, Garcia, Shin, & Kim, 2017).

3.2 Carbohydrate-based nanostructures

Natural biopolymers, polysaccharides, and proteins are structural assemblies that have been increasingly applied in food, medical, chemical, pharmaceutical, materials science, and nanotechnology. Proteins and polysaccharides typically exist in foods in many different aggregated states. Each biopolymer ingredient, under certain conditions, shows specific types of aggregation or self-assembly and interactions and intermolecular cross-linking during industrial processing. By tailoring the design of food-grade biopolymers, new structures can be generated with improved functionalities, which can be used as delivery vehicles for bioactive molecules. They can be also used to design packaging materials and for improving food quality and safety. Functional biopolymers can be carefully designed using food-grade proteins and carbohydrates based on several physicochemical principles, including complex formation, segregation, and gelation. In spite of the different advantages of carbohydrate and protein-based nanomaterials, their ability to scale up is a limiting factor, in comparison with lipid-based nanomaterials, due to uncontrollable chemical or heat treatments during processing (Fathi et al., 2012).

HPH has been used in formulating different delivery systems using various proteins, polysaccharides, and their complexes. Table 4 summarizes the HPH operating conditions for the formulation of nanoparticles using food-grade proteins and carbohydrates and their complexes for the delivery of different bioactive compounds. Carbohydrate-based nanoparticles, natural and modified polysaccharides, are promising inexpensive vehicles for micro- and nanoencapsulation of bioactive food ingredients. Polysaccharides, essentially found in plants, animals, algae, and microbial fermentation, are natural carbohydrate polymers that consist of monosaccharides with glycosidic bonds. Polysaccharides are considered the most suitable matrices for encapsulation of bioactives due to their large molecular structure, with different reactive groups located on them, and the presence of both hydrophobic and hydrophilic motifs in their molecular chains, which allows effective entrapment of

Table 4 High-pressure homogenization (HPH) operating conditions for the formulation of nanoparticles using food-grade proteins and carbohydrates and their complexes for the delivery of different bioactive compounds.

Biopolymer	Core formulation	Bioactive	HPH conditions	Purposes	References
Protein-based nanostructures					
Sodium caseinate	Stearin-rich milk fraction	α-Tocopherol	HPH 60 MPa, six cycles	Protecting from degradation	Relkin, Yung, Kalnin, & Ollivon, 2008
β-Lactoglobulin	Orange oil	β-Carotene	Microfluidization 62 MPa, three passes	Improved chemical stability	Qian, Decker, Xiao, & McClements, 2012
Casein Whey protein Soy protein	Corn oil	Quercetin	Microfluidization 82.7 MPa, three cycles	Emulsions with possessed favorable solubility, thermal stability, and quercetin bioaccessibility	Chen et al., 2018
Casein Hydrolyzed whey protein Soy protein Gelatin Combinations	Fish oil	ω3-Fatty acids	Microfluidic device	Stable formulations during 15 days under different stress conditions	Comunian, Ravanfar, Selig, & Abbaspourrad, 2018

Continued

Table 4 High-pressure homogenization (HPH) operating conditions for the formulation of nanoparticles using food-grade proteins and carbohydrates and their complexes for the delivery of different bioactive compounds.—cont'd

Biopolymer	Core formulation	Bioactive	HPH conditions	Purposes	References
Whey protein aggregates	Palm oil	α-Tocopherol	HPH 30 or 120 MPa 12 cycles	Encapsulation of α-tocopherol in hydrogenated palm oil-in-water nanoemulsions stabilized by whey protein aggregates (WPA), instead of only WPA presented a higher efficiency regarding stress processing conditions	Relkin, Shukat, & Moulin, 2014
Zein	Digestible lipid nanoparticles (corn O/W emulsion prepared by 1% Tween 80)	Curcumin loaded in zein nanoparticles prepared by antisolvent precipitation	Microfluidization 82.7 MPa for digestible lipid nanoparticles	Mixed colloidal particles as nanodelivery systems increased bioaccessibility of curcumin by forming mixed micelles in the small intestine	Zou et al., 2016
Zein (complexed with quercetagetin as alcohol-soluble flavonoid compound)	–	–	Microfluidization 50 MPa, two cycles	Together with quercetagetin, microfluidization improved thermal stability, reduced α-helix, increased β-sheets, and exhibited nanospheres with more compact structure and uniform particle distribution	Sun, Dai, Liu, & Gao, 2016

Soy protein isolate, SPI (complex of native and heat-denatured soy protein isolate with rutin)	Soybean oil	–	HPH 40 MPa, two passes	The addition of rutin resulted in a reduced interfacial tension, improved physical stability, and prohibited oxidation of SPI emulsions	Cui, Kong, et al., 2014; Cui, Chen, et al., 2014
Soy protein isolate	Soybean oil	–	UHPH 100–300 MPa	Higher oxidative stability compared with emulsions prepared by conventional homogenization at 15 MPa	Fernandez-Avila & Trujillo, 2016
Soy protein isolate	Soybean oil	Conjugated linoleic acid (CLA)	CH, 15 MPa UHPH 200 MPa	UHPH treatment, comparing with conventional homogenization, improved physical and oxidative stability of the emulsions, protected main biological CLA isomers, suggesting the potential use of UHPH technology for delivering CLA in functional foods	Fernandez-Avila & Trujillo, 2016

Continued

Table 4 High-pressure homogenization (HPH) operating conditions for the formulation of nanoparticles using food-grade proteins and carbohydrates and their complexes for the delivery of different bioactive compounds.—cont'd

Biopolymer	Core formulation	Bioactive	HPH conditions	Purposes	References
Soy protein isolate	Soybean oil	–	Microfluidization 40 MPa, one pass	Improved emulsification efficiency and emulsion stability have been achieved when increasing protein content used; however, high ionic strength and/or the heat pretreatment increased droplet flocculation and surface protein in emulsions	Shao & Tang, 2014
Soy protein isolate	Soybean oil	–	Microfluidization 40 and 80 MPa Two passes	The retention efficiency (RE) of the core material (soy oil) in the dry powders was higher at increased pressure level (80 MPa over 40 MPa)	Tang & Li, 2013
Soy glycinin nanoparticles	Soybean oil	–	Microfluidization 40 MPa, one pass	Increasing the ionic strength prior to microfluidization reduced average droplet size and interfacial surface coverage of the resultant emulsions, indicating improved emulsification performance	Liu & Tang, 2016

Acidic gliadin (wheat gluten cereal protein)	Algal oil	ω3-Fatty acids	Microfluidization 40, 80, and 120 MPa, one pass	High pressures formed physically and oxidatively stable algal oil structured emulsions through protein cross-linking. High pressure triggered the structure change of disulfide bonds that may reduce the IgE-mediated allergies	Liu, Liu, Guo, Yin, & Yang, 2017
Barely protein	Canola oil	β-Carotene	HPH –	Sufficient protection of β-carotene in the simulated gastric fluid. Controlled release in simulated intestinal fluid Improved solubility adsorption Low cytotoxicity through Caco-2 cell	
Pea proteins	Sunflower oil	–	UHPH 100 and 300 MPa, 1–20 passes	UHPH disrupted the protein disulfide bonds, exposing more hydrophobic segments which might improve protein encapsulation efficiency	Donsì et al., 2010

Continued

Table 4 High-pressure homogenization (HPH) operating conditions for the formulation of nanoparticles using food-grade proteins and carbohydrates and their complexes for the delivery of different bioactive compounds.—cont'd

Biopolymer	Core formulation	Bioactive	HPH conditions	Purposes	References
Legume protein isolates (kidney, red and mung beans; KPI, RPI, and MPI), comparing to SPI	Soybean oil	–	Microfluidization 80 MPa, two passes	Although the three legume proteins showed similar emulsifying properties, interfacial properties considerably varied, and microencapsulating properties and RE were much poorer than SPI. This was associated with the difference in interfacial properties between proteins	Liu, Chen, & Tang, 2014
Carbohydrate-based nanostructures					
Modified starches, alone or mixed with soy lecithin	Palm oil, sunflower oil	Terpene mixture, D-limonene	HPH 300 MPa, 10 cycles	Enhanced antimicrobial activity	Donsì, Sessa, et al., 2011
Sodium alginate	Lemongrass oil mixed with T ween 80	Lemongrass oil mixed with Tween 80	Microfluidization 150 MPa, three passes	Nanoemulsions with a minimum droplet size (6 nm), optical transparency, low viscosity, and high stability	

Starch nanoparticles (SNPs)	Span 80 and Tween 80 in cyclohexane	–	HPH cross-linking 10–60 MPa, 1–5 passes	W/O miniemulsion using with a uniform size distribution (300–500 nm) and good sphericity in shape	Shi, Li, Wang, Li, & Adhikari, 2011
Octenyl succinate starch (OSA) modified starches (derived from waxy maize blend with dried glucose syrup, waxy maize and tapioca)	Pressurized ethyl acetate	β-Carotene	HPH 6–6.5 MPa	Maximum encapsulation efficiency (70%–80%) and smaller particle sizes were achieved using a OSA starch refined from waxy maize	de Paz et al., 2014
OSA, whey protein isolate (WPI), Tween 20 and decaglycerol monolaurate (DML)	Medium chain triglyceride (MCT)	β-Carotene	HPH Two stage, different pressures Microfluidization –	Emulsions through microfluidizer had lower temperature and much smaller droplet sizes OSA starch and WPI were less effective for the formation of nanoemulsions with smaller droplets than Tween 20 and DML. WPI formed the most stable nanoemulsion and greatest capacity to protect β-carotene from degradation	Mao, Yang, Xu, Yuan, & Gao, 2010

Continued

Table 4 High-pressure homogenization (HPH) operating conditions for the formulation of nanoparticles using food-grade proteins and carbohydrates and their complexes for the delivery of different bioactive compounds.—cont'd

Biopolymer	Core formulation	Bioactive	HPH conditions	Purposes	References
Starch sodium octenyl succinate	Vitamin E acetate	Vitamin E	Microfluidization 20–300 MPa, 17 cycles	It was possible to produce vitamin E nanoparticle by UHPH and stabilized by microencapsulation with starch, which is a challenge. It was possible to avoid turbidity and physical stability problems encountered by other preparation methods	Chen & Wagner, 2004
Pectins	Itraconazole dissolved in chloroform (0.5%, w/w)	Itraconazole	HIH 100 MPa, 15 min	HPH created smaller particle size nanoemulsions than sonication Pectin nanoparticles maintained their solubility over 12 months of storage and improved redispersibility, dissolution, and stability of drug	Burapapadh, Takeuchi, & Sriamornsak, 2012
Apple pectin	–	Bifidobacteria and Lactobacilli	Microfluidization 155 MPa, six cycles	Microfluidization of pectin produced pectic oligosaccharides that increased the numbers of bifidobacteria and lactobacilli and decreased the number of bacteroides and clostridia	Chen et al., 2013

Mixed biopolymer-based nanostructures					
Soy proteins Soy polysaccharides	–	Folic acid	60 MPa for 2 min then heat treatment at 90°C for 1 h	HPH broke down aggregates of soy protein and helped binding protein with polysaccharide and folic acid Nanogels are a suitable delivery system of folic acid in food and beverages	Ding & Yao, 2013
Tween-20 (TW-20) and WPI	Medium chain triglycerides (MCTs)	β-Carotene	HPH Two stage, three cycles, different pressures	A blend of TW-20 and WPI (1:1) resulted in efficient surfactant–protein interaction and improved the stability of β-carotene nanoemulsion	Mao et al., 2009
Chitosan and alginate	–	(−)-Epigallocatechin-3-gallate (EGCG)	HPH 69 MPa, 0–5 cycles	The optimal conditions for preparing nano particles were a 1:1 alginate/chitosan ratio and three-cycle HPH. High encapsulation efficiency (80.1%) and DPPH radical scavenging activities	Park et al., 2015

Continued

Table 4 High-pressure homogenization (HPH) operating conditions for the formulation of nanoparticles using food-grade proteins and carbohydrates and their complexes for the delivery of different bioactive compounds.—cont'd

Biopolymer	Core formulation	Bioactive	HPH conditions	Purposes	References
β-Lactoglobulin and κ-carrageenan	Canola oil	(−)-Epigallocatechin-3-gallate (EGCG)	HPH 34.5, 69, and 104 MPa, six cycles	Higher pressure and number of cycles resulted in table emulsions. EGCG concentrations higher than 0.5% aggregated β-lactoglobulin and caused emulsion instability. Encapsulated EGCG showed an improved anticancer activity (in vitro) compared with the nonencapsulated EGCG	Ru et al., 2010
Sodium caseinate and alginates	Corn oil	β-Carotene	Microfluidization 120 MPa, 3 cycles	Microparticles exhibited better stability to pH, ionic environment and storage, and stability and bioaccessibility of β-carotene under in vitro gastrointestinal conditions	Liu, Wang, McClements, & Zou, 2018
Zein-propylene glycol alginate (PGA)	–	β-Carotene	Microfluidization 75 MPa, two cycles	β-Carotene in zein-PGA composite nanoparticles exhibited the improved physicochemical stability and the sustained release	Wei, Sun, Dai, Zhan, & Gao, 2018

Hydrolysates of SPI-maltodextrin	Fish oil	PUFA	HPH Two stage at 35 + 10 MPa	Limited hydrolysis of SPI resulted in stable emulsions; however, extensive hydrolysis affected the emulsion stability in a bad way	Zhang et al., 2014
Pea protein isolates and hydrolyzed starch	Miglyol 812 Neutraloel, a medium-chain triglyceride	–	HPH 50 MPa, three cycles	Using high DE value (28) resulted in an improved microencapsulation efficiency of dried emulsions with uniform microcapsules, which protected the interfacial protein film during the drying process	Gharsallaoui, Saurel, Chambin, & Voilley, 2012

bioactive compounds (Fathi et al., 2014). The structure of polysaccharides is greatly stable against harsh conditions, including temperature conditions and enzymatic degradation during digestion, but sensitive to microbial fermentation in the colon (Fathi et al., 2014).

Polysaccharides include starch, cellulose, pectin, guar gum, chitosan, alginate, dextrin, cyclodextrins, new sources of native gums, and their combinations and chemically modified forms.

Polysaccharides are classified, based on their intrinsic charge, into polyelectrolytes, which include cationic, anionic, and neutral subtypes, and non-polyelectrolytes (Liu, Jiao, Wang, Zhou, & Zhang, 2008; Liu, Tian, Zeng, & Chang, 2008). Chitosan is one of the most applicable cationic polyelectrolytes in the production of nanostructures based on polysaccharides, while alginate, heparin, pectin, and hyaluronic acid are examples of anionic polysaccharide nanoparticles. Finally, dextran, pullulan, and pectin represent the neutral polyelectrolyte (Salatin & Jelvehgarim, 2017).

The application of some polysaccharides, when they are present in the natural form, is limited for the encapsulation of bioactive compounds, due to their hydrophilicity like starch granules (Fathi et al., 2014) or low solubility as in cellulose fibers (Đorđević et al., 2016).

Carbohydrate-based delivery systems are formulated using various approaches including coacervation, spray-drying, supercritical fluid, electrospraying, electrospinning, reverse micelle, emulsion diffusion, emulsion-droplet coalescence, salting out, emulsification/solvent evaporation, HPH, and ultrasonication.

3.2.1 Starches

Starch, the second most common polysaccharide (after cellulose) in plants, normally exists as semicrystalline hydrophilic granules of widely varying sizes and shapes. Natural starches are mostly hydrophilic, which limits their application for the encapsulation of hydrophobic bioactive compounds (Fathi et al., 2014). Chemical modification, hydrolysis, and cross-linking of starches improve their functionality to protect volatile flavors during spray-drying, provide good protection against heat and oxidation, increase film formation ability, enhance hydrophobicity, and reduce the swelling ability (de Paz, Martín, Bartolomé, Largo, & Cocero, 2014; García-Tejeda, Salinas-Moreno, Hernández-Martínez, & Martínez-Bustos, 2016).

Further to chemical methods, physical modification of starches using high-energy methods such as HPH, microfluidization, and sonication has shown to break the hydrogen bonding inside crystalline structure of the large starch particles, resulting in improved functionality. Starch nanoparticles have been obtained by HPH with a nanoscale particle size (10 nm), narrow size distribution (Liu, Wu, Chen, & Chang, 2009), high dispersibility, spherical shape, and high ability to form transparent films with higher tensile strength. Starch nanoparticles have been developed using a variety of single and double emulsions with desirable structures and properties using HPH. Shi et al. (2011) employed HPH-cross-linking technique at 10–60 MPa, producing a W/O miniemulsion using sodium trimetaphosphate in which starch nanoparticles showed a uniform size distribution (300–500 nm) and good sphericity in shape. β-carotene-loaded nanoemulsions with a fine size distribution and high stability have been prepared by HPH using octenyl succinate starch, whey protein isolate, Tween 20, and decaglycerol monolaurate (Mao et al., 2009; Mao et al., 2010). UHPH was used to produce beverages fortified with vitamin E nanoparticles (100 nm) and stabilized by a starch coating (Chen & Wagner, 2004).

3.2.2 Celluloses

Cellulose is the most abundant biopolymer, found in the form of macroscopic fibers and microfibrillated aggregates in plant cell walls (Tingaut, Zimmermann, & Sebe, 2012). These fibers are long, flexible, and entangled and contain both amorphous and crystalline regions. Unmodified cellulose has limited use as an encapsulant in food and pharmaceutical industry due to its poor solubility. However, physical, chemical, or biochemical modification of cellulose enhances its functionality as a material for encapsulation (Đorđević et al., 2016). Acid hydrolysis can be implemented on microfibrils to remove the amorphous regions and obtain charged rodlike particles of microcrystalline and nanocrystalline cellulose particles, which have better functionality than original cellulose (Hemmati, Jafari, Kashaninejad, & Barani Motlagh, 2018). Cellulose nanocrystals and nanofibrillar cellulose, produced by acid hydrolysis, enhanced the encapsulation efficiency of curcumin and efficiency and stability of blueberry anthocyanins (Wang, Jung, & Zhao, 2017).

Mechanical methods such as HPH, sonication, grinding, and cryo-crushing have also been used to develop fine cellulose fibers. HPH is a simple, efficient, nontoxic technique that has been used to isolate fine fibrillated nanocellulose from cellulosic biomass. The fibers irreversibly disintegrate with alterations in the morphology and size by shearing and impact forces in the high-pressure valves and the successive and rapid opening and closing (Nakagaito & Yano, 2004). One of the limitations of this method is that cellulose is insoluble in water, and most of the organic solvents can clog the homogenizer valves, leading to nonuniform nanocellulose (Kaushik & Singh, 2011). Another disadvantage of using mechanical methods is the high-energy consumption.

3.2.3 Chitosan

Chitosan, the second most abundant natural polysaccharide after cellulose, can be obtained by the deacetylation of crustacean chitin. Chitosan is composed of β-(1–4)-linked D-glucosamine and *N*-acetyl-D-glucosamine (Hosseinnejad & Jafari, 2016). Chitosan is considered as nontoxic, biodegradable, biocompatible, antimicrobial, and efficient for its encapsulation ability (Azevedo, Bourbon, Vicente, & Cerqueira, 2014). Chitosan-based polyelectrolyte complexes have been prepared to encapsulate bioactive compounds and improve their encapsulation efficiencies, such as curcumin (Akolade, Oloyede, & Onyenekwe, 2017) or tea polyphenols (Liang et al., 2017). Chitosan nanoparticles have also been used as drug delivery systems (Wang et al., 2017). Due to its cationic character, chitosan can form a stable and complex matrix through cross-linking with other anionic agents such as cellulose, pectin, carrageenan, and alginate. For example, complexation by cross-linking of chitosan with cellulose nanocrystals resulted in improved encapsulation efficiency and stability of anthocyanins (Wang et al., 2017). Nanoliposomes containing vitamin C prepared by deposition of positive chitosan and negative sodium alginate were incorporated to mandarin juice (Liu et al., 2016), with improved organoleptic properties, oxidative stability, and vitamin C protection during 90 days of storage.

Due to chitosan poor solubility at physiological pH values, modified forms of chitosan, for example, carboxylated, thiolated, acylated, and conjugates have been reported to be potential alternatives (Elgadir et al., 2015). Chitosan nanoparticles (100–140 nm) were also produced using emulsification solvent diffusion method with a high entrapment efficiency of hydrophobic drugs (Singh & Mishra, 2013). In this method, O/W emulsions containing the drug were obtained by HPH.

3.2.4 Alginates

Alginate, obtained from marine brown algae by alkaline treatment, is an anionic hydrophilic polysaccharide (Tavassoli-Kafrani, Shekarchizadeh, & Masoudpour-Behabadi, 2016). Stability, delayed release, and thermal protection are all important characteristics of alginates when used as encapsulant. Alginate has gel-forming ability when it interacts with ions such as sodium or calcium, which results in beads or capsules suitable for micro- or nanoencapsulation of active compounds such as fats, acidulants and flavors, and a wide range of drug delivery applications (Mokhtari, Jafari, & Assadpour, 2017). Calcium alginate beads were used to protect antioxidant compounds from lemon balm extract (Najafi-Soulari, Shekarchizadeh, & Kadivar, 2016). Moreover, alginate is soluble at intestinal pH values, and it has been incorporated in chitosan and cellulose formulations to increase the release and stability of encapsulated antioxidant compounds, like vitamins, during intestinal transit (He, Zhang, Qi, Huang, & Su, 2015; Santa-Maria, Scher, & Jeoh, 2012). Riboflavin-loaded alginate-casein microspheres showed an encapsulation efficiency of 97.94% through process optimization (He et al., 2015).

HPH has been used to formulate alginate nanoparticles and encapsulate bioactive compounds such as essential oils (lemongrass oil) (Salvia-Trujillo et al., 2013) or β-carotene (Liu et al., 2018; Wei et al., 2018), improving their stability and bioaccessibility.

3.2.5 Pectins

Pectin is a complex polysaccharide rich in galacturonic acid (GalA) with at least 17 different types of monosaccharides (Voragen, Coenen, Verhoef, & Schols, 2009), and it is widely used in food and pharmaceutical industries. Pectin is extracted using strong mineral acids at high temperatures from citrus peels, apple pomace, and sugar beet pulp (Naqash, Masoodi, Rather, Wani, & Gani, 2017). Interestingly, pectin could be incorporated as an emulsifier in nanoemulsions. Formulation of nanoemulsions with essential oils (oregano, thyme, lemongrass, or mandarin) and high methoxyl pectin (HMP) showed a higher adsorption rate onto oil surface and improved stability, compared with Tween 80, a nonionic surfactant (Guerra-Rosas, Morales-Castro, Ochoa-Martínez, Salvia-Trujillo, & Martín-Belloso, 2016). The same authors tested the antimicrobial activity of these HMP nanoemulsions loaded with essential oils against *Escherichia coli* and *Listeria innocua*. Lemongrass-HMP nanoemulsion (11 and 13 nm) showed the highest antimicrobial activity. Pectin has also been employed in the encapsulation of peppermint (Wang, Soyama, & Luo, 2016) and clove (Sharma et al., 2017) essential oils in nanoemulsions to produce antimicrobial beverages. Clove formulations were stable to heat treatment and salt concentrations, except at pH 3–5 (Sharma et al., 2017). Burapapadh et al. (2012) formulated pectin-based nanoemulsions loaded with a model drug (itraconazole) using HPH and sonication. HPH produced smaller particle size nanoemulsions than sonication. Pectin nanoparticles maintained their solubility over 12 months of storage. High-pressure microfluidization has also shown its ability to increase the prebiotic potential of apple pomace pectin. In this sense, Chen et al. (2013) reported that microfluidization of pectin produced pectic oligosaccharides that increased the viability of bifidobacteria and lactobacilli and decreased the number of bacteroides and clostridia.

3.2.6 Gums

Gums and their derivatives have extensively been used due to their unique functionalities in food and pharmaceutical applications. Gums are classified according to their origin into plant-based gums, that is, plant exudate gums (e.g., gum Arabic), seed gums (e.g., guar gum), seaweed gums (e.g., carrageenans), and microbial exudate gums (e.g., xanthan gum) (Prajapati, Jani, Moradiya, Randeria, & Nagar, 2013). Gums have been employed, in combination with other encapsulants, to encapsulate flavors, aromas, phenolic compounds, antioxidants and nutraceutical compounds (Busch et al., 2017), fish oil (Botrel et al., 2017), curcumin (Tan, Xie, Zhang, Cai, & Xia, 2016), and drugs (Pinto, 2010).

Applications of HPH and microfluidization techniques have been investigated with gums, including xanthan gum (Gulrez et al., 2012; Harte & Venegas, 2010), guar gum (Villay et al., 2012), carrageenan (Harte & Venegas, 2010), tragacanth gum (Villay et al., 2012), gum Arabic (Villay et al., 2012), and cashew tree gum (Porto et al., 2015). The main purpose of these studies was to investigate the effect of HPH technology on the functional properties of gums such as viscosity reduction, molecular weight reduction, polydispersity, radius of gyration reduction, increased mass recovery, water uptake, and hydration rate reduction. Due to the high glass transition temperatures of gums, low-shear forces during conventional homogenization may not be sufficient to make structural changes in high-molecular-weight gum structures. HPH, on the other hand, may break the covalent bonds and change the structure due to the temperature rise during the treatment (Floury et al., 2002). HPH, as a pretreatment before spray-drying, was used to encapsulate red raspberry. Powders produced were less hygroscopic (3.45%), which consequently resulted in better encapsulation efficiency and retention of anthocyanins in the final powder (Syamaladevi et al., 2012).

Spray-dried fruit juices may show a high hygroscopicity and stickiness after drying. High-molecular-weight polymers, such as gum Arabic and maltodextrins, are added to overcome such challenges (Bhandari & Hartel, 2005). Further, they act as encapsulating carrier, which entrap and retain the flavor constituents and protect the bioactive and antioxidant compounds (Galmarini, Zamora, Baby, Chirife, & Mesina, 2008).

3.3 Protein-based nanoparticles

Food proteins are considered the most effective delivery systems for bioactive ingredients (Mohan, Rajendran, He, Bazinet, & Udenigwe, 2015). Proteins have the binding ability to hydrophobic and hydrophilic small molecules (Considine & Flanagan, 2008), surface activity, self-assembling properties, and good gelation characteristics (Livney, 2010). In general, proteins come from animal or plant origin. Animal proteins, including dairy proteins, gelatin, and egg proteins, are more commonly used for encapsulation than plant proteins (Can Karaca, Low, & Nickerson, 2015) due to their higher solubility, lower molecular weight, and flexibility. Plant proteins used for encapsulation include soy proteins, legume proteins (pea, lentil, or chickpea), and cereal proteins (zein and barley) (Can Karaca et al., 2015). Plant proteins are most preferred by some groups of consumers and are generally less expensive (Can Karaca et al., 2015).

Protein nanoparticles are simple to manufacture by just heating protein solutions and are compatible with the high-pressure emulsification process. Moreover, protein-based nanoparticles provide several advantages for the encapsulation of bioactive compounds, due to their suitability to be transformed and used without deleterious effects in the biological systems. Protein-based nanoparticles, on the other hand, have some drawbacks as carriers. These are associated with allergenicity after ingestion of egg, soybean, fish, and dairy proteins (Hefle, Nordlee, & Taylor, 1996). Protein nanoparticles can be formed with a high surface area that can cause rapid digestion and release in the intestinal tract, leading to adverse effects. They can also facilitate the adsorption of digestive or metabolic enzymes, interfering with their normal gastrointestinal function (McClements, Xiao, & Demokritou, 2017).

HPH has been applied to modify the physicochemical, functional, and structural properties of proteins such as thermal, conformational, and surface properties; hydrolytic stability; and protein aggregation state (Dissanayake & Vasiljevic, 2009; Donsì et al., 2010; Grácia-Juliá et al., 2008; Hu et al., 2011; Iordache & Jelen, 2003; Keerati-U-Rai & Corredig, 2009; Shen & Tang, 2012).

3.3.1 Milk proteins

In milk, casein micelles stabilize and deliver micronutrients such as calcium. Casein micelles have shown also a protective role for bioactive compounds such as vitamin D_2 and D_3 (Haham et al., 2012; Semo, Kesselman, Danino, & Livney, 2007) via reassembled casein micelles, which provided a protective effect against degradation of encapsulated vitamins. Sodium caseinate has been used to encapsulate β-carotene in an O/W emulsion using HPH with a high physical and oxidative stability (Yi, Li, Zhong, & Yokoyama, 2014).

Whey proteins have functional groups such as thiol/disulfide, carboxyl, and aromatic amino acids in their core. When whey proteins are heated or high-pressure homogenized, they undergo changes from high-ordered folded structures to denatured structures, and the functional groups become exposed to the surface to form different structures (Relkin, Eynard, & Launay, 1992). Denatured whey proteins are highly surface hydrophobic and can be self-assembled to form aggregates of different sizes and solubility under the influence of hydrophobic and covalent bonding (Keerati-U-Rai & Corredig, 2009). Alternatively, protein aggregation for the generation of nanoparticles can be achieved through cross-linking of proteins with enzymes such as peroxidase, transglutaminase, or tyrosinase (Xu, Teng, & Wang, 2016). Furthermore, these hydrophobic groups can bind to small-molecular-weight hydrophobic compounds such as fatty acids, retinol, α-tocopherol, phenolic compounds, phospholipids, aromatic compounds, and surfactants (Liang, Tremblay-Hébert, & Subirade, 2011; Relkin & Shukat, 2012; Zimet & Livney, 2009).

Whey proteins are normally used in the form of hydrogels or nanoparticles for encapsulation and controlled delivery of bioactive compounds (Abaee, Mohammadian, & Jafari, 2017). Whey protein hydrogels have been formed and successfully utilized to encapsulate and study the controlled release behavior of caffeine (Gunasekaran, Ko, & Xiao, 2007), iron (Remondetto, Beyssac, & Subirade, 2004), and probiotic bacteria such as *Lactobacillus rhamnosus* GG (Doherty et al., 2012).

Milk proteins can be used to formulate emulsion-based excipient foods (coingested with bioactive-rich foods) to improve the bioaccessibility of hydrophobic compounds such as quercetin. Chen, McClements, et al. (2018) and Chen, Shu, et al. (2018) used casein, whey protein, or soy protein to form excipient emulsions with possessed favorable solubility, thermal stability, and quercetin bioaccessibility. Excipient emulsions were obtained by HPH for three times at 80 MPa. In particular, casein was shown to be better at adsorbing onto oil-water interface and contributed to a more stable interfacial layer and improved bioaccessibility than whey and soy proteins.

The microfluidization method has been used to encapsulate and protect fish oil (Comunian et al., 2018) using dairy proteins. Fish oil-in-water emulsions encapsulated by casein, hydrolyzed whey proteins, soy proteins, and gelatin and their combinations were tested under different stress conditions. The formulations were stable during 15 days under different stress conditions; however, the best protection to the fish oil against oxidation was achieved when a combination of 0.5% (*w*/w) gelatin and casein (ratio 1:1) is used.

3.3.2 Plant protein-based nanoparticles

Zein, mainly obtained from corn meal by the method of solvent extraction, contains defined surface hydrophilic and hydrophobic domains that can form spherical colloidal nanoparticles by precipitating using the antisolvent precipitation method. Zein is considered as an ideal candidate for drugs and micronutrient delivery systems (Zhong & Jin, 2009) and has been applied also for the encapsulation of hydrophilic compounds, such as EGCG (Donsì, Voudouris, Veen, & Velikov, 2017), because zein is characterized by its high hydrophobicity and low water solubility but soluble in alcohol. Dynamic high-pressure microfluidization has been demonstrated to decrease α-helix content and increase the β-sheets of zein-derived nanoparticles (Sun et al., 2015), which leads to an increase in zein water solubility. When combining HPH (75 MPa) with heating, the synergistic effect can change the secondary structure of zein protein to a new shape (from sphericity to the dumbbell-like) and improve thermal stability (Sun et al., 2016).

Phenolic compounds can interact with proteins via noncovalent hydrophobic interactions, which may change the protein functional properties such as solubility, heat stability, and digestibility (Labuckas, Maestri, Perelló, Martínez, & Lamarque, 2008). Quercetagetin, as an alcohol-soluble flavonoid compound, was used to change the characteristics of zein nanoparticles by using high-pressure microfluidization, so that it will change the zein structure and the primary interactions with quercetagetin. Together with quercetagetin, microfluidization improved thermal stability, reduced α-helix, increased β-sheets, and exhibited nanospheres with more compact structures and uniform particle distribution (Sun et al., 2016). In this sense, Cui, Kong, Chen, Zhang, and Hua (2014) prepared a complex of native and heat-denatured soy protein isolate (SPI) with rutin in an O/W emulsion by hydrophobic interactions using HPH at 40 MPa during two passes.

The addition of rutin resulted in a reduced interfacial tension, improved physical stability, and prohibited oxidation of SPI emulsions, which was more obvious in emulsions formulated using native SPI, due to lower interfacial accumulation and physical location of rutin in heat-treated emulsions.

HPH has been also used to prepare other plant protein-based nanodelivery systems. Nanoemulsions based on pea proteins were formulated using UHPH (100–300 MPa) with a size of <200 nm (Donsì et al., 2010). The authors reported that UHPH disrupted the protein disulfide bonds, exposing more hydrophobic segments that improved protein encapsulation efficiency. In another study, increasing homogenization pressure during the emulsion preparation for micro −/nanoencapsulation decreased the emulsion viscosity and the oil droplet size (Yu, Wang, Yao, & Liu, 2007) improving microencapsulation and core material efficiency (Rusli, Sanguansri, & Augustin, 2006). This was related by Nesterenko, Alric, Silvestre, and Durrieu (2014) to the intense mechanical forces applied during HPH (50 MPa), which resulted in partial unfolding of protein chains and oil droplet dispersion and forming a protective layer on the oil droplets surface.

One of the methods to produce nanoemulsion delivery systems with improved encapsulation efficiency is using the emulsion cross-linking method, by cross-linking the protein at the oil-water interface using different types of cross-linkers (e.g., transglutaminase, laccase, and genipin). Even if this method can improve emulsion stability, the emulsion structure may adversely affect digestion and nutrition (Johnston, Nickerson, & Low, 2015). HPH was able to cross-link barley proteins for β-carotene encapsulation (Yang et al., 2014) and acidic gliadin for the encapsulation of edible algal oil (Liu et al., 2017). The nanoparticles produced provided sufficient protection of β-carotene in the simulated gastric fluid, with a complete release after 7 h in the simulated intestinal environments and bioaccessibility in in vitro studies using Caco-2 cells. Liu et al. (2017) reported that high pressures can form stable structured emulsions through protein cross-linking via intermolecular disulfide bonds. In their study, microfluidization was used in an innovative platform approach for the development of acidic gliadin-based delivery systems to design functional structured edible algal oils.

Macroscopic soy protein gels produced by enzymatic cross-linking using transglutaminase (Guo et al., 2016) or via heat treatment of soy proteins (Matsumiya & Murray, 2016) have been broken down to microgels by mechanical disruption. This approach resulted in small rigid particles (50–250 nm) that were highly effective as stabilizers for producing Pickering emulsions. Soy protein microgel particles produced by the breakage of a heat-treated macrogel using HPH were highly efficient to stabilize aerated systems and O/W emulsions (Matsumiya & Murray, 2016).

Soy proteins are functional and nutritional food ingredients derived from natural sources and considered as excellent carriers for bioactives over other food proteins (Abaee et al., 2017). Various delivery systems, including hydrogels, microspheres, and microemulsions, have been developed based on soy proteins (Liu, Jiao, et al., 2008; Liu, Tian, et al., 2008; Yu, Yan, Sun, & Gu, 2008). Highly stable nanoparticles (150–200 nm) have been successfully prepared using soy protein isolate (SPI) (Teng, Luo, & Wang, 2012; Zhang, Wu, Yang, He, & Wang, 2012), which exhibited a high encapsulation efficiency for hydrophobic compounds such as curcumin (Teng et al., 2012). UHPH as an emergent technology was found to cause conformational changes in SPI at the oil-water interface and result in fine and stable nanoemulsions based on SPI (<1 μm) (Fernandez-Avila & Trujillo, 2016).

It is well known that not only the encapsulating materials are important for obtaining good encapsulating effects of proteins and properties of spray-dried emulsions but also the optimization of the operating conditions, one of which is the homogenization condition used, prior to and during homogenization. For instance, increasing the ionic strength of the aqueous environment prior to high-pressure microfluidization to prepare soy glycinin nanoparticles was found to highly reduce average droplet size and interfacial surface coverage of the resultant emulsions (Liu & Tang, 2016). Shao and Tang (2014) studied the effect of varying protein concentrations (0.5%–4.0%, *w*/*v*) and ionic strength (0–300 mM NaCl) on the stability of native and preheated SPI-stabilized emulsions produced by high-pressure microfluidization (40 MPa). The authors reported that an improved emulsification efficiency and emulsion stability have been achieved when increasing protein content used; however, high ionic strength and/or the heat pretreatment increased droplet flocculation and surface protein in emulsions.

Protein type, nature, and structure can also affect emulsification and encapsulation efficiency. Liu et al. (2014) studied the microencapsulation capability of three legume protein isolates (kidney, red, and mung beans), comparing with SPI under HPH at 80 MPa for two passes. The authors reported that although the three legume proteins showed similar emulsifying properties, interfacial properties considerably varied, and microencapsulating properties and retention efficiency were much poorer than SPI. Microencapsulation efficiency varied even among the three legume proteins, which was associated by the authors with the difference in interfacial properties among proteins.

Increasing the time intensity of homogenization conditions in processing encapsulation systems containing proteins such as protein-stabilized emulsions is intended to extend physical stability against coalescence and is normally applied to reduce the particle size of the emulsion and adsorbing more protein at the interface of oil droplets. However, these intensive conditions can limit the use of a bioactive compound in a delivery system if it is incorporated in the melted lipid phase before homogenization and results in hindered shelf life. Relkin et al. (2014) used HPH (30 or 120 MPa for 12 cycles) to formulate matrix carriers consisted of whey protein aggregates, alone or stabilizing hydrogenated palm O/W nanoemulsions to encapsulate α-tocopherol. Encapsulation of the lipophilic α-tocopherol in fat droplets, instead of only whey protein aggregates presented a higher efficiency regarding stress processing conditions (30% degradation in protein aggregates at both pressures and 5% and 15% in nanoemulsions at 30 and 120 MPa, respectively). Fernández-Ávila et al. (2016) reported that SPI nanoemulsions prepared using UHPH technology at pressures up to 200 MPa were stable against lipid oxidation; however, using higher pressures (i.e., 300 MPa) resulted in higher temperatures at the homogenization valve exit, which caused the displacement of protein to rearrange into large aggregates. The microencapsulating properties of SPI by spray-drying were affected by pressure level of homogenization (Tang & Li, 2013), as the retention efficiency of the core material (soy oil) in the dry powders was higher at increased pressure levels (80 MPa over 40 MPa).

3.4 Nanostructures made of proteins-polysaccharide complexes or conjugates

Proteins and polysaccharides are generally good candidates for the encapsulation of bioactive ingredients. However, the encapsulated system is sensitive to environmental conditions such as temperature, salt, and pH (McClements, 2004); so, single protein or polysaccharide cannot be ideal (Cui, Chen, Kong, Zhang, & Hua, 2014). As an example, multilayers formed through the electrostatic interactions between proteins and polysaccharides could help minimize the deficiency of the protein or polysaccharide to stabilize emulsion systems (Dickinson, 2011a, 2011b).

The choice of wall material could affect not only the encapsulation efficiency of the bioactive compound but also its release behavior (Dias, Ferreira, & Barreiro, 2015). Matrix for encapsulation should be chosen carefully, as each bioactive compound has unique characteristics and different physicochemical behaviors. It has been revealed that conjugates prepared through Maillard reaction resulted in emulsions with better stability (Zhang et al., 2012). Tang and Li (2013) reported that blending SPI with lactose improved the microencapsulation efficiency, retention efficiency, and dissolution behavior; however, it decreased the stability of reconstituted emulsions and the glass transition temperature. Complex nanoparticles (162–243 nm), as a delivery systems for hydrophobic vitamin D_3, were also successfully produced using a mixture of SPI and carboxymethyl chitosan by cogelation using Ca^{2+} (Teng, Luo, & Wang, 2013). Although the complex nanoparticles resulted in a reduced vitamin release (42.3%) under simulated gastric conditions comparing with SPI nanoparticles (86.2%), vitamin release under intestinal conditions from complex nanoparticles was higher (36.0%) than pure SPI nanoparticles (8.2%). Other complexes have been used for the encapsulation of bioactive ingredients of which zein-chitosan complex nanoparticles were formulated by a low-energy phase separation method to protect vitamin D_3 and SPI-zein complex microparticles were prepared using Ca^{2+}-induced cold gelation method to protect riboflavin/vitamin B_{12} (Wan, Guo, & Yang, 2015). Coacervation process has been also used to encapsulate bioactive ingredients such as capsaicin by cross-linking gelatin with glutaraldehyde (Wang, Kluge, Leisk, & Kaplan, 2008), gelatin with acacia (Xing, Cheng, Yi, & Ma, 2005), and whey proteins with pectin for loading of limonene (Ghasemi, Jafari, Assadpour, & Khomeiri, 2017, 2018) to obtain nanocapsules in the size range of 100–600 nm. The authors reported an improvement in melting point and the thermal pyrolysis temperature, encapsulation efficiency, and dispersion properties.

HPH has been used to form nanoparticles from mixed biopolymers. Soy proteins and soy polysaccharide nanogels were produced using HPH at 60 MPa followed immediately by heating at 90°C during 1 h (pH = 4) to encapsulate folic acid (FA) (Ding & Yao, 2013). These nanogels were able to protect FA from decomposition by oxygen, heat, light, and at acidic conditions and helped the release of FA at neutral pH in the intestine. The authors mentioned that the purpose of applying HPH was to facilitate the binding of soy protein with soy polysaccharide and folic acid at pH = 4.0 by breaking down the original aggregates of soy protein.

HPH was also used for the encapsulation of *Quercus resinosa* leave infusion within a lactose-sodium caseinate matrix by spray-drying (Rocha-Guzmán et al., 2010) with an improved antioxidant activity. Paprika oleoresin emulsions (1 paprika oleoresin-four wall material) were prepared using HPH and gum Arabic (GA) or SPI as wall materials and then spray-dried (Rascón, Beristain, García, & Salgado, 2011). Microcapsules showed an increased carotenoid retention and protection against oxidation. However, in contrast to microcapsules prepared with SPI, GA microcapsules were unable to retain their structural integrity at water activities above 0.743. High-pressure microfluidization was used for encapsulation of D-limonene by mixing maltodextrins with a surface-active biopolymer, that is, whey protein concentrate (WPC), modified starch (Hi-cap), and Tween 20 as wall materials (Jafari et al., 2007). As reported by the authors, microfluidization resulted in small emulsion particle sizes (700–800 nm) with a narrow distribution and high retention (86%) of D-limonene.

Polyphenols, such as EGCG, have been shown to be biologically active, especially their antioxidant activity, when they are in an encapsulated form. Although single biopolymers, β-lactoglobulin for instance, have been used for their encapsulation, mixed biopolymers were also used. EGCG-loaded nanoparticles were fabricated through a complex anionic alginate and cationic chitosan by HPH with higher antioxidant activity (DPPH radical scavenging activity) (Park et al., 2015). HPH (34.5, 69, and 104 MPa for six cycles) was also successfully used to prepare O/W nanoemulsions stabilized by κ-carrageenan and β-lactoglobulin to encapsulate EGCG (Ru et al., 2010). The increase of homogenization pressure and the number of cycles decreased particle size and viscosity with minimal changes in particle size for 14 days, at EGCG concentration below 0.5%. However, higher EGCG concentrations than 0.5% bonded β-lactoglobulin and caused its aggregation and a significant increase in particle size and instability of the emulsions. Moreover, encapsulated EGCG showed an improved anticancer activity (in vitro) compared with the nonencapsulated EGCG.

Modification of proteins or the protein-polysaccharide conjugates by enzymatic hydrolysis before homogenization was reported to be an effective strategy to modify protein functional properties (Amarowicz, 2010) and to prepare an effective emulsifier. For instance, hydrolysates of sodium caseinate-maltodextrin conjugates were reported to be low-molecular-weight emulsifiers and stabilizers with more water solubility (O'Regan & Mulvihill, 2010). Maillard reaction combined with controlled enzymatic hydrolysis of the conjugates before HPH could result in improved emulsion stability. Zhang et al. (2012) described a method for conjugating soy β-conglycinin with dextran under dry-heated Maillard reaction followed by hydrolysis using trypsin and HPH at 30 MPa. The authors reported that hydrolysates of β-conglycinin-dextran conjugates with a degree of hydrolysis of 2.2% protected emulsions against bridging flocculation during homogenization and resulted in emulsions highly resistant to extreme pH values, salt, and heat. This was attributed by the authors to an increase of molecular flexibility of the conjugates and efficient adsorption and unfolding of peptide molecules at the oil droplet interface. Hydrolysates of SPI-maltodextrin conjugates have been shown to be a good carrier for PUFAs. Zhang et al. (2014) designed microcapsules of fish oil by HPH (two stage at 45 MPa total pressure) and freeze drying. Limited hydrolysis of SPI resulted in stable emulsions; however, extensive hydrolysis affected the emulsion stability in an unfavorable way. Gharsallaoui et al. (2012) produced microcapsules based on pea protein isolate and starch hydrolysates with a different dextrose equivalent (6–28) using HPH (50 MPa for three cycles) and subsequent spray-drying. Using high dextrose equivalent value (28) resulted in an improved microencapsulation efficiency of dried emulsions with uniform microcapsules, which protected the interfacial protein film during the drying process.

4 Applications of HPH in the preparation of bioactive-loaded nanostructures

HPH has found a wide range of applications as a physical-mechanical disruption process for the production, through a top-down approach, of bioactive-loaded nanostructures for several applications of industrial interest, exploiting the significant advantages it offers in terms of ease of operation, scalability, reproducibility, and high throughput (Donsì et al., 2013).

4.1 Phenolics and antioxidants

The strong driving force to the development of functional food products containing bioactive compounds, following the consumers' demand for health promotion and disease prevention through diet and nutrition, promoted the encapsulation especially of phenolics and antioxidant compounds. As an example, synthetic antioxidants such as butylated hydroxyanisole (BHA) and butylated hydroxytoluene (BHT), which are highly effective at controlling lipid oxidation, nowadays are considered undesired additives by consumers, which demand organic and natural/free food additives (Taghvaei et al., 2014; Taghvaei & Jafari, 2015). Therefore, several attempts to expand the number of naturally derived bioactive compounds of interest as food additives has promoted the screening of natural products, such as piceatannol, which is an analogue and metabolite of resveratrol

Table 5 Main antioxidant compounds encapsulated in nanodelivery systems prepared by high-pressure homogenization (HPH).

Antioxidant	Nanodelivery system	HPH processing conditions	Reference
Green tea catechins	O/W nanoemulsions, using Tween 80 as emulsifier and sunflower oil as lipid carrier	Two stage HPH (PANDA, GEA Niro Soavi), cooling system at 5°C, operated at 20–100 MPa for 1–6 passes	Gadkari, Shashidhar, & Balaraman, 2017
Quercetin	NLCs made of linseed oil and glyceryl monostearate, using Tween 80 as emulsifier	Hot emulsification by HPH (LiTu Mechanical equipment Engineering Co.), at 70°C, at 80 MPa for six passes	Huang, Wang, Li, Xia, & Xia, 2017
Naringenin	Polymeric micelles made of polaxamer 407, polaxamer 188, hydroxypropyl methyl cellulose	Micronization by HPH (Micron-LAB 40, APV) at 150 MPa for for 15 passes	Sumathi, Tamizharasi, & Sivakumar, 2017
Ferulic acid	NLCs made of cetyl palmitate and oleic acid, using poloxamer 188 and Tween 80 as emulsifiers	Hot emulsification by HPH (Micron-LAB 40, APV), at 75°C, at 50 MPa for 10 passes	Hassanzadeh, Arbabi, Atyabi, & Dinarvand, 2018
Resveratrol	O/W nanoemulsions, using glycerol monooleate and sugar ester as emulsifiers and peanut oil as lipid carrier	HPH (Nano DeBEE, Bee International, Inc.) at 300 MPa for 10 passes	Sessa et al., 2014
Curcumin	NLCs made of sunflower oil, medium-chain triglycerides, or palm oil, using monoolein or lecithin as emulsifiers	Hot preemulsification followed by HPH (orifice valve) at 5°C, at 200 MPa for three passes	Yucel et al., 2015

Table 5 Main antioxidant compounds encapsulated in nanodelivery systems prepared by high-pressure homogenization (HPH).—cont'd

Antioxidant	Nanodelivery system	HPH processing conditions	Reference
Lycopene	O/W nanoemulsions, using OSA modified starch as emulsifier and medium-chain triaglycerides as lipid carrier	HPH at 110 MPa for three passes	Li, Li, Xiao, Li, & Xie, 2018
Lutein	SLNs or NLCs made of cetyl palmitate, glyceryl tripalmitate, carnauba wax, caprylic/capric triglyceride, using caprylyl/capryl glycoside or Tween 80 as emulsifier	Hot HPH (Micron-LAB 40, APV) at 80°C, at 50 MPa for thee passes	Vishwanathan, Wilson, & Nicolosi, 2009

(Seyed, Jantan, Bukhari, & Vijayaraghavan, 2016). However, an alternative approach could be the use of the currently available "natural" antioxidants, by enhancing their activity through encapsulation.

Table 5 reports some examples of compounds, which have been encapsulated in nanometric delivery systems with the contribution of HPH processing in recent years, including the main flavonoid and nonflavonoid phenolic compounds as well as carotenoids. However, it must be highlighted that most of the studies reported in the literature that claim benefits from phenolic compounds encapsulation have been observed in vitro, such as the increase of bioaccessibility, or antioxidant activity on cell lines, whereas the benefits on human health upon consumption of functional food products have not been yet investigated in details (Arranz, Corredig, & Guri, 2016). Similarly the issues related to the interaction with other ingredients, effect of food processing, storage, and digestion, need further elucidation (Salvia-Trujillo, Soliva-Fortuny, Rojas-Graü, McClements, & Martín-Belloso, 2017).

In general, O/W nanoemulsions are especially suitable for the encapsulation of lipophilic compounds. However, if the payload compounds are scarcely soluble in oil (i.e., curcumin, quercetin, resveratrol, or lutein), the nanometric delivery systems require a more complex structure (i.e., SLNs or NLCs) to prevent the compound crystallization in the aqueous phase. Often, amphiphilic molecules are used to create a hydrophobic microenvironment around the bioactive molecules.

The attractive red color of some fruits and vegetables such as red raspberries is related to the anthocyanin molecules concentration. Anthocyanin pigments, responsible for red, blue, and violet color of many fruits, vegetables, and flowers, have several health-promoting properties. They also affect the sensory quality of the product. These pigments are sensitive to degradation by high temperature during processing (Mahdavee Khazaei, Jafari, Ghorbani, & Hemmati Kakhki, 2014; Mahdavi, Jafari, Ghorbani, & Assadpoor, 2014). Novel nonthermal treatments such as HPH are good candidates for the encapsulation of anthocyanins. HPH at 207 MPa for five passes was used to encapsulate red raspberry (*Rubus idaeus*) before spray-drying, which resulted in greater anthocyanins concentration (56.9 mmol Trolox equivalents = g DW), compared with (48.6 mmol Trolox equivalents = g DW) in nontreated samples (Syamaladevi et al., 2012). However, powders exhibited a lower brightness, redness, and yellowness. The greater antioxidant activity and better retention of anthocyanins are related to better coating and encapsulation of anthocyanins using high pressures, as a result of reduced particle size when HPH applied. These results have been also reported using other polymers to encapsulate other bioactive molecules such as limonene. Retention of limonene was greater at higher homogenization pressures in a freeze-dried gum Arabic-gelatin-sucrose matrix (Kaushik & Roos, 2007).

Although HPH was used to increase the encapsulation efficiency and retention of encapsulated raspberries, they exhibited lighter and less red color intensity compared with the encapsulation using conventional homogenization (Syamaladevi et al., 2012). Coating anthocyanins using HPH at high pressures could result in masking the surface color of the powder. As reported by the authors, as the HPH treatment resulted in temperature rise, the authors reported that ice cooling was needed to minimize the possible degradation of anthocyanins.

Several functional bioactive properties have also been described for plant extracts as rosemary supercritical extract, enriched in carnosic acid (Arranz et al., 2017). O/W emulsions containing lactoferrin or whey protein isolate were formulated by using HPH to encapsulated carnosic acid. Encapsulation protected carnosic acid from degradation during in vitro digestion. Moreover, emulsions enhanced bioavailability of carnosic acid in a cell model of the intestinal barrier.

4.2 Vitamins

Vitamin delivery exerts promising health benefits. Some important vitamins of interest to be encapsulated are water insoluble, such as vitamin A or D, therefore, less bioavailable (Katouzian & Jafari, 2016). Many dairy products in the market are fortified with such liposoluble vitamins, where the lipid phase can act as a solvent. Nonetheless the low liposoluble vitamins intake from our diet has brought out the need to boost their activity through delivery strategies using other food structures (Betoret, Betoret, Rocculi, & Dalla Rosa, 2015).

Food microstructure and HPH processing affect the in vitro bioaccessibility of the encapsulated bioactive compounds. In this sense, HPH (at varying pressure range) increased the in vitro bioaccessibility of α- and β-carotene in carrot O/W emulsions (5% olive oil), but not of lycopene in tomato emulsions (5% olive oil). In the carrot emulsions the mechanical disruption by HPH could be utilized to further increase the carotene bioaccessibility of an already preprocessed material, increasing its nutritional value (Svelander, Lopez-Sanchez, Pudney, Schumm, & Alminger, 2011). In a similar way, HPH (at varying pressures up to 140 MPa) produced barley nanoparticles of β-carotene-loaded vitamins, which were able to be internalized by Caco-2 cells and accumulated in the cytoplasm (Yang et al., 2014).

Many research studies have shown the applications of micelles as bioactive-loaded food delivery systems. In particular, casein micelles are considered as potential natural nanovehicles for encapsulation, protection, and delivery of unstable hydrophobic bioactive components (Semo et al., 2007). As an example the physicochemical stability and bioavailability of vitamin D have been improved by UHPH reassembled casein encapsulation, investigated in a human trial (Haham et al., 2012). As previously mentioned, not only the food microstructure but also the UHPH processing could be positively correlated with bioaccessibility. In fact, UHPH treatment of casein micelle dispersions at 200 MPa could modify casein micelle organization with an enhancement of the binding of α-tocopherol acetate, well known as vitamin E (Chevalier-Lucia et al., 2011).

4.3 Fatty acids

Nowadays, there is still a growing interest in the role of the essential fatty acids (mostly omega-3 polyunsaturated fatty acids and conjugated linoleic acid or CLA) and their addition to food products, which offers a wide range of health benefits and a well-considered marketing point of view too (Esfahani, Jafari, Jafarpour, & Dehnad, 2019; Ghorbanzade, Jafari, Akhavan, & Hadavi, 2017). However, those unsaturated fatty acids are more prone to lipid oxidation, which is the main reason for the deterioration of fats and oils due to their reaction with molecular oxygen, resulting in unsaturated hydroperoxides (Rodríguez-Alcalá, Alonso, & Fontecha, 2014). UHPH has improved the oxidative stability of emulsions containing soybean oil when compared with emulsions produced by conventional homogenization (Fernandez-Avila & Trujillo, 2016).

When focusing on in vitro digestion, no significant differences were seen for *trans*9*cis*11 and *cis*10*trans*12-CLA and their isomerization products (*cis*,*cis* and *trans*,*trans*-CLA) between different homogenization techniques (conventional and UHPH) and formulation treatments (Fernandez-Avila et al., 2016; Fernandez-Avila & Trujillo, 2017). However, matrix interactions in emulsions affected the CLA release, as the free CLA (nonemulsified control) showed a tendency to be oxidized. This could be explained by the small free fatty acid release percentages of emulsions during in vitro digestion (Fernandez-Avila & Trujillo, 2016), as well as the high amount of protein adsorbed at the interface in similar emulsions without CLA (Fernandez-Avila & Trujillo, 2016), which may decrease the extent of lipolysis. These results are in concordance with other studies in which lipid digestion is reduced when matrices are more viscous (Li et al., 2011; Li & McClements, 2011). In turn the reduction of the lipolysis rate can affect the release of bioactives and induce satiety (Guri & Corredig, 2014).

Similar percentages of CLA bioaccessibility (measured in the cell) and bioavailability (measured in the basolateral phase) in Caco-2 cells were found for conventional homogenization and UHPH emulsion treatments (Fernandez-Avila et al., 2016; Fernandez-Avila & Trujillo, 2017). These results are in line with those of Heo, Kim, Pan, and Kim (2016), who compared lecithin nanoemulsions containing both CLA forms (CLA triglycerides, TG, and CLA-FFA), where no differences in bioavailability were observed. Their emulsions also exhibited different particle size: 70 and 230 nm, respectively. However, they directly applied the emulsions to the Caco-2 cells, instead of submitting the emulsions through in vitro digestions for further application of the digestates. It was also found that UHPH treatment, compared with conventional homogenization, improved both physical and oxidative stability of the CLA emulsions (Fernandez-Avila & Trujillo, 2017).

The microfluidization method has been used to encapsulate and protect fish oil (Comunian et al., 2018) using dairy proteins. Fish O/W emulsions encapsulated by casein, hydrolyzed whey proteins, soy proteins, and gelatin and their combinations were tested under different stress conditions. The formulations were stable during 15 days under different stress conditions; however, the best protection to the fish oil against oxidation was achieved when a combination of 0.5% (*w*/w) gelatin and casein (ratio 1:1) was used. Liu et al. (2017) reported that high pressures can form stable structured emulsions through protein cross-linking via intermolecular disulfide bonds. In their study, microfluidization was used in an innovative platform approach for the development of acidic gliadin-based delivery system to design functional structured edible algal oil, highly rich in PUFA. Fernández-Ávila et al. (2016) produced SPI nanoemulsions using UHPH technology as a carrier for soybean oil rich in PUFA with a higher oxidative stability compared with conventional emulsions. Hydrolysates of SPI-maltodextrin conjugates have been shown to be a good carrier for PUFA. Zhang et al. (2014) designed microcapsules of fish oil by HPH (two stage at 45 MPa total pressure) and freeze drying. Limited hydrolysis of SPI resulted in stable emulsions; however, extensive hydrolysis affected the emulsion stability in a bad way.

4.4 Antimicrobials

The use of essential oils (EOs) as natural food preservatives against spoilage bacteria and foodborne pathogens has become increasingly popular. In addition, because of their nonspecific mechanism of action (Seow, Yeo, Chung, & Yuk, 2014), they prevent the development of microbial resistance.

Because of the lipophilic nature of EOs, which reduces their efficiency in the aqueous part of foods, where microorganisms proliferate, they require appropriate carriers, such as O/W nanoemulsions (Donsì, Annunziata, et al., 2011; Donsì, Sessa, et al., 2011; Donsì, Wang, et al., 2011). Nanoemulsions have been shown to enhance the antimicrobial activity and the physicochemical stability (Donsì & Ferrari, 2016). In particular, several studies have demonstrated that encapsulation of EOs in nanoemulsions significantly improves their dispersion in food, as well as masking their taste, because of the reduced partition into the aqueous phase (Donsì, Annunziata, et al., 2012; Donsì, Sessa, et al., 2012). In addition, the high reactivity of EO molecules, which might cause undesired interactions with food ingredients, is mitigated by their encapsulation and interaction with emulsifiers (Shah, Davidson, & Zhong, 2012).

EO-loaded nanoemulsions are routinely fabricated by colloid milling, ultrasonication, or HPH (Donsì et al., 2013). However, HPH is a continuous process and one of the most efficient techniques for EO emulsification on medium or large scale (Donsì, Annunziata, et al., 2012; Donsì, Sessa, et al., 2012; Stang, Schuchmann, & Schubert, 2001). In particular, HPH enables to achieve a uniform, narrow size distribution, for processing conditions in the range of 200–300 MPa (Donsì, Annunziata, et al., 2012; Donsì, Sessa, et al., 2012), even though for high emulsifier-to-oil ratios (≥ 1), significantly smaller pressures are required (Liang et al., 2012; Zahi et al., 2014).

However, several studies demonstrated that, in order to mitigate the effects of Ostwald ripening deriving from the nonnegligible solubility in water of many EOs components and increase the physical stability of EOs nanoemulsions, the addition of a low water solubility carrier oil, such as vegetable oils, is required.

Table 6 reports the formulation and processing conditions for the preparation of nanoemulsions encapsulating different EOs using HPH.

According to literature studies, nanoemulsions are likely to contribute to the EO interaction with the microbial cell membranes as a consequence of the increased surface area and passive transport via the porin proteins because of nanometric size (Donsì, Annunziata, et al., 2012; Donsì, Sessa, et al., 2012; Nazzaro, Fratianni, De Martino, Coppola, & De Feo, 2013); of the fusion with the membrane phospholipid bilayer promoted by the emulsifier, causing the EOs release at the desired sites; of the sustained release over time, due to the nanoemulsion droplets acting as nanotanks (Donsì, Annunziata, et al., 2012; Donsì, Sessa, et al., 2012); and of the eventual electrostatic interaction with the negatively charged microbial cell (Chang, McLandsborough, & McClements, 2015).

4.5 Food colorants

Color delivery is a topic of great interest to the food industry, especially when it involves natural colorants (i.e., bioactive molecules, such as anthocyanins). Similarly to what was discussed for antioxidants or antimicrobials, also color-bearing molecules need to be protected from interaction with other food ingredients and from processing, storage, and preparation conditions, to maintain unaffected their optical behavior. With this respect, nanostructured delivery systems might significantly contribute to protecting such molecules and to develop novel optical properties (Akhavan & Jafari, 2017).

Table 6 Main antimicrobial compounds encapsulated in nanodelivery systems prepared by high-pressure homogenization (HPH).

Essential oil/ essential oil component	Carrier oil	Emulsifier	HPH Conditions	Mean droplet size (nm)	Reference
Carvacrol	Peanut oil	Tween 20 and glycerol monooleate	HPH (orifice valve) at 200 MPa for five passes	120	Donsì et al., 2014
Cinnamaldehyde	–	Tween 80	Microfluidizer (Microfluidics Corporation) at 140 MPa for 10 passes	127	Jo et al., 2015
Eugenol	Bean oil	Tween 80 and sodium dodecylsulfate	Microfluidizer (Microfluidics Corporation) at 70 MPa for five passes	150	Li, Chen, et al., 2015; Li, Yang, et al., 2015
Lemongrass oil	–	Tween 80 + sodium alginate (1:1)	Microfluidizer (Microfluidics Corporation) at 100 MPa for 10 passes	6	Salvia-Trujillo, Rojas-Graü, Soliva-Fortuny, & Martín-Belloso, 2014
D-Limonene	–	Modified starch	HPH (Nano DeBEE, BEE International,) at 300 MPa for 10 passes	367	Donsì, Annunziata, et al., 2011; Donsì, Sessa, et al., 2011; Donsì, Wang, et al., 2011
D-Limonene	Sunflower oil	Tween 20 + monoolein	HPH (Nano DeBEE, BEE International,) at 300 MPa for 10 passes	131	Donsì, Annunziata, et al., 2011; Donsì, Sessa, et al., 2011; Donsì, Wang, et al., 2011
Peppermint oil	Medium chain tryglycerides	Modified starch	HPH (EmulsiFlex-C3, Avestin Inc.) at 100 MPa for 10 passes	200	Liang et al., 2012
Tea tree oil	–	Soy lecithin	HPH (Nano DeBEE, BEE International,) at 300 MPa for 10 passes	74	Donsì, Annunziata, et al., 2011; Donsì, Sessa, et al., 2011; Donsì, Wang, et al., 2011

Several studies have, in fact, reported that the colloidal properties of a dispersion, and, in particular, mean particle size, refractive index, and concentration of the disperse phase significantly influence the optical behavior of the system, by affecting light scattering and absorption. Moreover, when a chromophoric material is loaded in the dispersed phase, this effect is further enhanced (McClements, 2002). The interaction of the light with the dispersed phase results in absorption (or transmission), determining the chromatic properties of the dispersion, such as blueness, greenness, and redness, and scattering (or reflection), determining the turbidity, opacity, or lightness of the dispersion (McClements, 2002). The chromaticness of the medium is reported to increase, and hence the lightness to decrease, when the carrier particle size is increased, the color-bearing molecule concentration is increased, and the carrier particle concentration is decreased (Chanamai & McClements, 2001).

A disperse phase with a mean particle size below 60–100 nm causes the dispersion to be optically transparent, which is a desired property especially for clear beverages. In contrast, dispersions with a particle size above the 60–100 nm range are turbid or opaque (McClements, 2002). As a consequence the use of nanostructured delivery systems for chromophoric molecules requires the precise tailoring of the properties of the delivery system, including its particle size distribution, refractive index, concentration, and stability of the disperse phase.

To date, only a few examples have been reported to specifically address the issues of food colorant encapsulation in nanostructured delivery systems. For example, lutein or β-carotene-loaded nanoemulsions were proposed as a natural antioxidant, with an additional health-beneficial contribution towards the reduction of the incidences of certain chronic diseases (Rostamabadi, Falsafi, & Jafari, 2019). O/W nanoemulsions stabilized by β-lactoglobulin resulted efficiently in protecting β-carotene, which exhibited a reduced rate of color fading because of slower β-carotene degradation in comparison with Tween 20-stabilized nanoemulsions (Qian et al., 2012).

More recently, lutein-loaded nanoemulsions, formulated with Tween 80 and medium-chain triglycerides and obtained by spontaneous emulsification, were tested also against color fading over storage. However, despite the nanoemulsion stability at ambient temperature for up to 1 month, lutein degradation caused also some color fading (Surh, Decker, & McClements, 2017).

In another application the optical properties of food emulsions were investigated in a model system, consisting of structured O/W emulsions containing Nile red as a lipophilic colorant. Through the layer-by-layer electrostatic deposition, the emulsions were stabilized by interfacial protein-pectin complexes, subsequently covalently cross-linked through the use of laccase. Remarkably the emulsions prepared without laccase were the most stable to color fading, whereas in the presence of laccase, rapid color changes occurred (red to blue to white) (Selig et al., 2018).

Several studies focused on the use of biopolymeric colloidal particles to deliver colorant. For example, zein colloidal particles, in the nanometric range, were successfully exploited to encapsulate natural colorants, such as curcumin or indigocarmine, and therefore appear suitable to develop tunable strategies of color delivery for food applications (Patel, Heussen, Dorst, Hazekamp, & Velikov, 2013). In another application from the same authors, zein was used also to produce white colloidal particles. The key step was the purification of zein from colored impurities by washing with pure ethanol. The nanometric particles, obtained by antisolvent precipitation, could be exploited as an edible alternative for TiO_2 as white colorant in wet or liquid food products (de Boer, Kok, Imhof, & Velikov, 2018).

Optically active ethyl cellulose nanoparticles, also prepared by antisolvent precipitation, were instead loaded with organic UV filter molecules, such as oxybenzone, avobenzone, and octinoxate, to constitute the base of more environmentally friendly UV-protective coatings (Hayden, Kibbelaar, Imhof, & Velikov, 2018).

Interestingly the use of bioactive pigments, such as curcumin, indigocarmine, or β-carotene, as natural colorants, once they are stabilized into suitable delivery systems, is of great industrial interest because of an additional contribution to the health-beneficial properties of the foods.

Mixed colloidal particle nanodelivery systems containing both lipid and protein nanoparticles have been designed to increase the bioaccessibility of lipophilic bioactives. Digestible lipid nanoparticles produced using high-pressure microfluidization were mixed with zein nanoparticles containing curcumin produced by antisolvent precipitation method (Zou et al., 2016). As reported by the authors, curcumin bioaccessibility was increased by forming mixed micelles in the small intestine. Highly stable nanoparticles (150–200 nm) have been successfully prepared using soy protein isolate (SPI) (Teng et al., 2012), which exhibited high encapsulation efficiency for hydrophobic compounds such as curcumin.

Carotenoids, including β-carotene, lycopene, lutein, and astaxanthin, are natural pigments that can be found in fruits, vegetables, leaves, and sea products. Consumption of foods and beverages enriched with carotenoids may reduce the incidence of some chronic diseases. However, there are some limitations for using carotenoids in foods such as their poor water solubility, chemical degradation, high melting point, and low bioavailability (Rezaei, Fathi, & Jafari, 2019; Rostamabadi et al., 2019).

HPH has been successfully used to encapsulate carotenoids using different nanostructures, including lipid-based nanostructures, such as nanoemulsions, and protein and carbohydrate-based nanostructures and their conjugates or complexes.

Liang et al. (2013) used HPH to encapsulate β-carotene in an O/W nanoemulsion stabilized with modified starches and studied the emulsions stability and bioaccessibility of β-carotene. The authors reported an increase in bioaccessibility of the encapsulated β-carotene from 3.1% to 35.6%. Although modified starch produced a thick and rigid interfacial membrane around the oil droplets, which improved physical stability, it decreased the bioaccessibility of β-carotene.

Several factors affect the ability of HPH to produce highly stable encapsulated β-carotene nanoemulsions, one of which is the type of emulsifier or surfactant used. Nanoemulsions, O/W (60–140 nm), have been formulated using conventional homogenization and microfluidization, followed by solvent evaporation to encapsulate β-carotene (Tan & Nakajima, 2005). Homogenization pressure and cycle affected the emulsion particle size and formed fine and homogeneous system. It was demonstrated by the authors that emulsions oxidized faster, which was linked to the chosen emulsifier, suggesting that a special emulsifier system has to be utilized to stabilize chemically unstable β-carotene.In this sense, Mao et al. (2009) assessed different small and large molecular weight emulsifiers (Tween 20, octenyl succinate starch, decaglycerol monolaurate, whey protein isolate, and a mixture of Tween 20 and whey protein isolate) to obtain nanoemulsions for encapsulation of β-carotene using HPH. The authors reported different mechanisms of stabilization for each emulsifier. Lower particle size and interfacial tension were shown by Tween 20 and decaglycerol monolaurate; however, they aggregated easily. Octenyl succinate starch and whey protein isolate, on the other hand, resulted in emulsions with a strong interfacial membrane that had larger droplet size but showed higher stability. The best protection of β-carotene was provided by nanoemulsions stabilized with whey protein isolate.

Yuan, Gao, Zhao, and Mao (2008) studied the effect of emulsifier type (Tween 20, 40, 60, and 80) and concentration (4%–12%), homogenization pressure (two stage at 100/10 MPa), temperature (30–70°C), and number of homogenization cycles (three cycles) on the stability of β-carotene nanoemulsions. Smallest droplet size and narrowest size distribution were obtained using Tween 20. HPH (up to 100 MPa) and more cycles (up to 50°C) improved the physical stability of the nanoemulsions; however, emulsion stability deteriorated at higher temperatures. Although β-carotene nanoemulsions had good physical stability, significant chemical degradation of β-carotene occurred during storage.

A study evaluated stability and bioaccessibility astaxanthin loaded in O/W nanoemulsions prepared using two emulsifiers (modified lecithin [ML] and sodium caseinate [SC]) formulated using HPH (100 MPa) in four passes (Khalid et al., 2017). The authors mentioned that both emulsifiers resulted in stable emulsions containing astaxanthin; however, physicochemical stability of emulsions is strongly affected by environmental stresses and the emulsifier nature. Although SC-stabilized emulsions showed high physical and chemical stability (>70%) during 30 days of storage, bioaccessibility of astaxanthin was significantly higher (33%) in ML than in SC-stabilized nanoemulsions (6%). This may indicate a strong effect of emulsifier type on bioaccessibility. In two other studies of the same research group (Chen, McClements, et al., 2018; Chen, Shu, et al., 2018; Shu et al., 2018), ginseng and gypenoside saponins as natural emulsifiers were used for producing and stabilizing astaxanthin-loaded O/W nanoemulsions under the same conditions of homogenization pressure (100 MPa) and number of passes (four passes). The results of both studies using ginseng and gypenoside saponins demonstrated their high surface activity and showed the stability of the astaxanthin-loaded O/W nanoemulsions prepared by these emulsifiers against creaming. The results also have shown an improvement of heat stability of emulsions, which is a limiting factor for incorporating astaxanthin in food products. However, emulsions were more sensitive to other environmental conditions such as pH and ionic strength that markedly impacted the physical stability of nanoemulsions and caused droplet coalescence.

Another factor that could affect the stability of β-carotene-loaded nanoemulsions is the method used for emulsion preparation. For the sake of finding cheaper alternative methods to substitute high-energy methods such as HPH, Silva et al. (2011) used a high-energy emulsification-evaporation technique to prepare β-carotene nanoemulsions and study their stability. According to the authors, although this system was able to produce physically stable emulsions, the latter showed poor chemical stability during storage. Another limitation of this technique to be applied in the food industry is the use of an organic solvent as hexane during preparation. Similarly,

Anarjan et al. (2011) prepared astaxanthin nanodispersions (110–165 nm) using emulsification evaporation and reported that the most stable nanodispersion was produced using HPH at 30 MPa with three passes, followed by evaporation at 25°C.

Other carotenoids have been emulsified in O/W nanoemulsion systems using HPH including lycopene, lutein, and astaxanthin. Astaxanthin and lycopene are of industrial interest due to their potential health benefits, but their use is limited by their poor bioavailability due to sensitivity to chemical, thermal, and shelf stability towards processing and storage conditions. HPH up to 100 MPa was used to prepare linseed O/W nanoemulsions containing astaxanthin and lycopene with high bioaccessibility and enhanced physical and oxidative stabilities (Sotomayor-Gerding et al., 2016). Nanoemulsions prepared at 100 MPa were partially digested (66%) and showed high bioaccessibility (>70%), indicating the suitability of these emulsions to encapsulate carotenoids that can afford processing conditions and storage.

Nanoemulsions prepared by HPH for the encapsulation of carotenoids can increase the extent of digestion and improve bioaccessibility, as a result of the particle size reduction. However, on the other hand, one of the drawbacks of using nanoemulsions is the high oxidation rate with decreasing the particle size. Sodium caseinate (2%) O/W emulsions containing β-carotene (0.1%) were prepared using HP microfluidization (10–100 MPa) with droplet sizes (368–124 nm), with linear and inverse relation to homogenization pressures (Yi et al., 2014). It was reported by the authors that nanoemulsions (size <200 nm) were very stable to coalescence or flocculation over 30 days; however, β-carotene stability towards oxidation lowered as droplet diameter decreased to <200 nm. In vitro digestion increased with higher particle diameter, while the extent of lipolysis and bioaccessibility of β-carotene increased with decreasing particle diameter.

Plant protein nanoparticles have also been produced by HPH to protect carotenoids. Yang et al. (2014) were able to encapsulate β-carotene in barley protein nanoparticles with small sizes (90–150 nm) and narrow size distributions using HPH with no need of using cross-linking reagents or organic solvents. Barley nanoparticles provided sufficient protection of β-carotene during simulation of gastric fluid in addition to complete release in the simulated intestinal environments after 7 h. In vitro studies showed that barley protein nanoparticles could be internalized by Caco-2 cells and accumulated in the cytoplasm.

Several combinations of biopolymers have been used to produce carotenoid-loaded micro−/nanoparticles. β-Carotene-loaded microparticles were successfully produced using caseinate/alginate through microfluidization (Liu et al., 2018). Microparticles exhibited better stability to pH, ionic environment and storage, and stability and bioaccessibility of β-carotene under in vitro gastrointestinal conditions. Similarly, Wei et al. (2018) fabricated β-carotene-loaded zein-propylene glycol alginate composite nanoparticles using an emulsification-evaporation method by microfluidization treatment at 75 MPa for two cycles. The authors reported that β-carotene in zein-propylene glycol alginate composite nanoparticles exhibited improved physicochemical stability and sustained release. Paprika oleoresin emulsions (1 paprika oleoresin/4 wall materials) were prepared using HPH and gum Arabic or SPI as wall material and then spray-dried (Rascón et al., 2011). Microcapsules showed an increased carotenoid retention and protection against oxidation. However, in contrast to microcapsules prepared with SPI, gum Arabic microcapsules were unable to retain their structural integrity at water activities above 0.743.

4.6 Flavors and aromas

Flavor is one of the main properties of food products. Consumers' perception of food is greatly influenced by its flavor. Food flavor can be described as a combination of smell, taste, and/or trigeminal stimuli. Aroma is defined as molecules (volatiles and odor organic molecules) that interact with olfactory receptors in the nose (Zuidam & Heinrich, 2010). Aroma molecules have a small molecular weight (100–250 Da) and are often classified as hydrocarbons, alcohols, aldehydes, ketones, esters, or acids among others (Trifković et al., 2016). Since recent years, food flavor preservation during manufacturing, storage, packing, and controlled release of aroma attracted the attention of the food industry and related research.

One of the most efficient technologies to encapsulate flavors and aromas in food products is molecular inclusion in cyclodextrins (Madene, Jacquot, Scher, & Desobry, 2006). Another technique, spray-drying, is extensively applied to incorporate aromas in solid form and provides high loading efficiency. However, a high processing temperature could negatively affect and cause loss of volatile compounds. Multilayer emulsions have been described as an effective alternative to minimize loss of D-limonene during spray-drying (Burgos-Díaz et al., 2018).

Also, nanoencapsulation in emulsions has been investigated. However, one of the main disadvantages of nanoemulsions is the small droplet size, which facilitates undesirable rapid flavor diffusion (Trifković et al., 2016). The application of liposomes to encapsulate flavor has been described as an effective tool to prepare processed food targeting control release during cooking (Lengerich et al., 1991).

Even if extensive research on flavor encapsulation is focused on essential oils, bioactive compounds are also interesting for their antimicrobial or antioxidant properties. Yang et al. (2017) have described the effect of storage temperature on the aroma profile of emulsions containing citrus oil. O/W emulsions containing 10% (*w*/w) oil phase and Tween 80 (1%, *w*/*v*) were formulated and analyzed for aroma alteration using electric nose analysis. Preheating of citrus oil emulsions significantly affected the aroma, and the higher was the storage temperature, the more the aroma was modified. The control release of flavored nanoemulsions encapsulated in filled soluble hydrogels was recently investigated by Kwan and Davidov-Pardo (2018). O/W emulsions (5% orange oil, w/w) containing WPI or Tween 80 (0.5%–2%) were prepared using HPH at 200 MPa for five passes. Hydrogels were then prepared with nanoemulsions combined with low methoxyl pectin and WPI. In contact with artificial saliva, the soluble hydrogel released oil flavor. This alternative approach could be applied for bursting sensations.

4.7 Minerals

Essential minerals are elements of interest from a nutritional point of view. Fortification of food with essential minerals has attracted increasing attention (Gharibzahedi & Jafari, 2017a). Most performed research focuses on the incorporation of iron and calcium, two of the 16 essential minerals, in food products (Gharibzahedi & Jafari, 2017b).

Iron deficiency is a clinical condition, and it is the most common cause of anemia worldwide. Fortification of foods with iron could help to prevent associated anemia diseases. However, due to the high reactivity of iron, it may promote lipid oxidation in iron-fortified food products containing fat. The application of HPH to encapsulate iron has been recently described by Prichapan et al. (2018). Authors formulated W/O emulsions (20 wt%) containing ferrous sulfate. Incorporation of rice bran stearin crystals improved physical and oxidative stability and reduced particle size of emulsions. Successful encapsulation of ferrous sulfate in W/O emulsions could be applied in the formulation of aqueous-based food and beverage products.

However, further research should be conducted to evaluate the effect of pH and ionic strength of the external water phase on iron retention. W/O/W emulsions formulated with 15 wt% of whey protein isolate in the aqueous phase and 8 wt% polyglycerol polyricinoleate into corn oil had an iron loading capacity of >99% (Choi, Decker, & McClements, 2009). Despite iron was encapsulated in the inner water phase, it promoted lipid oxidation once added to fish oil emulsions.

The supplementation of food products, typically dairy and soy based, with calcium could mitigate deficient related diseases. Most commonly, calcium deficiency affects women, and it is associated with osteoporosis. Regardless of gender, most of the worldwide population does not meet the calcium recommended daily intake. The European Food Safety Authority (EFSA) has reported that in adults (≥25 years), calcium intake equals excretion at 715 mg/day, based on studies performed in North America. However, the encapsulation of calcium and incorporation into formulated food products can be technologically challenging. Supplementation of calcium in soy-based products may cause protein precipitation. Therefore the incorporation of calcium in emulsions could be an effective strategy. Márquez et al. (2018) incorporate calcium chloride in sunflower O/W emulsions (20% *w*/w) using HPH prepared with soybean flour dispersions. Not surprisingly, the incorporation of calcium decreased the solubility of the protein in the aqueous phase, which corroborates the importance of adjust calcium concentration in the formulations.

5 Conclusions and further remarks

Considerable interest has been recently given to nanostructures to protect food bioactive ingredients and improve their solubility, chemical stability against environmental stresses, and their release profile and biological efficacy in the human body. In this chapter, we focused on the development of different delivery systems using HPH or UHPH and microfluidization. In particular, we discussed the effect of applying these homogenization technologies on the functional properties of different wall materials, mainly lipids, carbohydrates, proteins, or mixed biopolymer ingredients, used to encapsulate different food ingredients. The application of these technologies to protect different compounds of interest and their effect on their biological (i.e., antimicrobial and antioxidant) activities, bioavailability and bioaccessibility were also discussed.

HPH technologies can change the physical state of biopolymers and improve their functionalities without the need of chemical modifications. It was noticed that most of the studies have been conducted using HPH with pressures up to 150 MPa; however, few studies have used ultrahigh pressures (300–400 MPa) to produce nanoparticles and study their encapsulation efficiency. More research is needed to investigate the application of HPH and UHPH in complexing different biopolymers to produce new functional nanostructures.

A synergistic effect between HPH and thermal treatments can produce modified form of the biopolymers (e.g., zein nanoparticles) with better structural properties and thermal behaviors for the development of efficient delivery systems for bioactive ingredients. HPH has also shown potential to cross-link biopolymers at the emulsion interface to design structures with better mechanical stability and sustained release. Future research is needed in this area to combine the effect of different technologies, such as HPH and sonication, to improve nanomaterial functionalities and encapsulation efficiency. Since it was found that even two nanostructures with the same droplet size produced by two different homogenization devices may give a completely different encapsulation efficiency, more research is required to study the molecular structure of biopolymers as affected by emulsification devices used. It is also of great importance to study the interaction between the encapsulant and encapsulated materials.

The main concern for the research studies is the delivery of the bioactive ingredient as HPH and other nanoencapsulation technologies are applied. However, little interest has been focused on improving the product quality, shelf life, organoleptic properties, and retarding oxidation, which should be the main goals of nanoencapsulation in the future.

HPH has shown the potential to encapsulate highly hydrophobic bioactive ingredients, such as vitamin E, starting from their powder form, and improve their solubility for possible application in beverage production with low turbidity and high physical stability. Future work has to be done with other hydrophobic bioactive powders to study their incorporation in beverages.

The potential of low-molecular-weight surfactants in nanostructure formulation alone or in combination with other biopolymers needs to be further studied. Although biopolymers can produce strong interfacial layers to protect encapsulated material, low-molecular-weight surfactants are able to reduce oil-water interfacial tension and produce nanoemulsions with a reduced particle size. Combining Tween 20 and whey protein isolate was found to improve the stability of nanoemulsions containing β-carotene. Low-molecular-weight surfactants (e.g., lecithin) can interact with biopolymers producing emulsions with high viscoelasticity, providing more benefits in nanoemulsion delivery systems.

Studies in the literature confirmed that HPH is an effective technique to produce β-carotene nanoemulsions physically stable, despite the significant degradation and oxidation during storage. It is also possible to apply HPH to produce stable emulsions to encapsulate flavors; however, severe homogenization conditions could adversely degrade flavor compounds. Emulsifier type and concentration and homogenization conditions were found to determine the stability and delivery of β-carotene and other bioactives. Future research is necessary to optimize the production conditions for the HPH technology to be commercialized. There is a gap in the literature on the use of HPH technology to formulate multiple and multilayered nanoemulsions by complexing of biopolymers and to study their stability and encapsulation efficiency.

Another important aspect that needs further studies concerns the fate upon consumption of food ingredient-loaded nanostructures. In vitro methods can be profitably employed to predict the fate in the gut of bioactives, without the need for in vivo methods.

Moreover, considerably more work will need to be done to determine the potential toxicology of nanostructures (<100 nm), especially in conjunction with HPH processing.

With all the scientific and technological progress in the design of carrier emulsions, it is imperative that these food structures are sustainable and produced by economically viable processing operations that can assure the physicochemical stability of the loaded bioactives during storage.

Nonetheless, it would be interesting to assess the effects of applying nanoemulsion technology on a wider range of functional ingredients and their potential interactions with the other food constituents to further improve their nutritional, physicochemical, and sensory characteristics.

References

Abaee, A., Mohammadian, M., & Jafari, S. M. (2017). Whey and soy protein-based hydrogels and nano-hydrogels as bioactive delivery systems. *Trends in Food Science & Technology*, *70* (Suppl. C), 69–81. https://doi.org/10.1016/j.tifs.2017.10.011.

Addo, C. N. A., & Ferragut, V. (2015). Evaluating the ultra-high pressure homogenization (UHPH) and pasteurization effects on the quality and shelf life of donkey milk. *International Journal of Food Studies*, *4*(1), 104–115. https://doi.org/10.7455/ijfs/4.1.2015.a9.

Aditya, N. P., Aditya, S., Yang, H., Kim, H. W., Park, S. O., & Ko, S. (2015). Co-delivery of hydrophobic curcumin and hydrophilic catechin by a water-in-oil-in-water double emulsion. *Food Chemistry, 173*, 7–13. https://doi.org/10.1016/j.foodchem.2014.09.131.

Akhavan, S., Assadpour, E., Katouzian, I., & Jafari, S. M. (2018). Lipid nano scale cargos for the protection and delivery of food bioactive ingredients and nutraceuticals. *Trends in Food Science & Technology, 74*, 132–146. https://doi.org/10.1016/j.tifs.2018.02.001.

Akhavan, S., & Jafari, S. M. (2017). Chapter 6—Nanoencapsulation of natural food colorants. In *Nanoencapsulation of Food Bioactive Ingredients* (pp. 223–260). London: Academic Press.

Akolade, J. O., Oloyede, H. O. B., & Onyenekwe, P. (2017). Encapsulation in chitosan-based polyelectrolyte complexes enhances antidiabetic activity of curcumin. *Journal of Functional Foods, 35*, 584–594. https://doi.org/10.1016/j.jff.2017.06.023.

Amador-Espejo, G. G., Suàrez-Berencia, A., Juan, B., Bárcenas, M. E., & Trujillo, A. J. (2014). Effect of moderate inlet temperatures in ultra-high-pressure homogenization treatments on physicochemical and sensory characteristics of milk. *Journal of Dairy Science, 97*(2), 659–671. https://doi.org/10.3168/jds.2013-7245.

Amarowicz, R. (2010). Modification of emulsifying properties of food proteins by enzymatic hydrolysis. *European Journal of Lipid Science and Technology, 112*(7), 695–696. https://doi.org/10.1002/ejlt.201000382.

Anarjan, N., Mirhosseini, H., Baharin, B. S., & Tan, C. P. (2011). Effect of processing conditions on physicochemical properties of sodium caseinate-stabilized astaxanthin nanodispersions. *LWT Food Science and Technology, 44*(7), 1658–1665. https://doi.org/10.1016/j.lwt.2011.01.013.

Ariyarathna, I. R., & Karunaratne, D. N. (2015). Use of chickpea protein for encapsulation of folate to enhance nutritional potency and stability. *Food and Bioproducts Processing, 95*, 76–82. https://doi.org/10.1016/j.fbp.2015.04.004.

Arranz, E., Corredig, M., & Guri, A. (2016). Designing food delivery systems: Challenges related to the in vitro methods employed to determine the fate of bioactives in the gut. *Food & Function*, 7(8), 3319–3336. https://doi.org/10.1039/C6FO00230G.

Arranz, E., Guri, A., Fornari, T., Mendiola, J. A., Reglero, G., & Corredig, M. (2017). In vitro uptake and immune functionality of digested Rosemary extract delivered through food grade vehicles. *Food Research International, 97*, 71–77. https://doi.org/10.1016/j.foodres.2017.03.033.

Assadpour, E., & Jafari, S. M. (2018). A systematic review on nanoencapsulation of food bioactive ingredients and nutraceuticals by various nanocarriers. *Critical Reviews in Food Science and Nutrition*, 1–47. https://doi.org/10.1080/10408398.2018.1484687.

Assadpour, E., Maghsoudlou, Y., Jafari, S. M., Ghorbani, M., & Aalami, M. (2016a). Optimization of folic acid nano-emulsification and encapsulation by maltodextrin-whey protein double emulsions. *International Journal of Biological Macromolecules, 86*, 197–207. https://doi.org/10.1016/j.ijbiomac.2016.01.064.

Assadpour, E., Maghsoudlou, Y., Jafari, S. M., Ghorbani, M., & Aalami, M. (2016b). Evaluation of folic acid nano-encapsulation by double emulsions. *Food and Bioprocess Technology, 9*(12), 2024–2032. https://doi.org/10.1007/s11947-016-1786-y.

Azevedo, M. A., Bourbon, A. I., Vicente, A. A., & Cerqueira, M. A. (2014). Alginate/chitosan nanoparticles for encapsulation and controlled release of vitamin B2. *International Journal of Biological Macromolecules, 71*, 141–146. https://doi.org/10.1016/j.ijbiomac.2014.05.036.

Betoret, E., Betoret, N., Rocculi, P., & Dalla Rosa, M. (2015). Strategies to improve food functionality: Structure–property relationships on high pressures homogenization, vacuum impregnation and drying technologies. *Trends in Food Science & Technology, 46*(1), 1–12. https://doi.org/10.1016/j.tifs.2015.07.006.

Bhandari, B. R., & Hartel, R. W. (2005). Phase transition during food powder production and powder stability. In C. Onwulate (Ed.), *Encapsulated and Powdered Foods* (pp. 261–292). Boca Raton, FL: Taylor & Francis.

Botrel, D. A., Borges, S. V., Fernandes, R. V. B., Antoniassi, R., de Faria-Machado, A. F., Feitosa, J. P. A., et al. (2017). Application of cashew tree gum on the production and stability of spray-dried fish oil. *Food Chemistry*, *221*, 1522–1529. https://doi.org/10.1016/j.foodchem.2016.10.141.

Bou, R., Cofrades, S., & Jiménez-Colmenero, F. (2014). Influence of high pressure and heating treatments on physical parameters of water-in-oil-in-water emulsions. *Innovative Food Science and Emerging Technologies*, *23*, 1–9. https://doi.org/10.1016/j.ifset.2014.04.001.

Burapapadh, K., Takeuchi, H., & Sriamornsak, P. (2012). Novel pectin-based nanoparticles prepared from nanoemulsion templates for improving in vitro dissolution and in vivo absorption of poorly water-soluble drug. *European Journal of Pharmaceutics and Biopharmaceutics*, *82*(2), 250–261. https://doi.org/10.1016/j.ejpb.2012.07.010.

Burgos-Díaz, C., Hernández, X., Wandersleben, T., Barahona, T., Medina, C., Quiroz, A., et al. (2018). Influence of multilayer O/W emulsions stabilized by proteins from a novel lupin variety AluProt-CGNA and ionic polysaccharides on D-limonene retention during spray-drying. *Colloids and Surfaces A: Physicochemical and Engineering Aspects*, *536*, 234–241. https://doi.org/10.1016/j.colsurfa.2017.04.032.

Busch, V. M., Pereyra-Gonzalez, A., Šegatin, N., Santagapita, P. R., Poklar Ulrih, N., & Buera, M. P. (2017). Propolis encapsulation by spray drying: Characterization and stability. *LWT Food Science and Technology*, *75*, 227–235. https://doi.org/10.1016/j.lwt.2016.08.055.

Can Karaca, A., Low, N. H., & Nickerson, M. T. (2015). Potential use of plant proteins in the microencapsulation of lipophilic materials in foods. *Trends in Food Science and Technology*, *42*(1), 5–12. https://doi.org/10.1016/j.tifs.2014.11.002.

Chanamai, R., & McClements, D. J. (2001). Prediction of emulsion color from droplet characteristics: Dilute monodisperse oil-in-water emulsions. *Food Hydrocolloids*, *15*, 83–91. https://doi.org/10.1016/S0268-005X(00)00055-2.

Chang, Y., McLandsborough, L., & McClements, D. J. (2015). Fabrication, stability and efficacy of dual-component antimicrobial nanoemulsions: Essential oil (thyme oil) and cationic surfactant (lauric arginate). *Food Chemistry*, *172*, 298–304. https://doi.org/10.1016/j.foodchem.2014.09.081.

Chen, J., Liang, R., Liu, W., Li, T., Liu, C., Wu, S., et al. (2013). Pectic-oligosaccharides prepared by dynamic high-pressure microfluidization and their in vitro fermentation properties. *Carbohydrate Polymers*, *91*(1), 175–182. https://doi.org/10.1016/j.carbpol.2012.08.021.

Chen, X., McClements, D. J., Zhu, Y., Chen, Y., Zou, L., Liu, W., et al. (2018). Enhancement of the solubility, stability and bioaccessibility of quercetin using protein-based excipient emulsions. *Food Research International*, *114*, 30–37. https://doi.org/10.1016/j.foodres.2018.07.062.

Chen, L., Remondetto, G. E., & Subirade, M. (2006). Food protein-based materials as nutraceutical delivery systems. *Trends in Food Science and Technology*, *17*(5), 272–283. https://doi.org/10.1016/j.tifs.2005.12.011.

Chen, F., Ren, C. G., Zhou, T., Wei, Y. J., & Dai, C. C. (2016). A novel exopolysaccharide elicitor from endophytic fungus *Gilmaniella* sp. AL12 on volatile oils accumulation in *Atractylodes lancea*. *Scientific Reports*, *6*, 34–35. https://doi.org/10.1038/srep34735.

Chen, Z., Shu, G., Taarji, N., Barrow, C. J., Nakajima, M., Khalid, N., et al. (2018). 'Gypenosides as natural emulsifiers for oil-in-water nanoemulsions loaded with astaxanthin: Insights of formulation, stability and release properties. *Food Chemistry*, *261*, 322–328. https://doi.org/10.1016/j.foodchem.2018.04.054.

Chen, C. C., & Wagner, G. (2004). Vitamin E nanoparticle for beverage applications. *Chemical Engineering Research and Design*, *82*(11), 1432–1437. https://doi.org/10.1205/cerd.82.11.1432.52034.

Chevalier-Lucia, D., Blayo, C., Gràcia-Julià, A., Picart-Palmade, L., & Dumay, E. (2011). Processing of phosphocasein dispersions by dynamic high pressure: Effects on the dispersion physico-chemical characteristics and the binding of α-tocopherol acetate to casein micelles. *Innovative Food Science and Emerging Technologies*, *12*(4), 416–425. https://doi.org/10.1016/j.ifset.2011.07.007.

Chevalier-Lucia, D., Cortés-Muñoz, M., Picart-Palmade, L., & Dumay, E. (2009). Submicron emulsions processed by ultra-high pressure homogenization. *High Pressure Research*, *29*(4), 732–738. https://doi.org/10.1080/08957950903417469.

Chi, E. Y., Krishnan, S., Randolph, T. W., & Carpenter, J. F. (2003). Physical stability of proteins in aqueous solution: Mechanism and driving forces in nonnative protein aggregation. *Pharmaceutical Research*, *20*(9), 1325–1336. https://doi.org/10.1023/A:102577142.

Choi, S. J., Decker, E. A., & McClements, D. J. (2009). Impact of iron encapsulation within the interior aqueous phase of water-in-oil-in-water emulsions on lipid oxidation. *Food Chemistry*, *116*(1), 271–276. https://doi.org/10.1016/j.foodchem.2009.02.045.

Codina-Torrella, I. (2014). *Optimización del proceso de elaboración y aplicación de la homogenización a ultra alta presión como tecnología de conservación de licuado de chufa*. Ph.D. thesis, Universitat Autònoma de Barcelona.

Codina-Torrella, I., Guamis, B., Ferragut, V., & Trujillo, A. J. (2017). Potential application of ultra-high pressure homogenization in thephysico-chemical stabilization of tiger nuts' milk beverage. *Innovative Food Science and Emerging Technologies*, *40*, 42–51. https://doi.org/10.1016/j.ifset.2016.06.023.

Comunian, T. A., Ravanfar, R., Selig, M. J., & Abbaspourrad, A. (2018). Influence of the protein type on the stability of fish oil in water emulsion obtained by glass microfluidic device. *Food Hydrocolloids*, 77, 96–106. https://doi.org/10.1016/j.foodhyd.2017.09.025.

Considine, T., Flanagan, J., Thompson, A., Boland, M., & Singh, H. B. T. -M. P. (2008). Chapter 13—Interaction between milk proteins and micronutrients. In *Food Science and Technology* (pp. 377–407). San Diego: Academic Press. https://doi.org/10.1016/B978-0-12-374039-7.00013-1.

Considine, T., Patel, H. A., Singh, H., & Creamer, L. K. (2007). Influence of binding conjugated linoleic acid and myristic acid on the heat-and high-pressure-induced unfolding and aggregation of β-lactoglobulin B. *Food Chemistry*, *102*(4), 1270–1280. https://doi.org/10.1016/j.foodchem.2006.07.022.

Cortés-Muñoz, M., Chevalier-Lucia, D., & Dumay, E. (2009). Characteristics of submicron emulsions prepared by ultra-high pressure homogenisation: Effect of chilled or frozen storage. *Food Hydrocolloids*, *23*(3), 640–654. https://doi.org/10.1016/j.foodhyd.2008.07.023.

Corzo-Martínez, M., Mohan, M., Dunlap, J., & Harte, F. (2015). Effect of ultra-high pressure homogenization on the interaction between bovine casein micelles and ritonavir. *Pharmaceutical Research*, *32*(3), 1055–1071.

Cruz, N., Capellas, M., Hernández, M., Trujillo, A. J., Guamis, B., & Ferragut, V. (2007). Ultra high pressure homogenization of soymilk: Microbiological, physicochemical and microstructural characteristics. *Food Research International*, *40*(6), 725–732. https://doi.org/10.1016/j.foodres.2007.01.003.

Cui, Z., Chen, Y., Kong, X., Zhang, C., & Hua, Y. (2014). Emulsifying properties and oil/water (O/W) interface adsorption behavior of heated soy proteins: Effects of heating concentration, homogenizer rotating speed, and salt addition level. *Journal of Agricultural and Food Chemistry*, *62*(7), 1634–1642. https://doi.org/10.1021/jf404464z.

Cui, Z., Kong, X., Chen, Y., Zhang, C., & Hua, Y. (2014). Effects of rutin incorporation on the physical and oxidative stability of soy protein-stabilized emulsions. *Food Hydrocolloids*, *41*, 1–9. https://doi.org/10.1016/j.foodhyd.2014.03.006.

Dai, L., Li, R., Wei, Y., Sun, C., Mao, L., & Gao, Y. (2018). Fabrication of zein and rhamnolipid complex nanoparticles to enhance the stability and in vitro release of curcumin. *Food Hydrocolloids*, 77, 617–628. https://doi.org/10.1016/j.foodhyd.2017.11.003.

de Boer, F. Y., Kok, R. N. U., Imhof, A., & Velikov, K. P. (2018). White zein colloidal particles: Synthesis and characterization of their optical properties on the single particle level and in concentrated suspensions. *Soft Matter*, *14*, 2870–2878. https://doi.org/10.1039/C7SM02415K.

de Paz, E., Martín, Á., Bartolomé, A., Largo, M., & Cocero, M. J. (2014). Development of water- soluble β-carotene formulations by high-temperature, high-pressure emulsification and antisolvent precipitation. *Food Hydrocolloids*, *37*, 14–24. https://doi.org/10.1016/j.foodhyd.2013.10.011.

de Vos, P., Faas, M. M., Spasojevic, M., & Sikkema, J. (2010). Encapsulation for preservation of functionality and targeted delivery of bioactive food components. *International Dairy Journal*, *20*, 292–302. https://doi.org/10.1016/j.idairyj.2009.11.008.

Degant, O & Schwechten, D 2002, Wheat flour with increased water binding capacity and process and equipment for its manufacture, German Patent DE10107885A1.

Deng, L., Kang, X., Liu, Y., Feng, F., & Zhang, H. (2017). Effects of surfactants on the formation of gelatin nanofibres for controlled release of curcumin. *Food Chemistry*, *231*(15), 70–77. https://doi.org/10.1016/j.foodchem.2017.03.027.

Dias, M. I., Ferreira, I. C. F. R., & Barreiro, M. F. (2015). Microencapsulation of bioactives for food applications. *Food & Function*, *6*(4), 1035–1052. https://doi.org/10.1039/C4FO01175A.

Dickinson, E. (2011a). Double emulsions stabilized by food biopolymers. *Food Biophysics*, *6*(1), 1–11. https://doi.org/10.1007/s11483-010-9188-6.

Dickinson, E. (2011b). Mixed biopolymers at interfaces: Competitive adsorption and multilayer structures. *Food Hydrocolloids*, *25*(8), 1966–1983. https://doi.org/10.1016/j.foodhyd.2010.12.001.

Diels, A. M., Callewaert, L., Wuytack, E. Y., Masschalck, B., & Michiels, C. W. (2005). Inactivation of *Escherichia coli* by high-pressure homogenisation is influenced by fluid viscosity but not by water activity and product composition. *International Journal of Food Microbiology*, *101*(3), 281–291. https://doi.org/10.1016/j.ijfoodmicro.2004.11.011.

Diels, A. M., & Michiels, C. W. (2006). High-pressure homogenization as a non-thermal technique for the inactivation of microorganisms. *Critical Reviews in Microbiology*, *32*(4), 201–216. https://doi.org/10.1080/10408410601023516.

Ding, S., Anton, N., Akram, S., Er-Rafik, M., Anton, H., Klymchenko, A., et al. (2017). A new method for the formulation of double nanoemulsions. *Soft Matter*, *13*(8), 1660–1669. https://doi.org/10.1039/c6sm02603f.

Ding, X., & Yao, P. (2013). Soy protein/soy polysaccharide complex nanogels: Folic acid loading, protection, and controlled delivery. *Langmuir*, *29*(27), 8636–8644. https://doi.org/10.1021/la401664y.

Dissanayake, M., & Vasiljevic, T. (2009). Functional properties of whey proteins affected by heat treatment and hydrodynamic high-pressure shearing. *Journal of Dairy Science*, *92*(4), 1387–1397. https://doi.org/10.3168/jds.2008-1791.

Doherty, S. B., Auty, M. A., Stanton, C., Ross, R. P., Fitzgerald, G. F., & Brodkorb, A. (2012). Survival of entrapped *Lactobacillus rhamnosus* GG in whey protein micro-beads during simulated ex vivo gastro-intestinal transit. *International Dairy Journal*, *22*(1), 31–43. https://doi.org/10.1016/j.idairyj.2011.06.009.

Donsì, F., Annunziata, M., & Ferrari, G. (2013). Microbial inactivation by high pressure homogenization: Effect of the disruption valve geometry. *Journal of Food Engineering*, *115*(3), 362–370. https://doi.org/10.1016/j.jfoodeng.2012.10.046.

Donsì, F., Annunziata, M., Sessa, M., & Ferrari, G. (2011). Nanoencapsulation of essential oils to enhance their antimicrobial activity in foods. *LWT-Food Science and Technology*, *44*, 1908–1914. https://doi.org/10.1016/j.lwt.2011.03.003.

Donsì, F., Annunziata, M., Vincensi, M., & Ferrari, G. (2012). Design of nanoemulsion-based delivery systems of natural antimicrobials: Effect of the emulsifier. *Journal of Biotechnology*, *159*(4), 342–350. https://doi.org/10.1016/j.jbiotec.2011.07.001.

Donsì, F., Cuomo, A., Marchese, E., & Ferrari, G. (2014). Infusion of essential oils for food stabilization: Unraveling the role of nanoemulsion-based delivery systems on mass transfer and antimicrobial activity. *Innovative Food Science and Emerging Technologies*, *22*, 212–220. https://doi.org/10.1016/j.ifset.2014.01.008.

Donsì, F., & Ferrari, G. (2016). Essential oil nanoemulsions as antimicrobial agents in food. *Journal of Biotechnology*, *233*, 106–120. https://doi.org/10.1016/j.jbiotec.2016.07.005.

Donsì, F., Ferrari, G., & Maresca, P. (2009). High-pressure homogenization for food sanitization. In *Global issues in food science and technology* (pp. 309–352). https://doi.org/10.1016/B978-0-12-374124-0.00019-3.

Donsì, F., Senatore, B., Huang, Q., & Ferrari, G. (2010). Development of novel pea protein-based nanoemulsions for delivery of nutraceuticals. *Journal of Agricultural and Food Chemistry*, *58*(19), 10653–10660. https://doi.org/10.1021/jf101804g.

Donsì, F., Sessa, M., & Ferrari, G. (2012). Effect of emulsifier type and disruption chamber geometry on the fabrication of food nanoemulsions by high pressure homogenization. *Industrial and Engineering Chemistry Research*, *51*(22), 7606–7618. https://doi.org/10.1021/ie2017898.

Donsì, F., Sessa, M., Mediouni, H., Mgaidi, A., & Ferrari, G. (2011). Encapsulation of bioactive compounds in nanoemulsion based delivery systems. *Procedia Food Science*, *1*, 1666–1671. https://doi.org/10.1016/j.profoo.2011.09.246.

Donsì, F., Voudouris, P., Veen, S. J., & Velikov, K. P. (2017). Zein-based colloidal particles for encapsulation and delivery of epigallocatechin gallate. *Food Hydrocolloids*, *63*, 508–517. https://doi.org/10.1016/j.foodhyd.2016.09.039.

Donsì, F., Wang, Y., & Huang, Q. (2011). Freeze-thaw stability of lecithin and modified starch-based nanoemulsions. *Food Hydrocolloids*, *25*(5), 1327–1336. https://doi.org/10.1016/j.foodhyd.2010.12.008.

Đorđević, Marinković, A. D., Nikolić, J. B., Drmanić, S. Ž., Rančić, M., Brković, D. V., et al. (2016). A study of the barrier properties of polyethylene coated with a nanocellulose/magnetite composite film. *Journal of the Serbian Chemical Society*, *81*(5), 589–605. https://doi.org/10.2298/JSC151217019D.

Dumay, E., Chevalier-Lucia, D., Picart-Palmade, L., Benzaria, A., Gràcia-Julià, A., & Blayo, C. (2013). Technological aspects and potential applications of (ultra) high-pressure homogenisation. *Trends in Food Science & Technology*, *31*(1), 13–26. https://doi.org/10.1016/j.tifs.2012.03.005.

Durand, E., Zhao, Y., Coupland, J. N., & Elias, R. J. (2015). Assessing interactions between lipophilic and hydrophilic antioxidants in food emulsions. *Journal of agricultural and food chemistry*, *63*(49), 10655–10661. https://doi.org/10.1021/acs.jafc.5b04152.

Elgadir, M. A., Uddin, M. S., Ferdosh, S., Adam, A., Chowdhury, A. J. K., & Sarker, M. Z. I. (2015). Impact of chitosan composites and chitosan nanoparticle composites on various drug delivery systems: A review. *Journal of Food and Drug Analysis*, *23*(4), 619–629. https://doi.org/10.1016/j.jfda.2014.10.008.

Esfahani, R., Jafari, S. M., Jafarpour, A., & Dehnad, D. (2019). Loading of fish oil into nanocarriers prepared through gelatin-gum Arabic complexation. *Food Hydrocolloids*, *90*, 291–298. https://doi.org/10.1016/j.foodhyd.2018.12.044.

Esfanjani, A. F., Jafari, S. M., Assadpoor, E., & Mohammadi, A. (2015). Nano-encapsulation of saffron extract through double-layered multipleemulsions of pectin and whey protein concentrate. *Journal of Food Engineering*, *165*, 149–155. https://doi.org/10.1016/j.jfoodeng.2015.06.022.

Esfanjani, A. F., Jafari, S. M., & Assadpour, E. (2017). Preparation of a multiple emulsion based on pectin-whey protein complex for encapsulation of saffron extract nanodroplets. *Food Chemistry*, *221*, 1962–1969. https://doi.org/10.1016/j.foodchem.2016.11.149.

Faridi Esfanjani, A., Assadpour, E., & Jafari, S. M. (2018). Improving the bioavailability of phenolic compounds by loading them within lipid-based nanocarriers. *Trends in Food Science & Technology*, *76*, 56–66. https://doi.org/10.1016/j.tifs.2018.04.002.

Fathi, M., Martín, Á., & McClements, D. J. (2014). Nanoencapsulation of food ingredients using carbohydrate based delivery systems. *Trends in Food Science & Technology*, *39*(1), 18–39. https://doi.org/10.1016/j.tifs.2014.06.007.

Fathi, M., Mozafari, M. R., & Mohebbi, M. (2012). Nanoencapsulation of food ingredients using lipid based delivery systems. *Trends Food Science and Technology*, *23*, 13–27. https://doi.org/10.1016/j.tifs.2011.08.003.

Fernandez-Avila, C., Arranz, E., Guri, A., Trujillo, A. J., & Corredig, M. (2016). Vegetable protein isolate-stabilized emulsions for enhanced delivery of conjugated linoleic acid in Caco-2 cells. *Food Hydrocolloids*, *55*, 144–154. https://doi.org/10.1016/j.foodhyd.2015.10.015.

Fernandez-Avila, C., & Trujillo, A. J. (2016). Ultra-high pressure homogenization improves oxidative stability and interfacial properties of soy protein isolate stabilized emulsions. *Food Chemistry*, *209*, 104–113. https://doi.org/10.1016/j.foodchem.2016.04.019.

Fernandez-Avila, C., & Trujillo, A. J. (2017). Enhanced stability of emulsions treated by ultra-high pressure homogenization for delivering conjugated linoleic acid in Caco-2 cells. *Food Hydrocolloids*, *71*, 271–281. https://doi.org/10.1016/j.foodhyd.2016.09.017.

Ferragut, V., Hernández-Herrero, M. M., Veciana-Nogués, M. T., Borras-Suarez, M., González-Linares, J., Vidal-Carou, M. C., et al. (2014). Ultra-high-pressure homogenization (UHPH) system for producing high-quality vegetable-based beverages: Physicochemical, microbiological, nutritional and toxicological characteristics. *Journal of the Science of Food and Agriculture*, *95*, 653–961. https://doi.org/10.1002/jsfa.6769.

Florence, A. T., & Whitehill, D. (1985). Stability and stabilization of water-in-oil-in water multiple emulsions. In D. O. Shah (Ed.), *Macro- and micro emulsions: Theory and application* (pp. 359–380). Washington: American Chemical Society.

Floury, J., Desrumaux, A., & Legrand, J. (2002). Effect of ultra-high-pressure homogenization on structure and on rheological properties of soy protein-stabilized emulsions. *Journal of Food Science*, *67*(9), 3388–3395. https://doi.org/10.1111/j.1365-2621.2002.tb09595.x.

Floury, J., Legrand, J., & Desrumaux, A. (2004). Analysis of a new type of high pressure homogeniser. Part B. Study of droplet break-up and recoalescence phenomena. *Chemical Engineering Science*, *59*(6), 1285–1294. https://doi.org/10.1016/j.ces.2003.11.025.

Gadkari, P. V., Shashidhar, M. G., & Balaraman, M. (2017). Delivery of green tea catechins through Oil-in-Water (O/W) nanoemulsion and assessment of storage stability. *J Food Eng*, *199*, 65–76. https://doi.org/10.1016/j.jfoodeng.2016.12.009.

Galmarini, M. V., Zamora, M. V., Baby, R., Chirife, J., & Mesina, V. (2008). Aromatic profiles of spray-dried encapsulated orange flavors: Influence of matrix composition on the aroma retention evaluated by sensory analysis and electronic nose techniques. *International Journal of Food Science and Technology*, *43*, 1569–1576.

Gao, Z. -M., Zhu, L. -P., Yang, X. -Q., He, X. -T., Wang, J. -M., Guo, J., et al. (2014). Soy lipophilic protein nanoparticles as a novel delivery vehicle for conjugated linoleic acid. *Food & Function*, *5*(6), 1286–1293. https://doi.org/10.1039/c3fo60497g.

García-Tejeda, Y. V., Salinas-Moreno, Y., Hernández-Martínez, A. R., & Martínez-Bustos, F. (2016). Encapsulation of purple maize anthocyanins in phosphorylated starch by spray drying. *Cereal Chemistry*, *93*(2), 130–137. https://doi.org/10.1094/CCHEM-04-15-0072-R.

Georget, E., Miller, B., Aganovic, K., Callanan, M., Heinz, V., & Mathys, A. (2014). Bacterial spore inactivation by ultra-high pressure homogenization. *Innovative Food Science & Emerging Technologies*, *26*, 116–123. https://doi.org/10.1016/j.ifset.2014.08.004.

Georget, E., Miller, B., Callanan, M., Heinz, V., & Mathys, A. (2014). (Ultra) high pressure homogenization for continuous high pressure sterilization of pumpable foods—A review. *Frontiers in Nutrition*, *1*, 1–6. https://doi.org/10.3389/fnut.2014.00015.

Gharibzahedi, S. M. T., & Jafari, S. M. (2017a). Chapter 9—Nanoencapsulation of minerals. In *Nanoencapsulation of Food Bioactive Ingredients* (pp. 333–400). London: Academic Press.

Gharibzahedi, S. M. T., & Jafari, S. M. (2017b). The importance of minerals in human nutrition: Bioavailability, food fortification, processing effects and nanoencapsulation. *Trends in Food Science & Technology*, *62*, 119–132. https://doi.org/10.1016/j.tifs.2017.02.017.

Gharibzahedi, S. M. T., & Jafari, S. M. (2018). Chapter 9—Fabrication of nanoemulsions by ultrasonication. In *Nanoemulsions* (pp. 233–285). London: Academic Press.

Gharsallaoui, A., Saurel, R., Chambin, O., & Voilley, A. (2012). Pea (*Pisum sativum*, L.) protein isolate stabilized emulsions: A novel system for microencapsulation of lipophilic ingredients by spray drying. *Food and Bioprocess Technology*, *5*(6), 2211–2221. https://doi.org/10.1007/s11947-010-0497-z.

Ghasemi, S., Jafari, S. M., Assadpour, E., & Khomeiri, M. (2017). Production of pectin-whey protein nano-complexes as carriers of orange peel oil. *Carbohydrate Polymers*, *177*(Suppl. C), 369–377. https://doi.org/10.1016/j.carbpol.2017.09.009.

Ghasemi, S., Jafari, S. M., Assadpour, E., & Khomeiri, M. (2018). Nanoencapsulation of D-limonene within nanocarriers produced by pectin-whey protein complexes. *Food Hydrocolloids*, *77*, 152–162. https://doi.org/10.1016/j.foodhyd.2017.09.030.

Ghorbanzade, T., Jafari, S. M., Akhavan, S., & Hadavi, R. (2017). Nano-encapsulation of fish oil in nano-liposomes and its application in fortification of yogurt. *Food Chemistry*, *216*, 146–152. https://doi.org/10.1016/j.foodchem.2016.08.022.

Grácia-Juliá, A., René, M., Cortés-Muñoz, M., Picart, L., López-Pedemonte, T., Chevalier, D., et al. (2008). Effect of dynamic high pressure on whey protein aggregation: A comparison with the effect of continuous short-time thermal treatments. *Food Hydrocolloids*, *22*(6), 1014–1032. https://doi.org/10.1016/j.foodhyd.2007.05.017.

Guamis, B., Trujillo, A. J., Ferragut, V., Quevedo, J. M., Lopez, T. J., & Buffa, M. N. (2015). *Continuous system and procedure of sterilization and physical stabilization of pumpable fluids by means of an ultra-high pressure homogenization*. UE patent application WO2012010284 A2.

Guerra-Rosas, M. I., Morales-Castro, J., Ochoa-Martínez, L. A., Salvia-Trujillo, L., & Martín-Belloso, O. (2016). Long-term stability of food-grade nanoemulsions from high methoxyl pectin containing essential oils. *Food Hydrocolloids*, *52*, 438–446. https://doi.org/10.1016/j.foodhyd.2015.07.017.

Gulrez, S. K. H., Al-Assaf, S., Fang, Y., Phillips, G. O., & Gunning, A. P. (2012). Revisiting the conformation of xanthan and the effect of industrially relevant treatments. *Carbohydrate Polymers*, *90*(3), 1235–1243. https://doi.org/10.1016/j.carbpol.2012.06.055.

Gunasekaran, S., Ko, S., & Xiao, L. (2007). Use of whey proteins for encapsulation and controlled delivery applications. *Journal of Food Engineering*, *83*(1), 31–40. https://doi.org/10.1016/j.jfoodeng.2006.11.001.

Gunstone, F. D., & Martini, S. (2010). Chemical and physical deterioration of bulk oils and shortenings, spreads and frying oils. In L. H. Skibsted, J. Risbo, & M. L. Andersen (Eds.), *Chemical deterioration and physical instability of food and beverages* (pp. 413–438). Boca Ratón: CRC Press. https://doi.org/10.1533/9781845699260.3.413.

Guo, J., Zhou, Q., Liu, Y. C., Yang, X. Q., Wang, J. M., Yin, S. W., et al. (2016). Preparation of soy protein-based microgel particles using a hydrogel homogenizing strategy and their interfacial properties. *Food Hydrocolloids*, *58*, 324–334. https://doi.org/10.1016/j.foodhyd.2016.03.008.

Guri, A., & Corredig, M. (2014). Dairy materials as delivery tools for bioactive components in dairy platforms. In M. Boland, M. Golding, & H. Singh (Eds.), *Food Structures, Digestion and Health* (pp. 465–488). London: Academi Press. https://doi.org/10.1016/B978-0-12-404610-8.00017-7.

Haham, M., Ish-Shalom, S., Nodelman, M., Duek, I., Segal, E., Kustanovich, M., et al. (2012). Stability and bioavailability of vitamin D nanoencapsulated in casein micelles. *Food & Function*, *3*(7), 737–744. https://doi.org/10.1039/C2FO10249H.

Harte, F., & Venegas, R. (2010). A model for viscosity reduction in polysaccharides subjected to high-pressure homogenization. *Journal of Texture Studies*, *41*(1), 49–61. https://doi.org/10.1111/j.1745-4603.2009.00212.x.

Hassanzadeh, P., Arbabi, E., Atyabi, F., & Dinarvand, R. (2018). Ferulic acid-loaded nanostructured lipid carriers: A promising nanoformulation against the ischemic neural injuries. *Life Sci*, *193*, 64–76. https://doi.org/10.1016/J.LFS.2017.11.046.

Hayden, D. R., Kibbelaar, H. V. M., Imhof, A., & Velikov, K. P. (2018). Size and optically tunable ethyl cellulose nanoparticles as carriers for organic UV filters. *ChemNanoMat*, *4*, 301–308. https://doi.org/10.1002/cnma.201700332.

He, Z., Zhang, X., Qi, W., Huang, R., & Su, R. (2015). Alginate-casein microspheres as bioactive vehicles for nutrients. *Transactions of Tianjin University*, *21*(5), 383–391. https://doi.org/10.1007/s12209-015-2692-5.

Hebishy, E., Buffa, M., Guamis, B., Blasco-Moreno, A., & Trujillo, A. J. (2015). Physical and oxidative stability of whey protein oil-in-water emulsions produced by conventional and ultra high-pressure homogenization: Effects of pressure and protein concentration on emulsion characteristics. *Innovative Food Science and Emerging Technologies*, *32*, 79–90. https://doi.org/10.1016/j.ifset.2015.09.013.

Hebishy, E., Buffa, M., Juan, B., Blasco-Moreno, A., & Trujillo, A. J. (2017). Ultra high-pressure homogenized emulsions stabilized by sodium caseinate: Effects of protein concentration and pressure on emulsions structure and stability. *LWT—Food Science and Technology*, *76*, 57–66. https://doi.org/10.1016/j.lwt.2016.10.045.

Hebishy, E., Zamora, A., Buffa, M., Blasco-Moreno, A., & Trujillo, A. J. (2017). Characterization of whey protein oil-in-water emulsions with different oil concentrations stabilized by ultra-high pressure homogenization. *Processes*, *5*(6), 1–19. https://doi.org/10.3390/pr5010006.

Hefle, S. L., Nordlee, J. A., & Taylor, S. L. (1996). Allergenic foods. *Critical Reviews in Food Science and Nutrition*, *36*(1), 69–89. https://doi.org/10.1080/10408399609527760.

Hemar, Y., Cheng, L. J., Oliver, C. M., Sanguansri, L., & Augustin, M. (2010). Encapsulation of resveratrol using water-in-oil-in-water double emulsions. *Food Biophysics*, *5*, 120–127. https://doi.org/10.1007/s11483-010-9152-5.

Hemmati, F., Jafari, S. M., Kashaninejad, M., & Barani Motlagh, M. (2018). Synthesis and characterization of cellulose nanocrystals derived from walnut shell agricultural residues. *International Journal of Biological Macromolecules*, *120*, 1216–1224. https://doi.org/10.1016/j.ijbiomac.2018.09.012.

Henry, J. V. L., Fryer, P. J., Frith, W. J., & Norton, I. T. (2010). The influence of phospholipids and food proteins on the size and stability of model sub-micron emulsions. *Food Hydrocolloids*, *24*, 66–71. https://doi.org/10.1016/j.foodhyd.2009.08.006.

Heo, W., Kim, J. H., Pan, J. H., & Kim, Y. J. (2016). Lecithin-based nano-emulsification improves the bioavailability of conjugated linoleic acid. *Journal of Agricultural and Food Chemistry*, *64*(6), 1355–1360. https://doi.org/10.1021/acs.jafc.5b05397.

Hosseinnejad, M., & Jafari, S. M. (2016). Evaluation of different factors affecting antimicrobial properties of chitosan. *International Journal of Biological Macromolecules*, *85*, 467–475. https://doi.org/10.1016/j.ijbiomac.2016.01.022.

Hu, X., Zhao, M., Sun, W., Zhao, G., & Ren, J. (2011). Effects of microfluidization treatment and transglutaminase cross-linking on physicochemical, functional, and conformational properties of peanut protein isolate. *Journal of Agricultural and Food Chemistry*, *59* (16), 8886–8894. https://doi.org/10.1021/jf201781z.

Huang, J., Wang, Q., Li, T., Xia, N., & Xia, Q. (2017). Nanostructured lipid carrier (NLC) as a strategy for encapsulation of quercetin and linseed oil: Preparation and in vitro characterization studies. *Journal of Food Engineering*, *215*, 1–12. https://doi.org/10.1016/J.JFOODENG.2017.07.002.

Huang, Q., Yu, H., & Ru, Q. (2010). Bioavailability and delivery of nutraceuticals using nanotechnology. *Journal of Food Science*, *75*(1), 50–57. https://doi.org/10.1111/j.1750-3841.2009.01457.x.

Innocente, N., Marchesini, G., & Biasutti, M. (2014). Effect of high-pressure homogenisation on the retention of selected aroma compounds in model dairy emulsions. *International Journal of Food Science & Technology*, *49*(9), 1992–2000. https://doi.org/10.1111/ijfs.12520.

Iordache, M., & Jelen, P. (2003). High pressure microfluidization treatment of heat denatured whey proteins for improved functionality. *Innovative Food Science & Emerging Technologies*, *4*(4), 367–376. https://doi.org/10.1016/S1466-8564(03)00061-4.

Jafari, S. M., Assadpoor, E., He, Y., & Bhandari, B. (2008). Re-coalescence of emulsion droplets during high-energy emulsification. *Food Hydrocolloids*, *22*, 1191–1202. https://doi.org/10.1016/j.foodhyd.2007.09.006.

Jafari, S. M., He, Y., & Bhandari, B. (2006). Nano-emulsion production by sonication and microfluidization—A comparison. *International Journal of Food Properties*, *9*(3), 475–485. https://doi.org/10.1080/10942910600596464.

Jafari, S. M., He, Y., & Bhandari, B. (2007). Production of sub-micron emulsions by ultrasound and microfluidization techniques. *Journal of Food Engineering*, *82*(4), 478–488. https://doi.org/10.1016/j.jfoodeng.2007.03.007.

Jafari, S. M., & McClements, D. J. (2017). Nanotechnology approaches for increasing nutrient bioavailability. In *Advances in food and nutrition research*. London, UK: Academic Press.

Jafari, M., Paknejad, Z., Rad, M. R., Motamedian, S. R., Eghbal, M. J., Nadjmi, N., et al. (2017). Polymeric scaffolds in tissue engineering: A literature review. *Journal of Biomedical Materials Research Part B: Applied Biomaterials*, *105*, 431–459. https://doi.org/10.1002/jbm.b.33547.

Jo, Y. J., Chun, J. Y., Kwon, Y. J., Min, S. G., Hong, G. P., & Choi, M. J. (2015). Physical and antimicrobial properties of trans-cinnamaldehyde nanoemulsions in water melon juice. *LWT—Food Science and Technology*, *60*(1), 444–451. https://doi.org/10.1016/j.lwt.2014.09.041.

Johnston, S. P., Nickerson, M. T., & Low, N. H. (2015). The physicochemical properties of legume protein isolates and their ability to stabilize oil-in-water emulsions with and without genipin. *Journal of Food Science and Technology*, *52*(7), 4135–4145. https://doi.org/10.1007/s13197-014-1523-3.

Katouzian, I., Faridi Esfanjani, A., Jafari, S. M., & Akhavan, S. (2017). Formulation and application of a new generation of lipid nano-carriers for the food bioactive ingredients. *Trends in Food Science & Technology*, *68*(Suppl. C), 14–25. https://doi.org/10.1016/j.tifs.2017.07.017.

Katouzian, I., & Jafari, S. M. (2016). Nano-encapsulation as a promising approach for targeted delivery and controlled release of vitamins. *Trends in Food Science & Technology*, *53*, 34–48. https://doi.org/10.1016/j.tifs.2016.05.002.

Kaushik, V., & Roos, Y. (2007). Limonene encapsulation in freeze-drying of gum Arabic–sucrose–gelatin systems. *LWT-Food Science and Technology*, *40*, 1381–1391. https://doi.org/10.1016/j.lwt.2006.10.008.

Kaushik, A., & Singh, M. (2011). Isolation and characterization of cellulose nanofibrils from wheat straw using steam explosion coupled with high shear homogenization. *Carbohydrate Research*, *346*(1), 76–85. https://doi.org/10.1016/j.carres.2010.10.020.

Keerati-U-Rai, M., & Corredig, M. (2009). Effect of dynamic high pressure homogenization on the aggregation state of soy protein. *Journal of Agricultural and Food Chemistry*, *57*(9), 3556–3562. https://doi.org/10.1021/jf803562q.

Khalid, N., Shu, G., Holland, B. J., Kobayashi, I., Nakajima, M., & Barrow, C. J. (2017). Formulation and characterization of O/W nanoemulsions encapsulating high concentration of astaxanthin. *Food Research International*, *102*, 364–371. https://doi.org/10.1016/j.foodres.2017.06.019.

Komaiko, J., Sastrosubroto, A., & McClements, D. J. (2015). Formation of oil-in-water emulsions from natural emulsifiers using spontaneous emulsification: Sunflower phospholipids. *Journal of Agricultural and Food Chemistry*, *63*(45), 10078–10088. https://doi.org/10.1021/acs.jafc.5b03824.

Kwan, A., & Davidov-Pardo, G. (2018). Controlled release of flavor oil nanoemulsions encapsulated in filled soluble hydrogels. *Food Chemistry*, *250*, 46–53. https://doi.org/10.1016/j.foodchem.2017.12.089.

Labrune, J. C., & Palmino, F. (2004). Nanowires. In C. Dupas, P. Houdy, & M. Lahmani (Eds.), *Nanoscience: Nanotechnologies and nanophysics* (pp. 325–379). Berlin: Springer.

Labuckas, D. O., Maestri, D. M., Perelló, M., Martínez, M. L., & Lamarque, A. L. (2008). Phenolics from walnut (*Juglans regia* L.) kernels: Antioxidant activity and interactions with proteins. *Food Chemistry*, *107*(2), 607–612. https://doi.org/10.1016/j.foodchem.2007.08.051.

Lee, S. H., Lefèvre, T., Subirade, M., & Paquin, P. (2009). Effects of ultra-high pressure homogenization on the properties and structure of interfacial protein layer in whey protein-stabilized emulsion. *Food Chemistry*, *113*, 191–195. https://doi.org/10.1016/j.foodchem.2008.07.067.

Lee, W. J., Liu, Y. C., Mwangi, F. K., Chen, W. H., Lin, S. L., Fukushima, Y., et al. (2011). Assessment of energy performance and air pollutant emissions in a diesel engine generator fueled with water-containing ethanol-biodiesel-diesel blend of fuels. *Energy*, *36*, 5591–5599. https://doi.org/10.1016/j.energy.2011.07.012.

Lee, L., & Norton, I. T. (2013). Comparing droplet breakup for a high-pressure valve homogeniser and a microfluidizer for the potential reduction of food grade nanoemulsions. *Journal of Food Engineering*, *114*, 158–163. https://doi.org/10.1016/j.jfoodeng.2012.08.009.

Lengerich, BV, Haynes, LC, Levine, H, Otterburn, MS, Mathewson, P, & Finley, J 1991, Extrusion baking of cookies having liposome encapsulated ingredients, US Patent 4999208A.

Li, W., Chen, H., He, Z., Han, C., Liu, S., & Li, Y. (2015). Influence of surfactant and oil composition on the stability and antibacterial activity of eugenol nanoemulsions. *LWT—Food Science and Technology*, *62*(1), 39–47. https://doi.org/10.1016/j.lwt.2015.01.012.

Li, Y., & McClements, D. J. (2011). Controlling lipid digestion by encapsulation of protein-stabilized lipid droplets within alginate–chitosan complex coacervates. *Food Hydrocolloids*, *25*(5), 1025–1033. https://doi.org/10.1016/j.foodhyd.2010.09.024.

Li, Y., Hu, M., Du, Y., Xiao, H., & McClements, D. J. (2011). Control of lipase digestibility of emulsified lipids by encapsulation within calcium alginate beads. *Food Hydrocolloids*, *25*(1), 122–130. https://doi.org/10.1016/j.foodhyd.2010.06.003.

Li, D., Li, L., Xiao, N., Li, M., & Xie, X. (2018). Physical properties of oil-in-water nanoemulsions stabilized by OSA-modified starch for the encapsulation of lycopene. *Colloids & Surfaces A: Physicochemical & Engineering Aspects*, *552*, 59–66. https://doi.org/10.1016/J.COLSURFA.2018.04.055.

Li, T., Yang, S., Liu, W., Liu, C., Liu, W., Zheng, H., et al. (2015). Preparation and characterization of nanoscale complex liposomes containing medium-chain fatty acids and vitamin C. *International Journal of Food Properties*, *18*(1), 113–124. https://doi.org/10.1080/10942912.2012.685683.

Liang, R., Huang, Q., Ma, J., Shoemaker, C. F., & Zhong, F. (2013). Effect of relative humidity on the store stability of spray-dried beta-carotene nanoemulsions. *Food Hydrocolloids*, *33*(2), 225–233. https://doi.org/10.1016/j.foodhyd.2013.03.015.

Liang, L., Tremblay-Hébert, V., & Subirade, M. (2011). Characterisation of the β-lactoglobulin/α-tocopherol complex and its impact on α-tocopherol stability. *Food Chemistry*, *126*(3), 821–826. https://doi.org/10.1016/j.foodchem.2010.12.029.

Liang, R., Xu, S., Shoemaker, C. F., Li, Y., Zhong, F., & Huang, Q. (2012). Physical and antimicrobial properties of peppermint oil nanoemulsions. *Journal of Agricultural and Food Chemistry*, *60*(30), 7548–7555. https://doi.org/10.1021/jf301129k.

Liang, J., Yan, H., Wang, X., Zhou, Y., Gao, X., Puligundla, P., et al. (2017). Encapsulation of epigallocatechin gallate in zein/chitosan nanoparticles for controlled applications in food systems. *Food Chemistry*, *231*, 19–24. https://doi.org/10.1016/j.foodchem.2017.02.106.

Liu, F., Chen, Z., & Tang, C. H. (2014). Microencapsulation properties of protein isolates from three selected Phaseolus legumes in comparison with soy protein isolate. *LWT—Food Science and Technology*, *55*(1), 74–82. https://doi.org/10.1016/j.lwt.2013.09.008.

Liu, Z., Jiao, Y., Wang, Y., Zhou, C., & Zhang, Z. (2008). Polysaccharides-based nanoparticles as drug delivery systems. *Advanced Drug Delivery Reviews*, *60*(15), 1650–1662. https://doi.org/10.1016/j.addr.2008.09.001.

Liu, X., Liu, Y. Y., Guo, J., Yin, S. W., & Yang, X. Q. (2017). Microfluidization initiated cross-linking of gliadin particles for structured algal oil emulsions. *Food Hydrocolloids*, *73*, 153–161. https://doi.org/10.1016/j.foodhyd.2017.07.001.

Liu, W., Liu, W., Ye, A., Peng, S., Wei, F., Liu, C., et al. (2016). Environmental stress stability of microencapsules based on liposomes decorated with chitosan and sodium alginate. *Food Chemistry*, *196*, 396–404. https://doi.org/10.1016/j.foodchem.2015.09.050.

Liu, F., & Tang, C. H. (2016). Soy glycinin as food-grade Pickering stabilizers: Part. I. Structural characteristics, emulsifying properties and adsorption/arrangement at interface. *Food Hydrocolloids*, *60*, 606–619. https://doi.org/10.1016/j.foodhyd.2015.04.025.

Liu, D., Tian, H., Zeng, J., & Chang, P. R. (2008). Core-shell nanoblends from soy protein/polystyrene by emulsion polymerization. *Macromolecular Materials and Engineering*, *293*(8), 714–721. https://doi.org/10.1002/mame.200800119.

Liu, W., Wang, J., McClements, D. J., & Zou, L. (2018). Encapsulation of β-carotene-loaded oil droplets in caseinate/alginate microparticles: Enhancement of carotenoid stability and bioaccessibility. *Journal of Functional Foods*, *40*, 527–535. https://doi.org/10.1016/j.jff.2017.11.046.Fwei.

Liu, D., Wu, Q., Chen, H., & Chang, P. R. (2009). Transitional properties of starch colloid with particle size reduction from micro- to nanometer. *Journal of Colloid and Interface Science*, *339*(1), 117–124. https://doi.org/10.1016/j.jcis.2009.07.035.

Livney, Y. D. (2010). Milk proteins as vehicles for bioactives. *Current Opinion in Colloid & Interface Science*, *15*(1), 73–83. https://doi.org/10.1016/j.cocis.2009.11.002.

Madene, A., Jacquot, M., Scher, J., & Desobry, S. (2006). Flavour encapsulation and controlled release–a review. *International Journal of Food Science & Technology*, *41*(1), 1–21. https://doi.org/10.1111/j.1365-2621.2005.00980.x.

Mahdavee Khazaei, K., Jafari, S. M., Ghorbani, M., & Hemmati Kakhki, A. (2014). Application of maltodextrin and gum Arabic in microencapsulation of saffron petal's anthocyanins and evaluating their storage stability and color. *Carbohydrate Polymers*, *105*(1), 57–62. https://doi.org/10.1016/j.carbpol.2014.01.042.

Mahdavi, S. A., Jafari, S. M., Ghorbani, M., & Assadpoor, E. (2014). Spray-drying microencapsulation of anthocyanins by natural biopolymers: A review. *Drying Technology*, *32*(5), 509–518. https://doi.org/10.1080/07373937.2013.839562.

Mao, L., Xu, D., Yang, J., Yuan, F., Gao, Y., & Zhao, J. (2009). Effects of small and large molecule emulsifiers on the characteristics of β-carotene nanoemulsions prepared by high pressure homogenization. *Food Technology and Biotechnology*, *47*(3), 336–342. https://hrcak.srce.hr/39852.

Mao, L., Yang, J., Xu, D., Yuan, F., & Gao, Y. (2010). Effects of homogenization models and emulsifiers on the physicochemical properties of β-carotene nanoemulsions. *Journal of Dispersion Science and Technology*, *31*(7), 986–993. https://doi.org/10.1080/01932690903224482.

Márquez, A. L., Wagner, J. R., & Palazolo, G. G. (2018). Effects of calcium content and homogenization method on the microstructure, rheology, and stability of emulsions prepared with soybean flour dispersions. *European Journal of Lipid Science and Technology*, *120*(7), 1700500–1700511. https://doi.org/10.1002/ejlt.201700500.

Mason, T. J. (1992). *Practical sonochemistry: User's guide to applications in chemistry and chemical engineering*. New York: Ellis Horwood Ltd.

Matsumiya, K., & Murray, B. S. (2016). Soybean protein isolate gel particles as foaming and emulsifying agents. *Food Hydrocolloids*, *60*, 206–215. https://doi.org/10.1016/j.foodhyd.2016.03.028.

McClements, D. J. (2002). Colloidal basis of emulsion color. *Current Opinion in Colloid & Interface Science*, 7(5–6), 451–455. https://doi.org/10.1016/S1359-0294(02)00075-4.

McClements, D. J. (2004). Protein-stabilized emulsions. *Current Opinion in Colloid & Interface Science*, *9*(5), 305–313. https://doi.org/10.1016/j.cocis.2004.09.003.

McClements, D. J. (2005). *Food emulsions: Principles, practices, and techniques* (2nd ed.). Boca Raton: CRC Press.

McClements, D. J. (2009). Biopolymers in food emulsions. In S. Kasapis, I. T. Norton, & J. B. Ubbink (Eds.), *Modern biopolymer science* (pp. 129–166). London, UK: Academic Press.

McClements, D. J. (2011). Edible nanoemulsions: Fabrication, properties, and functional performance. *Soft Matter*, 7(2), 297–316. https://doi.org/10.1039/C0SM00549E.

McClements, D. J. (2013). Edible lipid nanoparticles: Digestion, absorption, and potential toxicity. *Progress in Lipid Research*, *52*(4), 409–423. https://doi.org/10.1016/j.plipres.2013.04.008.

McClements, D. J., Xiao, H., & Demokritou, P. (2017). Physicochemical and colloidal aspects of food matrix effects on gastrointestinal fate of ingested inorganic nanoparticles. *Advances in Colloid and Interface Science*, *246*, 165–180. https://doi.org/10.1016/j.cis.2017.05.010.

Mehrnia, M., Jafari, S. M., Makhmal-Zadeh, B. S., & Maghsoudlou, Y. (2017). Rheological and release properties of double nano-emulsions containing crocin prepared with Angum gum, Arabic gum and whey protein. *Food Hydrocolloids*, *66*, 259–267. https://doi.org/10.1016/j.foodhyd.2016.11.033.

Mohammadi, A., Jafari, S. M., Assadpour, E., & Esfanjani, A. F. (2016). Nano-encapsulation of olive leaf phenolic compounds through WPC-pectin complexes and evaluating their release rate. *International Journal of Biological Macromolecules*, *82*, 816–822. https://doi.org/10.1016/j.ijbiomac.2015.10.025.

Mohan, A., Rajendran, S. R. C. K., He, Q. S., Bazinet, L., & Udenigwe, C. C. (2015). Encapsulation of food protein hydrolysates and peptides: A review. *RSC Advances*, *5* (97), 79270–79278. https://doi.org/10.1039/C5RA13419F.

Mokhtari, S., Jafari, S. M., & Assadpour, E. (2017). Development of a nutraceutical nano-delivery system through emulsification/internal gelation of alginate. *Food Chemistry*, *229*, 286–295. https://doi.org/10.1016/j.foodchem.2017.02.071.

Najafi-Soulari, S., Shekarchizadeh, H., & Kadivar, M. (2016). Encapsulation optimization of lemon balm antioxidants in calcium alginate hydrogels. *Journal of Biomaterials Science*, *27* (16), 1631–1644. https://doi.org/10.1080/09205063.2016.1226042.

Nakagaito, A. N., & Yano, H. (2004). The effect of morphological changes from pulp fiber towards nano-scale fibrillated cellulose on the mechanical properties of high-strength plant fiber based composites. *Applied Physics A*, *78*, 547–552. https://doi.org/10.1007/s00339-003-2453-5.

Naqash, F., Masoodi, F. A., Rather, S. A., Wani, S. M., & Gani, A. (2017). Emerging concepts in the nutraceutical and functional properties of pectin—A review. *Carbohydrate Polymers*, *168*, 227–239. https://doi.org/10.1016/j.carbpol.2017.03.058.

Naseri, N., Valizadeh, H., & Zakeri-Milani, P. (2015). Solid lipid nanoparticles and nanostructured lipid carriers: Structure, preparation and application. *Advanced Pharmaceutical Bulletin*, *5*(3), 305–313. https://doi.org/10.15171/apb.2015.043.

Nazzaro, F., Fratianni, F., De Martino, L., Coppola, R., & De Feo, V. (2013). Effect of essential oils on pathogenic bacteria. *Pharmaceuticals (Basel, Switzerland)*, *6*(12), 1451–1474. https://doi.org/10.3390/ph6121451.

Nesterenko, A., Alric, I., Silvestre, F., & Durrieu, V. (2014). Comparative study of encapsulation of vitamins with native and modified soy protein. *Food Hydrocolloids*, *38*, 172–179. https://doi.org/10.1016/j.foodhyd.2013.12.011.

O'Regan, J., & Mulvihill, D. M. (2010). Sodium caseinate–maltodextrin conjugate stabilized double emulsions: Encapsulation and stability. *Food Research International*, *43*(1), 224–231. https://doi.org/10.1016/j.foodres.2009.09.031.

Park, S. J., Garcia, C. V., Shin, G. H., & Kim, J. T. (2015). Fabrication and optimization of EGCG-loaded nanoparticles by high pressure homogenization. *Journal of Applied Polymer Science*. *133*(14). https://doi.org/10.1002/app.43269.

Park, S. J., Garcia, C. V., Shin, G. H., & Kim, J. T. (2017). Development of nanostructured lipid carriers for the encapsulation and controlled release of vitamin D3. *Food Chemistry*, *225*, 213–219. https://doi.org/10.1016/j.foodchem.2017.01.015.

Patel, A. R., Heussen, P. C. M., Dorst, E., Hazekamp, J., & Velikov, K. P. (2013). Colloidal approach to prepare colour blends from colourants with different solubility profiles. *Food Chemistry*, *141*, 1466–1471. https://doi.org/10.1016/j.foodchem.2013.03.082.

Pathakoti, K., Manubolu, M., & Hwang, H. (2017). Nanostructures: Current uses and future applications in food science. *Journal of Food and Drug Analysis*, *25*(2), 245–253. https://doi.org/10.1016/j.jfda.2017.02.004.

Pereda, J. (2009). *Utilización de la ultra alta presión por homogeneización como alternativa al tratamiento de pasteurización para la obtención de Leche de consumo*. Ph.D. thesis, Universitat Autònoma de Barcelona.

Pereda, J., Ferragut, V., Buffa, M., Guamis, B., & Trujillo, A. J. (2008). Proteolysis of ultra-high pressure homogenised treated milk during refrigerated storage. *Food Chemistry*, *111* (3), 696–702. https://doi.org/10.1016/j.foodchem.2008.04.040.

Pereda, J., Ferragut, V., Quevedo, J. M., Guamis, B., & Trujillo, A. J. (2007). Effects of ultra-high pressure homogenization on microbial and physicochemical shelf life of milk. *Journal of Dairy Science*, *90*(3), 1081–1093. https://doi.org/10.3168/jds.S0022-0302(07)71595-3.

Pereda, J., Ferragut, V., Quevedo, J. M., Guamis, B., & Trujillo, A. J. (2008). Effects of ultra-high-pressure homogenization treatment on the lipolysis and lipid oxidation of milk during refrigerated storage. *Journal of Agricultural and Food Chemistry*, *56*(16), 7125–7130. https://doi.org/10.1021/jf800972m.

Pereda, J., Ferragut, V., Quevedo, J. M., Guamis, B., & Trujillo, A. J. (2009). Heat damage evaluation in ultra-high pressure homogenized milk. *Food Hydrocolloids*, *23*(7), 1974–1979. https://doi.org/10.1016/j.foodhyd.2009.02.010.

Pereda, J., Jaramillo, D. P., Quevedo, J. M., Ferragut, V., Guamis, B., & Trujillo, A. J. (2008). Characterization of volatile compounds in ultra-high-pressure homogenized milk. *International Dairy Journal*, *18*(8), 826–834. https://doi.org/10.1016/j.idairyj.2007.12.002.

Phoon, P. Y., Paul, L. N., Burgner, J. W., San Martin-Gonzalez, M. F., & Narsimhan, G. (2014). Effect of cross-linking of interfacial sodium caseinate by natural processing on the oxidative stability of oil-in-water (O/W) emulsions. *Journal of Agricultural and Food Chemistry*, *62*, 2822–2829. https://doi.org/10.1021/jf403285z.

Pinnamaneni, S., Das, N. G., & Das, S. K. (2003). Comparison of oil-in-water emulsions manufactured by microfluidization and homogenization. *Pharmazie*, *58*, 554–558.

Pinto, J. F. (2010). Site-specific drug delivery systems within the gastro-intestinal tract: From the mouth to the colon. *International Journal of Pharmaceutics*, *395*(1), 44–52. https://doi.org/10.1016/j.ijpharm.2010.05.003.

Poliseli-Scopel, F. H., Hernández-Herrero, M., Guamis, B., & Ferragut, V. (2012). Comparison of ultra high pressure homogenization and conventional thermal treatments on the microbiological, physical and chemical quality of soymilk. *LWT—Food Science and Technology*, *46*(1), 42–48. https://doi.org/10.1016/j.lwt.2011.11.004.

Poliseli-Scopel, F. H., Hernández-Herrero, M., Guamis, B., & Ferragut, V. (2013). Characteristics of soymilk pasteurized by ultra high pressure homogenization. *Innovative Food Science & Emerging Technologies*, *20*, 73–80. https://doi.org/10.1016/j.ifset.2013.06.001.

Poliseli-Scopel, F. H., Hernández-Herrero, M., Guamis, B., & Ferragut, V. (2014). Sterilization and aseptic packaging of soymilk treated by ultra high pressure homogenization. *Innovative Food Science & Emerging Technologies*, *22*, 81–88. https://doi.org/10.1016/j.ifset.2014.01.001.

Porto, B. C., Augusto, P. E. D., Terekhov, A., Hamaker, B. R., & Cristianini, M. (2015). Effect of dynamic high pressure on technological properties of cashew tree gum (*Anacardium occidentale* L.). *Carbohydrate Polymers*, *129*, 187–193. https://doi.org/10.1016/j.carbpol.2015.04.052.

Prajapati, V. D., Jani, G. K., Moradiya, N. G., Randeria, N. P., & Nagar, B. J. (2013). Locust bean gum: A versatile biopolymer. *Carbohydrate Polymers*, *94*(2), 814–821. https://doi.org/10.1016/j.carbpol.2013.01.086.

PREDIMED Study Investigators (2015). Dietary fat intake and risk of cardiovascular disease and all-cause mortality in a population at high risk of cardiovascular disease. *The American Journal of Clinical Nutrition*, *102*(6), 1563–1573. https://doi.org/10.3945/ajcn.115.116046.

Prichapan, N., McClements, D. J., & Klinkesorn, U. (2018). Iron encapsulation in water-in-oil emulsions: Effect of ferrous sulfate concentration and fat crystal formation on oxidative stability. *Journal of Food Science*, *83*(2), 309–317.

Prombutara, P., Kulwatthanasal, Y., Supaka, N., Sramala, I., & Chareonpornwattana, S. (2012). Production of nisin-loaded solid lipid nanoparticles for sustained antimicrobial activity. *Food Control*, *24*(1–2), 184–190. https://doi.org/10.1016/j.foodcont.2011.09.025.

Qian, C., Decker, E. A., Xiao, H., & McClements, D. J. (2012). Physical and chemical stability of β-carotene-enriched nanoemulsions: Influence of PH, ionic strength, temperature, and emulsifier type. *Food Chemistry*, *132*(3), 1221–1229. https://doi.org/10.1016/j.foodchem.2011.11.091.

Qian, C., & McClements, D. J. (2011). Formation of nanoemulsions stabilized by model food-grade emulsifiers using high-pressure homogenization: Factors affecting particle size. *Food Hydrocolloids*, *25*, 1000–1008. https://doi.org/10.1016/j.foodhyd.2010.09.017.

Rafiee, Z., & Jafari, S. M. (2018). Application of lipid nanocarriers for the food industry. In J.-M. Mérillon & K. G. Ramawat (Eds.), *Bioactive molecules in food* (pp. 1–43). Cham: Springer International Publishing.

Rascón, M. P., Beristain, C. I., García, H. S., & Salgado, M. A. (2011). Carotenoid retention and storage stability of spray-dried encapsulated paprika oleoresin using gum Arabic and Soy protein isolate as wall materials. *LWT—Food Science and Technology*, *44*(2), 549–557. https://doi.org/10.1016/j.lwt.2010.08.021.

Relkin, P., Eynard, L., & Launay, B. (1992). Thermodynamic parameters of β-lactoglobulin and α-lactalbumin. A DSC study of denaturation by heating. *Thermochimica Acta*, *204*(1), 111–121. https://doi.org/10.1016/0040-6031(92)80320-V.

Relkin, P., & Shukat, R. (2012). Food protein aggregates as vitamin-matrix carriers: Impact of processing conditions. *Food Chemistry*, *134*(4), 2141–2148. https://doi.org/10.1016/j.foodchem.2012.04.020.

Relkin, P., Shukat, R., & Moulin, G. (2014). Encapsulation of labile compounds in heat- and high-pressure treated protein and lipid nanoparticles. *Food Research International*, *63*, 9–15. https://doi.org/10.1016/j.foodres.2014.03.038.

Relkin, P., Yung, J. M., Kalnin, D., & Ollivon, M. (2008). Structural behaviour of lipid droplets in protein-stabilized nano-emulsions and stability of α-tocopherol. *Food Biophysics*, *3*, 163–168. https://doi.org/10.1007/s11483-008-9064-9.

Remondetto, G. E., Beyssac, E., & Subirade, M. (2004). Iron availability from whey protein hydrogels: An in vitro study. *Journal of Agricultural and Food Chemistry*, *52*(26), 8137–8143. https://doi.org/10.1021/jf040286h.

Rezaei, A., Fathi, M., & Jafari, S. M. (2019). Nanoencapsulation of hydrophobic and low-soluble food bioactive compounds within different nanocarriers. *Food Hydrocolloids*, *88*, 146–162. https://doi.org/10.1016/j.foodhyd.2018.10.003.

Roach, A., & Harte, F. (2008). Disruption and sedimentation of casein micelles and casein micelle isolates under high-pressure homogenization. *Innovative Food Science & Emerging Technologies*, *9*(1), 1–8. https://doi.org/10.1016/j.ifset.2007.03.027.

Rocha-Guzmán, N. E., Gallegos-Infante, J. A., González-Laredo, R. F., Harte, F., Medina-Torres, L., Ochoa-Martínez, L. A., et al. (2010). Effect of high-pressure homogenization on the physical and antioxidant properties of *Quercus resinosa* infusions encapsulated by spray-drying. *Journal of Food Science*, *75*(5), N57–N61. https://doi.org/10.1111/j.1750-3841.2010.01653.x.

Rodríguez-Alcalá, L. M., Alonso, L., & Fontecha, J. (2014). Stability of fatty acid composition after thermal, high pressure, and microwave processing of cow milk as affected by polyunsaturated fatty acid concentration. *Journal of Dairy Science*, *97*(12), 7307–7315. https://doi.org/10.3168/jds.2013-7849.

Rodríguez-Alcalá, L. M., Harte, F., & Fontecha, J. (2009). Fatty acid profile and CLA isomers content of cow, ewe and goat milks processed by high pressure homogenization. *Innovative Food Science & Emerging Technologies*, *10*(1), 32–36. https://doi.org/10.1016/j.ifset.2008.10.003.

Roig-Sagués, A. X., Velázquez, R. M., Montealegre-Agramont, P., López-Pedemonte, T. J., Briñez-Zambrano, W. J., Guamis-López, B., et al. (2009). Fat content increases the lethality of ultra-high-pressure homogenization on *Listeria monocytogenes* in milk. *Journal of Dairy Science*, *92*(11), 5396–5402. https://doi.org/10.3168/jds.2009-2495.

Rostamabadi, H., Falsafi, S. R., & Jafari, S. M. (2019). Nanoencapsulation of carotenoids within lipid-based nanocarriers. *Journal of Controlled Release*, *298*, 38–67. https://doi.org/10.1016/j.jconrel.2019.02.005.

Ru, Q., Yu, H., & Huang, Q. (2010). Encapsulation of epigallocatechin-3-gallate (EGCG) using oil-in-water (O/W) submicrometer emulsions stabilized by ι-carrageenan and β-lactoglobulin. *Journal of Agricultural and Food Chemistry*, *58*(19), 10373–10381. https://doi.org/10.1021/jf101798m.

Rusli, J. K., Sanguansri, L., & Augustin, M. A. (2006). Stabilization of oils by microencapsulation with heated protein-glucose syrup mixtures. *Journal of the American Oil Chemists*, *83*(11), 965–972. https://doi.org/10.1007/s11746-006-5054-6.

Sabouri, S., Arranz, E., Guri, A., & Corredig, M. (2018). Sodium caseinate stabilized emulsions as a delivery system for epigallocatechin-gallate: Bioaccessibility, anti-proliferative activity and intestinal absorption. *Journal of Functional Foods*, *44*, 166–172. https://doi.org/10.1016/j.jff.2018.03.009.

Sagalowicz, L., & Leser, M. E. (2010). Delivery systems for liquid food products. *Current Opinion in Colloid and Interface Science*, *15*(1–2), 61–72. https://doi.org/10.1016/j.cocis.2009.12.003.

Salatin, S., & Jelvehgarim, M. (2017). Natural polysaccharide based nanoparticles for drug/gene delivery. *Pharmaceutical Sciences*, *23*, 84–94. https://doi.org/10.15171/PS.2017.14.

Saldo, J., Suárez-Jacobo, Á., Gervilla, R., Guamis, B., & Roig-Sagués, A. X. (2009). Use of ultra-high-pressure homogenization to preserve apple juice without heat damage. *High Pressure Research*, *29*(1), 52–56. https://doi.org/10.1080/08957950802715112.

Salvia-Trujillo, L., Rojas-Graü, M. A., Soliva-Fortuny, R., & Martín-Belloso, O. (2013). Effect of processing parameters on physicochemical characteristics of microfluidized lemongrass essential oil-alginate nanoemulsions. *Food Hydrocolloids*, *30*(1), 401–407. https://doi.org/10.1016/j.foodhyd.2012.07.004.

Salvia-Trujillo, L., Rojas-Graü, M. A., Soliva-Fortuny, R., & Martín-Belloso, O. (2014). Impact of microfluidization or ultrasound processing on the antimicrobial activity against *Escherichia coli* of lemongrass oil-loaded nanoemulsions. *Food Control*, *37*(1), 292–297. https://doi.org/10.1016/j.foodcont.2013.09.015.

Salvia-Trujillo, L., Soliva-Fortuny, R., Rojas-Graü, M. A., McClements, D. J., & Martín-Belloso, O. (2017). Edible nanoemulsions as carriers of active ingredients: A review. *Annual Review of Food Science and Technology*, *8*, 439–466. https://doi.org/10.1146/annurev-food-030216-025908.

Samer, C. J., & Schork, F. J. (1999). The role of high shear in continuous miniemulsion polymerization. *Industrial and Engineering Chemistry Research*, *38*(5), 1801–1807. https://doi.org/10.1021/ie9807203.

Sanguansri, P., & Augustin, M. A. (2006). Nanoscale meterials development: A food industry perspective. *Trends in Food Science and Technology*, *17*, 547–556. https://doi.org/10.1016/j.tifs.2006.04.010.

Santa-Maria, M., Scher, H., & Jeoh, T. (2012). Microencapsulation of bioactives in cross-linked alginate matrices by spray drying. *Journal of Microencapsulation*, *29*(3), 286–295. https://doi.org/10.3109/02652048.2011.651494.

Santana, R. C., Perrechil, F. A., & Cunha, R. L. (2013). High- and low-energy emulsifications for food applications: A focus on process parameters. *Food Engineering Reviews*, *5*, 107–122. https://doi.org/10.1007/s12393-013-9065-4.

Schultz, S., Wagner, G., Urban, K., & Ulrich, J. (2004). High-pressure homogenization as a process for emulsion formation. *Chemical Engineering and Technology*, *27*, 361–368. https://onlinelibrary.wiley.com/doi/abs/10.1002/ceat.200406111.

Selig, M. J., Mehrad, B., Zamani, H., Kierulf, A., Licker, J., & Abbaspourrad, A. (2018). Distribution of oil solubilized β-carotene in stabilized locust bean gum powders for the delivery of orange colorant to food products. *Food Hydrocolloids*, *84*, 34–37. https://doi.org/10.1016/J.FOODHYD.2018.05.027.

Semo, E., Kesselman, E., Danino, D., & Livney, Y. D. (2007). Casein micelle as a natural nano-capsular vehicle for nutraceuticals. *Food Hydrocolloids*, *21*(5), 936–942. https://doi.org/10.1016/j.foodhyd.2006.09.006.

Seow, Y. X., Yeo, C. R., Chung, H. L., & Yuk, H. -G. (2014). Plant essential oils as active antimicrobial agents. *Critical Reviews in Food Science and Nutrition*, *54*(5), 625–644. https://doi.org/10.1080/10408398.2011.599504.

Sessa, M., Balestrieri, M. L., Ferrari, G., Servillo, L., Castaldo, D., D'Onofrio, N., et al. (2014). Bioavailability of encapsulated resveratrol into nanoemulsion-based delivery systems. *Food Chemistry*, *147*, 42–50. https://doi.org/10.1016/j.foodchem.2013.09.088.

Setya, S., Talegaonkar, S., & Razdan, B. K. (2014). Nanoemulsions: Formulation methods and stability aspects. *World Journal of Pharmacy and Pharmaceutical Sciences*, *3*, 2214–2228. www.wjpps.com/download/article/1391535192.pdf.

Seyed, M. A., Jantan, I., Bukhari, S. N. A., & Vijayaraghavan, K. (2016). A comprehensive review on the chemotherapeutic potential of piceatannol for cancer treatment, with mechanistic insights. *Journal of Agricultural and Food Chemistry*, *64*(4), 725–737. https://doi.org/10.1021/acs.jafc.5b05993.

Shah, B., Davidson, P. M., & Zhong, Q. (2012). Encapsulation of eugenol using Maillard-type conjugates to form transparent and heat stable nanoscale dispersions. *LWT—Food Science and Technology*, *49*(1), 139–148. https://doi.org/10.1016/j.lwt.2012.04.029.

Shao, Y., & Tang, C. H. (2014). Characteristics and oxidative stability of soy protein-stabilized oil-in-water emulsions: Influence of ionic strength and heat pretreatment. *Food Hydrocolloids*, *37*, 149–158. https://doi.org/10.1016/j.foodhyd.2013.10.030.

Sharma, M., Mann, B., Sharma, R., Bajaj, R., Athira, S., Sarkar, P., et al. (2017). Sodium caseinate stabilized clove oil nanoemulsion: Physicochemical properties. *Journal of Food Engineering*, *212*, 38–46. https://doi.org/10.1016/j.jfoodeng.2017.05.006.

Shen, L., & Tang, C. H. (2012). Microfluidization as a potential technique to modify surface properties of soy protein isolate. *Food Research International*, *48*(1), 108–118. https://doi.org/10.1016/j.foodres.2012.03.006.

Shi, A., Li, D., Wang, L., Li, B., & Adhikari, B. (2011). Preparation of starch-based nanoparticles through high-pressure homogenization and miniemulsion cross-linking: Influence of various process parameters on particle size and stability. *Carbohydrate Polymers*, *83*(4), 1604–1610. https://doi.org/10.1016/j.carbpol.2010.10.011.

Shtay, R., Tan, C. P., & Schwarz, K. (2018). Development and characterization of solid lipid nanoparticles (SLNs) made of cocoa butter: A factorial design study. *Journal of Food Engineering*, *231*, 30–41. https://doi.org/10.1016/j.jfoodeng.2018.03.006.

Shu, G., Khalid, N., Chen, Z., Neves, M. A., Barrow, C. J., & Nakajima, M. (2018). Formulation and characterization of astaxanthin-enriched nanoemulsions stabilized using ginseng saponins as natural emulsifiers. *Food Chemistry*, *255*, 67–74. https://doi.org/10.1016/j.foodchem.2018.02.062.

Silva, H. D., Cerqueira, M. A., Souza, B. W. S., Ribeiro, C., Avides, M. C., Quintas, M. A. C., et al. (2011). Nanoemulsions of β-carotene using a high-energy emulsification–evaporation technique. *Journal of Food Engineering*, *102*(2), 130–135. https://doi.org/10.1016/j.jfoodeng.2010.08.005.

Singh, K., & Mishra, A. (2013). Water soluble chitosan nanoparticle for the effective delivery of lipophilic drugs: A review. *International Journal of Applied Pharmaceutics*, *5*(3), 1–6.

Sotomayor-Gerding, D., Oomah, B. D., Acevedo, F., Morales, E., Bustamante, M., Shene, C., et al. (2016). High carotenoid bioaccessibility through linseed oil nanoemulsions with enhanced physical and oxidative stability. *Food Chemistry*, *199*, 463–470. https://doi.org/10.1016/j.foodchem.2015.12.004.

Stang, M., Schuchmann, H., & Schubert, H. (2001). Emulsification in high-pressure homogenizers. *Engineering in Life Sciences*, *1*(4), 151–157. https://doi.org/10.1002/1618-2863(200110)1:4<151::AID-ELSC151>3.0.CO;2-D.

Sumathi, R., Tamizharasi, S., & Sivakumar, T. (2017). Formulation and evaluation of polymeric nanosuspension of naringenin. *International Journal of Applied Pharmaceutics*, *9*, 60–70.

Sun, C., Dai, L., Liu, F., & Gao, Y. (2016). Simultaneous treatment of heat and high pressure homogenization of zein in ethanol–water solution: Physical, structural, thermal and morphological characteristics. *Innovative Food Science & Emerging Technologies*, *34*, 161–170. https://doi.org/10.1016/j.ifset.2016.01.016.

Sun, C., & Gunasekaran, S. (2009). Effects of protein concentration and oil-phase volume fraction on the stability and rheology of menhaden oil-in-water emulsions stabilized by whey protein isolate with xanthan gum. *Food Hydrocolloids*, *23*(1), 165–174. https://doi.org/10.1016/j.foodhyd.2007.12.006.

Sun, C., Liu, F., Yang, J., Yang, W., Yuan, F., & Gao, Y. (2015). Physical, structural, thermal and morphological characteristics of zein quercetagetin composite colloidal nanoparticles. *Industrial Crops and Products*, *77*, 476–483. https://doi.org/10.1016/j.indcrop.2015.09.028.

Surh, J., Decker, A. E., & McClements, D. J. (2017). Utilisation of spontaneous emulsification to fabricate lutein-loaded nanoemulsion-based delivery systems: Factors influencing particle size and colour. *International Journal of Food Science & Technology*, *52*, 1408–1416. https://doi.org/10.1111/ijfs.13395.

Svelander, C. A., Lopez-Sanchez, P., Pudney, P. D., Schumm, S., & Alminger, M. A. (2011). High pressure homogenization increases the in vitro bioaccessibility of α-and β-carotene in carrot emulsions but not of lycopene in tomato emulsions. *Journal of Food Science*, *76*(9), H215–H225. https://doi.org/10.1111/j.1750-3841.2011.02418.x.

Syamaladevi, R. M., Insan, S. K., Dhawan, S., Andrews, P., & Sablani, S. S. (2012). Physicochemical properties of encapsulated red raspberry (*Rubus idaeus*) powder: Influence of high-pressure homogenization. *Drying Technology: An International Journal*, *30*(5), 484–493. https://doi.org/10.1080/07373937.2011.647369.

Taghvaei, M., & Jafari, S. M. (2015). Application and stability of natural antioxidants in edible oils in order to substitute synthetic additives. *Journal of Food Science and Technology*, *52*(3), 1272–1282. https://doi.org/10.1007/s13197-013-1080-1.

Taghvaei, M., Jafari, S. M., Mahoonak, A. S., Nikoo, A. M., Rahmanian, N., Hajitabar, J., et al. (2014). The effect of natural antioxidants extracted from plant and animal resources on the oxidative stability of soybean oil. *LWT—Food Science and Technology*, *56*(1), 124–130. https://doi.org/10.1016/j.lwt.2013.11.009.

Tahiri, I., Makhlouf, J., Paquin, P., & Fliss, I. (2006). Inactivation of food spoilage bacteria and *Escherichia coli* O157:H7 in phosphate buffer and orange juice using dynamic high pressure. *Food Research International*, *39*(1), 98–105. https://doi.org/10.1016/j.foodres.2005.06.005.

Tan, C. P., & Nakajima, M. (2005). β-Carotene nanodispersions: Preparation, characterization and stability evaluation. *Food Chemistry*, *92*(4), 661–671. https://doi.org/10.1016/j.foodchem.2004.08.044.

Tan, C., Xie, J., Zhang, X., Cai, J., & Xia, S. (2016). Polysaccharide-based nanoparticles by chitosan and gum arabic polyelectrolyte complexation as carriers for curcumin. *Food Hydrocolloids*, *57*, 236–245. https://doi.org/10.1016/j.foodhyd.2016.01.021.

Tang, C. H., & Li, X. R. (2013). Microencapsulation properties of soy protein isolate and storage stability of the correspondingly spray-dried emulsions. *Food Research International*, *52*(1), 419–428. https://doi.org/10.1016/j.foodres.2012.09.010.

Tavassoli-Kafrani, E., Shekarchizadeh, H., & Masoudpour-Behabadi, M. (2016). Development of edible films and coatings from alginates and carrageenans. *Carbohydrate Polymers*, *137*, 360–374. https://doi.org/10.1016/j.carbpol.2015.10.074.

Teng, Z., Luo, Y., & Wang, Q. (2012). Nanoparticles synthesized from soy protein: Preparation, characterization, and application for nutraceutical encapsulation. *Journal of Agricultural and Food Chemistry*, *60*(10), 2712–2720. https://doi.org/10.1021/jf205238x.

Teng, Z., Luo, Y., & Wang, Q. (2013). Carboxymethyl chitosan–soy protein complex nanoparticles for the encapsulation and controlled release of vitamin D_3. *Food Chemistry, 141*(1), 524–532. https://doi.org/10.1016/j.foodchem.2013.03.043.

Tingaut, P., Zimmermann, T., & Sebe, G. (2012). Cellulose nanocrystals and microfibrillated cellulose as building blocks for the design of hierarchical functional materials. *Journal of Materials Chemistry, 22*(38), 20105–20111.

Trifković, K., Đorđević, V., Balanč, B., Kalušević, A., Lević, S., Bugarski, B., et al. (2016). Novel approaches in nanoencapsulation of aromas and flavors. In A. M. Grumezescu (Ed.), *Encapsulations* (pp. 363–419). New York: Elsevier.

Trujillo, A. J., Capellas, M., Saldo, J., Gervilla, R., & Guamis, B. (2002). Applications of high-hydrostatic pressure on milk and dairy products: A review. *Innovative Food Science & Emerging Technologies, 3*(4), 295–307. https://doi.org/10.1016/S1466-8564(02)00049-8.

Valencia-Flores, D. C., Hernández-Herrero, M., Guamis, B., & Ferragut, V. (2013). Comparing the effects of ultra-high-pressure homogenization and conventional thermal treatments on the microbiological, physical, and chemical quality of almond beverages. *Journal of Food Science, 78*(2), E199–E205. https://doi.org/10.1111/1750-3841.12029.

Villay, A., Lakkis de Filippis, F., Picton, L., Le Cerf, D., Vial, C., & Michaud, P. (2012). Comparison of polysaccharide degradations by dynamic high-pressure homogenization. *Food Hydrocolloids, 27*(2), 278–286. https://doi.org/10.1016/j.foodhyd.2011.10.003.

Vincekovic, M., Viskic, M., Juric, S., Giacometti, J., Kovacevic, D. B., Putnik, P., et al. (2017). Innovative technologies for encapsulation of Mediterranean plants extracts. *Trends in Food Science & Technology, 69*, 1–12. https://doi.org/10.1016/j.tifs.2017.08.001.

Vishwanathan, R., Wilson, T. A., & Nicolosi, R. J. (2009). Bioavailability of a nanoemulsion of lutein is greater than a lutein supplement. *Nano Biomedecine & Engineering, 1*, 38–49.

Voragen, A. G. J., Coenen, G. J., Verhoef, R. P., & Schols, H. A. (2009). Pectin, a versatile polysaccharide present in plant cell walls. *Structural Chemistry, 20*(2), 263. https://doi.org/10.1007/s11224-009-9442-z.

Wan, Z. L., Guo, J., & Yang, X. Q. (2015). Plant protein-based delivery systems for bioactive ingredients in foods. *Food & Function, 6*(9), 2876–2889. https://doi.org/10.1039/C5FO00050E.

Wang, W., Jung, J., & Zhao, Y. (2017). Chitosan-cellulose nanocrystal microencapsulation to improve encapsulation efficiency and stability of entrapped fruit anthocyanins. *Carbohydrate Polymers, 157*, 1246–1253. https://doi.org/10.1016/j.carbpol.2016.11.005.

Wang, X., Kluge, J. A., Leisk, G. G., & Kaplan, D. L. (2008). Sonication-induced gelation of silk fibroin for cell encapsulation. *Biomaterials, 29*(8), 1054–1064. https://doi.org/10.1016/j.biomaterials.2007.11.003.

Wang, T., Soyama, S., & Luo, Y. (2016). Development of a novel functional drink from all natural ingredients using nanotechnology. *LWT—Food Science and Technology, 73*, 458–466. https://doi.org/10.1016/j.lwt.2016.06.050.

Wang, R., Tian, Z., & Chen, L. (2011). Nano-encapsulations liberated from barley protein microparticles for oral delivery of bioactive compounds. *International Journal of Pharmaceutics, 406*(1), 153–162. https://doi.org/10.1016/j.ijpharm.2010.12.039.

Waraho, T., Cardenia, V., Decker, E. A., & McClements, D. J. (2010). Lipid oxidation in emulsified food products. In E. A. Decker, R. J. Elias, & D. J. McClements (Eds.), *Oxidation in foods and beverages and antioxidant applications. Volume 2: Management in diferent industry sectors* (pp. 306–343). Cambridge: Woodhead Publishing Ltd.

Wei, Y., Sun, C., Dai, L., Zhan, X., & Gao, Y. (2018). Structure, physicochemical stability and in vitro simulated gastrointestinal digestion properties of β-carotene loaded zein-propylene glycol alginate composite nanoparticles fabricated by emulsification-evaporation method. *Food Hydrocolloids, 81*, 149–158. https://doi.org/10.1016/j.foodhyd.2018.02.042.

Wen, J., Chen, S., & Chen, G. (2018). Solid lipid nanoparticles. In S. Roohinejad, R. Greiner, I. Oey, & J. Wen (Eds.), *Emulsion-based systems for delivery of food active compounds* (pp. 121–138). Chichester, UK: John Wiley & Sons, Ltd. https://doi.org/10.1002/9781119247159.ch5

Windhab, E. J., Dressler, M., Feigl, K., Fischer, P., & Megias-Alguacil, D. (2005). Emulsion processing-from single-drop deformation to design of complex processes and products. *Chemical Engineering Science*, *60*(8–9), 2101–2113. https://doi.org/10.1016/j.ces.2004.12.003.

Wise, D. L. (2000). *Handbook of pharmaceutical controlled release technology*. New York: Taylor & Francis.

Xing, F., Cheng, G., Yi, K., & Ma, L. (2005). Nanoencapsulation of capsaicin by complex coacervation of gelatin, acacia, and tannins. *Journal of Applied Polymer Science*, *96*(6), 2225–2229. https://doi.org/10.1002/app.21698.

Xu, R., Teng, Z., & Wang, Q. (2016). Development of tyrosinase-aided crosslinking procedure for stabilizing protein nanoparticles. *Food Hydrocolloids*, *60*, 324–334. https://doi.org/10.1016/j.foodhyd.2016.04.009.

Yang, Y., Zhao, C., Tian, G., Lu, C., Zhao, S., Bao, Y., et al. (2017). Effects of preheating and storage temperatures on aroma profile and physical properties of citrus-oil emulsions. *Journal of Agricultural and Food Chemistry*, *65*(35), 7781–7789. https://doi.org/10.1021/acs.jafc.7b03270.

Yang, J., Zhou, Y., & Chen, L. (2014). Elaboration and characterization of barley protein nanoparticles as an oral delivery system for lipophilic bioactive compounds. *Food & Function*, *5*(1), 92–101. https://doi.org/10.1039/C3FO60351B.

Ye, R., & Harte, F. (2014). High pressure homogenization to improve the stability of casein–hydroxypropyl cellulose aqueous Systems. *Food hydrocolloids*, *35*, 670–677. https://doi.org/10.1016/j.foodhyd.2013.08.022.

Yi, J., Li, Y., Zhong, F., & Yokoyama, W. (2014). The physicochemical stability and in vitro bioaccessibility of beta-carotene in oil-in-water sodium caseinate emulsions. *Food Hydrocolloids*, *35*, 19–27. https://doi.org/10.1016/j.foodhyd.2013.07.025.

Yu, C., Wang, W., Yao, H., & Liu, H. (2007). Preparation of phospholipid microcapsule by spray drying. *Drying Technology*, *25*(4), 695–702. https://doi.org/10.1080/07373930701291017.

Yu, L., Yan, D., Sun, G., & Gu, L. (2008). Preparation and characterization of pH-sensitive hydrogel fibers based on hydrolyzed-polyacrylonitrile/soy protein. *Journal of Applied Polymer Science*, *108*(2), 1100–1108. https://doi.org/10.1002/app.27689.

Yuan, Y., Gao, Y., Zhao, J., & Mao, L. (2008). Characterization and stability evaluation of β-carotene nanoemulsions prepared by high pressure homogenization under various emulsifying conditions. *Food Research International*, *41*, 61–68. https://doi.org/10.1016/j.foodres.2007.09.006.

Yucel, C., Quagliariello, V., Iaffaioli, R. V., Ferrari, G., & Donsì, F. (2015). Submicron complex lipid carriers for curcumin delivery to intestinal epithelial cells: Effect of different emulsifiers on bioaccessibility and cell uptake. *International Journal of Pharmaceutics*, *494*, 357–369. https://doi.org/10.1016/j.ijpharm.2015.08.039.

Zahi, M. R., Wan, P., Liang, H., & Yuan, Q. (2014). Formation and stability of D-limonene organogel-based nanoemulsion prepared by a high-pressure homogenizer. *Journal of Agricultural and Food Chemistry*, *62*(52), 12563–12569. https://doi.org/10.1021/jf5032108.

Zamora, A., Ferragut, V., Jaramillo, P. D., Guamis, B., & Trujillo, A. J. (2007). Effects of ultra-high pressure homogenization on the cheese-making properties of milk. *Journal of Dairy Science*, *90*(1), 13–23. https://doi.org/10.3168/jds.S0022-0302(07)72604-8.

Zamora, A., & Guamis, B. (2015). Opportunities for ultra-high-pressure homogenisation (UHPH) for the food industry. *Food Engineering Reviews*, 7(2), 130–142. https://doi.org/10.1007/s12393-014-9097-4.

Zhang, Y., Tan, C., Abbas, S., Eric, K., Zhang, X., Xia, S., et al. (2014). The effect of soy protein structural modification on emulsion properties and oxidative stability of fish oil microcapsules. *Colloids and Surfaces. B: Biointerfaces*, *120*, 63–70. https://doi.org/10.1016/j.colsurfb.2014.05.006.

Zhang, J. B., Wu, N. N., Yang, X. Q., He, X. T., & Wang, L. J. (2012). Improvement of emulsifying properties of Maillard reaction products from β-conglycinin and dextran using controlled enzymatic hydrolysis. *Food Hydrocolloids*, *28*(2), 301–312. https://doi.org/10.1016/j.foodhyd.2012.01.006.

Zhong, Q., & Jin, M. (2009). Zein nanoparticles produced by liquid–liquid dispersion. *Food Hydrocolloids*, *23*(8), 2380–2387. https://doi.org/10.1016/j.foodhyd.2009.06.015.

Ziani, K., Fang, Y., & McClements, D. J. (2012). Fabrication and stability of colloidal delivery systems for flavor oils: Effect of composition and storage conditions. *Food Research International*, *46*, 209–216. https://doi.org/10.1016/j.foodres.2011.12.017.

Zimet, P., & Livney, Y. D. (2009). Beta-lactoglobulin and its nanocomplexes with pectin as vehicles for ω-3 polyunsaturated fatty acids. *Food Hydrocolloids*, *23*(4), 1120–1126. https://doi.org/10.1016/j.foodhyd.2008.10.008.

Zou, L., Zheng, B., Zhang, R., Zhang, Z., Liu, W., Liu, C., et al. (2016). Enhancing the bioaccessibility of hydrophobic bioactive agents using mixed colloidal dispersions: Curcumin-loaded zein nanoparticles plus digestible lipid nanoparticles. *Food Research International*, *81*, 74–82. https://doi.org/10.1016/j.foodres.2015.12.035.

Zuidam, N. J., & Heinrich, E. (2010). Encapsulation of aroma. In N. J. Zuidam & V. Nedovic (Eds.), *Encapsulation technologies for active food ingredients and food processing* (pp. 127–160). New York: Springer.

Further reading

Fathi, M., Donsì, F., & McClements, D. J. (2018). Protein-based delivery systems for the nanoencapsulation of food ingredients. *Comprehensive Reviews in Food Science and Food Safety*, *17*(4), 920–936. https://doi.org/10.1111/1541-4337.12360.

González-Ferrero, C., & Sáiz-Abajo, M. J. (2015). Characterization and stability studies of bioactive compounds and food matrices as evidence in support of health claims. *International Journal of Food Sciences and Nutrition*, *66*(Suppl. 1), S4–S12. https://doi.org/10.3109/09637486.2015.1042843.

Režek Jambrak, A., Vukušić, T., Donsì, F., Paniwnyk, L., & Djekic, I. (2018). Three pillars of novel nonthermal food technologies: Food safety, quality, and environment. *Journal of Food Quality*, *8619707*, 1–18. https://doi.org/10.1155/2018/8619707.

Syamaladevi, R. M., Andrews, P. K., Davies, N. M., Walters, T., & Sablani, S. S. (2011). Storage effects on anthocyanins, phenolics and antioxidant activity of thermally processed conventional and organic blueberries. *Journal of the Science of Food and Agriculture*, *92*(4), 916–924. https://doi.org/10.1002/jsfa.4670.

Zhou, H., Yue, Y., Liu, G., Li, Y., Zhang, J., Gong, Q., et al. (2010). Preparation and characterization of a lecithin nanoemulsion as a topical delivery system. *Nanoscale Research Letters*, *5*, 224–230. https://doi.org/10.1007/s11671-009-9469-5.

CHAPTER SEVEN

Production of food bioactive-loaded nanostructures by microfluidization

José Muñoz[a], M. Carmen Alfaro[b], Luis A. Trujillo-Cayado[b], Jenifer Santos[a], M. José Martín-Piñero[a]

[a]Chemical Engineering Department, Facultad de Química, C/Profesor García Gonzalez, University of Seville, Seville, Spain

[b]Chemical Engineering Department, Escuela Politécnica Superior, C/Virgen de África, University of Seville, Seville, Spain

Contents

1 Introduction

Today there is an increasing social interest in more natural, safer, and healthier products. With regard to food consumption, consumer habits are progressively changing. Nowadays it is a question not only of reducing our consumption of those foods whose intake in excess is harmful to our health but also of seeking those that yield healthy benefits and help us to delay

Nanoencapsulation of Food Ingredients by Specialized Equipment
https://doi.org/10.1016/B978-0-12-815671-1.00007-X

the occurrence of some diseases. For this reason the food industry is being challenged to develop formulations that incorporate natural and bioactive ingredients rather than synthetic compounds (Faridi Esfanjani, Assadpour, & Jafari, 2018; Katouzian & Jafari, 2016; Mahdavi, Jafari, Ghorbani, & Assadpoor, 2014). A bioactive component is a substance that is able to modulate one or more metabolic processes and, if consumed in sufficient quantity, provides health benefits both in the prevention and in the treatment of certain diseases (Guaadaoui, Benaicha, Elmajdoub, Bellaoui, & Hamal, 2014). Therefore they contribute to an improvement in the health of the brain, heart, and immune system, as well as reducing the risk of chronic diseases. One of the goals of the food industry nowadays is to increase the offer of foods enriched with food bioactive ingredients and nutraceuticals, such as prebiotics, probiotics, phenolic compounds and antioxidants, vitamins, essential fatty acids, enzymes, and minerals (Jafari, 2017). Many of these ingredients suffer destabilization when exposed to external environmental conditions, so nanoencapsulation is an effective way of applying a barrier to protect them from their external environment. From an engineering point of view, microfluidization turns out to be one of the more popular techniques to carry out nanoencapsulation of a fascinating variety of materials. In this chapter, the microfluidization technology is assessed such that operation modes, processing variables, and the applications of different available shearing chambers are analyzed. In addition, the role of microfluidization in the development of different types of nanostructures is discussed.

2 Microfluidization basics

Microfluidization, as a top-down methodology, is the state-of-the-art technology among high-energy methods, for the production of nanoemulsions, submicron emulsions, nanosuspensions, liposomes, and encapsulated materials, to promote deagglomeration of particle clusters and cell disruption. On the other hand, microfluidization can also be applied in a bottom-up methodology when used as a microfluidic reaction technology to yield crystals or precipitates.

Bench-scale Microfluidizers applies high pressures (3.4–275 MPa) with a pneumatic intensification pump, which oscillates between suction and compression strokes, following a trapezoidal profile. In this way the set point pressure is kept for longer than in conventional high-pressure valve homogenizers, which usually apply pressures according to a loosely triangular profile. In Microfluidizers, the inlet fluid is forced to flow into specifically designed microchannels, which are available with different geometries and characteristic diameters. These microchannels are the so-called interaction chambers (IXC). Fluids undergo laminar flow when transported through pipes towards and out of the IXC while undergoing turbulent flow and cavitation inside the microchannels (Villalobos-Castillejos et al., 2018). The reproducibility of batches produced with the microfluidization technique is quite good on account of the well-defined geometry of IXCs, uniform pressure profile, and limited amount (microliter order of magnitude) of samples pushed through the IXC after each compression stroke, which, in turn, promotes a higher energy-density transfer rate (Panagiotou & Fisher, 2012). These factors are also responsible for the narrow particle size distributions and low mean diameters, which can be achieved by using this technology.

Production-scale Microfluidizers can use two intensifier pistons to reduce the cycle lag between suction and compression strokes. In addition, the continuous electronic control of pumps allows a nearly constant pressure profile to be delivered, which enhances the reproducibility of industrial batches (Microfluidics, 2018).

At present, two kinds of geometries are commercially available for IXCs: the "Y-type" and the "Z-type" interaction chambers. The high pressure applied and sudden reduction in cross-sectional area when fluids flow from internal transport pipes to the inlet of interaction chambers yield high flow velocities. This along with changes of flow direction promotes high shear and extensional deformations, as well as turbulent flow and associated eddies. Products flowing through a "Z-type" IXC undergo huge impact forces when hitting the walls of the interaction chamber. On the other hand, collision with microchannel walls is not as important in "Y-type" IXC, so energy losses are minimized. In fact, size reduction is mainly achieved by the impingement of two identical product streams. Different nominal shear and flow rates can be reached, depending on the pressure applied and the cross-sectional area of the microchannels. The characteristic dimensions of "Y-type" IXCs range from 75 to 125 μm, while those of "Z-type" IXCs range from 87 to 300 μm (Microfluidics, 2018; Villalobos-Castillejos et al., 2018). The choice of the IXC and its size strongly depends on the application, the initial particle size, and the required final particle size distribution.

Interaction chambers will wear after extensive use, regardless of their constitutive material, which may be polycrystalline diamond or aluminum oxide. To delay wearing, samples with air trapped inside (i.e., the presence of foam and bubbles emptying the feeding reservoir completely) and inappropriate priming of the pump must be avoided. A further recommendation is to use an auxiliary processing module (APM), which is always a "Z-type" IXC. When using a "Y-type" IXC, the APM must be placed downstream to add backpressure and to stabilize the product flow. Conversely, if the main IXC is of "Z-type," the APM must be located upstream and consists of another "Z-type" chamber with a higher cross-sectional area. In this way, this APM works as a premixer or preprocessing unit that yields an initial particle size reduction before the sample is pushed to the lower-sized interaction chamber located downstream. This is a way of improving the efficiency of the second chamber and avoiding blockage of its microchannels.

"Y-type" IXCs are well suited to the production of O/W nanoemulsions, O/W submicron emulsions, and liposomes and for polymer encapsulation. On the other hand the "Z-type" is used for deagglomeration processes, W/O emulsions, and cell disruption. The usual reference to the characteristic IXC size is not sufficient when selecting a suitable one for specific applications, such as the preparation of nanoemulsions. For instance, both the F20Y and F12Y commercial chambers (Microfluidics, 2018) have a characteristic size of 75 μm, but the latter produces a higher shear on account of its smaller cross-sectional area. Therefore the F12Y is the IXC of choice to produce nanoemulsions; provided suitable emulsifiers are used.

Microfluidization can be carried out by either single-channel or dual-channel feeding. With regard to emulsification applications, the former requires that a coarse emulsion is previously formed, typically by using a rotor-stator homogenizer or a high-speed mixer. Subsequently the coarse dispersion is poured into a single feeding hopper and is forced to pass through the Microfluidizer. Therefore the role of a Microfluidizer is to further reduce the mean particle size to the submicron or nanoscales and to get a narrower particle size distribution (Rayner & Dejmek, 2015). Interestingly, if the previous coarse emulsion is prepared such that relatively small droplets are achieved thanks to a rapid and efficient adsorption of a sufficient quantity of emulsifier, the second microfluidization step will require a higher energy-density transfer rate. This is logical, considering that the Laplace pressure to overcome for further reduction of droplet sizes will increase as the mean droplet diameter decreases. In contrast, in the dual-channel homogenization method, the continuous-phase and dispersed-phase feed the homogenizer from two different reservoirs, so that the formation of the dispersion is carried out simultaneously, forcing the oil phase and aqueous phase through an air-driven high-pressure Microfluidizer.

The scale-up of microfluidization laboratory tests is quite straightforward since the microreactor design approach is used. However, the required number of parallel "Y-type" or "Z-type" IXCs may be determined to achieve specific product volumetric flow rates while maintaining the advantages of single multichannels (well-defined geometry, uniform pressure profile, and limited amount of sample pushed through each IXC, which guarantees a higher energy-density transfer rate). A parallel setup of microreactors makes maintenance operations easier since the whole processing unit can keep working, while one single chamber is stopped to solve blockage or other processing problems.

On the whole, one of the most important processing variables for microfluidization is the possibility of recycling products whenever the final quality specifications are not achieved. If the mean particle size is not low enough or the particle size distribution (PSD) is not narrow enough, multiple passes through the interaction chambers are currently made. The optimum performance of the microfluidization technique is to apply a sufficiently high pressure to avoid recycling products, since each pass through an interaction chamber implies an increase in energy consumption and temperature. This may be counterbalanced by cooling the output of the Microfluidizer by means of a shell and tube heat exchanger or a cooling coil placed inside a temperature-controlled bath. This raises the point of outlet temperature as one of the processing variables to control when using this technique. Excessive increase in temperature may even alter the performance of emulsifiers and the viscosity ratio (Panagiotou & Fisher, 2013). Decreasing the outlet temperature will slow down collisions among particles when producing emulsions or suspensions, which in turn will decrease the risk of coalescence and flocculation, respectively. A further problem related to recycling operations is the risk of recoalescence when dealing with emulsions (Jafari, Assadpoor, He, & Bhandari, 2008). Another processing variable, which is often overlooked, is the inlet temperature. Low inlet temperatures may compensate for the unavoidable heating effects caused by energy dissipation within the interaction chamber and, in other instances, must be optimized to ensure the right handling properties of emulsifiers.

Apart from temperature the main process variables are the choice of interaction chamber geometry, selection of upstream or downstream

APM, the interaction chamber characteristic size, number of cycles, and process pressure. Important process control variables are the power input and density of energy, which can be calculated as reported by Gavi, Kubicki, Padron, and Özcan-Taşkın (2018) and Heffernan, Kelly, Mulvihill, Lambrich, and Schuchmann (2011), respectively.

The power input, P, can be calculated as follows:

$$P\,(\mathrm{W}) = (V_{\mathrm{i}}/t_{\mathrm{i}}) \cdot \Delta P$$

where P (W) is the power input in watts, V_{i} is the volume of a single stroke of the pump (around $6 \cdot 10^{-6}$ m^3), t_{i} is the time of impulse (a few seconds), and ΔP is the operating pressure (Pa). However, it must be noted that the time of residence in interaction chambers is claimed to be even shorter by the Microfluidizer manufacturer (1–5 ms) (Microfluidics, 2018).

Another important processing parameter is the energy density, which can be calculated for multiple cycle operations as indicated by Heffernan et al. (2011):

$$E_{\mathrm{v}} = \Delta P \cdot N$$

where E_{V} is the energy density (i.e., energy/m^3), ΔP is the operating pressure, and N is number of cycles.

3 Development of submicron emulsions and nanoemulsions by microfluidization

At present a few functional foods incorporating bioactive components are available in the market, although two main challenges must be considered. Firstly, many of these bioactive agents are highly hydrophobic, exhibiting low solubility in aqueous systems (Akhavan, Assadpour, Katouzian, & Jafari, 2018; Rezaei, Fathi, & Jafari, 2019). This fact means that they cannot quickly be dispersed in aqueous-based systems. Secondly, some of them are physically and/or chemically highly unstable after their incorporation into some food products, due to processing, transport, and storage conditions (Rostamabadi, Falsafi, & Jafari, 2019). In addition, many bioactive compounds exhibit low biochemical stability, that is, they undergo modifications after consumption. In other cases, some bioactive agents provide undesirable flavors or are difficult to handle. All this reduces their bioavailability and limits their practical use. The encapsulation of bioactive compounds has been proposed as a way of stabilizing them, controlling their release, and increasing their bioavailability (Assadpour & Jafari, 2018;

Jafari & McClements, 2017). In principle, an emulsion could be a good way to encapsulate these compounds.

Emulsions are dispersions fabricated from two immiscible liquids (commonly oil and water), with one of the liquids being dispersed in the form of droplets (dispersed phase) in the other liquid (continuous phase) (McClements, 2016) and exhibiting mean droplet sizes ranging from 100 nm to 100 μm (Deen, Skovgaard, & Pedersen, 2016). They can be classified according to their morphology and composition. With regard to the latter, two major types of emulsions (oil-in-water (O/W) emulsions or water-in-oil (W/O) emulsions) can be formed. In W/O emulsions, water droplets are dispersed in the continuous oil phase. In contrast, in O/W emulsions, oil droplets are dispersed in the continuous aqueous phase. In both cases, bioactive molecules could be incorporated into the dispersed phase of emulsions (Jafari, Paximada, Mandala, Assadpour, & Mehrnia, 2017). In the food industry, most of the products are aqueous-based; hence we will focus here on O/W emulsions. Bioactive lipophilic components may be dissolved in the oil phase, acting as carrier or release system. However the particle size of the release system is an important parameter because the release rate of the bioactive agent is slower for large particle sizes (Tiwari & Amiji, 2006). Acosta (2009) found that bioactive compounds in droplets with a size below 500 nm exhibited greater absorption into the human body. This was attributed to the fact that small droplets can penetrate through the intestine epithelium allowing longer time for digestion and, therefore, enhanced absorption of the bioactive compound, that is, the increased permeability allows a rapid penetration of the bioactive component through the biological membranes (Odriozola-Serrano, Oms-Oliu, & Martín-Belloso, 2014). For this reason, conventional emulsions are not the option of choice. Fortunately, nanoemulsions do offer a good alternative due to their considerable potential application related to the development of novel functionalities. In particular the delivery of functional compounds, such as bioactive molecules, micronutrients, colorants, flavorings, or antimicrobial agents, into food and beverage products that naturally do not contain them is highly desired to increase product value by enhancing health benefits, nutritive profile, appearance, aroma, or shelf life (Donsì, 2018). The use of nanoemulsions is an efficient technique to protect the bioactive agents against environmental conditions and provide increased stability. Most importantly, nanoemulsions maintain their functional activity (Karthik, Ezhilarasi, & Anandharamakrishnan, 2017).

Nanoemulsions can be considered as conventional emulsions, which possess very small droplets. Regarding the droplet size of a nanoemulsion, different size ranges are considered in literature. Some authors consider a range between 50 and 500 nm (Golfomitsou, Mitsou, Xenakis, & Papadimitriou, 2018; Porras, Solans, Gonzalez, & Gutierrez, 2008) and other authors between 20 and 200 nm (Huang, Yu, & Ru, 2010; Jin et al., 2016), while others are more restrictive and propose that droplet sizes should range from 10 to 100 nm (Tadros, Izquierdo, Esquena, & Solans, 2004) or even between 1 and 100 nm (Mason, Wilking, Meleson, Chang, & Graves, 2006). Berg (2010) also proposed the latter range for nanoparticles in general. Therefore nanoemulsions can have a droplet size range between 1 and 500 nm (Yalçınöz & Erçelebi, 2018). For commercial applications, it is important to be able to attain emulsions with small droplets using efficient homogenization processes. Microfluidization is one of the best methods to produce small droplet sizes with a uniform droplet size distribution (Schuh et al., 2018).

The small size of the droplets in nanoemulsions grants them a series of advantages. On the one hand, they are more kinetically stable than emulsions against gravitational separation and droplet aggregation (Tadros et al., 2004; Wooster, Golding, & Sanguansri, 2008; McClements & Rao, 2011). Indeed, Brownian effects compensate for gravitational forces, and the attraction forces acting between droplets are small since decrease with decreasing droplet size (Silva, Cerqueira, & Vicente, 2012). On the other hand and depending on droplet size, nanoemulsions can be transparent or translucent (Tadros et al., 2004). For these reasons, nanoemulsions are likely to find more and more applications in commercial foods, such as salad dressings, or beverages, such as fortified water, soft drinks, or juices (Bai & McClements, 2016a; Guttoff, Saberi, & McClements, 2015; Pathakoti, Manubolu, & Hwang, 2017). They can also be used in sauces, yogurt, cream, etc. with new textures and a higher stability (De Francisco & García-Estepa, 2018). Incorporating bioactive agents into nanoemulsions has the advantage of protecting them from undesired interactions with external agents (oxygen, heat, ultraviolet light, and other incompatible food compounds). In this way, nanoemulsions enhance the stability and solubility of bioactive agents. However, nanoemulsions are thermodynamically unstable as emulsions, and therefore they will tend to break down over time. To obtain emulsions with such small droplet sizes (nanorange), different approaches can be used, high-energy methods being the most advantageous due to their higher efficiency and ease of scale-up (Salvia-Trujillo, Rojas-Graü, Soliva-Fortuny, & Martín-Belloso, 2013). Among them, microfluidization can generate very small droplet sizes, and narrow droplet size distributions provided the energy input that is optimized, and the right emulsifiers are selected. Thus we will focus on nanoemulsions obtained by microfluidization.

3.1 Formulation ingredients

All compounds of a nanoemulsion formulation must be nontoxic and suitable for human consumption as "generally recognized as safe" (GRAS) materials to be approved as a food-grade nanoemulsion. The basic formulation of a nanoemulsion consists of oil, water, and emulsifier although other compounds can be added to improve the stability, to modify physicochemical properties, or to increase the nutritional value or functionality of the nanoemulsion. We refer here to thickening agents, gelling agents, weighting agents, ripening inhibitors (McClements, 2016), or bioactive compounds. The following sections contain a description of these ingredients.

3.1.1 Oil/oil phase

To develop a nanoemulsion the first step consists of selecting the oil, which is the main component of the dispersed phase. Different oils with various properties (polarity, density, purity, or viscosity) can be used. Many of these properties influence the formation and stabilization of nanoemulsions. For example, the oil polarity determines the interfacial tension, which influences the properties of the interface formed, thus affecting the shelf life of the nanoemulsion. In addition, oil polarity may also alter the sensory and physicochemical properties by influencing the distribution of compounds such as antioxidants and colors between the oil and aqueous phases. The oils most frequently used to prepare nanoemulsions are acylglycerols, free fatty acids, flavoring oils, essential oils, mineral oils, fat substitutes, and waxes (Jin et al., 2016). In the food industry, mainly long or even medium- and short-chain triglycerides oils (LCT, MCT, and SCT) with different degrees of saturation are used due to their nontoxicity, low cost, and functional properties (McClements & Rao, 2011). Vegetable oils such as corn oil, palm oil, olive oil, sesame oil, soybean oil, hydrogenated soybean oil, peanut oil, safflower oil, sunflower oil, or canola oil are used as the dispersed phase, either alone or in combination with other oils (Singh et al., 2017). Furthermore, new oils, such as salmon oil, are raising increasing interest (Belhaj, Arab-Tehrany, & Linder, 2010).

Nevertheless the use of triglyceride oils to obtain nanoemulsions using microfluidization has a drawback. They have a high viscosity, which hinders the disruption of drops inside the interaction chamber and, therefore, hinders the production of droplets with very small sizes. This was demonstrated by Wooster, Golding, and Sanguansri (2008) who studied the formation of nanoemulsions of triglyceride oils, in particular peanut oil, and compared them with hexadecane oil nanoemulsions. In spite of this, it is possible to obtain nanoemulsions of small size by modifying the ratio of dispersed-phase viscosity/continuous-phase viscosity (η_D/η_C). Braginsky and Belevitskaya (1996) found that for turbulent shear, the η_D/η_C optimal range for a most efficient droplet rupture is between 0.1 and 5. On the other hand, there are oils that despite having a low viscosity and low interfacial tension, which facilitate the production of nanosizes, have high polarity and, therefore, high solubility in water. This causes a rapid destabilization of the nanoemulsion by Ostwald ripening or coalescence (McClements & Rao, 2011). This occurs with flavor oils or essential oils. With regard to essential oils, the great interest generated by their use as an antimicrobial agent to replace synthetic additives should be noted. The use of essential oils improves the idea of safety in foods for consumers and increasing the shelf life of the product. In addition to their known antiviral, antifungal, and antimicrobial properties, they possess medical properties such as antioxidant, anticarcinogenic, antiinflammatory, and antimutagenic effects (Shaaban, El-Ghorab, & Shibamoto, 2012). Despite these beneficial properties, essential oils have a high volatility and strong odor, and although they are considered natural, they are toxic if used at high concentrations. Thus their encapsulation into nanoemulsions provides a solution to these drawbacks while improving overall antimicrobial activity. Numerous essential oils (lemongrass, thyme, rosemary, lemon, tangerine, oregano, etc.) or compounds of essential oils (citral, cinnamaldehyde, carvacrol, D-limonene, etc.) have been encapsulated into nanoemulsions (Trujillo-Cayado, Santos, Alfaro, Calero, & Muñoz, 2016). They can be dissolved into common oils, used as carriers, or they can be the only oil in the oil phase. Some examples of oils used in the preparation of food-grade O/W nanoemulsions by microfluidization are shown in Table 1.

Another aspect to keep in mind is that the oil can be loaded with lipophilic bioactive components. In the encapsulation and release of these components, the oil is used as a solubilizing agent; hence the choice of oil is conditioned by the solubility of the active ingredients in the oil (Qadir, Faiyazuddin, Hussain, Alshammari, & Shakeel, 2016). Note that

Table 1 List of oils and emulsifiers used in food-grade nanoemulsions obtained by microfluidization

Oil	Emulsifier and classification	Reference
High-oleic palm oil	Whey powder (protein) Tween 20 (nonionic surfactant)	Ricaurte, de Jesús Perea-Flores, Martinez, & Quintanilla-Carvajal, 2016
Palm oil	Tween 20 (nonionic surfactant)	Gadkari & Balaraman, 2015
Sunflower oil	Tween 80 (nonionic surfactant) Lecithin (phospholipid)	
Corn oil	Whey protein isolate (protein) *Quillaja* saponin (surfactant)	Luo et al., 2017a
Corn oil	Whey protein (protein) Gum arabic (polysaccharide) Soy lecithin (phospholipid) *Quillaja* saponin (nonionic surfactant)	Bai et al., 2016
Corn oil	Tween 20 (nonionic surfactant)	Qian & McClements, 2011
Octadecane	Sodium dodecyl sulfate (anionic surfactant) Sodium caseinate (polysaccharide) β-Lactoglobulin (protein)	
Corn oil	Tween 20 (nonionic surfactant)	Jo & Kwon, 2014
Medium-chain triglyceride oil	Tween 80 (nonionic surfactant) Whey protein isolate (protein) Soy bean protein isolate (protein) Sodium caseinate (protein)	
Medium-chain triglyceride	Tween 80 (nonionic surfactant) Tween 85 (nonionic surfactant)	Yang, Marshall-Breton, Leser, Sher, & McClements, 2012
Fish oil	Mesquite gum (polysaccharide)	García-Márquez, Higuera-Ciapara, & Espinosa-Andrews, 2017
Fish oil	Sodium caseinate (protein) Modified starch (polysaccharide) Whey protein isolate (protein)	Jafari et al., 2006

Continued

Table 1 List of oils and emulsifiers used in food-grade nanoemulsions obtained by microfluidization—cont'd

Oil	Emulsifier and classification	Reference
Oleoresin capsicum	Tween 80 (nonionic surfactant)	Akbas et al., 2018
DHA algae oil	Tween 40 (nonionic surfactant) Sodium caseinate (protein) Soya lecithin (phospholipid)	Karthik & Anandharamakrishnan, 2016b
Neobee 1053 medium-chain triglycerides (caprylic/capric triglyceride oil)	Tween 20 (nonionic surfactant) Dodecyltrimethylammonium bromide (DTAB) (cationic surfactant) Sodium dodecyl sulfate (SDS) (anionic surfactant)	Silva, Cerqueira, & Vicente, 2015
Lemongrass, clove, tea tree, thyme, marjoram, palmarosa, rosewood, sage, mint	Tween 80 (nonionic surfactant)	Salvia-Trujillo, Rojas-Graü, Soliva-Fortuny, & Martín-Belloso, 2015
Essential oils (oregano, thyme, and lemongrass)	Tween 80 (nonionic surfactant)	Artiga-Artigas et al., 2018
Lemon oil	Sucrose monopalmitate (nonionic surfactant)	Rao & McClements, 2011
Lemon oil	Tween 80 (nonionic surfactant) Sucrose monopalmitate (nonionic surfactant)	Rao & McClements, 2012
Lemongrass	Tween 80 (nonionic surfactant) Sodium alginate (polysaccharide)	Salvia-Trujillo, Rojas-Graü, Soliva-Fortuny, & Martín-Belloso, 2014
Oregano, thyme, lemongrass, mandarin	High methoxyl pectin (polysaccharide) Tween 80 (nonionic surfactant)	Guerra-Rosas et al., 2016
Eugenol oil, canola oil	Modified starch (polysaccharide)	Majeed et al., 2016
Cinnamaldehyde	Tween 80 (nonionic surfactant)	Jo et al., 2014
Orange oil	Sucrose monopalmitate (nonionic surfactant) Lyso-lechitin (phospholipid)	Choi et al., 2011

Table 1 List of oils and emulsifiers used in food-grade nanoemulsions obtained by microfluidization—cont'd

Oil	Emulsifier and classification	Reference
D-Limonene	Modified starch (polysaccharide) Sodium caseinate (protein) Whey protein isolate (protein)	Jafari et al., 2007
Fish oil (ω-3 fatty acids) and lemon essential oil	Sodium alginate (polysaccharide)	Salvia-Trujillo, Decker, & McClements, 2016

incorporation of these compounds into the dispersed phase influences the formulation and properties of nanoemulsions. Indeed, it can provoke changes in droplet size and optical appearance. As mentioned the dispersed phase may contain bioactive agents that will be discussed later.

3.1.2 Emulsifiers

Once the oil is selected, the second stage is to choose the emulsifier, which is essential for nanoemulsion formation and stability, as well as for the selection of the homogenization method (Hasenhuettl, 2008; McClements & Rao, 2011; Yalçınöz & Erçelebi, 2018). A large interfacial area between oil and aqueous phases is present during the formation of nanoemulsions. Consequently, as aforementioned, they are thermodynamically unstable, and the addition of emulsifiers is necessary. Emulsifiers are surface-active agents that are able (a) to facilitate the disruption of the drops so that very fine drops can be obtained, thanks to their ability to reduce interfacial tension when they are adsorbed at the liquid-liquid interface; (b) to modify the interdroplet forces (usually by generating steric or electrostatic repulsive interactions), preventing or retarding the coalescence of newly formed droplets; and (c) to increase the shelf life, not only by slowing the rate of coalescence during storage (Narsimhan & Wang, 2008) but also by providing rheological properties to the liquid-liquid interface.

The most commonly used emulsifiers in the food industry are small-molecule surfactants, phospholipids, natural extracts, proteins, and polysaccharides (Guerra-Rosas, Morales-Castro, Ochoa-Martínez, Salvia-Trujillo, & Martín-Belloso, 2016; McClements & Jafari, 2018a).

According to their electrical properties, surfactants are classified as ionic, nonionic, and zwitterionic surfactants (Kralova & Sjöblom, 2009). Among them the most used in food industry are nonionic surfactants, since they have a low toxicity and are not irritants. Other surfactants worth mentioning are sugar esters such as sorbitan monooleate or sucrose monopalmitate, ethoxylated sorbitan esters such as Tweens and Spans, or polyoxyethylene ether surfactants such as Brijs (Yalçınöz & Erçelebi, 2018). Additionally, two natural widely used small-molecule surfactants are lecithins and saponins. Lecithins contain phospholipids and other lipids. The most common phospholipids in lecithin are phosphatidylcholine, phosphatidylethanolamine, and phosphatidylinositol. Their head groups are either anionic or zwitterionic. Lecithins stabilize W/O emulsions, so they are combined with other emulsifiers to improve the stability of O/W emulsions. Saponins are small-molecule surfactants contained in the extract from the bark of the *Quillaja saponaria* tree, and they possess sugar groups and phenolic groups (McClements, 2016).

Another criterion for the classification of surfactants is by using the HLB number concept. The HLB or hydrophilic/lipophilic balance value of an emulsifier indicates its degree of solubility either in the aqueous or in the lipophilic phase. According to its chemical structure, each emulsifier has an assigned HLB value. Emulsifiers with HLB values ranging from 0 to 10 show high lipophilicity, while values from 10 to 20 show high hydrophilicity (Kralova & Sjöblom, 2009; Setya, Talegaonkar, & Razdan, 2014). However, literature reveals that surfactants with HLB between 3 and 5 are the most suitable for the preparation of W/O nanoemulsions, and those with HLB between 10 and 12 for O/W nanoemulsions (McClements, 2016). Sometimes blends of surfactants are used to obtain the desired HLB value, and they are preferred because their use leads to better stability (McClements & Jafari, 2018b).

Proteins and some polysaccharides are emulsifiers considered as natural ingredients, and hence they are preferred by consumers over small-molecule surfactants, which are mostly synthetic. They provide stability through a combination of electrostatic repulsion and steric stabilization (Jo & Kwon, 2014). Specifically the proteins are usually electrically charged, so they are very sensitive to pH and ionic strength. Proteins used as emulsifiers in emulsion-based food products are milk, meat and fish, egg, and plant proteins. Among them the most commonly used proteins are whey protein, casein, β-lactoglobulin, ovoalbumin, soy protein isolate, and bovine serum albumin. Regarding polysaccharides, note that only some of them really act as emulsifiers, for example, gum arabic, modified starch, and some kinds of pectin or galactomannan. The surface activity they exhibit is due to either the presence of modified hydrophobic groups or the presence of protein residues (Dickinson, 2003).

The size of the droplets formed during the preparation of a nanoemulsion is dependent on the ability of the emulsifier to reduce the interfacial tension. Indeed the greater the reduction of interfacial tension, the smaller the droplets formed during emulsification. Consequently an emulsifier that sufficiently reduces the interfacial tension is necessary to reach a nanodrop size. On the other hand, in microfluidization, the time scale inside the interaction chamber is very small. Therefore, to obtain nanoemulsions by using this technique, not only is it necessary to use an emulsifier that decreases the interfacial tension sufficiently but also one that is adsorbed quickly at the interface of the fine droplets so that it can protect them against coalescence. It should be noted that low-molar-weight emulsifiers have a higher rate of adsorption (Danov, Kralchevsky, & Ivanov, 2001). For this reason, surfactants are usually more effective than other emulsifiers at forming nanoemulsions with small droplet sizes using high-energy methods (McClements & Rao, 2011; Yalçınöz & Erçelebi, 2018), and a smaller quantity of these emulsifiers is needed compared to the biopolymers (Öztürk, Argin, Ozilgen, & McClements, 2015).

When natural emulsifiers are used, the droplets are more difficult to break down within the homogenizer because these emulsifiers provide a higher interfacial tension than small-molecule surfactants (de Montes, Candal, & Herrera, 2017). Indeed, Qian and McClements (2011) compared two natural emulsifiers, namely, β-lactoglobulin (protein) and sodium caseinate (protein), with two small-molecule surfactants (Tween 20 and sodium dodecyl sulfate) to obtain food-grade nanoemulsions. They found that surfactants produced smaller sizes than biopolymers. This fact was attributed to differences in adsorption rates and interfacial properties. However, it is possible to obtain small droplet sizes from natural emulsifiers by using a dual-channel microfluidization method. Using this method the formation of a primary emulsion is not necessary as the process occurs in only one step, the device being fed with the dispersed and aqueous phases separately. Recently, Bai, Huan, Gu, and McClements (2016) formulated O/W nanoemulsions using natural emulsifiers (whey protein isolate (WPI), gum arabic, *Quillaja* saponins, and soy lecithin) and processed them with a dual-channel microfluidization method. They concluded that it was possible to prepare nanoemulsions, the effectiveness of *Quillaja* saponin and WPI is higher due to a faster adsorption, and there is a more marked decrease in the interfacial tension. Regarding hydrocolloids (gum arabic), it should be noted that the authors found a low reduction in interfacial tension and a loss of functionality when high pressure was applied during microfluidization.

Luo et al. (2017a) also fabricated O/W nanoemulsions containing natural emulsifiers (*Quillaja* saponin and WPI) and β-carotene as hydrophobic nutraceutical using high-pressure dual-channel microfluidization. Another recent work using two proteins, namely, sodium caseinate and pea protein isolate by high-pressure, is that of Yerramilli, Longmore, and Ghosh (2017). This work, additionally, highlights the relevance of molecular interactions to obtain stable systems. On the other hand, controlled enzymatic hydrolysis of whey proteins produces peptides that are smaller than intact protein (Christiansen, Vegarud, Langsrud, Ellekjaer, & Egelandsdal, 2004) and so have a higher rate of diffusion to the oil/water interface. According to Chu, Ichikawa, Kanafusa, and Nakajima (2007), peptides derived from whey and other proteins act as a nanoemulsifying agent. However the emulsification method used by these authors was not microfluidization but an emulsification-evaporation method.

In brief, there is a great interest in the use of natural emulsifiers, and numerous efforts are being made to produce stable nanoemulsions with them.

A list of emulsifiers used when food-grade nanoemulsions are obtained by microfluidization is shown in Table 1.

3.1.3 Aqueous phase

Many emulsifiers are dispersed into the aqueous phase. Water is the basic component of the aqueous phase but can contain a variety of solutes capable of modifying its properties, such as polarity (cosolvents), viscosity (polysaccharides or proteins), ionic strength (salts), pH (acids or bases), color (colorants), flavor (flavorings), and antimicrobial capacity (preservatives) (Jin et al., 2016; McClements & Rao, 2011). These changes can positively or negatively affect the stability of the nanoemulsion. It should be highlighted that some bioactive compounds, such as dietary fibers, which contribute to maintaining intestinal health, can also be introduced into the aqueous phase. An ingredient able to modify the texture and to improve the stability of nanoemulsions that is generally added into the aqueous phase is the so-called texture modifier. These are high-molecular-weight polymers that are extracted from either plants, seaweed, or animal collagen or are produced by microbial synthesis. Texture modifiers fall into two classes: thickening agents and gelling agents (McClements & Jafari, 2018a). Most of them are polysaccharides (alginates, carrageenans, pectins, galactomannans, tree exudates gums, xanthan gum, gellan gum, etc) or proteins (gelatin, caseins, and globular proteins). Sometimes blends of biopolymers are used to obtain the desired characteristics. In nanoemulsions, they modify the rheological properties; for example, they increase the viscosity, which hinders the movement of the droplets and, therefore, may retard or avoid their aggregation (Somogyi, 2000).

It should be taken into account that polysaccharides suffer modifications after the microfluidization process as a consequence of the high shear stresses applied, and these changes may affect the polysaccharide properties and interactions between the polysaccharide and the other components of the nanoemulsion. Consequently the modifications may affect the properties of the nanoemulsion. Hence the addition of these substances does not always lead to stable nanoemulsions, and the method of their incorporation is a key factor to obtain nanoemulsions with suitable properties. There are a number of studies concerning this issue, such as one recently published by Artiga-Artigas, Acevedo-Fani, and Martín-Belloso (2017), which investigates the effect of the incorporation of sodium alginate on the properties of corn oil/water nanoemulsions formulated with Tween 20 as emulsifier and processed by microfluidization. The work revealed that physicochemical properties of nanoemulsions can vary depending on their preparation procedure and that microfluidization impacted on both the degree of polymerization of alginate molecules and their disposition around oil droplets.

3.1.4 Bioactive lipophilic compounds

As mentioned earlier, nanoemulsions can be utilized in functional foods to encapsulate, protect, and deliver bioactive lipophilic compounds of which the most comprehensively studied and most usually incorporated are vitamins (A, D, E, and K), polyphenols (i.e., curcumin, resveratrol, catechin, tangeretin, nobiletin or quercetin, and $\propto$−tocopherol) (Assadpour, Jafari, & Esfanjani, 2017; Faridi Esfanjani et al., 2018), carotenoids (i.e., β-carotene, lycopene, lutein, and astaxanthin) (Rostamabadi et al., 2019), fatty acids (i.e., ω-3, linoleic acid, and butyric acid), coenzymes (Cho et al., 2014), and essential oils or compounds of essential oils (i.e., citral, D-limonene, and lemon oil) (McClements, Decker, & Weiss, 2007; Yalçınöz & Erçelebi, 2018). It should be highlighted that some bioactive compounds need to be heated to dissolve them in oil while others need to be diluted in a cosolvent (e.g., ethanol) before dissolving in oil.

Vitamins

Vitamins are either hydrophilic or lipophilic compounds with several biochemical functions, such as to protect vision, absorb and fix calcium in the body, prevent and treat cardiovascular and heart diseases, or help blood clotting and bone mineralization. Some examples of studies related to the encapsulation of vitamins into nanoemulsions found in the literature are vitamin E (Gong et al., 2012; Mayer, Weiss, & McClements, 2013; Saberi, Fang, & McClements, 2013) or vitamin D (Golfomitsou et al., 2018; Salvia-Trujillo, Fumiaki, Park, & McClements, 2017). We can find several examples of vitamins being incorporated using nanoemulsion delivery systems. Cheese, milk, ice cream, wine, bread, and juices are examples of food products enriched with minerals and vitamins (Dias, Botrel, Fernandes, & Borges, 2017; Öztürk, 2017). Nanoemulsions containing vitamins can be used in their original state or in a powdered form, the latter being far more stable (Katouzian & Jafari, 2017).

Polyphenols

Polyphenols are natural compounds that are found in a large number of fruits and vegetables. The most common polyphenols in food are flavonoids, phenolic acids, stilbenes, and lignans. They have antioxidant, antimicrobial, antiinflammatory, vasodilating, and prebiotic properties (Landete, 2012). In addition, flavonoids act against neurodegenerative disease, enhancing existing neuronal functions and stimulating neuronal regeneration. Quercetin, cocoa flavonoids, or tea catechins are able to reduce cardiovascular diseases. Polyphenols such as catechins, epicatechins, isoflavones, tannic acid, or saponins decrease the intestinal transport of glucose. Anthocyanin, resveratrol, or catechin have been reported to exert preventive effects against aging. Numerous studies of nanoemulsions fortified with polyphenols and stabilized with emulsifiers of different classes are found in the literature. By way of example, there are the studies by Ahmed, Li, McClements, and Xiao (2012), Zheng, Pen, Zhang, and McClements (2018), or Dammak and do Amaral Sobral (2017), where polyphenol-loaded nanoemulsions were obtained by microfluidization. Ahmed et al. (2012) investigated the effect of different lipids (corn oil (LCT), Miglyol (MCT), and tributyrin (SCT)) on the encapsulation of curcumin and, additionally, its bioaccessibility. β-Lactoglobulin (protein) as emulsifier was used to stabilize the nanoemulsions. It was necessary to heat the curcumin up to 60° C to dissolve it in the oil. Stable nanoemulsions were obtained except when using pure SCT. This nanoemulsion exhibited the lowest bioaccessibility of curcumin. In contrast the highest bioaccessibility was observed with medium-chain triglyceride oil.

Zheng et al. (2018) also encapsulated curcumin into corn oil/water nanoemulsions using a natural emulsifier (*Quillaja* saponin). Three different loading methods of curcumin were studied, namely, ambient temperature, 100°C for 15 min or traditional (60°C for 2 h). The first method gave the highest encapsulation efficiency, and all nanoemulsions gave similar bioaccessibilities to those of the best commercial formulation tested. Unlike the previous studies, Dammak and do Amaral Sobral (2017) investigated the encapsulation of rutin (flavonoid) into emulsions stabilized by a mixture of surfactants (Tween 80 and Span 80). Results demonstrated that 30-day physically stable rutin-loaded soybean oil/water emulsions with submicron droplet sizes can be obtained. The mean diameters were hardly affected by the concentration of rutin because it increased the viscosity of the dispersed phase. These diameters influenced the total rutin content after storage but around 22%–35% (*w*/w) were retained. Similar works related to the encapsulation of polyphenolic bioactives and their bioavailability but by methods other than microfluidization are resveratrol (Matos, Gutiérrez, Coca, & Pazos, 2014; Sessa et al., 2014; Sessa, Tsao, Liu, Ferrari, & Donsì, 2011), tangeretin (Ting, Chiou, Pan, Ho, & Huang, 2015), quercetin (Pool, Mendoza, Xiao, & McClements, 2013), epigallocatechin (Wang, Wang, & Huang, 2009), etc.

Carotenoids

Carotenoids are organic pigments, mainly found in vegetables and fruits. More than 1000 carotenoids are known (Yabuzaki, 2017). They have a polyene hydrocarbon chain that may have oxygen atoms attached (xanthophylls such as lutein or zeaxanthin) or may be oxygen-free (carotenes such as α-carotene, β-carotene, or lycopene). In recent years, researchers have proposed that carotenoids are beneficial to human health, protecting against and preventing cancer, cardiovascular disease, macular degeneration, and cataracts (Jo & Kwon, 2014). In addition, they exhibit antiinflammatory activity as well (Chuyen & Eun, 2017). β-Carotene, lycopene, and lutein are the most common carotenoids. Food fortification with these molecules is a way to ensure a daily intake. Nonetheless, carotenoids in foods are susceptible to chemical degradation, and therefore they are encapsulated into a nanoemulsion. Encapsulation of carotenoids enhances their stability and water solubility and improves their bioavailability (Rostamabadi et al., 2019). Several works have explored the encapsulation of β-carotene (pigment and antioxidant) into nanoemulsions by different methods, and in each study the influence of formulation and the processing variables on the stability of nanoemulsions as well as on the β-carotene degradation rate has been established. Among works that encapsulated β-carotene by microfluidization was that of Jo and Kwon (2014) and Luo et al. (2017a). Jo and Kwon (2014) studied both the nature of the emulsifier (surfactants vs proteins) and the processing variables (pressure and cycles) to obtain β-carotene-loaded nanoemulsions. At the same time, they studied the degradation rate of the bioactive agent. The authors demonstrated that it was possible to obtain nanoemulsions by microfluidization with a suitable physical stability. Additionally, they found smaller droplet sizes when a small-molecule surfactant was used as emulsifier. However, nanoemulsions stabilized with Tween 20 showed the lowest β-carotene retention. This result revealed that β-carotene degradation is very dependent on the droplet size. Later, Luo et al. (2017a) also produced nanoemulsions as earlier but now using dual-channel microfluidization. Natural emulsifiers were used to stabilize the nanoemulsion. Storage temperature was investigated. Regardless of the emulsifier used, there was chemical degradation of β-carotene, this being greater with higher temperature. A further interesting paper reports the encapsulation of β-carotene from emulgels prepared on the basis of a previous nanoemulsion. Emulgels proved to be a suitable matrix to significantly slow down β-carotene oxidation (Santos, Jimenez, Calero, Alfaro, & Muñoz, 2019). These works highlighted the need to continue research into optimizing the encapsulation of β-carotene in such a way that it can be stabilized against chemical degradation.

Encapsulation of other carotenoids has also been studied, examples being lutein by Surh, Decker, and McClements (2017); lycopene by Li, Li, Xiao, Li, and Xie (2018); or astaxanthin by Shu et al. (2018). In these studies, high-pressure and spontaneous emulsification were the techniques used to obtain the nanoemulsions.

Fatty acids

Omega fatty acids are essential to the normal functioning of the body, and it is necessary to introduce them into the diet. Polyunsaturated fatty acids are subdivided mainly into two groups depending on the location of the last double bond: ω-3 fatty acids and ω-6 fatty acids. On the one hand, ω-3 fatty acids mainly consist of $\propto -$linolenic acid, eicosapentaenoic acid (EPA), and docosahexaenoic acid (DHA). It has been reported that they decrease the risk of cardiovascular diseases (Colussi, Catena, Novello, Bertin, & Sechi, 2017), diseases related to either immune response disorders (Upala, Yong,

Theparee, & Sanguankeo, 2017), or even mental disorders (Hooper, Barreto, Pahor, Weiner, & Vellas, 2018).

On the other hand, ω-6 fatty acids mainly consist of linoleic acid, γ-linolenic acid, and arachidonic acid. Omega fatty acids are able to regulate the pressure and clotting of blood and act in the development and functioning of the nervous system and brain. Furthermore, they act in immune regulation and inflammation. Another fatty acid, but in this case monounsaturated, is ω-9. This fatty acid is able to increase metabolism and eliminate plaque from the arteries and, therefore, to contribute to the reduction of cardiovascular diseases. The high impact of fatty acid consumption on health justifies their incorporation into functional foods.

Several studies concerning the incorporation of omega fatty acids into nanoemulsions can be found in the literature. By way of example, there are the works of Uluata, McClements, and Decker (2015); Karthik and Anandharamakrishnan (2016a, 2016b); and Komaiko, Sastrosubroto, and McClements (2016). In all of them the nanoemulsions were obtained by microfluidization. Uluata and collaborators (2015) compared the effectiveness of different emulsifiers (natural and synthetic) in the nanoemulsification of fish oil. The droplet size increased in the order Tween 80 < lecithin < sodium dodecyl sulfate < *Quillaja* saponin. However, emulsions with Tween 80 and lecithin were less stable to thermal treatments. A similar study was carried out by Karthik and Anandharamakrishnan (2016a, 2016b). These authors firstly prepared DHA algae oil nanoemulsion using different high-energy techniques, and then they studied the role of different emulsifiers (Tween 40, sodium caseinate, and soya lecithin), using microfluidization, on the stability and in vitro digestibility of the nanoemulsions. They concluded that nanoemulsions stabilized with Tween 40 as emulsifier delivered bioactive ingredients more gradually than the other emulsifiers, enhancing the storage stability and bioavailability of ω-3 fatty acids. Komaiko et al. (2016) prepared fish oil/water nanoemulsions using four different sunflower phospholipids and demonstrated that they are a good alternative to synthetic surfactants in the stabilization of nanoemulsions.

3.1.5 Other components

Other components, such as cosolvents, cosurfactants, ripening retarders, weighting agents, and texture modifiers (thickening agents/stabilizers), can also be incorporated into nanoemulsions, in either the aqueous or oil phase.

Cosolvents are substances like alcohols and polyols that are added to solvents to improve their performance. In nanoemulsions formulated with natural emulsifiers and obtained by high-energy methods, such as microfluidization or high pressure, the incorporation of cosolvents makes it possible to obtain stable emulsions. This fact has been proved by Zeeb, Herz, McClements, and Weiss (2015) who studied systems stabilized with proteins. The incorporation of butanol optimized the dispersed-phase viscosity (η_D)/continuous-phase viscosity (η_C) ratio by reducing the viscosity of the oil, making it possible to obtain emulsions with nanosizes (nanoemulsions) (de Montes et al., 2017).

Cosurfactants are amphiphilic molecules such as ethanol or glycerol able to aid the emulsifier to decrease the interfacial tension and, therefore, to facilitate the formation of nanoemulsions. Alfaro, López-Tarruello, García, and Muñoz (2018) studied the effect of incorporating glycerol into the continuous phase of O/W nanoemulsions formulated with a mixture of $\propto$-pinene and *N*,*N*-dimethyldecanamide (30 wt%) and processed by microfluidization, on the properties of emulsions. The authors found that glycerol increased the physical stability and the shear strength.

A ripening inhibitor is a hydrophobic molecule (triglyceride oil, mineral oil, or ester gum) with a poor solubility in water. When it is added to the dispersed phase of a nanoemulsion whose oil has a high solubility in water (flavor oils and essential oils), it delays destabilization by Ostwald ripening (Pérez-Mosqueda, Ramírez, Trujillo-Cayado, Santos, & Muñoz, 2014, McClements & Jafari, 2018a).

A weighting agent is a substance that, when added into the dispersed phase, makes its density similar to that of the continuous phase, thus reducing the tendency of the emulsion to destabilize by gravitational separation (creaming or sedimentation). However, in nanoemulsions, due to the small droplet size, creaming is usually not a problem (Yalçınöz & Erçelebi, 2018).

Texture modifiers are agents that control the rheology of the aqueous phase in several types of dispersions. Due to their small droplet size, nanoemulsions may exhibit very different rheological properties to conventional emulsions with the same oil content. Thus there may be some novel applications of nanoemulsions to modify the texture of food products. It is well known that emulsions can show different rheological properties as viscous liquids, viscoelastic liquids, viscoelastic solids, viscoplastics depending on their composition, structure, and droplet interaction (McClements, 2011b). The addition of texture modifiers (e.g., polysaccharides, sugars, polyols, and proteins) to the aqueous phase may be useful to improve nanoemulsion viscosity and stability.

The usual method to incorporate a polysaccharidic stabilizer to a nanoemulsions is by "post addition," that is, avoiding that the food gum undergoes the high pressures and deformations occurring within the microchannels (which would result in biopolymer hydrolysis). In this way the first step consists of preparing the nanoemulsion by microfluidization and, subsequently, adding a concentrated food gum aqueous dispersion. The second step involves the use of a mixer under controlled rotational speed, as indicated by Santos, Calero, Muñoz, and Cidade (2018) when using an advanced-performance xanthan gum. This paper also emphasized the fact that the important point is not only to increase the viscosity of the aqueous phase but also to impart viscoelasticity. A certain grade of viscoelasticity enhances the physical stability of nanoemulsions as demonstrated by monitoring multiple light-scattering scans.

Further applications of nanoemulsions and biopolymers are emerging in the field of new materials and active packaging systems. Edible films including active ingredients can be used as an alternative to preserve food products and enhance their quality and their shelf life (Burgos, Mellinas, García-Serna, & Jiménez, 2017; Vahedikia et al., 2019). Films can be prepared from nanoemulsions by casting and subsequent drying. This new technology protects foods from dehydration and acts as a vapor barrier to the surrounding media. Edible films consist of biopolymer matrices and functional ingredients, which increase the shelf life of packaged food (Acevedo-Fani, Salvia-Trujillo, Rojas-Graü, & Martín-Belloso, 2015).

3.2 Influence of processing variables on nanoemulsion properties developed by microfluidization

Nanoemulsion properties, such as final mean diameters and droplet size distribution achieved or physical stability, depend in a complicated manner on many factors. To achieve better droplet sizes and physical stability, the optimization of the emulsification process by microfluidization covers two main classes of processing variables. The first one is finding the appropriate design of the microfluidizer device, and the second one is identifying an appropriate combination of processing variables, such as premixing, number of cycles, homogenization pressure, and temperature. These processing variables are not independent and often influence the final nanoemulsion properties.

Improved knowledge of the processing conditions that favor the required minimum droplet sizes and physical stability in nanoemulsions can lead to their optimized use in the food industry and thereby improve the value of the related products. For example, harsh processing conditions when using a microfluidizer device, such as high pressure or temperature or a large number of cycles, can lead to a deterioration of the proteins that are used as emulsifiers. Considering the importance of microfluidization operating conditions on the final nanoemulsion properties, this section presents recent findings concerning several factors that determine the development of these systems. This section would serve as a guideline to increasing the efficiency of the emulsification process as well as aiming to help in the selection of appropriate processing variables for the development of food nanoemulsion systems.

3.2.1 *Effect of the primary homogenization*

In general, when microfluidization is used to develop nanoemulsions, not much attention is given to primary homogenization, even though it has been shown that the method used to carry out this process influences the final properties of the microfluidized nanoemulsions. To carry out the primary homogenization, a well-accepted technology is a rotor-stator device. Using this technique the liquid is drawn into the space between the rotor, which is a fast spinning inner element, and a stationary outer sheath called the stator. Then a high shear is applied to the liquid before it comes out of the device.

Trujillo-Cayado, Alfaro, Santos, Calero, and Muñoz (2018) reported the influence of primary homogenization conditions on microfluidized emulsions formulated with thyme oil and a green surfactant derived from wheat biomass. An Ultra-Turrax T-50 and Silverson L5M rotor-stator homogenizer and three different homogenization rates were used for the development of the coarse emulsions (primary homogenization). Coarse emulsions showed a recoalescence process that was favored by higher homogenization speed and the use of Ultra-Turrax T-50. No differences in mean droplet sizes for preemulsions prepared using both rotor-stator devices at the same homogenization rate were observed. However, mean droplet diameters were mainly influenced by homogenization speeds and polydispersity by the rotor-stator geometry. Microfluidized emulsions showed a recoalescence process that was favored by higher homogenization speed and the use of Ultra-Turrax T-50 in the primary homogenization step. Therefore the droplet size distribution of the emulsions derived from the primary homogenization is a key variable that strongly affects the droplet size distribution and mean droplet sizes of the microfluidized nanoemulsions obtained after the secondary homogenization.

Higher droplet diameters for coarse emulsions could favor the breakup of the droplets inside the Microfluidizer M-110P chamber, allowing lower droplet diameters in microfluidized nanoemulsions to be obtained. In this way, Trujillo-Cayado, Alfaro, and Muñoz (2018) demonstrated that a tight control of the primary homogenization conditions in the development of microfluidized nanoemulsions is highly recommended. In this line, some research has been carried out in which the processing conditions in the primary homogenization are optimized as a previous step to microfluidization. Bitter fennel essential oil-in-water nanoemulsions were produced by using a rotor-stator device (Llinares, Ramírez, Carmona, Carrillo, & Muñoz, 2018). Physical stability, droplet sizes, and droplet size distribution of these systems were analyzed using response surface methodology (RSM). The studied variables were the surfactant concentration and the emulsification time. Taking into account the results obtained from RSM and the statistical analysis of the data, emulsification time had a little influence on mean droplet diameters and physical stability. However, in another investigation carried out by Trujillo-Cayado, Alfaro, and Muñoz (2018), it was shown that an increase in homogenization speed caused the droplet size distributions to shift towards lower values of droplet size. These coarse emulsions, which were formulated with D-limonene and a surfactant derived from rapeseed oil, were taken as a starting point for a second study to improve the emulsification method using a Microfluidizer M-110P. An increase in the microfluidization pressure allowed lower droplet sizes to be obtained, but only the emulsions prepared using a rather high pressure (172.4 MPa) significantly decreased the mean droplet diameter of the coarse emulsions.

3.2.2 Type of microfluidizer interaction chamber

There are two clearly differentiated types of interaction chambers in microfluidization commercial devices: Y-type and Z-type interaction chambers (Microfluidics Y and Z chambers). To obtain the maximum deformation level at different homogenization pressures for the development of nanoemulsions, a Y-type interaction chamber with the narrowest size of microchannels is preferred (Microfluidics, 2018).

According to Trujillo-Cayado, Santos, Ramírez, Alfaro, and Muñoz (2018), the use of series chambers (a Y-type chamber followed by a Z-type interaction chamber) demonstrated an increased capacity to obtain smaller mean droplet diameters than just one chamber (Y type). These nanoemulsions with very narrow droplet size distributions were obtained using a surfactant derived from coconut oil. Primary emulsification was carried out using a L5M emulsor screen rotor-stator unit (Silverson, UK). Subsequently the coarse emulsions were passed through a Microfluidizer M-110P at different pressures (34.5, 103.4, or 172.4 MPa) using two different chamber configurations (Y or Y + Z). The reduction in mean droplet diameters may be related to the increment in the shear produced by the use of the series configuration, since shear depends on not only homogenization pressure but also the chamber configuration. Therefore the droplet sizes were influenced by shear but not by homogenization pressure for these formulations.

3.2.3 *Microfluidization methods for developing nanoemulsions: Single-channel versus dual-channel microfluidization*

The single-channel microfluidization method is considered the traditional process for the development of nanoemulsions. Although historically it is the classic method for the development of emulsions by microfluidization, even today, single-channel microfluidization is still the more widely used method for the development of food nanoemulsions (Akbas, Soyler, & Oztop, 2018; Arredondo-Ochoa et al., 2017; Artiga-Artigas, Guerra-Rosas, Morales-Castro, Salvia-Trujillo, & Martín-Belloso, 2018; Pérez-Córdoba et al., 2018; Sedaghat Doost, Devlieghere, Dirckx, & Van der Meeren, 2018). However, a study reported by Bai and McClements (2016a) demonstrated that dual-channel microfluidization produced smaller droplets than single-channel microfluidization at similar homogenization pressures and surfactant concentrations. Monomodal droplet size distributions were obtained using the dual-channel microfluidization method, while using the single-channel method produced bimodal distributions, probably due to the differences in droplet disruption within the different methods.

In addition, the dual-channel method has two other advantages compared with the other single-channel microfluidization method: the possibility of producing emulsions with higher concentrations of dispersed phase and the reduction of time and cost that the primary homogenization entails. Several studies have been conducted recently into dual-channel microfluidization for the development of food-grade nanoemulsions (Bai & McClements, 2016a; Bai et al., 2016; Luo et al., 2017a, 2017b) and core-shell nanoparticles (Ebert, Koo, Weiss, & McClements, 2017).

3.2.4 *Microfluidization pressure and number of cycles*

The main factors affecting the properties of the nanoemulsions produced by microfluidization are probably the homogenization pressure used and the number of cycles. The effect of these variables on a nanoemulsion produced by a microfluidizer device is one of the major areas of study in high-pressure homogenization.

Usually, nanoemulsion droplet sizes produced by microfluidization can be decreased by increasing the homogenization pressure and/or number of cycles (Qian & McClements, 2011; Lee, Hancocks, Noble, & Norton, 2014; Bai & McClements, 2016b; Uluata, Decker, & McClements, 2016; Dammak & do Amaral Sobral, 2017; Gadkari, Shashidhar, & Balaraman, 2017; Trujillo-Cayado, Alfaro, & Muñoz, 2018; Yan, Park, & Balasubramaniam, 2017). Recently, Llinares, Santos, Trujillo-Cayado, Ramírez, and Muñoz (2018) described the variation of volumetric mean diameters as a function of the number of cycles for nanoemulsions formulated with rosemary essential oil and mixtures of Tween 80 and Span 80. It was observed that an increase in the number of cycles provoked a clear decrease in volumetric mean diameters. Rosemary oil-in-water nanoemulsions prepared using ten or more cycles showed mean diameters below 10 nm, but there were no significant differences in the 10–12 passes range at 137.9 MPa. However, several studies have shown that there is usually an optimum for homogenization pressure and/or number of cycles when processing such systems. For example, it has been reported that a microfluidizer is capable of producing palm oil-based nanoemulsions encapsulating curcumin (Raviadaran, Chandran, Shin, & Manickam, 2018) at an operating pressure of 350 bar and beyond five passes. Droplet mean diameters decreased with increased homogenization pressure until they reached an optimum level, whereas the number of cycles had no significant effect.

For some formulations an increase in homogenization pressures and number of passes led to overprocessing of the nanoemulsion droplets due to recoalescence (Jafari et al., 2008; Jafari, He, & Bhandari, 2007). This phenomenon is produced by the high shear, turbulence, and cavitation present in the microfluidizer device, which result in an increment of droplet size due to mechanical damage (Jafari, He, & Bhandari, 2006; Piorkowski & McClements, 2014). Trujillo-Cayado et al. (2016) compared the droplet size distribution, physical stability, and rheological properties of emulsions formulated with a mixture of green solvents and a polyoxyethylene glycerol fatty acid ester as emulsifier. After a short aging time, the influence of the microfluidization pressure on the droplet size distribution and physical stability of nanoemulsions was not significant. However, a second pass through the microfluidizer device led to a slight increase in the droplet sizes due to a recoalescence phenomenon. By way of example, see the works of Bai, Huan, Li, and McClements (2017), Santos et al. (2018), Santos, Jimenez, et al. (2019), and Trujillo-Cayado, Alfaro, García, and Muñoz (2017); Trujillo-Cayado, García, Santos, Carmona, and Alfaro (2017).

Several studies have concluded that, under certain conditions, the homogenization pressure and/or the number of passes does not affect droplet size distributions. However, this fact does not imply that these variables do not affect other factors such as the degree of flocculation or rheological properties. Trujillo-Cayado et al. (2016) compared the properties of nanoemulsions fabricated with two rotor-stator devices and two high-pressure homogenizers (high-pressure valve homogenizer vs microfluidization) using different processing variables (homogenization rates, homogenization pressures, and number of cycles). The influence of the pressure level (from 34.5 to 103.4 MPa) and the number of passes (1 and 2 pass) on droplet mean diameter were not significant, but a second pass through high-pressure devices resulted in greater volumetric mean diameter. The viscosity of the nanoemulsions decreased as the pressure increased, which is a consequence of the disruption of some droplet flocs. The most stable emulsion obtained was that prepared with Microfluidizer M-110P at the highest pressure and only one cycle. In an investigation reported by Santos, Trujillo-Cayado, Calero, Alfaro, and Muñoz (2016), a Microfluidizer device allowed the development of nanoemulsions regardless of the homogenization pressure used. In spite of the fact that all nanoemulsions did not show significant changes in the droplet size distributions, these emulsions exhibited different values of zero shear viscosity. Samples developed at higher homogenization pressures showed lower values of zero shear viscosity, which is related to a different degree of flocculation. This flocculation process led to droplet coalescence. Furthermore, slightly flocculated emulsions did not show a coalescence process, but a creaming process did take place. Consequently a moderate pressure of 103.4 MPa responded better than higher or lower pressures due to the lack of creaming and a lower coalescence.

3.2.5 Viscosities of continuous and dispersed phases

There is a direct relation between the viscosity of the dispersed phase and the final droplet size distribution of microfluidized nanoemulsions. If the viscosity of the dispersed phase increases, the breakup of the drops will be more difficult to carry out, and therefore the average diameters will be greater (Jafari et al., 2008). Qian and McClements (2011) concluded that beyond the homogenization pressure and number of passes used for the development of microfluidized nanoemulsions, the emulsifier type and concentration are very important factors affecting mean droplet diameters. This is due to greater and faster stabilization of the oil/water interfaces by the emulsifier and to the role of inertial to viscous forces in droplet breakup (Monroy-Villagrana, Alamilla-Beltrán, Hernández-Sánchez, & Gutiérrez-López, 2015).

3.2.6 Influence of processing temperature

Microfluidization requires a large amount of energy that is mostly converted into heat, causing an increase in the temperature of the sample. Furthermore, this temperature increase is proportional to the homogenization pressure used and the number of cycles. Since properties such as interfacial tension or viscosity of the continuous and dispersed phases depend on temperature, controlling this variable may be important for the properties of the prepared nanoemulsions. However, in the microfluidization process, a final cooling stage is usually incorporated by means of a water-cooled coil.

Bucci, Van Hekken, Tunick, Renye, and Tomasula (2018) investigated the use of mild heat treatments in combination with microfluidization to produce cheese. Pasteurized milk samples were processed at two inlet temperatures (42°C and 54°C) and two different homogenization pressures (125 and 170 MPa). Milk sample properties (droplet size distribution, microbiological properties, viscosity...) were highly influenced by processing pressure, which produced an increment in outlet temperature for all microfluidized samples. Conversely, Mao, Yang, Xu, Yuan, and Gao (2010) reported that droplet size distributions of β-carotene/sunflower O/W nanoemulsions were slightly affected by homogenization temperature. An increment from 25°C to 50°C resulted in a reduction of 13 nm in the mean droplet sizes.

4 Nanoparticles produced by microfluidization

4.1 Introduction to the nanoparticle definition

Nanoparticles are defined as colloidal particles with sizes ranging from 10 to 1000 nm. In the last three decades, nanoparticles have raised much attention due to their wide range of applications, such as in agriculture and food products as well as medical and pharmaceutical sciences. These applications are related to the ability of nanoparticles to perform as delivery systems for drug targeting (Anton, Benoit, & Saulnier, 2008). Some advantages of the use of nanoparticles when compared with other colloidal delivery systems are their high kinetic stability and rigidity. Many nanoparticles are produced through the previous preparation of nanoemulsions, which in turn are generated by either high-energy or low-energy methods. In this section, we are going to focus on nanoparticles obtained by microfluidization. However, it is important to note that the preparation of nanoemulsions can potentially affect bioactives and drugs encapsulated in nanoparticles.

Nanoparticles can be divided into two different types: nanospheres and nanocapsules. The difference between them lies in the structure of the particle. While nanospheres have a homogeneous structure, nanocapsules present a typical core-shell structure. Many techniques have been used to prepare nanoparticles and are generally divided into (i) methods involving the polymerization of monomers and (ii) methods that use preformed polymer from nanoemulsions.

4.2 Nanospheres

4.2.1 Polymerization of monomers

Polymeric nanospheres can be formed by in situ polymerization from nanoemulsions formed using high-energy methods, such as a microfluidizer. The chemical reactions involved are principally described as radical polymerization of droplets. Hence the droplets of nanoemulsions contain pure monomers and adsorbed surfactants in the interface. In this way the polymerization begins with the addition of initiator molecules, which can be an enzyme (Qi, Jones, & Schork, 2006), or with UV radiation (Craparo, Cavallaro, Bondì, Mandracchia, & Giammona, 2006) or ultrasonication (Bradley, Prescott, Schoonbrood, Landfester, & Grieser, 2005).

4.2.2 Preformed polymers

In this case the polymers are dissolved in the dispersed phase (usually an organic solvent) of a nanoemulsion. The preformed polymer process consists of removing the organic solvent from the nanoemulsion droplets, so that the polymer precipitates inside the droplets. This results in a polymer-based nanosphere suspension. This can be achieved by evaporation or diffusion shock. Using this methodology, not only synthetic polymers but also natural macromolecules (alginate, gelatin, chitosan, and other polysaccharides) can be used. This methodology is very common in the preparation of food products. Nevertheless, only a few studies have been carried out on the preparation of nanospheres using preformed polymers by microfluidization (Sun, Dai, Liu, & Gao, 2016; Sun et al., 2016). These researchers brought to light the fact that the homogenization pressure applied by a microfluidizer is a key parameter to be controlled in the preparation of nanospheres. This recent study reveals that the morphology of nanospheres formulated with zein (a plant protein) can be modified from their spherical initial state to needlelike shapes after treatment at 75 MPa. These nanospheres produced from microfluidized nanoemulsions by an antisolvent precipitation method present an intermediate transition state. However, if the homogenization pressure is higher (125 MPa), nanospheres and nanocapsules (core-shell structure) can be observed (Sun, Dai, et al., 2016). These authors suggest the existence of a critical value of pressure for the occurrence of morphological transformations. In addition, the use of the microfluidizer enhanced the thermal stability of these zein nanoparticles.

4.3 Polymeric and biopolymeric nanocapsules (NCs)

Nanocapsules consist of colloidal systems with a core-shell structure. On the one hand the core contains the food bioactive ingredient, and it is usually a lipophilic solvent (mainly an oil). On the other hand the shell is generally made of polymers or biopolymers. This delivery system presents some advantages: (i) high encapsulation efficiency due to the optimized food ingredient solubility in the hydrophobic core, (ii) low polymer concentration compared to polymeric nanospheres, and (iii) high protection of the food ingredient against degradation.

4.3.1 Production of polymeric nanocapsules by microfluidization

Kwon et al. (2002) produced polymethyl methacrylate (PMMA) nanoparticles containing coenzyme Q10 by using the microfluidization technique and solvent evaporation method. This study revealed the important role of sodium dodecyl sulfate (SDS) on the mean diameter of the nanoparticles prepared. As expected, mean diameters decreased with the number of passes through the Microfluidizer reaching a plateau value after three passes. The polydispersity followed the same tendency. Furthermore, no free coenzyme Q10 crystals were found in the aqueous suspension of PMMA nanoparticles, proving its encapsulation efficiency.

The selection of different polymers influences the production mechanism of nanoparticles produced by microfluidization. For example, the use of ethylcellulose (EC) instead of poly(lactic acid) (PLA) for the preparation of nanoparticles showed that EC emulsions underwent a coalescence process before the nucleation of the nanoparticles. This was due to the fact that EC molecules competed with the surfactant (SDS) at the interface inducing nanoemulsion instability, aggregation, and coalescence of the oil droplets (Desgouilles et al., 2003). Another parameter taken into account was the presence of additives. For example, Bejrapha, Surassmo, Choi, Nakagawa, and Min (2011) studied the effect of a freezing process in the absence or presence of gelatin and κ-carrageenan on the stability of capsicum-loaded nanocapsules produced by microfluidization. The aforementioned process is a common technique to remove water that can extend the shelf life of proteins, liposomes, and nanocapsules. Furthermore, it can reduce physicochemical dissociations to preserve heat-sensitive biological materials (Franks, 1998). Interestingly, they proved that gelatin and κ-carrageenan acted as cryoprotectants and lyoprotectants.

4.3.2 Production of biopolymeric nanocapsules by microfluidization

Many biopolymers can be used to prepare biopolymeric nanoparticles: lignin, chitosan, alginate, gelatin, proteins, etc. Alkanawati, Wurm, Thérien-Aubin, and Landfester (2018) were able to produce biopolymeric nanoparticles with a low polydispersity index (PDI) and small batch variability using microfluidization. They stated that there is a value of surfactant concentration earlier, which there is no improvement in the particle size of nanocapsules produced by microfluidization. In addition, these authors also demonstrated that homogenization pressure variation led to the same tendency. Hence, higher homogenization pressures do not always mean lower nanocapsule sizes. Although the size was quite similar above 69 MPa, the PDI increased with homogenization pressure above 69 MPa. This fact has also been reported in the preparation of core-shell protein nanoparticles formulated with casein and zein (Ebert et al., 2017). Obviously the specific homogenization pressure depends on the formulation used. In the latter case an increase in PDI and a stabilization of nanoparticle size were observed above 110 MPa. In spite of the increase in PDI, biopolymeric nanoparticles presented a rather narrow range of particle sizes (lower than 0.25 PDI) (Alkanawati et al., 2018; Chu et al., 2007; Ebert et al., 2017; Lin, Ma, Su, Bian, & Zhu, 2018).

In a recent study the comparison between different proteins cross-linked with TDI (2,4-toluene diisocyanate) to form nanoparticles using microfluidization was studied (Alkanawati et al., 2018). These authors prepared nanoparticles using bovine serum albumin, hydroxyethyl starch (HES), and lignin. HES nanoparticles showed bigger sizes than albumin or lignin nanoparticles. This was attributed to differences in the viscosity of the dispersions studied. Furthermore the influence of other cross-linking agents was reported. Interestingly the use of HES with adipic acid provoked the generation of nanospheres due to the high solubility of this acid in water.

Production of biopolymeric nanoparticles by microfluidizer reduces sizes to below 200 nm (Chu et al., 2007; Ebert et al., 2017; Lin et al., 2018). In the case of caseinate-zein nanoparticles (caseinate in the shell and zein in the core), the size was below 125 nm using dual-channel microfluidization (Ebert et al., 2017). This can be obtained by optimizing the processing parameters (number of passes and pressure) and formulation. For example, if the concentration of zein is higher than 3 wt%, there is not enough caseinate to cover the zein core, and the nanoparticle will be bigger.

4.4 Solid lipid nanoparticles

Solid lipid nanoparticles (SLNs) are considered to be a promising carrier system for the encapsulation of bioactive compounds (antioxidants, colorants, minerals, and essential oils) for food applications (Katouzian, Faridi Esfanjani, Jafari, & Akhavan, 2017). The use of this type of system provides some advantages, such as (i) flexible controlled release, (ii) higher encapsulation efficiency, and (iii) large-scale production without using an organic solvent. In fact the bioactive compounds can be incorporated into the matrix, the core, or the shell of SLNs.

SLNs can be achieved by two different approaches: a hot homogenization method or a cold homogenization method (Rafiee & Jafari, 2018). In the latter case, core materials are mixed with the melted lipid. Subsequently, this dispersion is cooled such that a phase transition to the solid state takes place. This material is ground by a mill and dispersed in a cold surfactant. Finally the nanoparticles are homogenized using a microfluidizer or another high-energy device. On the other hand, in the hot homogenization method, the lipid is heated to 10°C above its melting point. The core material and the surfactant are added to the melted lipid and the mixture homogenized using a microfluidizer at a controlled temperature. Finally the nanoemulsion is cooled, which leads to recrystallization.

Asumadu-Mensah, Smith, and Ribeiro (2013) prepared SLNs for functional ingredients in foods. They prepared dispersions, which contained SLNs or a mixture of lipid solid and sunflower oil (SL + O) by a hot melt microfluidization method. They used different types of lipids, namely, stearic acid, candelilla wax, and carnauba wax. The particle sizes were different when preparing SLNs and SL + O. The latter formulation presented the lowest mean diameters. In addition, particle sizes for the three lipids were stearic acid (195 nm), candelilla wax (178 nm), and carnauba wax (303 nm). TEM results revealed the morphology for these systems, showing circle, platelike, and needlelike structures.

Helgason, Salminen, Kristbergsson, McClements, and Weiss (2015) reported an interesting study concerning the influence of formulation and some microfluidization processing parameters on the droplet size of SLNs. The use of an interaction chamber with Y configuration was an advantage when compared with the use of Z configuration, resulting in SLNs with lower mean diameters. In addition, the surfactant concentration required was different depending on the homogenization pressure applied. For instance, more than 3 wt% of SDS was needed to avoid the aggregation of SLNs above 28,500 psi.

5 Nanoliposomes produced by microfluidization

Another technology for the encapsulation of food ingredients is the production of liposomes. This structure is also present in the formulation of many pharmaceutical, cosmetic, and nutraceutical products. Liposomes are closed and continuous structures that form bilayers and are composed mainly of lipids and/or phospholipids (Ghorbanzade, Jafari, Akhavan, & Hadavi, 2017; Tavakoli, Hosseini, Jafari, & Katouzian, 2018). Furthermore, nanoliposomes can be prepared by using high-energy devices and an appropriate formulation (Danaei et al., 2018; Mozafari, Johnson, Hatziantoniou, & Demetzos, 2008).

These delivery systems have attracted a lot of attention in the dairy food industry. One of the first applications of liposomes in food products was in cheese production (Law & King, 1985). Nowadays, one of the best-known applications of nanoliposomes in the food industry is their use in the encapsulation and delivery of carotenoids (Livney, 2015; Tan et al., 2014). This is because carotenoids are relatively unstable within food matrices and susceptible to oxygen, light, and autooxidation (Xianquan, Shi, Kakuda, & Yueming, 2005).

Liposomes are prepared using surfactants, mainly phospholipids, with intermediate HLB numbers and optimum curvatures close to zero (Israelachvili, 2011). The main sources of phospholipids in the food industry are soy, egg, sunflower, or milk lecithins (Mozafari et al., 2008; Singh, Thompson, Liu, & Corredig, 2012; Taylor, Weiss, Davidson, & Bruce, 2005). The lipid nature of the phospholipid is due to the long-chain fatty acids. The properties of phospholipids, and hence liposomes, mainly depend on the fatty acid part of phospholipid molecule. There are different types of liposomes: small unilamellar vesicles (SUVs), large unilamellar vesicles (LUVs), multilamellar vesicles (MLVs), and multivesicular vesicles (MVVs).

While unilamellar vesicles are formed of just one bilayer, the multilamellar vesicles are composed of multiple concentric bilayers. In addition, SUVs are considered as unilamellar liposomes whose diameter is below 100 nm.

Despite the fact that bilayers can be formed spontaneously when phospholipids are dispersed in an aqueous medium by the hydrophilic interaction of the lipid head groups with water, this is not usual. Normally, specific fabrication methods are needed to prepare liposomes. They can be classified as (a) solvent evaporation, (b) solvent displacement, (c) surfactant displacement, and (d) based on homogenization procedures. In the latter, phospholipids and the aqueous phase are mixed together and then forced to pass through a high-pressure homogenizer, microfluidizer, membranes, or sonicator.

Similar to nanoemulsion preparation, homogenization pressure and the number of cycles through a microfluidizer are the essential parameters that control the size distributions of the nanoliposomes. The use of microfluidization reduces the liposome sizes to 100 nm or less. Even at low pressures, the size reduction is noticeable. In addition, the production of unimodal nanoliposomes is usually related to high pressures and more than three passes through a microfluidizer (Barnadas-Rodrı´guez & Sabés, 2001; Takahashi et al., 2007). Furthermore, Takahashi et al. (2007) produced curcumin-loaded SUVs using a low-cost commercial lecithin. They controlled the size of the nanoliposomes by adjusting the pressure and number of passes. The trapping efficiency obtained was 85% with nanoliposomes of 100 nm (Takahashi et al., 2007).

To encapsulate ascorbic acid in nanoliposomes, Farhang, Kakuda, and Corredig (2012) produced nanoliposomes by varying the concentration of phospholipids derived from milk fat globule membranes using a microfluidization technique. Using low phospholipid concentrations, five passes were needed to obtain nanoliposomes of 100 nm. In addition, these authors revealed that a higher number of cycles through a microfluidizer are required when higher phospholipid concentrations are used to reach the same nanoliposome sizes. The encapsulation efficiency was linearly dependent on phospholipid concentration up to 10 wt%. Above this critical concentration a plateau zone was observed. Interestingly, MLVs were observed by transmission electronic microscopy at the highest phospholipid concentration studied in this research, suggesting a relation between the microstructure and the phospholipid concentration. In contrast the transformation of MLVs to SUVs using a microfluidizer after 5–10 passes has been reported by some authors (Khanniri et al., 2016; Vemuri & Rhodes, 1995).

Although nanoliposomes are excellent carriers for the controlled release (Mozafari, 2006; Mozafari et al., 2008) of bioactive ingredients, they present some limitations, such as a tendency to leak and lose encapsulated components with aging time, in addition to gradual coalescence (Laye, McClements, & Weiss, 2008). To improve the bioavailability of nisin-loaded nanoliposomes, Imran et al. (2012) prepared nanoliposomes embedded into a polymer matrix to formulate biodegradable films. They prepared nanoliposomes with soy lecithin at 2000 bar and five cycles, reaching a size of 150 nm with an encapsulation efficiency of 50%.

6 Dietary nanofibers produced by microfluidization

Dietary fibers have recently attracted attention due to their health benefits, such as anticancer properties, cholesterol reduction, and adjustment of blood sugar level, as well as their effects on human intestines (Anderson, 1987; Lindström et al., 2006; Murakami et al., 2007). This has provoked an increase in the consumption of fiber-based products and hence the necessity to search for new sources of fibers. Some dietary fibers, which are normally used as additives in food products, are cellulose, hemicellulose, lignin, pectin, and gums. To obtain these nanofibers from raw materials, different techniques based on mechanical methods such as refining, grinding, or high-pressure homogenization can be used.

In the last 10 years, some studies concerning the production of food nanofibers using microfluidization have noted the importance of this tendency for the future. Nanofibers not only confer a high added value to the food product but also may play the role of stabilizers. Ketenoglu, Mert, and Tekin (2014) produced orange vest and corn bran nanofibers using microfluidization for their subsequent incorporation into food emulsions. The size of the nanofibers is very important for the stability of some food products. In fact a reduction in nanofiber size is achieved after using microfluidization, resulting in a larger area to cover the oil/water interface, which in turn enhances the stability of the droplets (Ketenoglu et al., 2014). In addition, higher concentrations of nanofibers in food products yield higher viscoelastic properties, which confer enhanced stability to food emulsions.

Charani, Dehghani-Firouzabadi, Afra, and Shakeri (2013), who prepared microfibrillated cellulose gels using the microfluidization technique, also reported the same phenomena. In addition, these authors revealed that nanofiber suspensions can be prepared using microfluidization up to 5.5 wt%

nanofibers. Also, cellulose nanofibers have been considered better stabilizers than cellulose nanoparticles, and they can be produced in a range of 10–100nm using the microfluidization technique (Gestranius, Stenius, Kontturi, Sjöblom, & Tammelin, 2017; Hettrich, Pinnow, Volkert, Passauer, & Fischer, 2014; Spence, Venditti, Rojas, Habibi, & Pawlak, 2011). Microfibrillation using a microfluidizer has been shown to provide food products with better physical and optical properties, as well as water interaction, compared with other homogenization techniques (Spence et al., 2011). Some researchers prepared cellulose nanofibers derived from olive tree pruning (Fillat et al., 2018). They proved the theory proposed by Missoum, Belgacem, and Bras (2013), which states that the smaller the chamber size, the higher the degree of defibrillation. Interestingly, Fillat et al. (2018) related the number of cycles through the microfluidizer with the nanofibrillation yield. Furthermore, a parameter related to the number of cycles and defibrillation time also influenced the quality of nanofibers produced. Pacaphol and Aht-Ong (2017) prepared hemp nanofibers at different defibrillation times, and they concluded that at a high defibrillation time, nanofibers were smaller and more uniform. Recently, Willemsen et al. (2017, 2018) have shown that suspensions of microfluidized lemon peel fibers show enhanced viscoelasticity, the results being dependent on pH and salt concentration.

7 Concluding remarks

Microfluidization is increasingly used for the preparation of a wide variety of nanostructures, which find applications not only in the food industry but also in health sciences (medicine and pharmacy), personal care, cosmetics, agricultural chemistry, and chemical and materials engineering. Therefore this technology is the key step of the production process of submicron emulsions, nanoemulsions, nanospheres, nanocapsules, nanoliposomes, solid lipid nanoparticles, and nano- and microfibers.

Microfluidization can be carried out by either single-channel or dual-channel feeding. The former implies that all ingredients and additives must be premixed before being forced to flow through the microchannels. Conversely, dual-channel feeding allows part of the feeding materials to remain separated from each other within independent inlet reservoirs until they are homogenized inside the microchannels. Further process variables are microchannel geometry and size, location of an auxiliary processing module (if used), pressure, number of cycles, and inlet and outlet temperature.

Important engineering functions to compare different process conditions are the power input and the energy density. The choice of microchannel geometry is of paramount importance regarding the nanostructure to be produced. Thus Y-shaped interaction chambers are the choice for producing O/W submicron emulsions, O/W nanoemulsions, and liposomes and for polymer encapsulation. On the other hand, *Z*-shaped interaction chambers must be used if the targets are W/O emulsions, cell disruption, or deagglomeration of particles.

Nanoemulsions obtained by single- or double-channel microfluidization must exhibit a very low mean droplet diameter. However, there is not an agreement among different authors regarding the maximum size that could be associated to the term nanoemulsion. The mean diameter achieved by microfluidization will depend on the dispersed-phase concentration, emulsifier/oil mass ratio, and processing variables. Formulation of nanoemulsions has been reviewed in this chapter. The role of (a) different oils and hydrophobic solubilized materials, (b) low- and high-molecular-weight emulsifiers, (c) bioactive lipophilic compounds (vitamins, polyphenols, carotenoids, and fatty acids), (d) hydrophilic materials, and (e) minor components (cosolvents, cosurfactants, ripening retarders, weighting agents, and texture modifiers [thickening agents/stabilizers]) has been assessed.

Microfluidization plays a relevant role in the production of nanoparticles (nanospheres and either polymeric or biopolymeric nanocapsules) as well as in solid lipid nanoparticles, liposomes, and dietary fibers as reviewed in this chapter. Since nanoparticles produced by microfluidization are based on the previous formation of a nanoemulsion, their production is significantly influenced by the aforementioned processing variables for nanoemulsions. However, the pressure applied is able to modify not only the droplet size of nanoparticles but also their microstructure. In addition, the use of Y configuration presented some advantages for the production of solid lipid nanoparticles. Concerning nanoliposomes produced by microfluidization, enhanced results are obtained using high pressures and more than three cycles. On top of that the use of microfluidization for the preparation of food nanofibers is a clear tendency for the future of this field.

From an engineering point of view, the microfluidization technique is likely to be improved by following the principles of process intensification. The evolution of commercial devices, integrating vacuum pumps in the inlet reservoir to get rid of bubbles or designing external cooling systems for interaction chambers, to avoid high temperatures during fabrication, would enhance the production of many bioactive-loaded nanostructures. In addition the use of microfluidization following a bottom-up methodology is likely to contribute new interesting nanostructures.

References

Acevedo-Fani, A., Salvia-Trujillo, L., Rojas-Graü, M. A., & Martín-Belloso, O. (2015). Edible films from essential-oil-loaded nanoemulsions: physicochemical characterization and antimicrobial properties. *Food Hydrocolloids*, *47*, 168–177.

Acosta, E. (2009). Bioavailability of nanoparticles in nutrient and nutraceutical delivery. *Current Opinion in Colloid & Interface Science*, *14*(1), 3–15.

Ahmed, K., Li, Y., McClements, D. J., & Xiao, H. (2012). Nanoemulsion-and emulsion-based delivery systems for curcumin: encapsulation and release properties. *Food Chemistry*, *132*(2), 799–807.

Akbas, E., Soyler, B., & Oztop, M. H. (2018). Formation of capsaicin loaded nanoemulsions with high pressure homogenization and ultrasonication. *LWT- Food Science and Technology*, *96*, 266–273.

Akhavan, S., Assadpour, E., Katouzian, I., & Jafari, S. M. (2018). Lipid nano scale cargos for the protection and delivery of food bioactive ingredients and nutraceuticals. *Trends in Food Science & Technology*, *74*, 132–146.

Alfaro, M. C., López-Tarruello, M., García, M. C., & Muñoz, J. (2018). Influencia de la adición de glicerol en la distribución de tamaños de gotas, diámetros medios y estabilidad física de emulsiones O/W ecológicas procesadas con tecnología Microfluidizer. *Afinidad*, *75*, 175–181.

Alkanawati, M. S., Wurm, F. R., Thérien-Aubin, H., & Landfester, K. (2018). Large-scale preparation of polymer nanocarriers by high-pressure microfluidization. *Macromolecular Materials and Engineering*, *303*(1), 1700505.

Anderson, J. W. (1987). Dietary fiber, lipids and atherosclerosis. *American Journal of Cardiology*, *60*(12), G17–G22.

Anton, N., Benoit, J. P., & Saulnier, P. (2008). Design and production of nanoparticles formulated from nano-emulsion templates—a review. *Journal of Controlled Release*, *128*(3), 185–199.

Arredondo-Ochoa, T., García-Almendárez, B. E., Escamilla-García, M., Martín-Belloso, O., Rossi-Márquez, G., Medina-Torres, L., et al. (2017). Physicochemical and antimicrobial characterization of beeswax–starch food-grade nanoemulsions incorporating natural antimicrobials. *International Journal of Molecular Sciences*, *18*(12), 2712.

Artiga-Artigas, M., Acevedo-Fani, A., & Martín-Belloso, O. (2017). Effect of sodium alginate incorporation procedure on the physicochemical properties of nanoemulsions. *Food Hydrocolloids*, *70*, 191–200.

Artiga-Artigas, M., Guerra-Rosas, M. I., Morales-Castro, J., Salvia-Trujillo, L., & Martín-Belloso, O. (2018). Influence of essential oils and pectin on nanoemulsion formulation: a ternary phase experimental approach. *Food Hydrocolloids*, *81*, 209–219.

Assadpour, E., & Jafari, S. M. (2018). A systematic review on nanoencapsulation of food bioactive ingredients and nutraceuticals by various nanocarriers. *Critical Reviews in Food Science and Nutrition*, *8*, 1–47.

Assadpour, E., Jafari, S. M., & Esfanjani, A. F. (2017). Protection of phenolic compounds within nanocarriers. *CAB Reviews*, *12*(057), 1–8.

Asumadu-Mensah, A., Smith, K. W., & Ribeiro, H. S. (2013). Solid lipid dispersions: potential delivery system for functional ingredients in foods. *Journal of Food Science*, *78*(7), E1000–E1008.

Bai, L., Huan, S., Gu, J., & McClements, D. J. (2016). Fabrication of oil-in-water nanoemulsions by dual-channel microfluidization using natural emulsifiers: saponins, phospholipids, proteins, and polysaccharides. *Food Hydrocolloids*, *61*, 703–711.

Bai, L., Huan, S., Li, Z., & McClements, D. J. (2017). Comparison of emulsifying properties of food-grade polysaccharides in oil-in-water emulsions: gum Arabic, beet pectin, and corn fiber gum. *Food Hydrocolloids*, *66*, 144–153.

Bai, L., & McClements, D. J. (2016a). Development of microfluidization methods for efficient production of concentrated nanoemulsions: comparison of single-and dual-channel microfluidizers. *Journal of Colloid and Interface Science*, *466*, 206–212.

Bai, L., & McClements, D. J. (2016b). Formation and stabilization of nanoemulsions using biosurfactants: rhamnolipids. *Journal of Colloid and Interface Science*, *479*, 71–79.

Barnadas-Rodrı´guez, R., & Sabés, M. (2001). Factors involved in the production of liposomes with a high-pressure homogenizer. *International Journal of Pharmaceutics*, *213*(1–2), 175–186.

Bejrapha, P., Surassmo, S., Choi, M. J., Nakagawa, K., & Min, S. G. (2011). Studies on the role of gelatin as a cryo-and lyo-protectant in the stability of capsicum oleoresin nanocapsules in gelatin matrix. *Journal of Food Engineering*, *105*(2), 320–331.

Belhaj, N., Arab-Tehrany, E., & Linder, M. (2010). Oxidative kinetics of salmon oil in bulk and in nanoemulsion stabilized by marine lecithin. *Process Biochemistry*, *45*(2), 187–195.

Berg, J. C. (2010). *An introduction to interfaces & colloids. The bridge to nanoscience*. Singapore: World Scientific.

Bradley, M. A., Prescott, S. W., Schoonbrood, H. A., Landfester, K., & Grieser, F. (2005). Miniemulsion copolymerization of methyl methacrylate and butyl acrylate by ultrasonic initiation. *Macromolecules*, *38*(15), 6346–6351.

Braginsky, L. M., & Belevitskaya, M. A. (1996). Kinetics of droplets breakup in agitated vessels. In N. N. Kulov (Ed.), *Liquid–liquid systems* (p. 1). Commack, NY: Nova Science.

Bucci, A. J., Van Hekken, D. L., Tunick, M. H., Renye, J. A., & Tomasula, P. M. (2018). The effects of microfluidization on the physical, microbial, chemical, and coagulation properties of milk. *Journal of Dairy Science*, *101*, 6990–7001.

Burgos, N., Mellinas, A. C., García-Serna, E., & Jiménez, A. (2017). Nanoencapsulation of flavor and aromas in food packaging. In *Food packaging* (pp. 567–601). London: Academic Press.

Charani, P. R., Dehghani-Firouzabadi, M., Afra, E., & Shakeri, A. (2013). Rheological characterization of high concentrated MFC gel from kenaf unbleached pulp. *Cellulose*, *20*(2), 727–740.

Cho, H. T., Salvia-Trujillo, L., Kim, J., Park, Y., Xiao, H., & McClements, D. J. (2014). Droplet size and composition of nutraceutical nanoemulsions influences bioavailability of long chain fatty acids and Coenzyme Q10. *Food Chemistry*, *156*, 117–122.

Choi, S. J., Decker, E. A., Henson, L., Popplewell, L. M., Xiao, H., & McClements, D. J. (2011). Formulation and properties of model beverage emulsions stabilized by sucrose monopalmitate: influence of pH and lyso-lecithin addition. *Food Research International*, *44*(9), 3006–3012.

Christiansen, K. F., Vegarud, G., Langsrud, T., Ellekjaer, M. R., & Egelandsdal, B. (2004). Hydrolyzed whey proteins as emulsifiers and stabilizers in high-pressure processed dressings. *Food Hydrocolloids*, *18*(5), 757–767.

Chu, B. S., Ichikawa, S., Kanafusa, S., & Nakajima, M. (2007). Preparation of protein-stabilized β-carotene nanodispersions by emulsification–evaporation method. *Journal of the American Oil Chemists' Society*, *84*(11), 1053–1062.

Chuyen, H. V., & Eun, J. B. (2017). Marine carotenoids: bioactivities and potential benefits to human health. *Critical Reviews in Food Science and Nutrition*, *57*(12), 2600–2610.

Colussi, G., Catena, C., Novello, M., Bertin, N., & Sechi, L. A. (2017). Impact of omega-3 polyunsaturated fatty acids on vascular function and blood pressure: relevance for cardiovascular outcomes. *Nutrition, Metabolism, and Cardiovascular Diseases*, *27*(3), 191–200.

Craparo, E. F., Cavallaro, G., Bondì, M. L., Mandracchia, D., & Giammona, G. (2006). PEGylated nanoparticles based on a polyaspartamide. Preparation, physico-chemical characterization, and intracellular uptake. *Biomacromolecules*, 7(11), 3083–3092.

Dammak, I., & do Amaral Sobral, P. J. (2017). Formulation and stability characterization of rutin-loaded oil-in-water emulsions. *Food and Bioprocess Technology*, *10*(5), 926–939.

Danaei, M., Dehghankhold, M., Ataei, S., Hasanzadeh Davarani, F., Javanmard, R., Dokhani, A., et al. (2018). Impact of particle size and polydispersity index on the clinical applications of lipidic nanocarrier systems. *Pharmaceutics*, *10*(2), 57.

Danov, K. D., Kralchevsky, P. A., & Ivanov, I. B. (2001). Dynamic processes in surfactant-stabilized emulsions. In *Encyclopedic handbook of emulsion technology* (pp. 621–659). New York: Marcel Dekker.

De Francisco, E. V., & García-Estepa, R. M. (2018). Nanotechnology in the agrofood industry. *Journal of Food Engineering*, *238*, 1–11.

Montes de Oca-Ávalos, J. M. M., Candal, R. J., & Herrera, M. L. (2017). Nanoemulsions: stability and physical properties. *Current Opinion in Food Science*, *16*, 1–6.

Deen, G. R., Skovgaard, J., & Pedersen, J. S. (2016). Formation and properties of nanoemulsions. In Grumezescu (Ed.), *Emulsions* (pp. 193–226). London: Academic Press.

Desgouilles, S., Vauthier, C., Bazile, D., Vacus, J., Grossiord, J. L., Veillard, M., et al. (2003). The design of nanoparticles obtained by solvent evaporation: a comprehensive study. *Langmuir*, *19*(22), 9504–9510.

Dias, D. R., Botrel, D. A., Fernandes, R. V. D. B., & Borges, S. V. (2017). Encapsulation as a tool for bioprocessing of functional foods. *Current Opinion in Food Science*, *13*, 31–37.

Dickinson, E. (2003). Hydrocolloids at interfaces and the influence on the properties of dispersed systems. *Food Hydrocolloids*, *17*(1), 25–39.

Donsì, F. (2018). Applications of nanoemulsions in foods. In S. M. Jafari & D. J. McClements (Eds.), *Nanoemulsions* (pp. 349–377). London: Elsevier, Academic Press.

Ebert, S., Koo, C. K., Weiss, J., & McClements, D. J. (2017). Continuous production of core-shell protein nanoparticles by antisolvent precipitation using dual-channel microfluidization: caseinate-coated zein nanoparticles. *Food Research International*, *92*, 48–55.

Farhang, B., Kakuda, Y., & Corredig, M. (2012). Encapsulation of ascorbic acid in liposomes prepared with milk fat globule membrane-derived phospholipids. *Dairy Science & Technology*, *92*(4), 353–366.

Faridi Esfanjani, A., Assadpour, E., & Jafari, S. M. (2018). Improving the bioavailability of phenolic compounds by loading them within lipid-based nanocarriers. *Trends in Food Science & Technology*, *76*, 56–66.

Fillat, Ú., Wicklein, B., Martin-Sampedro, R., Ibarra, D., Ruiz-Hitzky, E., Valencia, C., et al. (2018). Assessing cellulose nanofiber production from olive tree pruning residue. *Carbohydrate Polymers*, *179*, 252–261.

Franks, F. (1998). Freeze-drying of bioproducts: putting principles into practice. *European Journal of Pharmaceutics and Biopharmaceutics*, *45*(3), 221–229.

Gadkari, P. V., & Balaraman, M. (2015). Extraction of catechins from decaffeinated green tea for development of nanoemulsion using palm oil and sunflower oil based lipid carrier systems. *Journal of Food Engineering*, *147*, 14–23.

Gadkari, P. V., Shashidhar, M. G., & Balaraman, M. (2017). Delivery of green tea catechins through oil-in-water (O/W) nanoemulsion and assessment of storage stability. *Journal of Food Engineering*, *199*, 65–76.

García-Márquez, E., Higuera-Ciapara, I., & Espinosa-Andrews, H. (2017). Design of fish oil-in-water nanoemulsion by microfluidization. *Innovative Food Science & Emerging Technologies*, *40*, 87–91.

Gavi, E., Kubicki, D., Padron, G. A., & Özcan-Taşkın, N. G. (2018). Breakup of nanoparticle clusters using microfluidizer M110-P. *Chemical Engineering Research and Design*, *132*, 902–912.

Gestranius, M., Stenius, P., Kontturi, E., Sjöblom, J., & Tammelin, T. (2017). Phase behaviour and droplet size of oil-in-water pickering emulsions stabilised with plant-derived nanocellulosic materials. *Colloids and Surfaces A: Physicochemical and Engineering Aspects*, *519*, 60–70.

Ghorbanzade, T., Jafari, S. M., Akhavan, S., & Hadavi, R. (2017). Nano-encapsulation of fish oil in nano-liposomes and its application in fortification of yogurt. *Food Chemistry*, *216*, 146–152.

Golfomitsou, I., Mitsou, E., Xenakis, A., & Papadimitriou, V. (2018). Development of food grade O/W nanoemulsions as carriers of vitamin D for the fortification of emulsion based food matrices: a structural and activity study. *Journal of Molecular Liquids*, *268*, 734–742.

Gong, Y., Wu, Y., Zheng, C., Fan, L., Xiong, F., & Zhu, J. (2012). An excellent delivery system for improving the oral bioavailability of natural vitamin E in rats. *AAPS PharmSciTech*, *13*(3), 961–966.

Guaadaoui, A., Benaicha, S., Elmajdoub, N., Bellaoui, M., & Hamal, A. (2014). What is a bioactive compound? A combined definition for a preliminary consensus. *International Journal of Nutrition and Food Sciences*, *3*(3), 174–179.

Guerra-Rosas, M. I., Morales-Castro, J., Ochoa-Martínez, L. A., Salvia-Trujillo, L., & Martín-Belloso, O. (2016). Long-term stability of food-grade nanoemulsions from high methoxyl pectin containing essential oils. *Food Hydrocolloids*, *52*, 438–446.

Guttoff, M., Saberi, A. H., & McClements, D. J. (2015). Formation of vitamin D nanoemulsion-based delivery systems by spontaneous emulsification: factors affecting particle size and stability. *Food Chemistry*, *171*, 117–122.

Hasenhuettl, G. L. (2008). Synthesis and commercial preparation of food emulsifiers. In *Food emulsifiers and their applications* (pp. 11–37). New York, NY: Springer.

Heffernan, S. P., Kelly, A. L., Mulvihill, D. M., Lambrich, U., & Schuchmann, H. P. (2011). Efficiency of a range of homogenisation technologies in the emulsification and stabilization of cream liqueurs. *Innovative Food Science & Emerging Technologies*, *12*(4), 628–634.

Helgason, T., Salminen, H., Kristbergsson, K., McClements, D. J., & Weiss, J. (2015). Formation of transparent solid lipid nanoparticles by microfluidization: influence of lipid physical state on appearance. *Journal of Colloid and Interface Science*, *448*, 114–122.

Hettrich, K., Pinnow, M., Volkert, B., Passauer, L., & Fischer, S. (2014). Novel aspects of nanocellulose. *Cellulose*, *21*(4), 2479–2488.

Hooper, C., Barreto, P. D. S., Pahor, M., Weiner, M., & Vellas, B. (2018). The relationship of omega 3 polyunsaturated fatty acids in red blood cell membranes with cognitive function and brain structure: a review focussed on Alzheimer's disease. *The Journal of Prevention of Alzheimer's Disease*, *5*(1), 78–84.

Huang, Q., Yu, H., & Ru, Q. (2010). Bioavailability and delivery of nutraceuticals using nanotechnology. *Journal of Food Science*, *75*(1), R50–R57.

Imran, M., Revol-Junelles, A. M., René, N., Jamshidian, M., Akhtar, M. J., Arab-Tehrany, E., et al. (2012). Microstructure and physico-chemical evaluation of nano-emulsion-based antimicrobial peptides embedded in bioactive packaging films. *Food Hydrocolloids*, *29*(2), 407–419.

Israelachvili, J. (2011). *Intermolecular and surface forces* (3rd ed.). London: Academic Press.

Jafari, S. M. (2017). *Nanoencapsulation of food bioactive ingredients. Principles and applications.* London: Academic Press-Elsevier.

Jafari, S. M., Assadpoor, E., He, Y., & Bhandari, B. (2008). Re-coalescence of emulsion droplets during high-energy emulsification. *Food Hydrocolloids*, *22*(7), 1191–1202.

Jafari, S. M., He, Y., & Bhandari, B. (2006). Nano-emulsion production by sonication and microfluidization—a comparison. *International Journal of Food Properties*, *9*(3), 475–485.

Jafari, S. M., He, Y., & Bhandari, B. (2007). Optimization of nano-emulsions production by microfluidization. *European Food Research and Technology*, *225*(5–6), 733–741.

Jafari, S. M., & McClements, D. J. (2017). Nanotechnology approaches for increasing nutrient bioavailability. In *Advances in food and nutrition research*. Academic Press.

Jafari, S. M., Paximada, P., Mandala, I., Assadpour, E., & Mehrnia, M. A. (2017). 2—Encapsulation by nanoemulsions. In *Nanoencapsulation technologies for the food and nutraceutical industries* (pp. 36–73). Academic Press.

Jin, W., Xu, W., Liang, H., Li, Y., Liu, S., & Li, B. (2016). Nanoemulsions for food: properties, production, characterization, and applications. In A. M. Grumezescu (Ed.), *Emulsions* (pp. 1–36). London: Elsevier. Academic Press.

Jo, Y. J., & Kwon, Y. J. (2014). Characterization of β-carotene nanoemulsions prepared by microfluidization technique. *Food Science and Biotechnology*, *23*(1), 107–113.

Karthik, P., & Anandharamakrishnan, C. (2016a). Fabrication of a nutrient delivery system of docosahexaenoic acid nanoemulsions via high energy techniques. *RSC Advances*, *6*(5), 3501–3513.

Karthik, P., & Anandharamakrishnan, C. (2016b). Enhancing omega-3 fatty acids nanoemulsion stability and in-vitro digestibility through emulsifiers. *Journal of Food Engineering*, *187*, 92–105.

Karthik, P., Ezhilarasi, P. N., & Anandharamakrishnan, C. (2017). Challenges associated in stability of food grade nanoemulsions. *Critical Reviews in Food Science and Nutrition*, *57*(7), 1435–1450.

Katouzian, I., Faridi Esfanjani, A., Jafari, S. M., & Akhavan, S. (2017). Formulation and application of a new generation of lipid nano-carriers for the food bioactive ingredients. *Trends in Food Science & Technology*, *68*(Suppl C), 14–25.

Katouzian, I., & Jafari, S. M. (2016). Nano-encapsulation as a promising approach for targeted delivery and controlled release of vitamins. *Trends in Food Science & Technology*, *53*, 34–48.

Katouzian, I., & Jafari, S. M. (2017). Nanoencapsulation of vitamins. *Nanoencapsulation of food bioactive ingredients*. (pp. 145–181). London, UK: Academic Press

Ketenoglu, O., Mert, B., & Tekin, A. (2014). Effects of microfluidized dietary fibers on stability properties of emulsions. *Journal of Texture Studies*, *45*(4), 295–306.

Khanniri, E., Bagheripoor-Fallah, N., Sohrabvandi, S., Mortazavian, A. M., Khosravi-Darani, K., & Mohammad, R. (2016). Application of liposomes in some dairy products. *Critical Reviews in Food Science and Nutrition*, *56*(3), 484–493.

Komaiko, J., Sastrosubroto, A., & McClements, D. J. (2016). Encapsulation of ω-3 fatty acids in nanoemulsion-based delivery systems fabricated from natural emulsifiers: sunflower phospholipids. *Food Chemistry*, *203*, 331–339.

Kralova, I., & Sjöblom, J. (2009). Surfactants used in food industry: a review. *Journal of Dispersion Science and Technology*, *30*(9), 1363–1383.

Kwon, S. S., Nam, Y. S., Lee, J. S., Ku, B. S., Han, S. H., Lee, J. Y., et al. (2002). Preparation and characterization of coenzyme Q10-loaded PMMA nanoparticles by a new emulsification process based on microfluidization. *Colloids and Surfaces A: Physicochemical and Engineering Aspects*, *210*(1), 95–104.

Landete, J. M. (2012). Updated knowledge about polyphenols: functions, bioavailability, metabolism, and health. *Critical Reviews in Food Science and Nutrition*, *52*(10), 936–948.

Law, B. A., & King, J. S. (1985). Use of liposomes for proteinase addition to Cheddar cheese. *Journal of Dairy Research*, *52*(1), 183–188.

Lee, L., Hancocks, R., Noble, I., & Norton, I. T. (2014). Production of water-in-oil nanoemulsions using high pressure homogenisation: a study on droplet break-up. *Journal of Food Engineering*, *131*, 33–37.

Laye, C., McClements, D. J., & Weiss, J. (2008). Formation of biopolymer-coated liposomes by electrostatic deposition of chitosan. *Journal of Food Science, 73*(5), 7–15.
Li, D., Li, L., Xiao, N., Li, M., & Xie, X. (2018). Physical properties of oil-in-water nanoemulsions stabilized by OSA-modified starch for the encapsulation of lycopene. *Colloids and Surfaces A: Physicochemical and Engineering Aspects, 552*, 59–66.
Lin, C., Ma, Q., Su, Q., Bian, H., & Zhu, J. Y. (2018). Facile synthesis of highly hydrophobic cellulose nanoparticles through post-esterification microfluidization. *Fibers, 6*(2), 22.
Lindström, J., Peltonen, M., Eriksson, J. G., Louheranta, A., Fogelholm, M., Uusitupa, M., et al. (2006). High-fibre, low-fat diet predicts long-term weight loss and decreased type 2 diabetes risk: the Finnish diabetes prevention study. *Diabetologia, 49*(5), 912–920.
Livney, Y. D. (2015). Nanostructured delivery systems in food: latest developments and potential future directions. *Current Opinion in Food Science, 3*, 125–135.
Llinares, R., Ramírez, P., Carmona, J., Carrillo, F., & Muñoz, J. (2018). Formulation and optimization of emulsions based on bitter fennel essential oil and EO/BO block copolymer surfactant. *Colloids and Surfaces A: Physicochemical and Engineering Aspects, 536*, 142–147.
Llinares, R., Santos, J., Trujillo-Cayado, L. A., Ramírez, P., & Muñoz, J. (2018). Enhancing rosemary oil-in-water microfluidized nanoemulsion properties through formulation optimization by response surface methodology. *LWT- Food Science and Technology, 97*, 370–375.
Luo, X., Zhou, Y., Bai, L., Liu, F., Deng, Y., & McClements, D. J. (2017a). Fabrication of β-carotene nanoemulsion-based delivery systems using dual-channel microfluidization: physical and chemical stability. *Journal of Colloid and Interface Science, 490*, 328–335.
Luo, X., Zhou, Y., Bai, L., Liu, F., Zhang, R., Zhang, Z., et al. (2017b). Production of highly concentrated oil-in-water emulsions using dual-channel microfluidization: use of individual and mixed natural emulsifiers (saponin and lecithin). *Food Research International, 96*, 103–112.
Mahdavi, S. A., Jafari, S. M., Ghorbani, M., & Assadpoor, E. (2014). Spray-drying microencapsulation of anthocyanins by natural biopolymers: a review. *Drying Technology, 32* (5), 509–518.
Majeed, H., Antoniou, J., Hategekimana, J., Sharif, H. R., Haider, J., Liu, F., et al. (2016). Influence of carrier oil type, particle size on in vitro lipid digestion and eugenol release in emulsion and nanoemulsions. *Food Hydrocolloids, 52*, 415–422.
Mao, L., Yang, J., Xu, D., Yuan, F., & Gao, Y. (2010). Effects of homogenization models and emulsifiers on the physicochemical properties of β-carotene nanoemulsions. *Journal of Dispersion Science and Technology, 31*(7), 986–993.
Mason, T. G., Wilking, J. N., Meleson, K., Chang, C. B., & Graves, S. M. (2006). Nanoemulsions: formation, structure, and physical properties. *Journal of Physics: Condensed Matter, 18*(41), R635.
Matos, M., Gutiérrez, G., Coca, J., & Pazos, C. (2014). Preparation of water-in-oil-in-water (W1/O/W2) double emulsions containing trans-resveratrol. *Colloids and Surfaces A: Physicochemical and Engineering Aspects, 442*, 69–79.
Mayer, S., Weiss, J., & McClements, D. J. (2013). Vitamin E-enriched nanoemulsions formed by emulsion phase inversion: factors influencing droplet size and stability. *Journal of Colloid and Interface Science, 402*, 122–130.
McClements, D. J. (2011). Edible nanoemulsions: fabrication, properties, and functional performance. *Soft Matter*, 7(6), 2297–2316.
McClements, D. J. (2016). *Food emulsions: Principles, practices, and techniques*. Boca Ratón, FL. USA: CRC press.
McClements, D. J., Decker, E. A., & Weiss, J. (2007). Emulsion-based delivery systems for lipophilic bioactive components. *Journal of Food Science, 72*(8), R109–R124.
McClements, D. J., & Jafari, S. M. (2018a). General aspects of nanoemulsions and their formulation. In *Nanoemulsions* (pp. 3–20). London: Elsevier. Academic Press.

McClements, D. J., & Jafari, S. M. (2018b). Improving emulsion formation, stability and performance using mixed emulsifiers: a review. *Advances in Colloid and Interface Science*, *251*, 55–79.

McClements, D. J., & Rao, J. (2011). Food-grade nanoemulsions: formulation, fabrication, properties, performance, biological fate, and potential toxicity. *Critical Reviews in Food Science and Nutrition*, *51*(4), 285–330.

Microfluidics (2018). *Microfluidizer processor user guide*. Newton, MA: Microfluidics.

Missoum, K., Belgacem, M. N., & Bras, J. (2013). Nanofibrillated cellulose surface modification: a review. *Materials*, *6*(5), 1745–1766.

Monroy-Villagrana, A., Alamilla-Beltrán, L., Hernández-Sánchez, H., & Gutiérrez-López, G. F. (2015). Hydrodynamic characterization of the formation of alpha-tocopherol nanoemulsions in a microfluidizer. In *Food nanoscience and nanotechnology* (pp. 163–175). London: Elsevier. Academic Press.

Mozafari, M. R. (Ed.), (2006). *Nanocarrier technologies: Frontiers of nanotherapy*. Dordrecht, Netherlands: Springer.

Mozafari, M. R., Johnson, C., Hatziantoniou, S., & Demetzos, C. (2008). Nanoliposomes and their applications in food nanotechnology. *Journal of Liposome Research*, *18*(4), 309–327.

Murakami, K., Sasaki, S., Okubo, H., Takahashi, Y., Hosoi, Y., & Itabashi, M. (2007). Dietary fiber intake, dietary glycemic index and load, and body mass index: a cross-sectional study of 3931 Japanese women aged 18–20 years. *European Journal of Clinical Nutrition*, *61* (8), 986.

Narsimhan, G., & Wang, Z. (2008). Guidelines for processing emulsion-based foods. In *Food emulsifiers and their applications* (pp. 349–394). New York, NY: Springer.

Odriozola-Serrano, I., Oms-Oliu, G., & Martín-Belloso, O. (2014). Nanoemulsion-based delivery systems to improve functionality of lipophilic components. *Frontiers in Nutrition*, *1*, 24.

Öztürk, B. (2017). Nanoemulsions for food fortification with lipophilic vitamins: production challenges, stability, and bioavailability. *European Journal of Lipid Science and Technology*, *119*(7), 1500539.

Öztürk, B., Argin, S., Ozilgen, M., & McClements, D. J. (2015). Formation and stabilization of nanoemulsion-based vitamin E delivery systems using natural biopolymers: whey protein isolate and gum arabic. *Food Chemistry*, *188*, 256–263.

Pacaphol, K., & Aht-Ong, D. (2017). Preparation of hemp nanofibers from agricultural waste by mechanical defibrillation in water. *Journal of Cleaner Production*, *142*, 1283–1295.

Panagiotou, T., & Fisher, R. (2012). Improving product quality with entrapped stable emulsions: from theory to industrial application. *Challenges*, *3*, 84–113.

Panagiotou, T., & Fisher, R. (2013). Producing micron- and nano- size formulations for functional foods applications. *Functional Foods in Health and Disease*, *3*(7), 274–289.

Pathakoti, K., Manubolu, M., & Hwang, H. M. (2017). Nanostructures: current uses and future applications in food science. *Journal of Food and Drug Analysis*, *25*(2), 245–253.

Pérez-Córdoba, L. J., Norton, I. T., Batchelor, H. K., Gkatzionis, K., Spyropoulos, F., & Sobral, P. J. (2018). Physico-chemical, antimicrobial and antioxidant properties of gelatin-chitosan based films loaded with nanoemulsions encapsulating active compounds. *Food Hydrocolloids*, *79*, 544–559.

Pérez-Mosqueda, L. M., Ramírez, P., Trujillo-Cayado, L. A., Santos, J., & Muñoz, J. (2014). Development of ecofriendly submicron emulsions stabilized by a bio-derived gum. *Colloids and Surfaces B: Biointerfaces*, *123*, 797–802.

Piorkowski, D. T., & McClements, D. J. (2014). Beverage emulsions: recent developments in formulation, production, and applications. *Food Hydrocolloids*, *42*, 5–41.

Pool, H., Mendoza, S., Xiao, H., & McClements, D. J. (2013). Encapsulation and release of hydrophobic bioactive components in nanoemulsion-based delivery systems: impact of physical form on quercetin bioaccessibility. *Food & Function, 4*(1), 162–174.

Porras, M., Solans, C., Gonzalez, C., & Gutierrez, J. M. (2008). Properties of water-in-oil (W/O) nano-emulsions prepared by a low-energy emulsification method. *Colloids and Surfaces A: Physicochemical and Engineering Aspects, 324*(1–3), 181–188.

Qadir, A., Faiyazuddin, M. D., Hussain, M. T., Alshammari, T. M., & Shakeel, F. (2016). Critical steps and energetics involved in a successful development of a stable nanoemulsion. *Journal of Molecular Liquids, 214*, 7–18.

Qi, G., Jones, C. W., & Schork, F. J. (2006). Enzyme-initiated miniemulsion polymerization. *Biomacromolecules*, 7(11), 2927–2930.

Qian, C., & McClements, D. J. (2011). Formation of nanoemulsions stabilized by model food-grade emulsifiers using high-pressure homogenization: factors affecting particle size. *Food Hydrocolloids, 25*(5), 1000–1008.

Rafiee, Z., & Jafari, S. M. (2018). Application of lipid nanocarriers for the food industry. In J. -M. Mérillon & K. G. Ramawat (Eds.), *Bioactive molecules in food* (pp. 1–43). Cham: Springer International Publishing.

Rao, J., & McClements, D. J. (2011). Food-grade microemulsions, nanoemulsions and emulsions: fabrication from sucrose monopalmitate & lemon oil. *Food Hydrocolloids, 25*(6), 1413–1423.

Rao, J., & McClements, D. J. (2012). Food-grade microemulsions and nanoemulsions: role of oil phase composition on formation and stability. *Food Hydrocolloids, 29*(2), 326–334.

Raviadaran, R., Chandran, D., Shin, L. H., & Manickam, S. (2018). Optimization of palm oil in water nano-emulsion with curcumin using microfluidizer and response surface methodology. *LWT- Food Science and Technology, 96*, 58–65.

Rayner, M., & Dejmek, P. (2015). *Engineering aspects of food emulsification and homogenization.* Boca Raton, FL. USA: CRC Press.

Rezaei, A., Fathi, M., & Jafari, S. M. (2019). Nanoencapsulation of hydrophobic and low-soluble food bioactive compounds within different nanocarriers. *Food Hydrocolloids, 88*, 146–162.

Ricaurte, L., de Jesús Perea-Flores, M., Martinez, A., & Quintanilla-Carvajal, M. X. (2016). Production of high-oleic palm oil nanoemulsions by high-shear homogenization (microfluidization). *Innovative Food Science & Emerging Technologies, 35*, 75–85.

Rostamabadi, H., Falsafi, S. R., & Jafari, S. M. (2019). Nanoencapsulation of carotenoids within lipid-based nanocarriers. *Journal of Controlled Release, 298*, 38–67.

Saberi, A. H., Fang, Y., & McClements, D. J. (2013). Fabrication of vitamin E-enriched nanoemulsions: factors affecting particle size using spontaneous emulsification. *Journal of Colloid and Interface Science, 391*, 95–102.

Salvia-Trujillo, L., Decker, E. A., & McClements, D. J. (2016). Influence of an anionic polysaccharide on the physical and oxidative stability of omega-3 nanoemulsions: antioxidant effects of alginate. *Food Hydrocolloids, 52*, 690–698.

Salvia-Trujillo, L., Fumiaki, B., Park, Y., & McClements, D. J. (2017). The influence of lipid droplet size on the oral bioavailability of vitamin D 2 encapsulated in emulsions: an in vitro and in vivo study. *Food & Function, 8*(2), 767–777.

Salvia-Trujillo, L., Rojas-Graü, M. A., Soliva-Fortuny, R., & Martín-Belloso, O. (2013). Effect of processing parameters on physicochemical characteristics of microfluidized lemongrass essential oil-alginate nanoemulsions. *Food Hydrocolloids, 30*(1), 401–407.

Salvia-Trujillo, L., Rojas-Graü, M. A., Soliva-Fortuny, R., & Martín-Belloso, O. (2014). Impact of microfluidization or ultrasound processing on the antimicrobial activity against *Escherichia coli* of lemongrass oil-loaded nanoemulsions. *Food Control, 37*, 292–297.

Salvia-Trujillo, L., Rojas-Graü, A., Soliva-Fortuny, R., & Martín-Belloso, O. (2015). Physicochemical characterization and antimicrobial activity of food-grade emulsions and nanoemulsions incorporating essential oils. *Food Hydrocolloids, 43*, 547–556.

Santos, J., Calero, N., Muñoz, J., & Cidade, M. T. (2018). Development of food emulsions containing an advanced performance xanthan gum by microfluidization technique. *Food Science and Technology International, 24*(5), 373–381.

Santos, J., Jimenez, M., Calero, N., Alfaro, M. C., & Muñoz, J. (2019). Influence of a shear post-treatment on rheological properties, microstructure and physical stability of emulgels formed by rosemary essential oil and a fumed silica. *Journal of Food Engineering, 241*, 136–148.

Santos, J., Trujillo-Cayado, L. A., Calero, N., Alfaro, M. C., & Muñoz, J. (2016). Development of eco-friendly emulsions produced by microfluidization technique. *Journal of Industrial and Engineering Chemistry, 36*, 90–95.

Schuh, R. S., Poletto, É., Fachel, F. N., Matte, U., Baldo, G., & Teixeira, H. F. (2018). Physicochemical properties of cationic nanoemulsions and liposomes obtained by microfluidization complexed with a single plasmid or along with an oligonucleotide: implications for CRISPR/Cas technology. *Journal of Colloid and Interface Science, 530*, 243–255.

Sedaghat Doost, A., Devlieghere, F., Dirckx, A., & Van der Meeren, P. (2018). Fabrication of Origanum compactum essential oil nanoemulsions stabilized using Quillaja Saponin biosurfactant. *Journal of Food Processing and Preservation, 42*, e13668.

Sessa, M., Balestrieri, M. L., Ferrari, G., Servillo, L., Castaldo, D., D'Onofrio, N., et al. (2014). Bioavailability of encapsulated resveratrol into nanoemulsion-based delivery systems. *Food Chemistry, 147*, 42–50.

Sessa, M., Tsao, R., Liu, R., Ferrari, G., & Donsì, F. (2011). Evaluation of the stability and antioxidant activity of nanoencapsulated resveratrol during in vitro digestion. *Journal of Agricultural and Food Chemistry, 59*(23), 12352–12360.

Setya, S., Talegaonkar, S., & Razdan, B. K. (2014). Nanoemulsions: formulation methods and stability aspects. *World Journal of Pharmacy and Pharmaceutical Sciences, 3*(2), 2214–2228.

Shaaban, H. A., El-Ghorab, A. H., & Shibamoto, T. (2012). Bioactivity of essential oils and their volatile aroma components. *Journal of Essential Oil Research, 24*(2), 203–212.

Shu, G., Khalid, N., Chen, Z., Neves, M. A., Barrow, C. J., & Nakajima, M. (2018). Formulation and characterization of astaxanthin-enriched nanoemulsions stabilized using ginseng saponins as natural emulsifiers. *Food Chemistry, 255*, 67–74.

Silva, H. D., Cerqueira, M. Â., & Vicente, A. A. (2012). Nanoemulsions for food applications: development and characterization. *Food and Bioprocess Technology, 5*(3), 854–867.

Silva, H. D., Cerqueira, M. A., & Vicente, A. A. (2015). Influence of surfactant and processing conditions in the stability of oil-in-water nanoemulsions. *Journal of Food Engineering, 167*, 89–98.

Singh, Y., Meher, J. G., Raval, K., Khan, F. A., Chaurasia, M., Jain, N. K., et al. (2017). Nanoemulsion: concepts, development and applications in drug delivery. *Journal of Controlled Release, 252*, 28–49.

Singh, H., Thompson, A., Liu, W., & Corredig, M. (2012). Liposomes as food ingredients and nutraceutical delivery systems. In *Encapsulation technologies and delivery systems for food ingredients and nutraceuticals* (pp. 287–318). Sawston, Cambridge, UK: Woodhead Publishing.

Somogyi, L. P. (2000). Food additives. In *Kirk-Othmer Encyclopedia of Chemical Technology* (pp. 1-59). Wiley & Sons.

Spence, K. L., Venditti, R. A., Rojas, O. J., Habibi, Y., & Pawlak, J. J. (2011). A comparative study of energy consumption and physical properties of microfibrillated cellulose produced by different processing methods. *Cellulose*, *18*(4), 1097–1111.
Sun, C., Dai, L., Liu, F., & Gao, Y. (2016). Dynamic high pressure microfluidization treatment of zein in aqueous ethanol solution. *Food Chemistry*, *210*, 388–395.
Sun, C., Yang, J., Liu, F., Yang, W., Yuan, F., & Gao, Y. (2016). Effects of dynamic high-pressure microfluidization treatment and the presence of quercetagetin on the physical, structural, thermal, and morphological characteristics of zein nanoparticles. *Food and Bioprocess Technology*, *9*(2), 320–330.
Surh, J., Decker, E. A., & McClements, D. J. (2017). Utilisation of spontaneous emulsification to fabricate lutein-loaded nanoemulsion-based delivery systems: factors influencing particle size and colour. *International Journal of Food Science & Technology*, *52*(6), 1408–1416.
Tadros, T., Izquierdo, P., Esquena, J., & Solans, C. (2004). Formation and stability of nano-emulsions. *Advances in Colloid and Interface Science*, *108*, 303–318.
Takahashi, M., Inafuku, K. I., Miyagi, T., Oku, H., Wada, K., Imura, T., et al. (2007). Efficient preparation of liposomes encapsulating food materials using lecithins by a mechanochemical method. *Journal of Oleo Science*, *56*(1), 35–42.
Tan, C., Xue, J., Lou, X., Abbas, S., Guan, Y., Feng, B., et al. (2014). Liposomes as delivery systems for carotenoids: comparative studies of loading ability, storage stability and in vitro release. *Food & Function*, *5*(6), 1232–1240.
Tavakoli, H., Hosseini, O., Jafari, S. M., & Katouzian, I. (2018). Evaluation of physicochemical and antioxidant properties of yogurt enriched by olive leaf phenolics within nanoliposomes. *Journal of Agricultural and Food Chemistry*, *66*(35), 9231–9240.
Taylor, T. M., Weiss, J., Davidson, P. M., & Bruce, B. D. (2005). Liposomal nanocapsules in food science and agriculture. *Critical Reviews in Food Science and Nutrition*, *45*(7–8), 587–605.
Ting, Y., Chiou, Y. S., Pan, M. H., Ho, C. T., & Huang, Q. (2015). In vitro and in vivo anticancer activity of tangeretin against colorectal cancer was enhanced by emulsion-based delivery system. *Journal of Functional Foods*, *15*, 264–273.
Tiwari, S. B., & Amiji, M. M. (2006). Nanoemulsion formulations for tumor-targeted delivery. In *Nanotechnology for cancer therapy* (pp. 724–740). Boca Ratón, FL. USA: CRC Press.
Trujillo-Cayado, L. A., Alfaro, M. C., García, M. C., & Muñoz, J. (2017). Physical stability of N, N-dimethyldecanamide/α-pinene-in-water emulsions as influenced by surfactant concentration. *Colloids and Surfaces B: Biointerfaces*, *149*, 154–161.
Trujillo-Cayado, L. A., Alfaro, M. C., & Muñoz, J. (2018). Effects of ethoxylated fatty acid alkanolamide concentration and processing on D-limonene emulsions. *Colloids and Surfaces A: Physicochemical and Engineering Aspects*, *536*, 198–203.
Trujillo-Cayado, L. A., Alfaro, M. C., Santos, J., Calero, N., & Muñoz, J. (2018). Influence of primary homogenization step on microfluidized emulsions formulated with thyme oil and Appyclean 6548. *Journal of Industrial and Engineering Chemistry*, *66*, 203–208.
Trujillo-Cayado, L. A., García, M. C., Santos, J., Carmona, J. A., & Alfaro, M. C. (2017). Progress in the formulation of concentrated ecological emulsions for agrochemical application based on environmentally friendly ingredients. *ACS Sustainable Chemistry & Engineering*, *5*(5), 4127–4132.
Trujillo-Cayado, L. A., Santos, J., Alfaro, M. C., Calero, N., & Muñoz, J. (2016). A further step in the development of oil-in-water emulsions formulated with a mixture of green solvents. *Industrial & Engineering Chemistry Research*, *55*(27), 7259–7266.
Trujillo-Cayado, L. A., Santos, J., Ramírez, P., Alfaro, M. C., & Muñoz, J. (2018). Strategy for the development and characterization of environmental friendly emulsions by microfluidization technique. *Journal of Cleaner Production*, *178*, 723–730.

Uluata, S., Decker, E. A., & McClements, D. J. (2016). Optimization of nanoemulsion fabrication using microfluidization: role of surfactant concentration on formation and stability. *Food Biophysics*, *11*(1), 52–59.

Uluata, S., McClements, D. J., & Decker, E. A. (2015). Physical stability, autoxidation, and photosensitized oxidation of ω-3 oils in nanoemulsions prepared with natural and synthetic surfactants. *Journal of Agricultural and Food Chemistry*, *63*(42), 9333–9340.

Upala, S., Yong, W. C., Theparee, T., & Sanguankeo, A. (2017). Effect of omega-3 fatty acids on disease severity in patients with psoriasis: a systematic review. *International Journal of Rheumatic Diseases*, *20*(4), 442–450.

Vahedikia, N., Garavand, F., Tajeddin, B., Cacciotti, I., Jafari, S. M., Omidi, T., et al. (2019). Biodegradable zein film composites reinforced with chitosan nanoparticles and cinnamon essential oil: physical, mechanical, structural and antimicrobial attributes. *Colloids and Surfaces B: Biointerfaces*, *177*, 25–32.

Vemuri, S., & Rhodes, C. T. (1995). Preparation and characterization of liposomes as therapeutic delivery systems: a review. *Pharmaceutica Acta Helvetiae*, *70*(2), 95–111.

Villalobos-Castillejos, F., Granillo-Guerrero, V. G., Leyva-Daniel, D. E., Alamilla-Beltrán, L., Gutiérrez-López, G. F., Monroy-Villagrana, A., et al. (2018). Fabrication of nanoemulsions by microfluidization. In *Nanoemulsions. Formulation, applications, and characterization* (pp. 207–232). London: Academic Press. Elsevier.

Wang, X., Wang, Y. W., & Huang, Q. (2009). Enhancing stability and oral bioavailability of polyphenols using nanoemulsions. *ACS symposium series*. (pp. 198–212). Oxford. UK: Oxford University Press.

Willemsen, K. L. D. D., Panozzo, A., Moelants, K., Debon, S. J. J., Desmet, C., Cardinaels, R., et al. (2017). Physico-chemical and viscoelastic properties of high pressure homogenized lemon peel fiber fraction suspensions obtained after sequential pectin extraction. *Food Hydrocolloids*, *72*, 358–371.

Wooster, T. J., Golding, M., & Sanguansri, P. (2008). Impact of oil type on nanoemulsion formation and Ostwald ripening stability. *Langmuir*, *24*(22), 12758–12765.

Xianquan, S., Shi, J., Kakuda, Y., & Yueming, J. (2005). Stability of lycopene during food processing and storage. *Journal of Medicinal Food*, *8*(4), 413–422.

Yabuzaki, J. (2017). Carotenoids database: structures, chemical fingerprints and distribution among organisms. *Database*, *2017*, .

Yalçınöz, Ş., & Erçelebi, E. (2018). Potential applications of nano-emulsions in the food systems: an update. *Materials Research Express*, *5*(6)062001.

Yan, B., Park, S. H., & Balasubramaniam, V. M. (2017). Influence of high pressure homogenization with and without lecithin on particle size and physicochemical properties of whey protein-based emulsions. *Journal of Food Process Engineering*, *40*(6), e12578.

Yang, Y., Marshall-Breton, C., Leser, M. E., Sher, A. A., & McClements, D. J. (2012). Fabrication of ultrafine edible emulsions: comparison of high-energy and low-energy homogenization methods. *Food Hydrocolloids*, *29*(2), 398–406.

Yerramilli, M., Longmore, N., & Ghosh, S. (2017). Improved stabilization of nanoemulsions by partial replacement of sodium caseinate with pea protein isolate. *Food Hydrocolloids*, *64*, 99–111.

Zeeb, B., Herz, E., McClements, D. J., & Weiss, J. (2015). Reprint of: impact of alcohols on the formation and stability of protein-stabilized nanoemulsions. *Journal of Colloid and Interface Science*, *449*, 13–20.

Zheng, B., Pen, S., Zhang, X., & McClements, D. J. (2018). Impact of delivery system type on curcumin bioaccessibility: comparison of curcumin-loaded nanoemulsions with commercial curcumin supplements. *Journal of Agricultural and Food Chemistry*, *66*(41), 10816–10826.

Further reading

Ding, W. K., & Shah, N. P. (2009). Effect of homogenization techniques on reducing the size of microcapsules and the survival of probiotic bacteria therein. *Journal of Food Science*, *74* (6), M231–M236.

García, M. C., Cabral, R., Muñoz, J., & Alfaro, M. C. (2017). Aplicación de alta presión sobre disoluciones acuosas de goma Diutan: influencia sobre las propiedades reológicas. *Afinidad*, *74*(577), 30–35.

Jo, Y. J., Chun, J. Y., Kwon, Y. J., Min, S. G., Hong, G. P., & Choi, M. J. (2015). Physical and antimicrobial properties of trans-cinnamaldehyde nanoemulsions in water melon juice. *LWT- Food Science and Technology*, *60*(1), 444–451.

Liu, F., Zhu, Z., Ma, C., Luo, X., Bai, L., Decker, E. A., et al. (2016). Fabrication of concentrated fish oil emulsions using dual-channel microfluidization: impact of droplet concentration on physical properties and lipid oxidation. *Journal of Agricultural and Food Chemistry*, *64*(50), 9532–9541.

Narsimhan, G., & Goel, P. (2001). Drop coalescence during emulsion formation in a high-pressure homogenizer for tetradecane-in-water emulsion stabilized by sodium dodecyl sulfate. *Journal of Colloid and Interface Science*, *238*(2), 420–432.

Santos, J., Alfaro, M. C., Trujillo-Cayado, L. A., Calero, N., & Muñoz, J. (2019). Encapsulation of β-carotene in emulgels-based delivery systems formulated with sweet fennel oil. *LWT—Food Science and Technology*, *100*, 189–195.

Willemsen, K. L. D. D., Panozzo, A., Moelants, K., Cardinaels, R., Wallecan, J., Moldenaers, P., et al. (2018). Effect of pH and salts on microstructure and viscoelastic properties of lemon peel acid insoluble fiber suspensions upon high pressure homogenization. *Food Hydrocolloids*, *82*, 144–154.

CHAPTER EIGHT

Production of food bioactive-loaded nanostructures by ultrasonication

Roya Koshani[a,b] **and Seid Mahdi Jafari**[b]
[a]Department of Chemistry, Quebec Centre for Advanced Materials, Pulp and Paper Research Centre, McGill University, Montreal, QC, Canada
[b]Department of Food Materials and Process Design Engineering, Faculty of Food Science and Technology, Gorgan University of Agricultural Sciences and Natural Resources, Gorgan, Iran

Contents

1 Introduction

Recent advances in the design and production of specialized nanostructures for nanoencapsulation of bioactive compounds have been directed attention toward the development of green and facile manufacturing techniques, which provide control over size, shape, and surface chemistry (Jafari, 2017; Ezhilarasi, Karthik, Chhanwal, & Anandharamakrishnan, 2013).

Nanoencapsulation of Food Ingredients by Specialized Equipment
https://doi.org/10.1016/B978-0-12-815671-1.00008-1

It has been well established that the employed fabrication procedures can influence and modulate the overall properties of nanoengineered delivery systems (de Souza Simões et al., 2017; Livney, 2015). The biochemical and technofunctional properties of such systems are dictated to a great extent by size of structure-forming particles (Tamjidi, Shahedi, Varshosaz, & Nasirpour, 2013). Indeed, due to reduced particle size and increased active surface area, nanostructured carriers exhibit a significant improvement in solubility, dissolution, bioavailability, controlled release, and targeted delivery of encapsulated constituents compared with macro- and microsized counterparts. They are also able to minimize undesirable chemical reactions and protect bioactive components against degradation while being passed through different gastrointestinal compartments (Joye, Davidov-Pardo, & McClements, 2014; Huang, Yu, & Ru, 2010).

In the food and nutraceutical sectors, various techniques are being applied in order to manufacture nanosized structures and systems for encapsulating bioactive agents based on "bottom-up" or top-down" strategies. The "bottom-up" approach generates highly ordered two- and three-dimensional nanoscale structures by spontaneously assembling atoms and/or molecules of materials. In fact, these techniques (this technique does not) do not depend on a high-energy system. By contrast, "top-down" approach is engineered by scaling down a complex entity of the materials into its component parts through high-energy devices. This downstream route requires mechanical forces that allow size reduction and shaping of the structure (Joye & McClements, 2014; Steed, Turner, & Wallace, 2007). Nevertheless a combination of these pathways can be more advantageous in nanocarrier synthesis.

Ultrasound technique is commonly used to form different types of nanostructures suitable for nanoencapsulation, which represents an outstanding example of "top-down" approach and a potential trigger of "bottom-up" approach. Ultrasonically generated cavitations are principally responsible for many physicochemical effects of ultrasound waves particularly size reduction of materials (Patil & Pandit, 2007). The role of cavitation is not only limited to synthesize but also expanded to control morphological features of nanomaterials and processing techniques such as nanoencapsulation, coating, and nanocomposites. More details are given in the next section.

Among existing technologies, ultrasonication owing to intrinsic advantages including safety, being environmentally friendly, straightforward operation, energy efficiency, and simple implementation, as well as scale-up potential, has been receiving a great deal of attention for >100 years. However, the interest in the employment of this technique for nanomaterial production was born only recently (Manickam & Rana, 2011; Soria & Villamiel, 2010).

Selection of the most appropriate technique to fabricate nanostructures is a vital step in nanoencapsulation purposes, because physicochemical and biological properties of encased and encasing components are closely dependent upon preparation strategies and the key parameters effect on the condition of process (Anandharamakrishnan, 2014).

This chapter provides an outlook on the fundamentals of ultrasonication and equipment settings. Different kinds of bioactive bearing nanostructures prepared by means of sonication, along with their advantages and disadvantages, are examined. Effect of ultrasonic operational parameters on yielded nanostructures, nanoencapsulation, and release process are also discussed. Last section includes some notes about potential future developments of sonication in different aspects of nanoencapsulation.

2 Ultrasonication: A versatile technique for nanoencapsulation

2.1 General principles (principal mechanisms)

Ultrasound is a series of sound waves (also known as acoustic waves) with frequencies above the human hearing range (>16 kHz), which can be classified into different frequency and intensity ranges. Firstly, low-intensity (high-frequency) ultrasound passes through a medium without causing any destructive physicochemical changes. In spite of operating on high frequencies ranging at 1–10 MHz, its power intensity is too low ($<1\ W/cm^2$) to alter the material. Secondly, high-intensity (low-frequency) ultrasound is characterized by high-power intensities of $10–1000\ W/cm^2$ and frequencies of 16–100 kHz. Therefore, low-intensity ultrasound is commonly employed for detection/diagnosis purposes in the medium (such as sonography), while high-intensity ultrasound is being increasingly used for such material-altering purposes as depolymerizing the macromolecules, downsizing particles to nanoscale, making emulsions (homogenization), deflocculating droplets and so on, especially in laboratories and food industries (Bermúdez-Aguirre, Mobbs, & Barbosa-Cánovas, 2011; Patist & Bates, 2008; Lorimer & Mason, 1987).

Since acoustically generated energy cannot be absorbed by molecules, it should be converted to a chemically usable form via the indirect and complex phenomenon recognized as acoustic cavitation (Kardos & Luche,

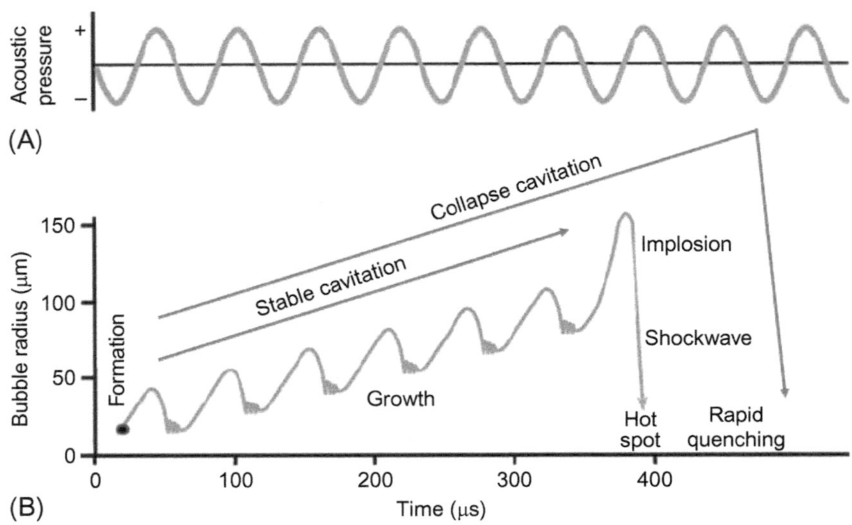

Fig. 1 Schematic representation of stable and collapse acoustic cavitations. *(Adapted from Suslick, K. S. (1995). Applications of ultrasound to materials chemistry. MRS Bulletin, 20(4), 29–34 with permission).*

2001). Acoustic waves passing through a liquid comprise of expansion and compression waves, which are considered to be pressure waves (Fig. 1A). Induced pressure by expansion waves can break the cohesion of liquid media and form a microbubble (a microcavity). The compression wave comes along and compresses this cavity; then, another expansion wave re-expands it. Eventually, an oscillating bubble going back and forth with the speed of 20,000 times a second comes out. The oscillating bubbles are capable of accumulating ultrasonic energy and simultaneously growing to a certain size (typically tens of micrometers) (Bang & Suslick, 2010).

It should be noted that collapse conditions are mainly dependent on the intensity and energy of ultrasound waves, along with medium viscosity, surface tension, vapor pressure, nature and concentration of dissolved gas, presence of solid particles, temperature, and pressure of the treatment (Soria & Villamiel, 2010; Brennen, 2014). Bubbles can only repeatedly oscillate without any implosion at low ultrasound intensities, which is named "stable cavitation." As ultrasound intensity goes up, bubbles begin to oscillate nonlinearly and implosively collapse, releasing the concentrated energy within the bubble on a very short time (Fig. 1B). Extremely high localized temperatures, high pressures, free radicals, and shock waves are the consequences of this type of bubble activity labeled to "collapse," "transient," or "inertial" cavitation, which provides the primary mechanism for sonochemical effects (Richardson, Pitt, & Woodbury, 2007; Suslick & Price, 1999).

Many physical effects of high-intensity ultrasound on different structures such as size distribution of particles and structural manipulations are attributed to cavitations, particularly collapse-based form (Richardson et al., 2007). Moreover, sonochemically generated free radicals by sonolysis of water accounts for chemical effects, which can react with and modify other molecules (Bang & Suslick, 2010; Soria & Villamiel, 2010).

Although acoustic cavitations play a key role in the production of nanomaterials, efforts are being made to find out the driving forces responsible for nanoparticle generation when a solution containing reactants is irradiated with ultrasound. The sonochemical reactions can be occurred in three different regions of a media as schematically shown in Fig. 2. The first place is the central environment of the bubble where the temperature and pressure are extremely high because of implosive collapse. The second region is interfacial zone where the conditions are lower than that of the core of the bubble and enough to induce a chemical reaction. The third region is the bulk liquid including ambient conditions.

One possible explanation of nanostructures formed by sonication is the fast kinetics associated with the rapid intense collapse conditions of the bubble, which impedes the growth of the nuclei and subsequently structural

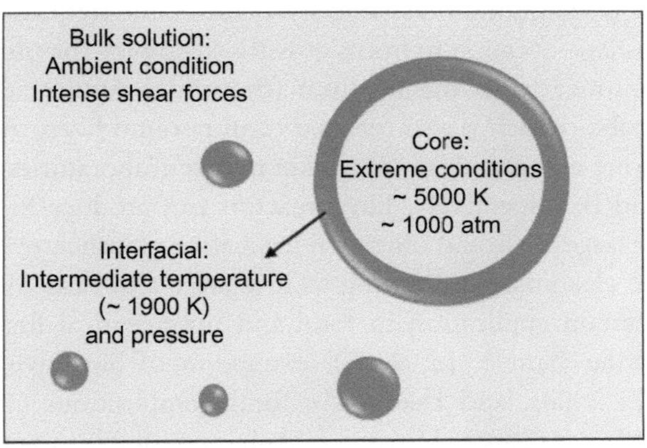

Fig. 2 Schematic representation of different regions occurring sonochemical reaction. *(Redrawn from Manickam, S., & Rana, R. K. (2011). Production of nanomaterials using ultrasonic cavitation—A simple, energy efficient and technological approach. In* Ultrasound technologies for food and bioprocessing *(pp. 405–444). Springer, New York, NY, with permission).*

organization or crystallization. This may yield to the nanostructured products. The reaction can as well happen in the interfacial region when the precursor is nonvolatile and unable to enter into the bubble. Nanoamorphous or nanocrystalline products are formed depending on the temperature of this region. Formation of nanostructures is even feasible in the third region, albeit slowly (Gedanken, 2004; Manickam & Rana, 2011).

2.2 Ultrasound operating systems

Sonication alone or in combination with most nanoencapsulation methods is applied for breaking up the aggregates of particles and polymeric structures, thereby reducing their size to nanometer scales (Fathi, Martin, & McClements, 2014). Given that the energy provided by cavitation is approximately 10–100 kJ/mol, only hydrogen bonds can be targeted by this technique (Tischer, Sierakowski, Westfahl, & Tischer, 2010). Equipment settings play an important role in the operation and optimization of sonication process for achieving these purposes.

A variety of specialized apparatuses with different setups are commercially available for sonochemical reactions. Examples include ultrasonic cleaning baths, direct immersion ultrasonic horns, and flow reactors for continuous processes. A typical ultrasonic equipment mainly consists of (i) the electrical generator producing electrical current with a specified power rating, (ii) a single/multiple piezoelectric transducer converting electrical energy into sound energy by vibrating mechanically at ultrasonic frequencies, and (iii) an emitter in shape of titanium horns or baths to convey the ultrasonic wave from the transducer into the medium (Bermúdez-Aguirre et al., 2011). Ultrasonic probe, which is also referred to ultrasound horn, and ultrasonic cleaning bath are extensively employed in research laboratories, as displayed in Fig. 3A and B, respectively. Flow reactors can produce high ultrasonic intensities for large-scale and industrial irradiations (Suslick & Price, 1999).

Ultrasonic cleaning baths owing to low price and early availability are applied for certain application in food and nutraceutical fields including freezing/thawing (Sun & Li, 2003), extraction of bioactive compounds (Aleixo et al., 2004), and changes in food compositions (Joyce, Phull, Lorimer, & Mason, 2003). However, their undesirable functions such as power loss over time, nonuniform transmission of ultrasound, unadjustable power, and difficult control of temperature in the bath have led to low reproducible results (Moreira, Morais, & Delerue-Matos, 2017).

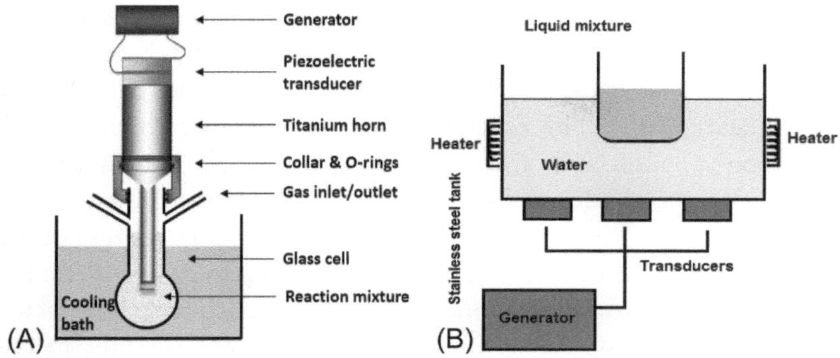

Fig. 3 Schematic representation of probe (A) and cleaning bath (B) ultrasounic systems. *(Adapted from Bang, J. H., & Suslick, K. S. (2010). Applications of ultrasound to the synthesis of nanostructured materials.* Advanced Materials, *22(10), 1039–1059 and redrawn from Mason, T. J. (2002). High powered ultrasound in physical and chemical processing. In New acoustics. Selected topics. CSIC Madrid with permission, respectively).*

Note that cleaning baths have an insufficient intensity to produce nanostructured materials (Bang & Suslick, 2010). On the contrary the commonly used probe-based sonicators show a reproducible manner and permit parameters such as sonication power, frequency, and time to be selected as required (Moreira et al., 2017). Nonetheless, the high-power sonication horns can result in the generation of considerable heat and shear forces, which can chemically destabilize sensitive and labile bioactive components (or molecules) via thermal decomposition or the development of free radicals. The rise in temperature can be feasibly controlled by using a cooling jacket, operating in cycles, and reducing the sonication time (Gharibzahedi & Jafari, 2018).

High-intensity ultrasonic horns mostly operate at a frequency of 20 or 40 kHz, while frequencies close to 40 kHz are provided for cleaning baths. There are also different ultrasound setups having the capability to work in the frequency regimes of few hundred kilohertz to few megahertz. Previous studies have shown that cavitational heating of collapsing bubbles can be occurred over the full frequency range, whereas ultrasonically induced physical effects, for instance, emulsification process or nanomaterial generation, are dominantly appeared in lower frequencies (Bang & Suslick, 2010).

3 Lipid-based and surfactant-based nanostructures produced by ultrasonication

Lipid-based nanoengineered structures are rapidly expanding as the best candidates for nanoencapsulation systems due to their exceptional bioattributes including abilities to enhance gastrointestinal solubilization and oral bioavailability of biological active compounds, as well as biocompatibility (Cerqueira et al., 2014) (Akhavan, Assadpour, Katouzian, & Jafari, 2018; Rafiee & Jafari, 2018). The majority of food-grade bioactive components including polyphenols, fatty acids, vitamins and minerals, aromas, and antimicrobial agents are hydrophobic in nature (Faridi Esfanjani, Assadpour, & Jafari, 2018; Rezaei, Fathi, & Jafari, 2019). For the accommodation of these agents, lipid-based nanocarriers are required enabling to protect them from adverse process/environmental conditions and physicochemical interactions through gastrointestinal tract (McClements, 2012; McClements, Decker, Park, & Weiss, 2009).

Moreover, extensive research has been conducted on vesicular nanostructures as a new class of nanocapsules, which the bioactive agents can be incorporated into a cavity surrounded by a lipid or surfactant-based layer (Anandharamakrishnan, 2014).

Several types of delivery systems consisting of lipid and surfactant molecules such as nanoliposomes, nanoemulsions, nanostructured lipid carriers, niosomes, cubosomes, and hexosomes that are mostly employed for nanoencapsulation will be explained in the upcoming subsections. However, due to similarities in constituents forming these nanovehicles, a quite distinct classification based on surfactant and/or lipid can become somewhat tricky.

3.1 Nanoliposomes

Nanometric version of liposomes referred to as nanoliposomes is vesicular structures produced by increasing the amount of energy input to the same combination of constituents of liposomes in an aqueous environment (Ghorbanzade, Jafari, Akhavan, & Hadavi, 2017; Tavakoli, Hosseini, Jafari, & Katouzian, 2018). The main component of such lipid vesicles are amphiphilic phospholipids having a propensity to form bilayer-shaped structures by self-assembly process (Mozafari, Johnson, Hatziantoniou, & Demetzos, 2008). The interior aqueous phase of nanoliposomal carriers acts as a reservoir for hydrophilic compounds, whereas hydrocarbon tail-bearing bilayers offer a suitable environment for the incorporation of low solubility and hydrophobic nutrients (Briuglia, Rotella, McFarlane, & Lamprou, 2015). Nanoliposomes have many physicochemical similarities to liposomes, albeit with higher bioavailability and smaller mean diameter (Demirci, Caglar, Cakir, & Gülseren, 2017).

Resemblance to natural cell membranes, simultaneous delivery of both hydrophilic and lipophilic components, long blood circulation time, biodegradability, biocompatibility, and relatively nontoxicity are other potential advantages of liposomes at micro and nanometer scales (Livney, 2015). Moreover, reports exist about the biologically beneficial impacts of phospholipids and related complex lipids of liposomes and nanoliposomes for humans including liver protection, memory enhancement, canner treatment, and reduction of intestinal cholesterol absorption (Mozafari et al., 2008). Interestingly, nanoliposomes can be fabricated by using food-grade raw materials such as egg, soy, or milk as a rich source of phospholipids (Thompson, Mozafari, & Singh, 2007). These capabilities have encouraged food and nutrition scientists to employ and develop them extensively for the protection and delivery of bioactive agents. However, sensitivity to shear and environmental stresses, for instance, osmotic pressure differences and difficulties in scaling up, are disadvantages of liposomal systems (Livney, 2015).

Different methodologies are described in the literature for the fabrication of liposomes: solvent evaporation, electroformation, thin-film dehydration/rehydration, proliposome, membrane extrusion, dialysis, and high-pressure homogenization (Isailović et al., 2013; Mozafari et al., 2008; Taylor, Weiss, Davidson, & Bruce, 2005). Sonication or exposure to ultrasound is the most widely used technique for reducing the size of liposomes and producing nanoliposomes. Nevertheless, little information is available about the exact mechanism of the effects of ultrasound on liposome dispersions. Liposomes can be appeared as unilamellar or multilamellar vesicles ranging from approximately 200 nm to several microns. If the size is <150 nm, then they are called nanoliposome (Jafari, 2017; Livney, 2015).

After making micrometric multilamellar vesicle type liposomes via conventional methods, a probe sonicator can be employed by inserting the tip in the flask containing vesicles for 10–15 min with on and off intervals (cycles) to obtain nanoliposomes. Ultrasonically generated nanoliposomes are predominantly in the form of small unilamellar vesicles (Panahi et al., 2017). Besides, immersing the flask in a bath sonicator stowed with water makes nanoliposome formation possible but in 30–45 min (Davaran et al., 2014). It is worthy to note that extensive sonication is required to reach minimum size limit of small unilamellar vesicles in bath systems, which is not always possible (Akhavan et al., 2018). It has been accepted that encapsulation efficiency and loading content are dependent on the size and specific surface areas of the liposomes (Feller, Gawrisch, & MacKerell, 2002). High-intensity ultrasound irradiation can impose its influence on liposomal dispersions by extreme pressure and temperature condition as well as highly turbulent flow arising from collapse cavitation. High shear forces cause disruption of the multilamellar vesicle structures, which in turn leads to leakage of encased components from liposomes. Moreover, the resulting small unilamellar vesicles provide less space for entrapment of lipophiles like resveratrol into the bilayer membrane and hydrophiles in the cores of ultrasonically degraded liposomes (Isailović et al., 2013; Taylor et al., 2005). A study reported a greater encapsulation efficiency of nanoliposomes carrying polyunsaturated fatty acids in comparison with liposomes. Indeed, size reduction increases the surface area of the particles, which can raise the chance of contact and bond between the fatty acids and phospholipids in the bilayers (Rasti, Jinap, Mozafari, & Yazid, 2012).

Nanoliposome preparation by probe-based sonication has some possible disadvantages such as probable degradation of bioactive materials owing to localized overheating, constant cooling requirement, low storage stability, and metal pollution resulting from probe tip (Akhavan et al., 2018; Yu & Tang, 2016). Despite all difficulties, a probing process system still has superiority to other equipment-based techniques because it can be employed directly on hydrated multilamellar vesicles and led to form more homogenous dispersion of small unilamellar vesicles (Torchilin & Weissig, 2003). Table 1 summarizes process conditions and characteristics of different nanoliposomes bearing bioactive compounds produced by sonication probe systems.

3.2 Nanoemulsions

Nanoemulsion-based delivery systems are one of the most widely used vehicles for encapsulating bioactive compounds due to their relatively simple preparation with commonly available processing equipment (Raikos & Ranawana, 2017; McClements, Decker, & Weiss, 2007). Nanoemulsions are metastable oil-in-water (O/W) or water-in-oil (W/O) dispersions with droplet diameters ranging from 50 to 200 nm (transparent) and up to 500 nm (translucent) (McClement, 2012). Exclusive advantages of nanoemulsions over ordinary emulsions including high physical stability against sedimentation and creaming, high bioavailability, lower emulsifier requirement, low turbidity, and scale-up potential endow them with a great potential for application in food, cosmetic and pharmaceutical industries (Jafari, Fathi, & Mandala, 2015; Ghosh, Mukherjee, & Chandrasekaran, 2013).

Table 1 Some selected studies on nanoencapsulation of food bioactive compounds by ultrasonically prepared nanoliposomal systems.

Bioactive ingredient	Sonication parameters	Encapsulation results	Carrier size (nm)	Ref.
Curcumin	Frequency: 40 kHz Power: 40% of full power Time: 6 min	• Encapsulation efficiency: ~56.2% • Increased the solubility and the bioavailability of curcumin • High stability for >1 month • Strong candidate for prevention or treatment of neurodegenerative disorders	~ 114.37	Hasan et al. (2018)
Fish oil	Frequency: 20 kHz Power: 160 W Time: 7 min Temperature: 25°C	• Yogurt fortification • Significant reduction in acidity, syneresis, and peroxide value of yogurt • High content and stability of DHA and EPA after 21 days storage	~ 300–500	Ghorbanzade et al. (2017)
Nisin and Garlic extract	Frequency: 20 kHz Power: 250 W Time: 5 min	• Encapsulation efficiency: > ~80% • Improved antimicrobial activity	~ 130–180	Pinilla and Brandelli (2016)
Carotenoids	Power: 240 W Time: 10 min	• Encapsulation efficiency: > ~80% • High stability during preparation and gastrointestinal incubation	~ 60–80	Tan et al. (2014)
Resveratrol	Power: 500 W Amplitude: 40% Time: 60 min	• Encapsulation efficiency: 44%–55% • Prolonged release	~ 120–290	Isailović et al. (2013)
Polyunsaturated fatty acids (docosahexaenoic acid and eicosapentaenoic acid)	Frequency: 20 kHz Power: 120 W Time: 7 min	• Encapsulation efficiency: > ~80% • High physical and oxidative stability	~ 50–200	Rasti et al. (2012)
Vitamin C	Power: 20 W Time: 2 min	• Encapsulation efficiency: (~ 96%) • Coated with chitosan • Improved storage stability for 15 weeks • Enhanced protection against oxidation (over 85% of vitamin C)	~ 96–97	Liu and Park (2010)

Hydrophilic bioactive agents like polyphenols (Mohammadi, Jafari, Assadpour, & Esfanjani, 2016) can be incorporated in W/O nanoemulsion, while a O/W nanoemulsion is capable to encapsulate lipophilic agents such as sterols, carotenoids (Sotomayor-Gerding et al., 2016), oil-soluble vitamins (Lv et al., 2018), and essential fatty acids (Lane, Li, Smith, & Derbyshire, 2014).

Being nonequilibrium systems, nanoemulsions cannot be formed spontaneously and therefore need the energy input provided by specialized mechanical devices or from the chemical potential of compounds (Gutiérrez et al., 2008). Having less control over the particle size distribution and stability of nanoemulsions as well as high-energy consumption poses limitations against employing high-speed agitators and high-pressure homogenizers. On the other side, ultrasound-assisted emulsification is documented as a simple, non-toxic, reproducible, environmentally friendly, and efficient high-energy method for producing nanoemulsions enabling to break the interface and downsize the emulsion droplets to nanometric scale (Ramisetty, Pandit, & Gogate, 2015; Shanmugam & Ashokkumar, 2014; Rao & McClements, 2011). Ultrasound-assisted emulsification is believed to happen by two mechanisms: first, mechanical vibrations arising from transferring sound waves through a liquid medium and, second, the intense turbulence and microjets as a result of collapse cavitation phenomenon, which cause to the disruption of the coarse droplets and the eruption of the oil phase into the aqueous medium (Ghosh et al., 2013; Kentish et al., 2008). Numerous studies have shown that a variety of factors impact the efficiency of ultrasound devices and consequently the physicochemical properties of yielded nanoemulsions, including the applied frequency, power, amplitude, time, plus the setup of ultrasonic processor (Shao et al., 2018; Aday, Temizkan, Büyükcan, & Caner, 2013). Recently developed configurations of ultrasound equipment along with their action mechanism and advantages have been presented in a chapter of published book by Gharibzahedi and Jafari (2018). In general, batch ultrasonic systems are effective on just $<2\%$ of the liquid volume and long homogenizing time required to achieve nanoemulsion with desired droplet diameter. Therefore, continuous-flow homogenizers are mostly preferred especially in industrial settings because of shorter exposure time of fluids to acoustic fields.

A time-based study was carried out on ultrasonic-assisted formation of eucalyptus oil nanoemulsions by Sugumar, Ghosh, Nirmala, Mukherjee, and Chandrasekaran (2014). Frequency of 20 kHz, maximum power of

750 W, amplitude of 40%, and different sonication times with 5-min intervals from 0 to 30 min applied on nanoemulsions revealed a decrease in droplet size and polydispersity index as well as the alteration in white milky appearance to transparent by increasing sonication time. No phase separation, creaming, and change in the droplet diameter were occurred in nanoemulsions subjected to sonication for 25 and 30 min (Sugumar et al., 2014). Moreover, the droplet size can be decreased in an ultrasonic homogenizer by increasing the amplitude in shorter times similar to that one exposed to lower amplitude for longer time periods. In fact, more energy dissipates into the system at higher amplitudes, which remarkably elevates cavitational collapse pressure, thereby providing minimum droplet size with a high stability (O'Sullivan & Norton, 2016; Jadhav, Holkar, Karekar, Pinjari, & Pandit, 2015). The amplitude of 40% after 5 min led to the emergence of completely stable nanoemulsion, while the frequency and power were set on 20 kHz and 750 W, respectively. However, stable nanoemulsions appeared at the amplitudes of 30% and 20% within 10 and 15 min of sonication time, respectively (Carpenter & Saharan, 2017).

A similar trend has been reported that higher-power levels at shorter times facilitate the fabrication of stable nanoemulsions in both bath (Kaltsa, Michon, Yanniotis, & Mandala, 2013; Delmas et al., 2011) and continuous-based (O'Sullivan & Norton, 2016) ultrasound systems. It must be taken into consideration that extremely high levels of power may actually enlarge the droplet size owing to the enhanced temperature and/or the reduced functionality of surfactants. More importantly, only physical effects of ultrasonication were dominant during emulsification process under above conditions (Carpenter & Saharan, 2017). However, some studies have reported the chemical effects of free radicals released in acoustic cavitation on produced emulsions, which caused to the lipid oxidation (Tang, Shridharan, & Sivakumar, 2013; Chemat, Grondin, Sing, & Smadja, 2004; Pandit & Joshi, 1993). Ultrasound is also utilized to donate antimicrobial property to nanoemulsions or intensify it. Larger active surface area allows rapid incorporation of antimicrobial active compounds and great interaction of droplets with bacterial membranes leading to faster delivery of encased bioactives with a high ratio (Noori, Zeynali, & Almasi, 2018; Gul, Saricaoglu, Besir, Atalar, & Yazici, 2018; Sugumar et al., 2014). Table 2 shows some recent researches on ultrasonically prepared nanoemulsions carrying bioactive materials.

Table 2 Process conditions and functionalities of ultrasound-assisted nanoemulsions carrying bioactive compounds.

Bioactive compound	Ultrasonic setup	Formulation parameters: Oil phase (%)	Formulation parameters: Surfactant/ stabilizer (%)	Operating parameters[b]: UF (kHz)	UP (W)	UA (%)	ST (min)	UT (°C)	Droplet size (nm)	Functionalities	Ref.
Mustard oil	Ultrasonic probe (batch type)	10	Tween 80 Span 80 (8%)	20	750	40	30	< 5	87.38	• High stability against mechanical and thermal stresses • Long-term stability (up to 3 months) • No chemical effects on oil structure	Carpenter and Saharan (2017)
Nettle essential oil (NEO)	Ultrasonic processor with a titanium sonotrode probe	10 (Canola and NEO)	Tween 40 (20% wt)	24	165.87		23.64	–	86.75	• No substantial difference in the droplets growth between 4°C and 25°C • The highest killing effect on the yeast • High antibacterial potential against the Gram-negative • Antioxidant activity	Ghorbanzade et al. (2017)

Ginger essential oil (GEO)	Sonotrode containing a piezoelectric crystal	5		Tween 80 (30% wt of ginger)	20	200	–	5	$\Delta T^a < 10$	57.4	• Increase the shelf life of chicken breast by significant decreasing of total aerobic psychrophilic bacteria and molds and yeasts • The highest total acceptance of nanoemulsion-coated fillets	Noori et al. (2018)
Clove essential oil (CEO) Hazelnut protein	Ultrasonic probe (batch type)		Glycerol (30% *w*/w of protein content) + 3% (CEO)	Tween 80 (2% *v*/v of essential oil)	–	750	50 75 100	2 4 6	25	160–248	• Capable of film formation • Increased tensile strength of films with ultrasound treatment • More homogeneous structure • Improved mechanical, barrier, and antimicrobial properties	Gul et al. (2018)

Continued

Table 2 Process conditions and functionalities of ultrasound-assisted nanoemulsions carrying bioactive compounds.—cont'd

		Formulation parameters		Operating parameters							
Bioactive compound	**Ultrasonic setup**	**Oil phase (%)**	**Surfactant/ stabilizer (%)**	**UF (kHz)**	**UP (W)**	**UA (%)**	**ST (min)**	**UT (°C)**	**Droplet size (nm)**	**Functionalities**	**Ref.**
Eugenol	Ultrasonic apparatus (batch type)	10	Tween 20 (1% wt/v)	–	450	–	6	–	80–100	• Great storage stability and thermal stability • Excellent antioxidant capacity and antimicrobial activity	Shao et al. (2018)

[a]Temperature difference between initial coarse emulsions to final nanoemulsion.
[b]UF, ultrasound frequency; UP, ultrasound power; UA, ultrasound amplitude; UT, ultrasound temperature; ST, sonication time.

3.3 Nanostructured lipid carriers (NLCs)

NLCs are consisted of a mixture of solid and liquid lipids plus the bioactive compounds dispersed in a surfactant solution, which ultimately remains in the solid state at room/body temperature. As a result of partial crystallization of oil droplets, many amorphous crystals are immersed within the inner core of NLCs creating larger empty spaces for bioactive compounds. Besides, these less ordered structures reduce the expulsion rate of accommodated components over the storage period. The fluidity of lipids in the central region results in better solubilization of bioactive ingredients and consequently higher payload capacity of NLCs. The enclosed compounds are protected from physical, chemical, and microbial degradations by confinement of them in a solid lipid matrix (Katouzian, Esfanjani, Jafari, & Akhavan, 2017; Tamjidi et al., 2013).

The superiority of NLCs over solid lipid nanoparticles (SLNs) is the anti-recrystallization characteristics of lipids forming NLCs minimizing the expulsion of bioactive payload and higher available space for bioactive compounds (Müller, Petersen, Hommoss, & Pardeike, 2007). As an efficient alternative to conventional nanovehicles (nanoliposomes, nanoemulsions, and biopolymeric nanoparticles), NLCs offer several exclusive features including straightforward and safe procedure, higher colloidal stability (because of the higher density of solid lipid), dispersibility in aqueous-based formulations, good physicochemical stability due to the less mobility of guest molecules in their solid matrix, slow and controlled release, entrapment of both lipophilic and hydrophilic nutraceuticals, biocompatible and biodegradable lipid matrices, and low cytotoxicity. However, raw material supply and environmental determinants are considered the obstacles of applying NLCs in the food sectors (Akhavan et al., 2018; Park, Garcia, Shin, & Kim, 2017).

Different types of NLCs are obtained depending on the formulation composition and method of preparation, which are classified into (I) the imperfect type; (II) the amorphous type; and (III) the multiple types with different particle sizes ranging from 50 to 1000 nm, as displayed in Fig. 4. Since the ongoing crystallization process increases the expulsion rate of bioactive molecules in lipid nanostructures, the amorphous type of NLCs can dispel this problem by avoiding crystallization (Jaiswal, Gidwani, & Vyas, 2016).

High-pressure homogenization (Lv et al., 2016), high shear homogenization (Manea, Vasile, & Meghea, 2014), microemulsion technique (Joshi, Pathak, Sharma, & Patravale, 2008), and solvent diffusion method

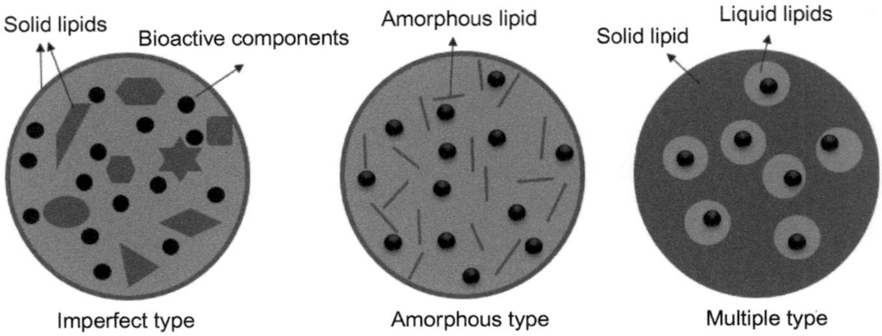

Fig. 4 Different types of bioactive compound-loaded NLCs (light and dark browns illustrate the liquid oil phase and the solid lipid, respectively). *(Redrawn from Katouzian, I., Esfanjani, A. F., Jafari, S. M., & Akhavan, S. (2017). Formulation and application of a new generation of lipid nano-carriers for the food bioactive ingredients. Trends in Food Science & Technology, 68, 14–25 with permission).*

(Hejri, Khosravi, Gharanjig, & Hejazi, 2013) are various developed techniques for NLC fabrication. Ultrasonic emulsification is another method used for the preparation of lipid nanospheres that besides being a simple procedure, it has no need to organic solvents (Zhao et al., 2010). In order to produce NLCs, at first, bioactive materials are dissolved in a blend of melted lipids and then dispersed in a hot surfactant solution. The resulted pre-emulsion is homogenized by mechanical stirring or ultrasonication to obtain a hot O/W nanoemulsion. The emulsion droplets are imperfectly crystallized during cooling step. Ultrasonically provided energy leads to the particle size reduction of the core lipid emulsion (Puglia et al., 2008; Müller et al., 2007). Indeed, it breaks the lipid mixture into small droplets, which lead to higher solubility of the oil in the solid. These fine oily compartments precipitate in the solid lipid matrix during the cooling process. The horn tip erosion and corresponding metal release into solution should be carefully regarded in terms of safety-related concerns and quality deterioration of NLCs arising from lipid oxidation during storage (Liu & Wu, 2010; Schenk et al., 2004). When NLC-encapsulated bioactives are prone to the attack of radicals produced in lipid oxidation reactions, increasing the ratio of saturated lipids (solid matrix) to unsaturated ones can be effective by firmer confinement and more protection of them (Boon, McClements, Weiss, & Decker, 2010). It is also reported that probe sonication operating as a batch processing system generally yields NLCs with higher polydispersity index than those fabricated through continuous setups, for instance, high-pressure homogenizer. The values of polydispersity index

(PDI) varying between 0.1 and 0.25 express an acceptable narrow size distribution for the long-term stability of nanodispersions (Tamjidi, Shahedi, Varshosaz, & Nasirpour, 2014).

In recent years, NLCs have been employed for improvement of chemical stability and bioavailability of many food bioactive compounds. Table 3 gives information regarding some of the bioactive bearing NLCs produced utilizing ultrasonication. By virtue of ultrasound homogenization, Lasoń, Sikora, Ogonowski, Tabaszewska, and Skoczylas (2016) manufactured terpene-loaded NLCs using modified beeswax as a solid lipid and caprylic/capric triglyceride as a liquid lipid. The average particle size for NLCs incorporating mono-(geranic acid), di-(forskolin), and tri-(ursolic acid) terpenes was 101.1, 106.4, and 97.6 nm, respectively. The obtained zeta potential values (close to −30 mV) proved a high physical stability of NLCs. In addition, in vitro release studies revealed that terpenes with lower lipophilicity are able to remain within the lipid nanoparticles for a longer time, whereas higher lipophilic terpenes are released easier from the NLC matrix. Another stable NLC dispersion incorporating lutein was prepared by sonication technique. The formulation comprised solid lipids (Precirol, Myverol, and Pluronic), lutein in corn oil, and emulsifiers. The obtained NLCs indicated an imperfect crystalline lattice and a spherical morphology along with 90% entrapment efficiency. As the time of ultrasonication increased from 2 to 10 min, the particle size of the NLCs decreased from 228 to 130 nm. NLCs were capable of effectively protecting the encapsulated lutein in simulated gastric fluid and slowly released lutein in simulated intestinal fluid (Liu & Wu, 2010).

3.4 Niosomes

Nonionic surfactant-based vesicular nanocarriers, the so-called niosomes, are mainly formed from the self-association of nonionic amphiphiles in aqueous media resulted in bilayer structures similar to liposomes. Cholesterol and its derivatives as one of the most important additive agents attach to the hydrophilic head of surfactants by hydrogen bonds (Fig. 5A), thereby facilitating the self-assembly process (Moghassemi & Hadjizadeh, 2014; Biswal, Murthy, Sahu, Sahoo, & Amir, 2008). Niosomes were discovered by the L'Oreal cosmetic company in the late 1970s (Handjani-Vila, Ribier, Rondot, & Vanlerberghie, 1979), but their applicability as nanocarrier systems in nutraceutical and pharmaceutical sectors has been broadly exploring. They are physically analogous to phospholipid vesicles (liposomes)

Table 3 Food bioactive compounds encapsulated within ultrasound-assisted SLNs and NLCs.

Bioactive agents[a]	Method of preparation	Composition		Ultrasound conditions		Particle size (nm)	Achievements	Ref.
				Configuration	Process parameters[a]			
Lycopene	High shear homogenization and ultrasonication	Glyceryl distearate/ caprylic–capric triglyceride	Tween 80	Ultrasound generator	Power output: 25% ST: 225 s	~<200	• Encapsulation efficiency (>64%) • Any chemical reaction with coating materials • Good physical stability • No remarkable leakage during storage • No difference between fortified sample with lycopene nanocarriers and blank orange drink	Zardini, Mohebbi, Farhoosh, and Bolurian (2018)
Epigallocatechin-3 gallate and Folate	High shear homogenization and ultrasonication	Precirol ATO 5/Mygliol 812	Tween 60	Probe sonicator Vibra-Cell	UP: 130 W UA: 70% ST: 5 min	~ 350	• Encapsulation efficiency (>80%) • Enhanced oral absorption • Increased bioavailability • Storage stability up to 8 weeks • Enhanced nutraceuti cal effects	Granja et al. (2017)

β sitosterol	High-speed homogenization and ultrasonication	Propolis wax/ pomegranate seed oil	Tween 80	Probe sonicator	UP: 250 W ST: 8 UT ~ melting point of lipids	~ 100	• Narrow particle size distribution • Effective in lowering the risk for cardio-vascular diseases	Soleimanian, Goli, Varshosaz, and Maestrelli (2018)
Sesamol	High-speed homogenization and ultrasonication	Compritol® 888 ATO/sesame oil	Lutrol F68	Probe sonicator (lab scale for larger samples)	UP: 400 W UF: 24 kHz	~ 200	• High encapsulation efficiency (>90%) • Enhanced and prolonged antioxidant activity of sesamol • Transformation of sesamol crystal to an amorphous form dispersed in the lipid matrix	Puglia et al. (2017)

Continued

Table 3 Food bioactive compounds encapsulated within ultrasound-assisted SLNs and NLCs.—cont'd

Bioactive agents	Method of preparation	Composition		Ultrasound conditions		Particle size (nm)	Achievements	Ref.
				Configuration	Process parameters			
Astaxanthin	Melt-emulsification and ultrasonication	Glyceryl behenate/oleic acid	Tween 80	Bath sonicator Followed by probe sonicator	ST: 15 min UP: 100 W UA: 50% ST: 4 min (2 s on—2 s off)	~ 94	• Physical and storage stability • New crystalline lattice with low crystallinity for NLC • An appropriate carrier system for the delivery of astaxanthin and other lipophilic compounds into aqueous-based foods and transparent beverages	Tamjidi et al. (2014)
Mono, di and triterpene	Ultrasound homogenization	Modified beeswax/caprylic–capric triglyceride	Alkyl polyglucoside	Probe-type sonicator	ST: 5 min	~ 100	• Terpene- loaded NLCs with high physical stability • Suitable nanocarriers for burst and sustained release delivery systems of terpenes	Lasoń et al. (2016)

Curcumin	Ultrasound homogenization	Shea butter	Tween-80 and Span 80	Probe-type sonicator	ST: 6 min (1 s On- 1 s Off) 40% of maximum power	~ 50–230	• A natural, cheap, and easily prepared nanovector • Curcumin-enriched shell model • Higher encapsulation efficiency for SLNs with smaller diameter and larger surface area	Hajj Ali et al. (2016)

[a]UF, ultrasound frequency; UP, ultrasound power; UA, ultrasound amplitude; UT, ultrasound temperature; ST, sonication time.

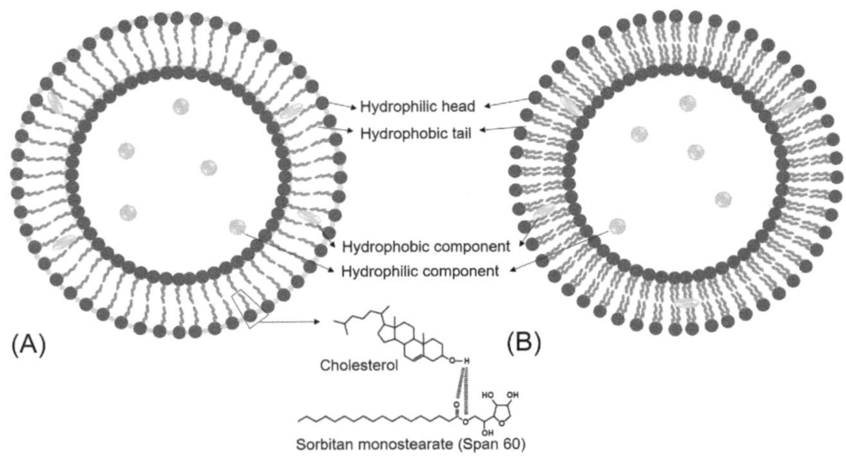

Fig. 5 Schematic representation of niosomes along with the position of cholesterol molecules in their bilayer membrane (A) compared with liposomes (B). *(Inspired from Desmet, E., Van Gele, M., & Lambert, J. (2017). Topically applied lipid-and surfactant-based nanoparticles in the treatment of skin disorders. Expert Opinion on Drug Delivery, 14(1), 109–122).*

(Fig. 5) and also capable of incorporating both hydrophilic and hydrophobic substances owing to their unique structure. In fact, hydrophiles can be adsorbed within vesicular aqueous core, while the lipophiles are localized by partitioning into the lipophilic domain of the bilayers (Abaee & Madadlou, 2016; Pardakhty & Moazeni, 2013). However, the main structural component is entirely different for niosomes and liposomes including double-chain phospholipids and single-chain surfactants, respectively (Desmet, Van Gele, & Lambert, 2017).

Besides having all the advantages of liposomes, lower price of surfactants, more osmotically active, and higher chemical and physical stability due to the nonionic surfactants instead of phospholipids that are more vulnerable to heat and oxidation degradation plus longer storage time of niosomes have led to their exploitation as alternatives to phospholipid-based liposomes (Tarekegn, Joseph, Palani, Zacharia, & Ayenew, 2010; Uchegbu & Vyas, 1998). In spite of being a superb vehicle for delivery purposes, niosomes might be at the risk of several physicochemical instabilities such as aggregation, fusion, leakage, and hydrolysis of encapsulated constituents that can be counted as drawbacks (Abaee & Madadlou, 2016; Verma, Singh, Syan, Mathur, & Valecha, 2010). Therefore, complementary processes are required to modulate the characteristics of bioactive compound-conveying niosomes.

The particle size has a profound impact on biological fate of vesicles and their entrapped molecules, as well as their physical properties, particularly, membrane permeability, penetration, stability, and toxicity (Essa, 2014; Goyal et al., 2005; Litzinger, Buiting, van Rooijen, & Huang, 1994). Vesicles, irrespective of their types, are categorized in three major forms according to their size range: (I) small unilamellar vesicles (SUV) (10–100 nm), (II) large unilamellar vesicles (LUV) (100–3000 nm), and (III) multilamellar vesicles (MLV) where more than one bilayer is formed (Kumar & Rajeshwarrao, 2011). Although most niosomes are in the nano- or submicron size range, the "nanoniosome" or "nanovesicle" term has not been used in published articles, which is possibly because of the introduction of new nanotechnology-associated phrases during the past few years (Pardakhty & Moazeni, 2013; Steed et al., 2007).

It is noteworthy that niosomes, similar to the aforementioned lipidic nanostructures, cannot be made spontaneously and therefore require an energy input for the arrangement of surfactant and/or lipid molecules. The widely reported preparation techniques of niosomes are thin-film hydration (Nematollahi, Pardakhty, Torkzadeh-Mahanai, Mehrabani, & Asadikaram, 2017), hand shaking (Marwa, Omaima, Hanaa, & Mohammed, 2013), the bubble (Verma et al., 2010), ether injection (Kumar & Rai, 2011), reverse phase evaporation (Abdelkader et al., 2012), heating (Mozafari, Reed, & Rostron, 2007), freezing/thawing (Abdelkader, Ismail, Kamal, & Alany, 2011), dehydration/rehydration (Uchegbu & Vyas, 1998), and proniosome technology (Hu & Rhodes, 2000). Ultrasonication is one of the most popular equipment-based methods for fabricating a population of lipid/surfactant vesicles in the nanometric scale. In this procedure an aliquot of bioactive compound-containing solution is added to the surfactant/cholesterol mixture and then sonicated at the surfactant transition temperature for minutes to yield niosomes (Mehta, Raval, Patel, & Pandit, 2018; Alam et al., 2013). The principal effect of sonication is corresponded to the cavitation especially its collapse type, which has been revealed to be responsible for many physical effects of ultrasound on lipid membranes. Moreover, mechanical or thermal energy supplied by niosome preparation approaches causes bilayer sheets to separate and form MLVs, which can subsequently be broken down into SUVs with a high radius of curvature using sonication. Thus often a size reduction step should be followed by the initial hydration step in the niosome production procedure (Basiri, Rajabzadeh, & Bostan, 2017).

Controlling the size of niosomal vesicles is a possibility by optimizing the processing variables such as power, frequency, amplitude, and time and selecting an appropriate setup of ultrasonic systems. The particle size of niosomal suspensions sonicated for different time intervals by a temperature-controlled bath sonicator showed a significant reduction trend by increasing the exposure time up to 30 min (Essa, 2014). The maximum magnitude of size reduction was related to the first 10 min, while the decline in size was minimum after 30 min of exposure. It was found that the ability of ultrasound waves to generate cavitation bubbles in the niosome suspension medium decreases by passing time (Essa, 2014). A study on film method-prepared niosomes exhibited the formation of the most homogenous and smallest spherical-shaped vesicles after adding sonication to regular vortex mixing. In contrast, niosomes obtained by classic film hydration method were observed as large vesicles with tubular shape even after 10 min vortexing. The results confirmed the influence of ultrasound on size and geometry of niosomal nanostructures and also emphasized the importance of agitation method in preparation process (Bayindir & Yuksel, 2010). Similarly a film hydration-probe ultrasound method has been used for generating niosomes, appeared as unilamellar vesicular nanostructures with uniform sizes and shapes. It is noticeable that no conglutination of vesicles was observed (Wang et al., 2012). The formation of niosomes using ultrasonic processing system has also been compared with traditional thin-film hydration method (Khan, Madni, Hirvonen, & Peltonen, 2017). Both techniques produced well-defined spherical vesicles, but the ultrasonically prepared vesicles were smaller and monodisperse than those fabricated by thin-film hydration method. Encapsulation efficiency also showed an increase trend by increasing the amplitude from 50% to 70%, whereas it began to decrease with going toward 90% amplitude. Too long sonication time led to the loss of encapsulation efficiency, happening due to the reformulation of larger vesicles and consequently the leakage of enclosed compounds from niosomes. The optimum conditions of sonication process for small niosomes with low polydispersity values and high entrapment efficiencies were obtained in amplitude of 70% and sonication time of 45 min.

Sonication method as an eco-friendly alternative has been utilized, alone or in combination with other techniques, for the production of a variety of bioactive component-loaded niosomes, as summarized in Table 4. A kind of niosome prepared from sorbitan monostearate (Span 60) as a food-grade synthetic surfactant and lauryl alcohol (dodecanol) as a stabilizer using a probe-based sonication, subsequent to the thin-film hydration, was used for nanoencapsulating resveratrol for yogurt enrichment (Pando, Beltran, Gerone, Matos, & Pazos, 2015).

Table 4 A summary of recent researches on niosomes carrying bioactive components produced by ultrasound processing systems.

Bioactive agents[a]	Initial method of preparation	Ultrasound conditions: Setup	Ultrasound conditions: Process parameters	Type of surfactants	Particle size (nm)	Efficiency (%)	Reference
α-Tocopherol	Thin-film hydration	Probe sonicator	UP: 100 W UF: 20 kHz 4 cycles (180 s on/60 s off)	Tween 60 Span 60	~ 106.8–190	~ 83–99	Basiri et al. (2017)
α-Tocopherol	Heating	Probe sonicator	UP: 100 W UF: 20 kHz 5 cycles (40 s on/10 s off)	Tween 60 Span 60	~ 70.9–186	~ 61–98	Basiri et al. (2017)
Resveratrol	Thin-film hydration	Probe sonicator	UP: 500 W UF: 20 kHz UA: 45% ST: 30 min	Span 60	~ 183	~ 42	Pando et al. (2015)
Resveratrol α-tocopherol curcumin	Thin-film hydration	Ultrasonic bath	ST: 10 min UT: 60°C 3 cycles	Tween 60	~ 471–565	–	Tavano, Muzzalupo, Picci, and de Cindio (2014)
Capsaicin	Thin-film hydration	Ultrasonic bath	ST: 30 min UT: 60°C	Tween 80 Span 80	~ 331–479	~ 62–76	Tavano, Alfano, Muzzalupo, and de Cindio (2011)
Thai flower extracts	Thin-film hydration	Probe sonicator	UA: 35% ST: 10 min (3 s on/1 s off)	Tween 61	~ 159.3–545	–	Manosroi et al. (2013)

[a]UF, ultrasound frequency; UP, ultrasound power; UA, ultrasound amplitude; UT, ultrasound temperature; ST, sonication time.

3.5 Cubosomes and hexosomes

Amphiphilic molecules are able to spontaneously self-assemble to create a variety of well-defined, thermodynamically stable liquid-crystalline structures offering long-range order in one, two, or three dimensions. Such structures not only are constrained to the lamellar phase (like liposomes) but also involve nonlamellar phases, which are together recognized as lyotropic liquid-crystalline systems. More recently, inverse bicontinuous cubic phase at low temperatures along with its colloidal nanoparticles named as cubosomes and inverse hexagonal phases at high temperatures along with its colloidal nanoparticles named as hexosomes have drawn a great attention as nanocarriers and delivery systems for pharmaceutical, cosmetic, and food applications (Urandur et al., 2018; Akbar, Anwar, Ayish, Elliott, & Squires, 2017). Particularly, they are considered promising contenders in the food industry to improve solubilization and protection of bioactive components, to be used as reactors, to control aroma and flavor release, or to generate nanostructured food products (Garti, Libster, & Aserin, 2012; Ubbink, Burbidge, & Mezzenga, 2008; Sagalowicz, Leser, Watzke, & Michel, 2006).

When cubic and hexagonal phases are dispersed in water with the extra hydrophobic compounds (such as oleic acid and vitamin E acetate) and surfactants (such as Pluronic F 127, Pluronic F 108, and Myrj series), cubosomes and hexosomes are formed, respectively (Urandur et al., 2018; Lancelot, Sierra, & Serrano, 2014). Various methods used for the production of these nanoparticles are categorized into the top-down and bottom-up approaches. In top-down route, self-assembled systems are made in a viscous bulk phase, resulted from hydration of structure-forming lipid and stabilizers. The bulk is then dispersed into an aqueous solution by either mechanical shearing or high-throughput techniques such as high-pressure homogenization (Wörle et al., 2007) or ultrasonication (Akhlaghi, Ribeiro, Boyd, & Loh, 2016). Due to being a simple, rapid, and low-cost technique, sonication is often applied for the preparation of uniform nanodispersions of cubosomes and hexosomes with a particle size below 200 nm and low polydispersity (Chong, Mulet, Boyd, & Drummond, 2015). However, bioactive components, mainly those based on proteins and peptides, are sensitive to high temperature, which may be induced by high amount of energy provided during preparation.

In the bottom-up route, water-insoluble lipids can be dissolved to generate single-phase liquid precursors in the presence of a hydrotrope (for instance, ethanol and chloroform) for the fabrication of nanoparticles using less energy input. It should be noted that formation of vesicles cannot be effectively prevented in this method (Karami & Hamidi, 2016). A comparative investigation between top-down and bottom-up approaches indicated that ultrasonication as one of the best candidates can be easily employed for the fabrication of cubosome formulations with less aggregates, low liposome content, small particle size, and high storage stability (Akhlaghi et al., 2016).

A small number of amphiphilic lipids are currently used for the formulation of cubsome and hexosome nanoparticles including glyceryl monooleate, glyceryl monolinoleate, and phytantriol (Serieye, Méducin, Milošević, Fu, & Guillot, 2017). The phase diagram of the phytantriol/water binary systems, studied first by Barauskas and Landh (2003), and the representative 3-D models of the liquid-crystalline phases have been displayed in Fig. 6. By adding water at room temperature, lamellar crystalline phase (L_{α}) in coexistence with an inverse micellar (L_{II}) and the inverted

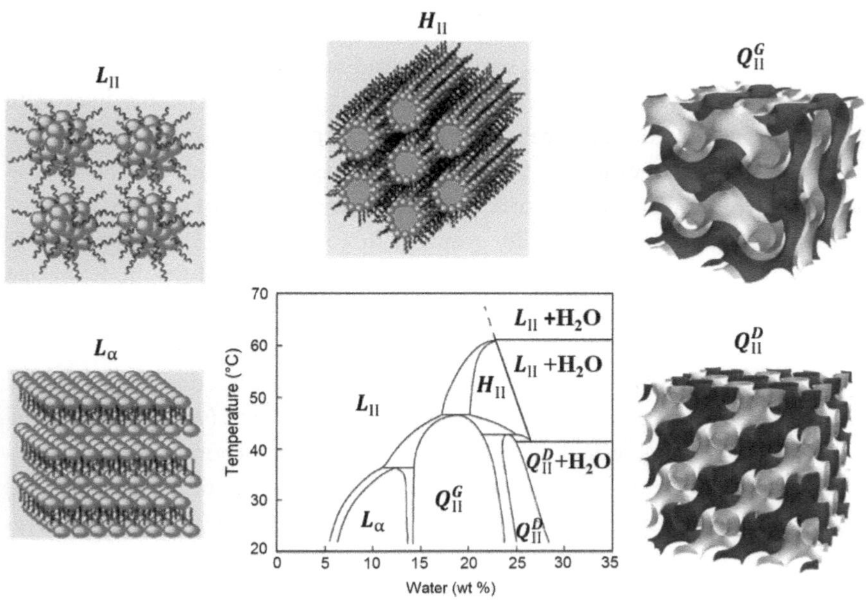

Fig. 6 Phase diagram of the phytantriol/water binary systems **as a function of** temperature and water concentration and the representative 3-D models of the liquid-crystalline phases. *(Adapted from* Akbar, S., Anwar, A., Ayish, A., Elliott, J.M., & Squires, A. M. (2017). Phytantriol based smart nano-carriers for drug delivery applications. European Journal of Pharmaceutical Sciences, 101, 31–42, *with permission).*

bicontinuous cubic phases (Q_{II}^{G}, double gyroid and Q_{II}^{D}, double diamond) appeared. A transition to the reversed hexagonal phase (H_{II}) followed by the L_{II} *phase* is also observed *at elevated temperatures.*

It is known that temperature changes, in addition to water content and lipid/surfactant composition, play a critical role in the arrangement of lyotropic liquid-crystalline structures. Phytantriol-based curcumin-loaded dispersions prepared by ultrasonication after heat treatment at 55°C showed the structural transformation of H_{II} *phases to* Q_{II} *phases while by reducing temperature the* Q_{II} *phases were prevailed* (Tu et al., 2014).

In a study by Dong et al. (2010), the lyotropic-phase behavior of liquid-crystalline disperse and bulk systems was investigated. Disperse systems with particle size of 200–300 nm and polydispersity index <0.25 were achieved by ultrasonication at a maximum amplitude of 40% for 20 min in pulse mode of 0.5 s on/0.5 s off. Disperse systems showed a greater magnitude of supercooling phenomenon as a result of the limited crystal growth after nuclei formation. In addition, no H_{II} phase was detected in the dispersion systems in comparison with the bulk nondispersed systems. In fact, submicron size of dispersed particles and limited number of unit cells within each particle could not provide the packing stress required for the formation of H_{II} phase in an equilibrium condition. On the other hand, dispersed systems due to the large surface area-to-volume ratio of the submicrometer-sized particles have a high water influx efficiency, affecting the phase transition and the release behavior. Recent studies reported that the burst release of bioactive molecules happens from ultrasonically prepared hexosomes (Wei, Li, Guo, Liu, & Wang, 2018) and cubosomes (Tu et al., 2014) because of the high surface area-to-volume ratio, which also confirms that the water transport is rapid for these nanoparticles.

The attractiveness of Q_{II} and H_{II} systems is corresponded to the structural tunability in aqueous environment, high interfacial area (specific interfacial area ~ 400 m^2/g), efficient solubilization and incorporation of various hydrophilic, hydrophobic and amphiphilic moieties in their well-defined internal nanostructures, and sustained or controlled release properties, as well as nanostructural versatility, compatibility, digestibility, and bioadhesive properties of involved lipid constituents (Serieye et al., 2017; Azmi, Moghimi, & Yaghmur, 2015).

The Q_{II} and H_{II} systems are three- (3-D) and two-dimensional (2D) complexes consisted of aqueous nanochannels embedded in continuous hydrophobic matrices (Yaghmur & Glatter, 2009). Hydrophobic bioactive agents intrinsically tend to reside within the lipid bilayer of matrices, hydrophilic agents prefer to reside in the water channels, and the amphiphiles are located in the interface. Location of incorporated/solubilized molecules plus the size and polarity of nanoparticulate systems are important factors affecting on diffusion mechanism and release kinetics of payload (Clogston & Caffrey, 2005).

In spite of all the advantageous mentioned for liquid-crystalline nanoparticles, some issues limit their application especially in the food sector such as high viscosity, large amount of surfactants required for their formation, low taste acceptability of final product when concentration exceeds threshold, and changes in the inner structure of the dispersed particles when they are mixed with an emulsion, being not anymore cubic or hexagonal (Sagalowicz & Leser, 2010).

It is also important to note that the functionality of these nonlamellar phases is different and prominently dependent upon their nanostructure. Release of entrapped molecules can be influenced by switching from one phase to another one like temperature-induced phase transition from cubic to hexagonal (Fong, Hanley, & Boyd, 2009). The H_{II} phase showed a more sustained release than the Q_{II} phase with larger aqueous channels. It has been demonstrated that the glucose diffusion coefficient in H_{II} phase is almost 10 times lower than in Q_{II} phase (Fong et al., 2009; Boyd, Whittaker, Khoo, & Davey, 2006).

To date, ultrasonicated liquid-crystalline nanoparticles have been exploited for encapsulating various bioactive compounds, mostly aimed at using in pharmaceutical field, as given in Table 5. An interesting example is curcumin, a bioactive phenolic compound derived from turmeric that has antitumor properties and therapeutic activity in malignant tumors and Alzheimer's disease, but its usage is limited because of its poor bioavailability, pH sensitivity, and poor tissue absorption. Besides, encapsulation of curcumin in lipidic and polymeric delivery systems is a great challenge owing to its low aqueous solubility. It was reported that an ultrasonicated hexagonal formulation based on soybean lecithin improved solubilization and sustained release behavior of curcumin. Hexosomes bearing curcumin as well revealed good elastic properties and better antioxidant activity (Wei et al., 2018). In another study by Tu et al. (2014), curcumin was incorporated in piperine-loaded cubosomes via ultrasonication in order to improve oral bioavailability and tissue distribution of curcumin.

Another bioactive compound incorporated into colloidal nonlamellar liquid-crystalline nanoparticles and their corresponding phases was resveratrol, a phenolic compound that naturally exists in grapes and peanuts and

Table 5 Representative examples of the ultrasonicated cubosomes and hexosomes carrying bioactive components.

Type of dispersion	Bioactive agent[a]	Ultrasound conditions		Particle size (nm)	Achievements	Reference
		Setup	Process parameters			
Hexosome	Curcumin	Probe sonicator	UP: 400 W ST: 18 min Pulse mode: 10 s on/5 s off	~ 100–150	• Better structural stability • Biphase release (fast and then sustained release) • Good elastic properties and thus suitable candidate for oral drug • Desired antioxidant activity	Wei et al. (2018)
Cubosome	Antimicrobial peptides	Probe sonicator	40% of maximum power ST: 5 min Pulse mode: 3 s on/7 s off	~ 191	• Fewer number of vesicles • Slower release rate • High peptide association efficacy to the particles • High proteolytic stability • More effective in bacterial killing	Boge et al. (2017)
Cubosome	Resveratrol	Probe sonicator	ST: 15 min	~ 161	• High entrapment efficiency (> 83%) • Controlled and sustains release • Higher trans-nasal permeation and better distribution to brain	Ahirrao and Shrotriya (2017)
Cubosome	Curcumin piperine	Probe sonicator	(1) 80% of maximum power ST: 30 min (2) 40% of maximum power ST: 10 min Pulse mode: 0.5 s on/0.5 s off	~ 153	• High entrapment efficiency • Slower elimination rate • Biphase release (rapid and then sustained release) • Higher bioavailability • Higher level of curcumin in tissues	Tu et al. (2014)

[a]UF, ultrasound frequency; UP, ultrasound power; UA, ultrasound amplitude; UT, ultrasound temperature; ST, sonication time.

possess anticancer, antioxidant, immunomodulatory, and cardioprotective activities (Ahirrao & Shrotriya, 2017). Also, Boge et al. (2017) studied nanoencapsulation of antimicrobial peptides into cubosomes to enhance its solubility and stability in the physiological medium.

4 Biopolymeric and polymeric nanostructures engineered by ultrasonication

Biopolymeric and polymeric nanostructures are often developed and used as nanosized carrier-based delivery systems, which can be classified into nanoparticle-based carriers and nanocomplex-based carriers, theoretically with dimensions between 1 and 100 nm. Practically, nanostructured carriers also consist of submicron particles with sizes below 1000 nm (Jafari & McClements, 2017; Fu, Sarkar, Bhunia, & Yao, 2016). Due to the nanometer scale and high surface area-to-volume ratio, polymeric and biopolymeric nanostructures show unique physicochemical features such as vastly modifiable surfaces, high loading capacity, solubility enhancer of bioactive agents particularly lipophilics, simple engineering of targeted carriers, easy design, and preparation. Moreover, based on the material used in the preparation, carrier-based delivery systems can be categorized into lipid-based systems, examined in above subsections, and biopolymer-based systems. The most important fact is the biosafety and biocompatibility of the used biopolymers and preparation methods, which are critical characteristics in the field of edible delivery systems (Amoabediny et al., 2018).

Nanostructure-based carriers can be produced from biopolymers (such as proteins and polysaccharides) and synthetic polymers (such as polylactic acid, polycaprolactone, polyacrylonitrile, $mPEG_{5000}$-b-p, and paclitaxel) using a simple and safe sonication technique. The size, shape, charge, and internal organization of structures and their corresponding nanoparticles depend considerably on the method and materials used to fabricate them, thereby enabling them to be tailored for intended applications (Van der Sman, 2012). Promisingly enough the mechanism underlying the ultrasonication-induced production of nanostructures is the result of a physical rather than a chemical process associated with acoustic cavitation, which may produce secondary effects (Chatani et al., 2009). It has also been reported that high-intensity ultrasound can decrease the molecular weight of chitosan being mainly due to mechanical forces while not affecting the degree of acetylation (Wu, Zivanovic, Hayes, & Weiss, 2008). Different food bioactive compounds that have been encapsulated into nanoparticle and nanocomplex-based carriers, created by ultrasound contribution, are discussed in the succeeding texts.

4.1 Nanoparticle-based carriers

Polymer nanoparticles are typically defined as solid, colloidal particles in the range of 10–1000 nm, which can be assembled in the form of nanospheres and nanocapsules using various fabrication methods (Fig. 7). Polymeric nanospheres are matrix systems in which the bioactive molecules can be adsorbed either at the sphere surface or entrapped within the sphere. Unlike nanospheres, nanocapsules are vesicular systems, acting as a reservoir in which the bioactive molecules are confined to an aqueous core surrounded by the polymeric shell around them (Anandharamakrishnan, 2014). Given that biopolymers have the ability to entrap both hydrophilic and hydrophobic bioactive agents within their networks and be formed under relatively mild preparation conditions without using harsh or harmful chemicals, biopolymer-based nanoparticle systems have been successfully developed for encapsulating, protecting, and delivering bioactive agents (Hu, Liu, Zhang, & Zeng, 2017; Li, Jiang, Xu, & Gu, 2015). In fact the biopolymer-based nanoparticles increase the absorption and bioavailability of bioactive molecules mainly via different routes: (1) protection of the bioactive molecule from the harsh environment of the gastrointestinal tract, (2) prolongation of the residence time in the gut by mucoadhesion, (3) endocytosis of the particles, and/or (4) permeabilizing effect of the polymer (Bhushani, Harish, & Anandharamakrishnan, 2017; des Des Rieux, Fievez, Garinot, Schneider, & Préat, 2006). The unique structure; surface

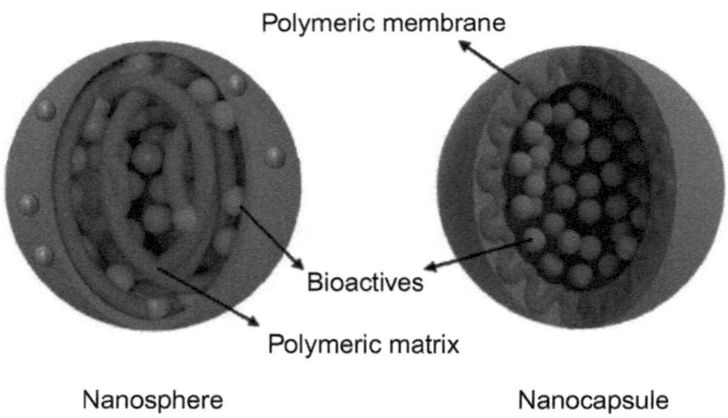

Fig. 7 Schematic illustration of polymeric-based nanocapsules and nanospheres.

reactive sites as hydroxyl, carboxyl, and amino groups; abundance; low cost; biodegradability; and biocompatibility are other general advantages of selecting polysaccharides and proteins as suitable structural building blocks for bioactive carriers (El-Salam & El-Shibiny, 2016; Zhang, Chan, Moretti, & Uhrich, 2015). For the purpose of oral consumption and diminishing carrier-induced cytotoxicity in the delivery of food bioactive compounds, there is no better option than utilization of food-grade polymers that have been approved as safe, being suitable for developing such delivery systems.

There are various approaches for producing the nanoparticles from individual biopolymers, which can be broadly classified as top-down, bottom-up, or combinational procedures such as desolvation, precipitation, gelation using thermal treatment, and self-assembly (Joye & McClements, 2014; Jones & McClements, 2011). Ultrasonication is of particular interest in the nanoparticle fabrication owing to several technically benefits including simplicity, versatility, environmentally friendly, and by-product free (Tsai, Bai, & Chen, 2008; Wu et al., 2008). Interestingly a combination of two opposing effects of ultrasonication including the formation and the breakdown can occur during the production process of nanostructures. It was stated by Chatani et al. (2009) that the production of minimal and monodispersed amyloid fibrils using ultrasonication pulses is attainable by minimum free energy under competition between formation and breakdown processes of the fibrils.

Some researches have been done on the development of biopolymer-derived and ultrasonically produced nanoparticles as nanovehicles for hydrophilic and hydrophobic bioactive compounds, which are examined in the succeeding text. The water solubility and pH stability of curcumin were greatly enhanced after entrapping in ultrasound-treated β-lactoglobulin. Moreover, encapsulated curcumin was found to be resistant to pepsin digestion but sensitive to trypsin. In the Caco-2 cell model, the digested curcumin-loaded β-lactoglobulin significantly improved the permeation rate of curcumin (Li, Cui, Ngadi, & Ma, 2015). Conjugated linoleic acids are a group of linoleic isomers having several diverse health effects such as reducing body fat, lowering serum cholesterol level, atherosclerosis reduction, and immune response enhancement. To prevent the oxidative deterioration, they were incorporated into soy lipophilic protein nanoparticles by probe-mediated ultrasonication at an amplitude of 30% for 5 min. Nanoparticles with the mean diameter of 170 nm indicated a high loading capacity of 26.3%, oxidation protection, and a sustained releasing profile in vitro for conjugated linoleic acids. Moreover, coating the surface of the soy lipophilic protein nanoparticles carrying linoleic acids with sodium caseinate improved their colloidal stability (Gao et al., 2014).

A production protocol of casein micelle nanocapsules was optimized in different levels of sonication treatment and alkaline pH in a way that the original structure and morphological characteristics were well maintained. It was hereby found that the step of sonication processing plays a leading role in the enhancement of the encapsulation efficiency of casein micelles. The highest efficiency of encapsulation belonged to the sonicated samples at an amplitude of 50%, power of 600 W, and frequency of 25 kHz for 2 min. Additionally the nanocapsules exhibited an superb oxidative protection against UV light. Ultrasonication in alkaline environments probably leads to more accessibility of the hydrophobic moieties of casein micelles facilitating the penetration of oil particles in their inner regions. Therefore ultrasound technology introduces an effective way for encapsulation and delivery of hydrophobic bioactive compounds at the nanoscale, for example, unsaturated fatty acids in food products and pharmaceutics, aiming to promote the society's health (Ghasemi & Abbasi, 2014).

Regarding to combining ultrasound with other nanoencapsulation techniques, the encapsulation of blueberry bioactive compounds via ultrasonic-mediated spray drying provided better control over capsule size and shape than those of conventional spray drying and freeze drying. The usage of ultrasonic technique in spray drying strengthened the retention of phenolic compounds, anthocyanins, and storage stability (Tatar Turan, Cengiz, Sandıkçı, Dervisoglu, & Kahyaoglu, 2016). Ultrasound-assisted atomization system emits high-frequency sound waves or vibrations by piezoelectric transducers during the atomization of a liquid that creates liquid droplets with a uniform size. The vibrational energy of ultrasonic-mediated atomization system introduces less mechanical stress decreasing the mechanical degradation of the bioactive compounds (Shishir & Chen, 2017).

Antisolvent precipitation as well is an efficient technique in nanoencapsulation process particularly for the fabrication of nanoparticles using polysaccharides and/or proteins. Its application is often limited in terms of particle agglomeration happening during the process, arising from the lack of sufficient repulsion only after particle generation (Joye & McClements, 2014). In this case, implementing an ultrasound processing system is thought to be productive. It can be assembled with the speedy mixing device to improve the mixing process, increase the nucleation rates, and minimize agglomeration (Thorat & Dalvi, 2012).

In spite of the safety concerns on the synthetic polymers and the unwillingness to use them for food applications, the polyester families such as polylactic acid, poly caprolactone, and poly glycolic acid are favorite polymers in the biomedical sectors owing to their good biocompatibility and biodegradability. Poly (D,L-lactide-*co*-glycolide), recognized as GRAS for pharmaceutical applications, breaks down into its monomers (lactic acid and glycolic acids), which are eliminated as harmless waste products, carbon dioxide and water (Murugesan, Mishra, & Jain, 2008). The degradation profile of polymeric compounds affects the release profile of entrapped drugs or bioactive agents. Polymers with a higher molecular weight degrade slower associated with a prolonged constant bioactive release over time. Also, nanoparticles with a higher amount of lactide monomers demonstrated a slower release rate for hydrophobic agents. In essence the more hydrophobic the polymer, the more strong hydrophobic interaction between the polymer and entrapped agent and thus a slower release profile. On the other hand, when the bioactive molecules are mostly localized in the center of the nanoparticle, diffusion processes depend on the polymer molecular weight (Sabliov, Chen, & Yada, 2015; Astete, Dolliver, Whaley, Khachatryan, & Sabliov, 2011).

Dendrimers are other unique class of polymeric nanostructured carriers creating a different avenue for nanodelivery of different bioactive compounds. The word dendrimer originates from the two Greek words dendri- (tree branch-like) and meros (part of) (Gardikis, Hatziantoniou, Viras, Wagner, & Demetzos, 2006). Dendrimers are considered as highly branched and star-shaped macromolecules with nanometer-scale dimensions. The well-defined structure, monodispersity of size, surface functionalization capability, and stability render dendrimers attractive nanocarrier candidates. Nonetheless a suitable dendrimer as bioactive delivery system must as well be nontoxic, nonimmunogenic, and biodegradable (Khosravi-Darani, Pardakhty, Honarpisheh, Rao, & Mozafari, 2007). The first complete dendrimer family that has been synthesized, characterized, and commercialized as a safe and nonimmunogenic nanocarrier is the poly(amidoamine) (PAMAM) dendrimer (Tomalia & Frechet, 2001). Bioactive molecules can be binded to dendrimers via either complexation or encapsulation as schematically displayed in Fig. 8. In this way, Madadlou, Sheehan, Emam-Djomeh, and Mousavi (2011) nanoencapsulated a fungal protease within nanoparticles yielded via silicification of polyamidoamine dendrimer template generation 0. It was used to hydrolyze the main milk protein (casein) and generate angiotensin-converting enzyme

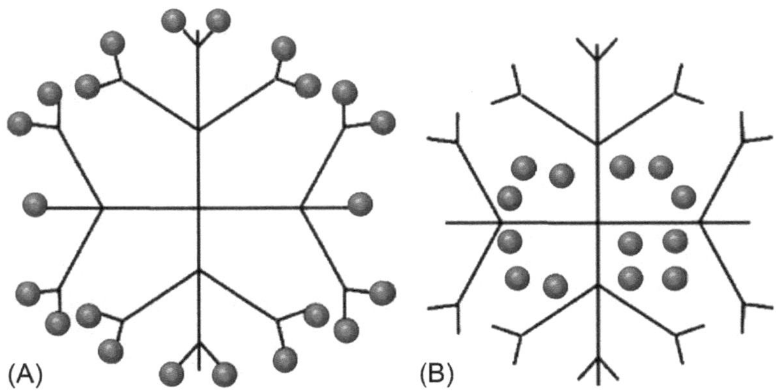

Fig. 8 A simple schematic representation of the dendrimer structure carrying bioactive agents. (A) Complexation: covalent attachment to end groups. (B) Encapsulation: bioactive entrapment inside the core of dendrimer. *(Redrawn from (Khosravi-Darani, K., Pardakhty, A., Honarpisheh, H., Rao, V. M., & Mozafari, M. R. (2007). The role of high-resolution imaging in the evaluation of nanosystems for bioactive encapsulation and targeted nanotherapy. Micron: The International Research and Review Journal for Microscopy, 38(8), 804–818.) with permission).*

(ACE)-inhibitory bioactive peptides in the absence or presence of sonication. An extreme mass transfer limitation was obtained for the supramolecular casein particles through the encapsulating matrix. A much longer time was required to achieve maximum ACE-inhibitory activity in the absence of ultrasound, while sonication boosted the production of ACE-inhibitory peptides and shortened the time for bioprocessing. The mechanism for enhancement of enzymatic hydrolysis after sonication was attributed to an increase in diffusion rates as a result of cavitation. Besides, sonication is able to disrupt the self-assembled casein nanoparticles in the solution and readily expose the hidden moieties to enzymatic lysis. More importantly the dendrimer-mediated encapsulated protease retained 100% of its origin activity even at the ultrasound exposure with an acoustic power of 4.5 W.

4.2 Nanocomplex-based carriers

Complexes can be formed by a variety of physical and chemical interactions including van der Waals, electrostatic, covalent, hydrophobic, and hydrogen bonding. Many classes of biopolymers are capable of forming molecular complexes through the assembly of lipophilic and hydrophilic entities. The bioactive molecules can be nanoencapsulated and highly protected by virtue of protein-protein, polysaccharide-polysaccharide, and protein-polysaccharide (nano) complexes (Jafari, 2017). It is worthy to note that submicron complexes made of soft colloidal particles are intrinsically more advantageous than larger complexes/precipitates for encapsulation and controlled release purposes, because of more stability over sedimentation, as well as owning a larger surface area for molecular release of the captured bioactives (Kurukji, Norton, & Spyropoulos, 2016).

One of the most common methods of assembling proteins and polysaccharides into functional nanocomplexes is based on electrostatic attraction between oppositely charged groups under appropriate solution conditions (Spada, Marczak, Tessaro, & Cardozo, 2015; Zimet & Livney, 2009). This process generally comprises three steps: the dissolution of biopolymers in the solvent, the mixing of polymers in the required ratio, and an acidification stage. The acidification step has a significant effect on the formed complex size, carrying out either before mixing two different polymers termed preblending acidification or after mixing the two polymers termed as postblending acidification (Ghasemi, Jafari, Assadpour, & Khomeiri, 2017, 2018). Postblending acidification has been proposed to produce smaller particles in comparison with preblending acidification (Bédié, Turgeon, & Makhlouf, 2008). However, subjecting the formed particles to high shear forces in order to reduce their sizes overcomes the drawback associated with preblending acidification.

Electrostatic attraction between groups with opposite charges in suitable solution conditions is considered as one of the most common ways of assembling proteins and polysaccharides into functional nanocomplexes (Spada et al., 2015; Zimet & Livney, 2009). This process generally comprises three stages: dissolving the biopolymers in the solvent, mixing the biopolymers in the required ratio, and finally acidifying them (Ghasemi et al., 2017, 2018). The acidification step has a significant effect on the formed complex size, carrying out either before mixing two different polymers termed preblending acidification or after mixing the two polymers termed as postblending acidification (Assadpour, Jafari, & Maghsoudlou, 2017; Assadpour, Maghsoudlou, Jafari, Ghorbani, & Aalami, 2016). By comparison with preblending acidification, the postblending methodology has been proposed to produce very tiny particles (Bédié et al., 2008). However, subjecting the formed particles to high shear forces in order to reduce their sizes overcomes the drawbacks connected with the preblending acidification.

In this regard, Kurukji et al. (2016) have successfully diminished the size of the sodium caseinate-chitosan complexes that were prepared by employing a preblending acidification approach using sonication. In the absence of chitosan in the formulation, the particle size did not reduce to the colloidal scale, and protein precipitates were appeared after sonication because of the self-association of sodium caseinate near to its isoelectric point (IEP). It is a well-known fact that dispersing protein particles (to a colloidal size) is a thermodynamically unfavorable process and extremely hard to obtain even by means of typical high-energy processing techniques. Addition of chitosan could compensate this fundamental problem by the level of charge repulsion introduced between the individual complexes. More importantly, rhodamine B and riboflavin as bioactive models could be incorporated after complex formation providing a desirable strategy for nanoencapsulation of temperature- or shear-sensitive bioactive molecules (Kurukji et al., 2016).

The effect of applying high-intensity ultrasound to a kappa-carrageenan solution prior to complexing with β-lactoglobulin has previously been examined by Hosseini et al. (2013). With increasing amplitude and time of the sonication process, the viscosity of polysaccharide solutions decreased due to the formation of smaller polysaccharide chains. Hence, solutions of sonicated kappa-carrageenan and protein complexes showed a lower turbidity than intact ones. The affinity constant between β-lactoglobulin and kappa-carrageenan was considerably decreased at pH = 4.2 by the sonication. This phenomenon was corresponded to the reduction of the kappa-carrageenan reactivity, resulting from structural changes and some heterogeneous sonochemical interactions. As a result, the reactive sites present along the kappa-carrageenan backbone may be eliminated, or the interaction between kappa-carrageenans may be promoted, which reduces the number of binding sites for the β-lactoglobulin molecules. Sonochemical interactions reduced the charge density of sonicated kappa-carrageenan, which was detected by lower zeta potential values of sonicated nanocomplexes than untreated ones. Besides, sonication had a positive impact on the homogenization of nanoparticles in the mixed dispersion. The highest amount of interactions between β-lactoglobulin and kappa-carrageenan was achieved at pH values of 1.0–2.0. The dissociation of complexes did not occur even at lower pH values. Isothermal titration calorimetry results exhibited exothermic sequences for the interactions of β-lactoglobulin with kappa-carrageenan, which is associated with electrostatic neutralization of biopolymers.

Rahaiee, Shojaosadati, Hashemi, Moini, and Razavi (2015) developed chitosan-sodium alginate nanocomplexes carrying crocin, a water-soluble carotenoid present in saffron. The mixture of chitosan, sodium alginate, and bioactive ingredient was sonicated at 20-kHz frequency with 100%

amplitude for 3-min and 6-min time periods to achieve particles at nanometer scale. FTIR spectroscopy confirmed that the major driving force for crocin entrapping into the polysaccharide nanocomplex is hydrogen bonding. The nanoencapsulation efficiency and loading capacity were obtained approximately 38% and 31%, respectively. In addition, this type of nanocarrier based on carbohydrate complexes enhanced the crocin stability under unfavorable environmental conditions.

It should be taken into consideration that complexes formed by electrostatic interactions are susceptible to dissociation or degradation under specific environmental conditions, for example, pH, temperature, and ionic strength. However, dissociation of complexes, thanks to the changes in the environmental factors, can make them, in addition to a kind of protective carriers, very versatile as building blocks for developing stimuli-sensitive carriers being dedicated to the targeted and controlled delivery of bioactive constituents throughout the GIT (Semenova et al., 2014). Complexation process mostly happens in low pH values, and hence, the resulted structures will be resistant to acidic conditions, which is very helpful in protecting the bioactive agents in the stomach acidic conditions. However, dissociation can take place in the alkaline conditions of intestine, followed by releasing and absorbing entrapped bioactives (Jones, Lesmes, Dubin, & McClements, 2010).

Covalent complexation such as polysaccharide glycosylated proteins prepared via the Maillard reaction is a suitable alternative to overcome the limitation of dissociation or degradation of electrostatic complexes. This makes the fabricated particles more resistant to the environmental changes and also well suited for carrying bioactive components. However, high temperatures required in dry- or wet-heating conditions typically causes protein aggregation and lower degree of glycation. A combination of high-intensity ultrasound and polyethylene glycol as a crowding agent with high molecular mass was designed by Perusko, Al-Hanish, Velickovic, and Stanic-Vucinic (2015) to be applied to model systematic Maillard reaction, which considerably enhanced whey protein glycation in the presence of arabinose. An increase in glycation efficiency leads only to slight structural change of whey proteins. Solubility at different pH values, thermal stability, and antioxidative capacity of glycated whey protein were increased in comparison with ultrasonified nonglycated proteins. Therefore, ultrasonication under crowding conditions can be employed in Maillard reactions involving labile compounds such as proteins and bioactives because of the ability of reaction enhancement under lower temperatures and short times. In another research study carried out by Stanic-Vucinic, Prodic, Apostolovic, Nikolic, & Velickovic, 2013, they observed that high-intensity ultrasound can efficiently promote glycation of β-lactoglobulin by Maillard reaction and that resulting glycoconjugates possess an improved antioxidative capacity, with a negligible effect on protein's secondary and tertiary structure. Several studies proved that high-intensity ultrasound can promote Maillard reaction by accelerating early (Corzo-Martinez et al., 2014), intermediate, and final stages of Maillard reaction (Stanic-Vucinic et al., 2013; Guan et al., 2011).

5 Bioattributes of sonicated and nanoencapsulated food bioactive components

Nanostructures as nanoscale delivery systems due to the subcellular size offer promising means of increasing the bioavailability and bioaccessibility of nutraceutical ingredients, especially poorly soluble ones including functional lipids (e.g., carotenoids, phytosterols, and fatty acids), natural antioxidants, and many other compounds that are extensively consumed as bioactive ingredients in various food products (McClements, 2014). In principal, bioavailability represents the proportion of ingested nutrient or bioactive compound that is absorbed from the intestine content and reaches the systemic circulation (Jafari, Esfanjani, Katouzian, & Assadpour, 2017). The complex process of bioavailability encompasses different stages such as digestion or liberation of food matrix, absorption, metabolism, tissue distribution, and elimination phase (Eastwood, 2013). Differently, bioaccessibility is described as the quantity or the proportion of the bioactive ingredients capable of releasing from the food matrix in the GIT and being readily available for absorption process (Heaney, 2001). Bioavailability of a compound is evaluated in vivo in animals or humans, estimated after the administration of an acute or chronic dose of an isolated component or a food containing same component (Rein et al., 2013), while bioaccessibility is usually determined by in vitro digestion protocols, simulating gastric and small intestinal digestion plus sometimes absorption with the help of Caco-2 cells (Courraud et al., 2013).

The delivery of any bioactive ingredient to various sites into the body is greatly influenced by particle size. Ultrasonication reduces the size of encapsulates to the nanoscale, thereby improving bioadhesiveness in the mucus covering the intestinal epithelium and consequently prolonging compound residence time throughout the gastrointestinal tract. In addition, large surface contact area available per volume can lead to the rapid dissolution of bioactive agents within the gastrointestinal tract and more interaction of bioactive-loaded nanostructures with the biological support (Borel & Sabliov, 2014).

It has been elucidated that nanoparticle-based carriers derived from biomacromolecules can promote the absorption and bioavailability of bioactive molecules mostly through the following pathways: (I) enhancing their solubility, (II) protecting the bioactive molecules from the harsh conditions of the GIT, (III) prolonging the intestinal transit time by mucoadhesion, (IIII) elevating the permeability across the small intestine, and (V) boosting the amount of bioactive components in the bloodstream. For the oral administration path, the food macromolecular-derived nanoparticles should first be able to tolerate the severe pH and digestive enzymes through the GIT, especially in the stomach; preserve the loading bioactives; and reach the targeted absorption site in the small intestine; otherwise, the oral delivery of food macromolecular-based nanoparticles will be unattainable (Hu et al., 2017). Release can be stabilized using the biopolymers such as alginate or chitosan and a layer-by-layer technique to avoid a burst release (Quintanilla-Carvajal et al., 2010). Also, modulations of surface properties of nanostructures make it feasible to achieve the targeted release of bioactive compounds (Bouwmeester et al., 2009).

In the case of niosome, as a surfactant-based nanostructured carrier, often a complementary downstream process especially ultrasonication is applied during the niosome production subsequent to the initial hydration step because of the importance of the vesicle size in vesicle biodistribution (Basiri et al., 2017). For instance, sub-200-nm vesicles have been demonstrated to prevent splenic but not liver uptake (Litzinger et al., 1994). Tu et al. (2014) reported that piperine-loaded cubosomes produced by ultrasonication for carrying curcumin improved the oral bioavailability and tissue distribution of curcumin.

Highly hydrophobic bioactive agents (like vitamin E, vitamin D, curcumin, and carotenoids), typically possess poor oral bioaccessibility and bioavailability owing to their low water solubility (Rafiee, Nejatian, Daeihamed, & Jafari, 2018; Rostamabadi, Falsafi, & Jafari, 2019). Nonetheless, they are still able to dissolve in aqueous gastrointestinal fluids due to the presence of mixed micelles and vesicles that have hydrophobic regions to solubilize them. In addition, their bioaccessibility can be enhanced considerably by coingesting them with a digestible lipid that generates mixed micelles (Jafari & McClements, 2017). In principal, during digestion in the small intestine, the triacylglycerols in the small oil droplets are hydrolyzed into free fatty acids and monoacylglycerols resulting from the action of lipase. Then, these monoacylglycerols and free fatty acids are combined with phospholipids and bile salts in the gastrointestinal fluids to form mixed micelles. During digestion in the small intestine, the hydrolysis of triacylglycerols in the small oil droplets into monoacylglycerol and free fatty acids is basically accomplished by the hydrolytic action of lipase. Then, the resulting products are combined with phospholipids and bile salts in the gastrointestinal fluids to form mixed micelles. In the meanwhile, the released lipophilic vitamins from the oil droplets solubilize inside the hydrophobic entities of the mixed micelles, enabling to transverse the mucus layer and reach the epithelial intestinal cells to be absorbed and then packed into chylomicrons (Lv et al., 2018). It has been observed that the bioaccessibility of lipophilic bioactives in nanoscale delivery systems is significantly dependent on the dimensions of the particles encapsulating them. Experiments simulating the GIT conditions showed that the bioaccessibility of β-carotene entrapped into O/W nanoemulsions increased when the particle size becomes smaller (Yi, Li, Zhong, & Yokoyama, 2014; Salvia-Trujillo, Qian, Martín-Belloso, & McClements, 2013). Walia, Dasgupta, Ranjan, Chen, and Ramalingam (2017) encapsulated vitamin D in fish oil, which was formulated by ultrasonication. Nanoemulsions passed through simulated GIT revealed an increased bioavailability than nonencapsulated vitamin. These results have been proved with coenzyme Q10, a type of lipophilic bioactive component, by animal feeding tests (Cho et al., 2014). The fact that the smaller the particle size, the more rapid and complete the digestion of the lipophilic compounds followed by a better bioaccessibility and bioavailability can be inferred from these results.

6 Recent advances and further perspectives

The major challenges of ultrasound application include issues of scale-up and energy efficiency. Although laboratory-scale ultrasonic devices are readily commercially available for scientific investigations, larger-scale equipment for industrial applications still remains relatively uncommon. The utilization of pilot-scale sonicators and the optimization of process variables can assist manufactures in designing and scaling up for industrial facilities. Continuous-flow high-power processor is one of the recent advancements in the ultrasound technologies to produce nanosized structures, which have a great potential for commercial applications. A number of technological challenges still need to be examined for the optimized fabrication of nanostructure-based delivery systems using sonication. In the meanwhile, health and safety factors should be prioritized before their widespread application. While encapsulation of bioactive compounds in different nanostructures manufactured by sonication has been studied, there is no enough information on the toxicity, oxidative stability, and bioavailability rate of incorporated ingredients. More clinical and in vivo tests are needed on the gastrointestinal fate of nanoencapsulated nutrients and their effects on human health. With regard to the current limitations, future researches should be directed to the development of new precursor materials and more systematical control over operational parameters. A successful evolution in this area will be the exploration of smart multifunctional delivery systems containing two or several hydrophobic and hydrophilic bioactive ingredients with the ability of controlled release of their cargoes in responding external stimuli. Commercial ultrasonic devices are developing, and no novel process for the application of ultrasound in laboratories and industries is possible without ultrasonic equipment designers and engineers willing to establish new configurations in consistent with the requirements of users. It is eventually inferred that while the technology has great promises, it will be carefully developed and scaled up for every unique application.

7 Conclusions

Much research has been done on the fabrication and development of specialized nanostructures for the encapsulation of food-grade bioactive ingredients such as vitamins, phenolic compounds, antioxidants, fatty acids, enzymes, and probiotics. However, much less work is related to find application in commercial products because the nanostructures are often made of ingredients or processing operations that are too expensive and/or unsuitable for food applications. In addition, they pose adverse effects to the properties of the end product such as being physically instable to aggregation or sedimentation and chemically instable over the product shelf life and unfavorably changing appearance or mouthfeel.

Hence, this chapter has provided an overview of ultrasonication as a facile and green nanoencapsulation technique encompassing the fundamentals and different setups. High-intensity ultrasound irradiation provides unique reaction conditions via acoustic cavitations, namely, "stable" and "collapse" cavitation. Extremely highly localized temperatures and pressures, free radicals plus shock waves arising from collapse cavitation are mainly responsible for its physical and chemical effects. The utilization of ultrasound, alone or in combination with other encapsulation approaches, alongside biopolymers, can lead to a rapid generation of edible carrier nanostructures to fortify food products with the desired bioactives. Ultrasonically prepared nanostructures can be derived from (1) lipids or surfactants (such as nanoliposomes, nanoemulsions, NLCs, cubosomes, and hexosomes) and (2) biopolymers or synthetic polymers (such as nanoparticle carriers and nanocomplexes carriers).

Mechanism of ultrasound action and reaction parameters involved in the formation of nanostructures along with advantages and disadvantages were discussed from the viewpoint of safety, stability, droplet size, and process impacts on some functionalities.

Ultrasonication as a facile, rapid, safe, and cost-effective method provides manufactures with an opportunity to develop nanoengineered delivery systems with high stability, small droplet sizes, and narrow polydispersity. These nanosized systems are used to improve technofunctional properties of encapsulated bioactive agents such as solubility, dissolution, availability, controlled, and targeted release profile in vitro and in vivo plus the antioxidant and antimicrobial capacities.

References

Abaee, A., & Madadlou, A. (2016). Niosome-loaded cold-set whey protein hydrogels. *Food Chemistry*, *196*, 106–113.

Abdelkader, H., Ismail, S., Hussein, A., Wu, Z., Al-Kassas, R., & Alany, R. G. (2012). Conjunctival and corneal tolerability assessment of ocular naltrexone niosomes and their ingredients on the hen's egg chorioallantoic membrane and excised bovine cornea models. *International Journal of Pharmaceutics*, *432*(1–2), 1–10.

Abdelkader, H., Ismail, S., Kamal, A., & Alany, R. G. (2011). Design and evaluation of controlled-release niosomes and discomes for naltrexone hydrochloride ocular delivery. *Journal of Pharmaceutical Sciences*, *100*(5), 1833–1846.

Aday, M. S., Temizkan, R., Büyükcan, M. B., & Caner, C. (2013). An innovative technique for extending shelf life of strawberry: Ultrasound. *LWT-Food Science and Technology*, *52* (2), 93–101.

Ahirrao, M., & Shrotriya, S. (2017). In vitro and in vivo evaluation of cubosomal in situ nasal gel containing resveratrol for brain targeting. *Drug Development and Industrial Pharmacy*, *43* (10), 1686–1693.

Akbar, S., Anwar, A., Ayish, A., Elliott, J. M., & Squires, A. M. (2017). Phytantriol based smart nano-carriers for drug delivery applications. *European Journal of Pharmaceutical Sciences*, *101*, 31–42.

Akhavan, S., Assadpour, E., Katouzian, I., & Jafari, S. M. (2018). Lipid nano scale cargos for the protection and delivery of food bioactive ingredients and nutraceuticals. *Trends in Food Science & Technology*, *74*, 132–146.

Akhlaghi, S. P., Ribeiro, I. R., Boyd, B. J., & Loh, W. (2016). Impact of preparation method and variables on the internal structure, morphology, and presence of liposomes in phytantriol-Pluronic® F127 cubosomes. *Colloids and Surfaces B: Biointerfaces*, *145*, 845–853.

Alam, M., Zubair, S., Farazuddin, M., Ahmad, E., Khan, A., Zia, Q., et al. (2013). Development, characterization and efficacy of niosomal diallyl disulfide in treatment of disseminated murine candidiasis. *Nanomedicine: Nanotechnology, Biology and Medicine*, *9*(2), 247–256.

Aleixo, P. C., Júnior, D. S., Tomazelli, A. C., Rufini, I. A., Berndt, H., & Krug, F. J. (2004). Cadmium and lead determination in foods by beam injection flame furnace atomic absorption spectrometry after ultrasound-assisted sample preparation. *Analytica Chimica Acta*, *512*(2), 329–337.

Amoabediny, G., Haghiralsadat, F., Naderinezhad, S., Helder, M. N., Akhoundi Kharanaghi, E., Mohammadnejad Arough, J., et al. (2018). Overview of preparation methods of polymeric and lipid-based (niosome, solid lipid, liposome) nanoparticles: a comprehensive review. *International Journal of Polymeric Materials and Polymeric Biomaterials*, *67*(6), 383–400.

Anandharamakrishnan, C. (2014). *Techniques for nanoencapsulation of food ingredients*. New York: Springer.

Assadpour, E., Jafari, S. -M., & Maghsoudlou, Y. (2017). Evaluation of folic acid release from spray dried powder particles of pectin-whey protein nano-capsules. *International Journal of Biological Macromolecules*, *95*, 238–247.

Assadpour, E., Maghsoudlou, Y., Jafari, S. -M., Ghorbani, M., & Aalami, M. (2016). Optimization of folic acid nano-emulsification and encapsulation by maltodextrin-whey protein double emulsions. *International Journal of Biological Macromolecules*, *86*, 197–207.

Astete, C. E., Dolliver, D., Whaley, M., Khachatryan, L., & Sabliov, C. M. (2011). Antioxidant poly (lactic-co-glycolic) acid nanoparticles made with α-tocopherol–ascorbic acid surfactant. *ACS Nano*, *5*(12), 9313–9325.

Azmi, I. D., Moghimi, S. M., & Yaghmur, A. (2015). Cubosomes and hexosomes as versatile platforms for drug delivery. *Therapeutic Delivery*, *6*(12), 1347–1364.

Bang, J. H., & Suslick, K. S. (2010). Applications of ultrasound to the synthesis of nanostructured materials. *Advanced Materials*, *22*(10), 1039–1059.

Barauskas, J., & Landh, T. (2003). Phase behavior of the phytantriol/water system. *Langmuir*, *19*(23), 9562–9565.

Basiri, L., Rajabzadeh, G., & Bostan, A. (2017). Physicochemical properties and release behavior of span 60/tween 60 niosomes as vehicle for α-tocopherol delivery. *LWT-Food Science and Technology*, *84*, 471–478.

Bayindir, Z. S., & Yuksel, N. (2010). Characterization of niosomes prepared with various nonionic surfactants for paclitaxel oral delivery. *Journal of Pharmaceutical Sciences*, *99*(4), 2049–2060.

Bédié, G. K., Turgeon, S. L., & Makhlouf, J. (2008). Formation of native whey protein isolate–low methoxyl pectin complexes as a matrix for hydro-soluble food ingredient entrapment in acidic foods. *Food Hydrocolloids*, *22*(5), 836–844.

Bermúdez-Aguirre, D., Mobbs, T., & Barbosa-Cánovas, G. V. (2011). Ultrasound applications in food processing. In *Ultrasound technologies for food and bioprocessing* (pp. 65–105). New York, NY: Springer.

Bhushani, A., Harish, U., & Anandharamakrishnan, C. (2017). Nanodelivery of nutrients for improved bioavailability. In *Nutrient delivery* (pp. 369–411). Academic Press.

Biswal, S., Murthy, P. N., Sahu, J., Sahoo, P., & Amir, F. (2008). Vesicles of non-ionic surfactants (niosomes) and drug delivery potential. *International Journal of Pharmaceutical Sciences and Nanotechnology*, *1*(1), 1–8.

Boge, L., Umerska, A., Matougui, N., Bysell, H., Ringstad, L., Davoudi, M., et al. (2017). Cubosomes post-loaded with antimicrobial peptides: characterization, bactericidal effect and proteolytic stability. *International Journal of Pharmaceutics*, *526*(1–2), 400–412.

Boon, C. S., McClements, D. J., Weiss, J., & Decker, E. A. (2010). Factors influencing the chemical stability of carotenoids in foods. *Critical Reviews in Food Science and Nutrition*, *50*(6), 515–532.

Borel, T., & Sabliov, C. M. (2014). Nanodelivery of bioactive components for food applications: types of delivery systems, properties, and their effect on ADME profiles and toxicity of nanoparticles. *Annual Review of Food Science and Technology*, *5*, 197–213.

Bouwmeester, H., Dekkers, S., Noordam, M. Y., Hagens, W. I., Bulder, A. S., De Heer, C., et al. (2009). Review of health safety aspects of nanotechnologies in food production. *Regulatory Toxicology and Pharmacology*, *53*(1), 52–62.

Boyd, B. J., Whittaker, D. V., Khoo, S. M., & Davey, G. (2006). Lyotropic liquid crystalline phases formed from glycerate surfactants as sustained release drug delivery systems. *International Journal of Pharmaceutics*, *309*(1–2), 218–226.

Brennen, C. E. (2014). *Cavitation and bubble dynamics*. Cambridge University Press.

Briuglia, M. L., Rotella, C., McFarlane, A., & Lamprou, D. A. (2015). Influence of cholesterol on liposome stability and on in vitro drug release. *Drug Delivery and Translational Research*, *5*(3), 231–242.

Carpenter, J., & Saharan, V. K. (2017). Ultrasonic assisted formation and stability of mustard oil in water nanoemulsion: effect of process parameters and their optimization. *Ultrasonics Sonochemistry*, *35*, 422–430.

Cerqueira, M. A., Pinheiro, A. C., Silva, H. D., Ramos, P. E., Azevedo, M. A., Flores-López, M. L., et al. (2014). Design of bio-nanosystems for oral delivery of functional compounds. *Food Engineering Reviews*, *6*(1–2), 1–19.

Chatani, E., Lee, Y. H., Yagi, H., Yoshimura, Y., Naiki, H., & Goto, Y. (2009). Ultrasonication-dependent production and breakdown lead to minimum-sized amyloid fibrils. *Proceedings of the National Academy of Sciences*, *106*(27), 11119–11124.

Chemat, F., Grondin, I., Sing, A. S. C., & Smadja, J. (2004). Deterioration of edible oils during food processing by ultrasound. *Ultrasonics Sonochemistry*, *11*(1), 13–15.

Cho, H. T., Salvia-Trujillo, L., Kim, J., Park, Y., Xiao, H., & McClements, D. J. (2014). Droplet size and composition of nutraceutical nanoemulsions influences bioavailability of long chain fatty acids and coenzyme Q10. *Food Chemistry*, *156*, 117–122.

Chong, J. Y., Mulet, X., Boyd, B. J., & Drummond, C. J. (2015). Steric stabilizers for cubic phase lyotropic liquid crystal nanodispersions (cubosomes). In *Advances in planar lipid bilayers and liposomes* (pp. 131–187). Vol. 21. Academic Press.

Clogston, J., & Caffrey, M. (2005). Controlling release from the lipidic cubic phase. Amino acids, peptides, proteins and nucleic acids. *Journal of Controlled Release*, *107*(1), 97–111.

Davaran, S., Rezaei, A., Alimohammadi, S., Khandaghi, A. A., Nejati-Koshki, K., Nasrabadi, H. T., et al. (2014). Synthesis and physicochemical characterization of biodegradable star-shaped poly lactide-co-glycolide-β-cyclodextrin copolymer nanoparticles containing albumin. *Advances in Nanoparticles*, *3*(01), 14.

de Souza Simões, L., Madalena, D. A., Pinheiro, A. C., Teixeira, J. A., Vicente, A. A., & Ramos, O. L. (2017). Micro-and nano bio-based delivery systems for food applications: in vitro behavior. *Advances in Colloid and Interface Science*, *243*, 23–45.

Demirci, M., Caglar, M. Y., Cakir, B., & Gülseren, İ. (2017). Encapsulation by nanoliposomes. In *Nanoencapsulation technologies for the food and nutraceutical industries* (pp. 74–113). .

Delmas, T., Piraux, H., Couffin, A. C., Texier, I., Vinet, F., Poulin, P., et al. (2011). How to prepare and stabilize very small nanoemulsions. *Langmuir*, *27*(5), 1683–1692.

Des Rieux, A., Fievez, V., Garinot, M., Schneider, Y. J., & Préat, V. (2006). Nanoparticles as potential oral delivery systems of proteins and vaccines: a mechanistic approach. *Journal of Controlled Release*, *116*(1), 1–27.

Desmet, E., Van Gele, M., & Lambert, J. (2017). Topically applied lipid-and surfactant-based nanoparticles in the treatment of skin disorders. *Expert Opinion on Drug Delivery*, *14*(1), 109–122.

Dong, Y. D., Tilley, A. J., Larson, I., Lawrence, M. J., Amenitsch, H., Rappolt, M., et al. (2010). Nonequilibrium effects in self-assembled mesophase materials: unexpected supercooling effects for cubosomes and hexosomes. *Langmuir*, *26*(11), 9000–9010.

Eastwood, M. A. (2013). *Principles of human nutrition*. Springer.

El-Salam, M. H. A., & El-Shibiny, S. (2016). Natural biopolymers as nanocarriers for bioactive ingredients used in food industries. In *Encapsulations* (pp. 793–829). Academic Press.

Essa, E. A. (2014). Effect of formulation and processing variables on the particle size of sorbitan monopalmitate niosomes. *Asian Journal of Pharmaceutics (AJP)*, *4*(4), 227–233.

Ezhilarasi, P. N., Karthik, P., Chhanwal, N., & Anandharamakrishnan, C. (2013). Nanoencapsulation techniques for food bioactive components: a review. *Food and Bioprocess Technology*, *6*(3), 628–647.

Faridi Esfanjani, A., Assadpour, E., & Jafari, S. M. (2018). Improving the bioavailability of phenolic compounds by loading them within lipid-based nanocarriers. *Trends in Food Science & Technology*, *76*, 56–66.

Fathi, M., Martin, A., & McClements, D. J. (2014). Nanoencapsulation of food ingredients using carbohydrate based delivery systems. *Trends in Food Science & Technology*, *39*(1), 18–39.

Feller, S. E., Gawrisch, K., & MacKerell, A. D. (2002). Polyunsaturated fatty acids in lipid bilayers: Intrinsic and environmental contributions to their unique physical properties. *Journal of the American Chemical Society*, *124*(2), 318–326.

Fong, W. K., Hanley, T., & Boyd, B. J. (2009). Stimuli responsive liquid crystals provide 'on-demand'drug delivery in vitro and in vivo. *Journal of Controlled Release*, *135*(3), 218–226.

Fu, Y., Sarkar, P., Bhunia, A. K., & Yao, Y. (2016). Delivery systems of antimicrobial compounds to food. *Trends in Food Science & Technology*, *57*, 165–177.

Gao, Z. M., Zhu, L. P., Yang, X. Q., He, X. T., Wang, J. M., Guo, J., et al. (2014). Soy lipophilic protein nanoparticles as a novel delivery vehicle for conjugated linoleic acid. *Food & Function*, *5*(6), 1286–1293.

Gardikis, K., Hatziantoniou, S., Viras, K., Wagner, M., & Demetzos, C. (2006). A DSC and Raman spectroscopy study on the effect of PAMAM dendrimer on DPPC model lipid membranes. *International Journal of Pharmaceutics*, *318*(1–2), 118–123.

Garti, N., Libster, D., & Aserin, A. (2012). Lipid polymorphism in lyotropic liquid crystals for triggered release of bioactives. *Food & Function*, *3*(7), 700–713.

Gedanken, A. (2004). Using sonochemistry for the fabrication of nanomaterials. *Ultrasonics Sonochemistry*, *11*(2), 47–55.

Gharibzahedi, S. M., & Jafari, S. M. (2018). Fabrication of Nanoemulsions by Ultrasonication. In *Nanoemulsions* (pp. 233–285). Academic Press.

Ghasemi, S., & Abbasi, S. (2014). Formation of natural casein micelle nanocapsule by means of pH changes and ultrasound. *Food Hydrocolloids*, *42*, 42–47.

Ghasemi, S., Jafari, S. M., Assadpour, E., & Khomeiri, M. (2017). Production of pectin-whey protein nano-complexes as carriers of orange peel oil. *Carbohydrate Polymers*, *177*(Supplement C), 369–377.

Ghasemi, S., Jafari, S. M., Assadpour, E., & Khomeiri, M. (2018). Nanoencapsulation of d-limonene within nanocarriers produced by pectin-whey protein complexes. *Food Hydrocolloids*, *77*, 152–162.

Ghorbanzade, T., Jafari, S. M., Akhavan, S., & Hadavi, R. (2017). Nano-encapsulation of fish oil in nano-liposomes and its application in fortification of yogurt. *Food Chemistry*, *216*, 146–152.

Ghosh, V., Mukherjee, A., & Chandrasekaran, N. (2013). Ultrasonic emulsification of food-grade nanoemulsion formulation and evaluation of its bactericidal activity. *Ultrasonics Sonochemistry*, *20*(1), 338–344.

Goyal, P., Goyal, K., Kumar, S. V., Singh, A., Katare, O. P., & Mishra, D. N. (2005). Liposomal drug delivery systems–clinical applications. *Acta Pharmaceutica*, *55*(1), 1–25.

Granja, A., Vieira, A. C., Chaves, L. L., Nunes, C., Neves, A. R., Pinheiro, M., et al. (2017). Folate-targeted nanostructured lipid carriers for enhanced oral delivery of epigallocatechin-3-gallate. *Food Chemistry*, *237*, 803–810.

Guan, Y. G., Zhang, B. S., Yu, S. J., Wang, X. R., Xu, X. B., Wang, J., et al. (2011). Effects of ultrasound on a glycin–glucose model system—a means of promoting maillard reaction. *Food and Bioprocess Technology*, *4*(8), 1391–1398.

Gul, O., Saricaoglu, F. T., Besir, A., Atalar, I., & Yazici, F. (2018). Effect of ultrasound treatment on the properties of nano-emulsion films obtained from hazelnut meal protein and clove essential oil. *Ultrasonics Sonochemistry*, *41*, 466–474.

Gutiérrez, J. M., González, C., Maestro, A., Solè, I. M. P. C., Pey, C. M., & Nolla, J. (2008). Nano-emulsions: new applications and optimization of their preparation. *Current Opinion in Colloid & Interface Science*, *13*(4), 245–251.

Hajj Ali, H., Michaux, F., Bouelet Ntsama, I. S., Durand, P., Jasniewski, J., & Linder, M. (2016). Shea butter solid nanoparticles for curcumin encapsulation: Influence of nanoparticles size on drug loading. *European Journal of Lipid Science and Technology*, *118* (8), 1168–1178.

Handjani-Vila, R. M., Ribier, A., Rondot, B., & Vanlerberghie, G. (1979). Dispersions of lamellar phases of non-ionic lipids in cosmetic products. *International Journal of Cosmetic Science*, *1*(5), 303–314.

Hasan, M., Latifi, S., Kahn, C., Tamayol, A., Habibey, R., Passeri, E., et al. (2018). The positive role of curcumin-loaded salmon nanoliposomes on the culture of primary cortical neurons. *Marine Drugs*, *16*(7), 218.

Heaney, R. P. (2001). Factors influencing the measurement of bioavailability, taking calcium as a model. *The Journal of nutrition*, *131*(4), 1344S–1348S.

Hejri, A., Khosravi, A., Gharanjig, K., & Hejazi, M. (2013). Optimisation of the formulation of β-carotene loaded nanostructured lipid carriers prepared by solvent diffusion method. *Food Chemistry*, *141*(1), 117–123.

Hosseini, S. M. H., Emam-Djomeh, Z., Razavi, S. H., Moosavi-Movahedi, A. A., Saboury, A. A., Mohammadifar, M. A., et al. (2013). Complex coacervation of β-lactoglobulin–κ-Carrageenan aqueous mixtures as affected by polysaccharide sonication. *Food chemistry*, *141*(1), 215–222.

Hu, B., Liu, X., Zhang, C., & Zeng, X. (2017). Food macromolecule based nanodelivery systems for enhancing the bioavailability of polyphenols. *Journal of Food and Drug Analysis*, *25*(1), 3–15.

Hu, C., & Rhodes, D. G. (2000). Erratum to 'proniosomes: a novel drug carrier preparation'. *International Journal of Pharmaceutics*, *206*, 109–122.

Huang, Q., Yu, H., & Ru, Q. (2010). Bioavailability and delivery of nutraceuticals using nanotechnology. *Journal of Food Science*, *75*(1), R50–R57.

Isailović, B. D., Kostić, I. T., Zvonar, A., Đorđević, V. B., Gašperlin, M., Nedović, V. A., et al. (2013). Resveratrol loaded liposomes produced by different techniques. *Innovative Food Science & Emerging Technologies*, *19*, 181–189.

Jadhav, A. J., Holkar, C. R., Karekar, S. E., Pinjari, D. V., & Pandit, A. B. (2015). Ultrasound assisted manufacturing of paraffin wax nanoemulsions: Process optimization. *Ultrasonics Sonochemistry*, *23*, 201–207.

Jafari, S. M. (2017). An overview of nanoencapsulation techniques and their classification. In *Nanoencapsulation technologies for the food and nutraceutical industries* (pp. 1–34). Academic Press.

Jafari, S. M., Esfanjani, A. F., Katouzian, I., & Assadpour, E. (2017). Release, characterization, and safety of nanoencapsulated food ingredients. In *Nanoencapsulation of food bioactive ingredients* (pp. 401–453). Academic Press. chapter 10.

Jafari, S. M., Fathi, M., & Mandala, I. (2015). Emerging product formation. In *Food waste recovery* (pp. 293–317). Academic Press.

Jafari, S. M., & McClements, D. J. (2017). Nanotechnology approaches for increasing nutrient bioavailability. In *Advances in food and nutrition research* (pp. 1–30). Vol. 81. Academic Press.

Jaiswal, P., Gidwani, B., & Vyas, A. (2016). Nanostructured lipid carriers and their current application in targeted drug delivery. *Artificial Cells, Nanomedicine, and Biotechnology, 44* (1), 27–40.

Jones, O. G., Lesmes, U., Dubin, P., & McClements, D. J. (2010). Effect of polysaccharide charge on formation and properties of biopolymer nanoparticles created by heat treatment of β-lactoglobulin–pectin complexes. *Food Hydrocolloids, 24*(4), 374–383.

Jones, O. G., & McClements, D. J. (2011). Recent progress in biopolymer nanoparticle and microparticle formation by heat-treating electrostatic protein–polysaccharide complexes. *Advances in Colloid and Interface Science, 167*(1–2), 49–62.

Joshi, M., Pathak, S., Sharma, S., & Patravale, V. (2008). Design and in vivo pharmacodynamic evaluation of nanostructured lipid carriers for parenteral delivery of artemether: nanoject. *International Journal of Pharmaceutics, 364*(1), 119–126.

Joyce, E., Phull, S. S., Lorimer, J. P., & Mason, T. J. (2003). The development and evaluation of ultrasound for the treatment of bacterial suspensions. A study of frequency, power and sonication time on cultured Bacillus species. *Ultrasonics Sonochemistry, 10*, 315–318.

Joye, I. J., Davidov-Pardo, G., & McClements, D. J. (2014). Nanotechnology for increased micronutrient bioavailability. *Trends in Food Science & Technology, 40*(2), 168–182.

Joye, I. J., & McClements, D. J. (2014). Biopolymer-based nanoparticles and microparticles: fabrication, characterization, and application. *Current Opinion in Colloid & Interface Science, 19*(5), 417–427.

Kaltsa, O., Michon, C., Yanniotis, S., & Mandala, I. (2013). Ultrasonic energy input influence on the production of sub-micron o/w emulsions containing whey protein and common stabilizers. *Ultrasonics Sonochemistry, 20*(3), 881–891.

Karami, Z., & Hamidi, M. (2016). Cubosomes: remarkable drug delivery potential. *Drug Discovery Today, 21*(5), 789–801.

Kardos, N., & Luche, J. L. (2001). Sonochemistry of carbohydrate compounds. *Carbohydrate Research, 332*(2), 115–131.

Katouzian, I., Esfanjani, A. F., Jafari, S. M., & Akhavan, S. (2017). Formulation and application of a new generation of lipid nano-carriers for the food bioactive ingredients. *Trends in Food Science & Technology, 68*, 14–25.

Kentish, S., Wooster, T. J., Ashokkumar, M., Balachandran, S., Mawson, R., & Simons, L. (2008). The use of ultrasonics for nanoemulsion preparation. *Innovative Food Science & Emerging Technologies, 9*(2), 170–175.

Khan, M. I., Madni, A., Hirvonen, J., & Peltonen, L. (2017). Ultrasonic processing technique as a green preparation approach for diacerein-loaded niosomes. *AAPS PharmSciTech, 18* (5), 1554–1563.

Khosravi-Darani, K., Pardakhty, A., Honarpisheh, H., Rao, V. M., & Mozafari, M. R. (2007). The role of high-resolution imaging in the evaluation of nanosystems for bioactive encapsulation and targeted nanotherapy. *Micron: The International Research and Review Journal for Microscopy, 38*(8), 804–818.

Kumar, K., & Rai, A. K. (2011). Development and evaluation of proniosome-encapsulated curcumin for transdermal administration. *Tropical Journal of Pharmaceutical Research, 10*(6), 697–703.

Kumar, G. P., & Rajeshwarrao, P. (2011). Nonionic surfactant vesicular systems for effective drug delivery—an overview. *Acta Pharmaceutica Sinica B, 1*(4), 208–219.

Kurukji, D., Norton, I., & Spyropoulos, F. (2016). Fabrication of sub-micron protein-chitosan electrostatic complexes for encapsulation and pH-modulated delivery of model hydrophilic active compounds. *Food Hydrocolloids, 53*, 249–260.

Lancelot, A., Sierra, T., & Serrano, J. L. (2014). Nanostructured liquid-crystalline particles for drug delivery. *Expert Opinion on Drug Delivery, 11*(4), 547–564.

Lane, K. E., Li, W., Smith, C., & Derbyshire, E. (2014). The bioavailability of an omega-3-rich algal oil is improved by nanoemulsion technology using yogurt as a food vehicle. *International Journal of Food Science & Technology, 49*(5), 1264–1271.

Lasoń, E., Sikora, E., Ogonowski, J., Tabaszewska, M., & Skoczylas, Ł. (2016). Release study of selected terpenes from nanostructured lipid carriers. *Colloids and Surfaces A: Physicochemical and Engineering Aspects, 510*, 87–92.

Li, M., Cui, J., Ngadi, M. O., & Ma, Y. (2015). Absorption mechanism of whey-protein-delivered curcumin using Caco-2 cell monolayers. *Food Chemistry, 180*, 48–54.

Li, Z., Jiang, H., Xu, C., & Gu, L. (2015). A review: Using nanoparticles to enhance absorption and bioavailability of phenolic phytochemicals. *Food Hydrocolloids, 43*, 153–164.

Litzinger, D. C., Buiting, A. M., van Rooijen, N., & Huang, L. (1994). Effect of liposome size on the circulation time and intraorgan distribution of amphipathic poly (ethylene glycol)-containing liposomes. *Biochimica et Biophysica Acta (BBA)-Biomembranes, 1190*(1), 99–107.

Liu, N., & Park, H. J. (2010). Factors effect on the loading efficiency of Vitamin C loaded chitosan-coated nanoliposomes. *Colloids and Surfaces B: Biointerfaces, 76*(1), 16–19.

Liu, C. H., & Wu, C. T. (2010). Optimization of nanostructured lipid carriers for lutein delivery. *Colloids and Surfaces A: Physicochemical and Engineering Aspects, 353*(2–3), 149–156.

Livney, Y. D. (2015). Nanostructured delivery systems in food: latest developments and potential future directions. *Current Opinion in Food Science, 3*, 125–135.

Lorimer, J. P., & Mason, T. J. (1987). Sonochemistry. Part 1—the physical aspects. *Chemical Society Reviews, 16*, 239–274.

Lv, S., Gu, J., Zhang, R., Zhang, Y., Tan, H., & McClements, D. J. (2018). Vitamin E encapsulation in plant-based nanoemulsions fabricated using dual-channel microfluidization: formation, stability and bioaccessibility. *Journal of Agricultural and Food Chemistry, 66*, 10532–10542.

Lv, F., Hasan, M., Dang, H., Hassan, S. G., Meng, W., Deng, Y., et al. (2016). Optimized luteolin loaded solid lipid nanoparticle under stress condition for enhanced bioavailability in rat plasma. *Journal of Nanoscience and Nanotechnology, 16*(9), 9443–9449.

Madadlou, A., Sheehan, D., Emam-Djomeh, Z., & Mousavi, M. E. (2011). Ultrasound-assisted generation of ACE-inhibitory peptides from casein hydrolyzed with nanoencapsulated protease. *Journal of the Science of Food and Agriculture, 91*(11), 2112–2116.

Manea, A. M., Vasile, B. S., & Meghea, A. (2014). Antioxidant and antimicrobial activities of green tea extract loaded into nanostructured lipid carriers. *Comptes Rendus Chimie, 17*(4), 331–341.

Manickam, S., & Rana, R. K. (2011). Production of nanomaterials using ultrasonic cavitation—A simple, energy efficient and technological approach. In *Ultrasound technologies for food and bioprocessing* (pp. 405–444). New York, NY: Springer.

Manosroi, A., Kietthanakorn, B. O., Chankhampan, C., Khositsuntiwong, N., Manosroi, W., Abe, M., et al. (2013). Physical characteristics and biological activities of Thai flower extracts loaded in niosomes. *Month, 40*(4), 603–617.

Marwa, A., Omaima, S., Hanaa, E. G., & Mohammed, A. S. (2013). Preparation and in-vitro evaluation of diclofenac sodium niosomal formulations. *International Journal of Pharmaceutical Sciences and Research, 4*(5), 1757–1765.

McClements, D. J. (2012). Advances in fabrication of emulsions with enhanced functionality using structural design principles. *Current Opinion in Colloid & Interface Science*, *17*(5), 235–245.

McClements, D. J. (2014). *Nanoparticle- and microparticle-based delivery systems: Encapsulation, protection and release of active compounds*. CRC Press.

McClements, D. J., Decker, E. A., Park, Y., & Weiss, J. (2009). Structural design principles for delivery of bioactive components in nutraceuticals and functional foods. *Critical Reviews in Food Science and Nutrition*, *49*(6), 577–606.

McClements, D. J., Decker, E. A., & Weiss, J. (2007). Emulsion-based delivery systems for lipophilic bioactive components. *Journal of Food Science*, *72*(8), R109–R124.

Mehta, H. R., Raval, A. G., Patel, Y. K., & Pandit, V. D. (2018). Niosome: a controlled and novel targeted drug delivery system. *World Journal of Pharmacy and Pharmaceutical Sciences*, 7(9), 1320–1340.

Moghassemi, S., & Hadjizadeh, A. (2014). Nano-niosomes as nanoscale drug delivery systems: an illustrated review. *Journal of Controlled Release*, *185*, 22–36.

Mohammadi, A., Jafari, S. M., Assadpour, E., & Esfanjani, A. F. (2016). Nano-encapsulation of olive leaf phenolic compounds through WPC–pectin complexes and evaluating their release rate. *International Journal of Biological Macromolecules*, *82*, 816–822.

Moreira, M. M., Morais, S., & Delerue-Matos, C. (2017). Environment-friendly techniques for extraction of bioactive compounds from fruits. In *Soft chemistry and food fermentation* (pp. 21–47). Academic Press.

Mozafari, M. R., Johnson, C., Hatziantoniou, S., & Demetzos, C. (2008). Nanoliposomes and their applications in food nanotechnology. *Journal of Liposome Research*, *18*(4), 309–327.

Mozafari, M. R., Reed, C. J., & Rostron, C. (2007). Cytotoxicity evaluation of anionic nanoliposomes and nanolipoplexes prepared by the heating method without employing volatile solvents and detergents. *Die Pharmazie*, *62*(3), 205–209.

Müller, R. H., Petersen, R. D., Hommoss, A., & Pardeike, J. (2007). Nanostructured lipid carriers (NLC) in cosmetic dermal products. *Advanced Drug Delivery Reviews*, *59*(6), 522–530.

Murugesan, S., Mishra, P., & Jain, N. K. (2008). Development of folate-conjugated PEGylated poly (d, l-lactide-co-glycolide) nanoparticulate carrier for docetaxel. *Current Nanoscience*, *4*(4), 402–408.

Nematollahi, M. H., Pardakhty, A., Torkzadeh-Mahanai, M., Mehrabani, M., & Asadikaram, G. (2017). Changes in physical and chemical properties of niosome membrane induced by cholesterol: a promising approach for niosome bilayer intervention. *RSC Advances*, 7(78), 49463–49472.

Noori, S., Zeynali, F., & Almasi, H. (2018). Antimicrobial and antioxidant efficiency of nanoemulsion-based edible coating containing ginger (*Zingiber officinale*) essential oil and its effect on safety and quality attributes of chicken breast fillets. *Food Control*, *84*, 312–320.

O'Sullivan, J. J., & Norton, I. T. (2016). Novel ultrasonic emulsification technologies. *Gums and stabilisers for the food industry 18: Hydrocolloid functionality for affordable and sustainable global food solutions* (p. 89). *18*, (p. 89). .

Panahi, Y., Farshbaf, M., Mohammadhosseini, M., Mirahadi, M., Khalilov, R., Saghfi, S., et al. (2017). Recent advances on liposomal nanoparticles: synthesis, characterization and biomedical applications. *Artificial Cells, Nanomedicine, and Biotechnology*, *45*(4), 788–799.

Pandit, A. B., & Joshi, J. B. (1993). Hydrolysis of fatty oils: Effect of cavitation. *Chemical Engineering Science*, *48*(19), 3440–3442.

Pando, D., Beltran, M., Gerone, I., Matos, M., & Pazos, C. (2015). Resveratrol entrapped niosomes as yoghurt additive. *Food Chemistry*, *170*, 281–287.

Pardakhty, A., & Moazeni, E. (2013). Nano-niosomes in drug, vaccine and gene delivery: a rapid overview. *Nanomedicine Journal*, *1*(1), 1–12.
Park, S. J., Garcia, C. V., Shin, G. H., & Kim, J. T. (2017). Development of nanostructured lipid carriers for the encapsulation and controlled release of vitamin D3. *Food Chemistry*, *225*, 213–219.
Patil, M. N., & Pandit, A. B. (2007). Cavitation–a novel technique for making stable nano-suspensions. *Ultrasonics Sonochemistry*, *14*(5), 519–530.
Patist, A., & Bates, D. (2008). Ultrasonic innovations in the food industry: From the laboratory to commercial production. *Innovative Food Science & Emerging Technologies*, *9*(2), 147–154.
Perusko, M., Al-Hanish, A., Velickovic, T. C., & Stanic-Vucinic, D. (2015). Macromolecular crowding conditions enhance glycation and oxidation of whey proteins in ultrasound-induced Maillard reaction. *Food Chemistry*, *177*, 248–257.
Pinilla, C. M. B., & Brandelli, A. (2016). Antimicrobial activity of nanoliposomes co-encapsulating nisin and garlic extract against gram-positive and gram-negative bacteria in milk. *Innovative Food Science & Emerging Technologies*, *36*, 287–293.
Puglia, C., Blasi, P., Rizza, L., Schoubben, A., Bonina, F., Rossi, C., et al. (2008). Lipid nanoparticles for prolonged topical delivery: an in vitro and in vivo investigation. *International Journal of Pharmaceutics*, *357*(1–2), 295–304.
Puglia, C., Lauro, M. R., Offerta, A., Crascì, L., Micicchè, L., Panico, A. M., et al. (2017). Nanostructured lipid carriers (NLC) as vehicles for topical administration of sesamol: In vitro percutaneous absorption study and evaluation of antioxidant activity. *Planta Medica*, *83*(5), 398–404.
Quintanilla-Carvajal, M. X., Camacho-Díaz, B. H., Meraz-Torres, L. S., Chanona-Pérez, J. J., Alamilla-Beltrán, L., Jimenéz-Aparicio, A., et al. (2010). Nanoencapsulation: a new trend in food engineering processing. *Food Engineering Reviews*, *2*(1), 39–50.
Rafiee, Z., & Jafari, S. M. (2018). Application of lipid nanocarriers for the food industry. In J. -M. Mérillon & K. G. Ramawat (Eds.), *Bioactive molecules in food* (pp. 1–43). Cham: Springer International Publishing.
Rafiee, Z., Nejatian, M., Daeihamed, M., & Jafari, S. M. (2018). Application of different nanocarriers for encapsulation of curcumin. *Critical Reviews in Food Science and Nutrition*, 1–77.
Rahaiee, S., Shojaosadati, S. A., Hashemi, M., Moini, S., & Razavi, S. H. (2015). Improvement of crocin stability by biodegradeble nanoparticles of chitosan-alginate. *International Journal of Biological Macromolecules*, *79*, 423–432.
Raikos, V., & Ranawana, V. (2017). Designing emulsion droplets of foods and beverages to enhance delivery of lipophilic bioactive components—A review of recent advances. *International Journal of Food Science & Technology*, *52*(1), 68–80.
Ramisetty, K. A., Pandit, A. B., & Gogate, P. R. (2015). Ultrasound assisted preparation of emulsion of coconut oil in water: Understanding the effect of operating parameters and comparison of reactor designs. *Chemical Engineering and Processing: Process Intensification*, *88*, 70–77.
Rao, J., & McClements, D. J. (2011). Food-grade microemulsions, nanoemulsions and emulsions: Fabrication from sucrose monopalmitate & lemon oil. *Food Hydrocolloids*, *25*(6), 1413–1423.
Rasti, B., Jinap, S., Mozafari, M. R., & Yazid, A. M. (2012). Comparative study of the oxidative and physical stability of liposomal and nanoliposomal polyunsaturated fatty acids prepared with conventional and Mozafari methods. *Food Chemistry*, *135*(4), 2761–2770.
Rein, M. J., Renouf, M., Cruz-Hernandez, C., Actis-Goretta, L., Thakkar, S. K., & da Silva Pinto, M. (2013). Bioavailability of bioactive food compounds: a challenging journey to bioefficacy. *British Journal of Clinical Pharmacology*, *75*(3), 588–602.

Rezaei, A., Fathi, M., & Jafari, S. M. (2019). Nanoencapsulation of hydrophobic and low-soluble food bioactive compounds within different nanocarriers. *Food Hydrocolloids, 88*, 146–162.

Richardson, E. S., Pitt, W. G., & Woodbury, D. J. (2007). The role of cavitation in liposome formation. *Biophysical Journal, 93*(12), 4100–4107.

Rostamabadi, H., Falsafi, S. R., & Jafari, S. M. (2019). Nanoencapsulation of carotenoids within lipid-based nanocarriers. *Journal of Controlled Release, 298*, 38–67.

Sabliov, C., Chen, H., & Yada, R. (Eds.), (2015). *Nanotechnology and functional foods: Effective delivery of bioactive ingredients*. John Wiley & Sons.

Sagalowicz, L., & Leser, M. E. (2010). Delivery systems for liquid food products. *Current Opinion in Colloid & Interface Science, 15*(1–2), 61–72.

Sagalowicz, L., Leser, M. E., Watzke, H. J., & Michel, M. (2006). Monoglyceride self-assembly structures as delivery vehicles. *Trends in Food Science & Technology, 17*(5), 204–214.

Salvia-Trujillo, L., Qian, C., Martín-Belloso, O., & McClements, D. J. (2013). Influence of particle size on lipid digestion and β-carotene bioaccessibility in emulsions and nanoemulsions. *Food Chemistry, 141*(2), 1472–1480.

Schenk, D. J., Jones, A. N., Braun, M. P., Yao, D., Wallace, M. A., Marques, R., et al. (2004). Ultrasonic particle size reduction of radiolabeled pharmaceuticals. *Journal of Labelled Compounds and Radiopharmaceuticals, 47*(7), 399–406.

Semenova, M. G., Moiseenko, D. V., Grigorovich, N. V., Anokhina, M. S., Antipova, A. S., Belyakova, L. E., et al. (2014). Protein–polysaccharide interactions and digestion of the complex particles. In *Food structures, digestion and health* (pp. 169–192). Academic Press.

Serieye, S., Méducin, F., Milošević, I., Fu, L., & Guillot, S. (2017). Interface tuning and stabilization of monoglyceride mesophase dispersions: food emulsifiers and mixtures efficiency. *Journal of Colloid and Interface Science, 496*, 26–34.

Shanmugam, A., & Ashokkumar, M. (2014). Ultrasonic preparation of stable flax seed oil emulsions in dairy systems—Physicochemical characterization. *Food Hydrocolloids, 39*, 151–162.

Shao, Y., Wu, C., Wu, T., Li, Y., Chen, S., Yuan, C., et al. (2018). Eugenol-chitosan nanoemulsions by ultrasound-mediated emulsification: Formulation, characterization and antimicrobial activity. *Carbohydrate Polymers, 193*, 144–152.

Shishir, M. R. I., & Chen, W. (2017). Trends of spray drying: a critical review on drying of fruit and vegetable juices. *Trends in Food Science & Technology, 65*, 49–67.

Soleimanian, Y., Goli, S. A. H., Varshosaz, J., & Maestrelli, F. (2018). Propolis wax nanostructured lipid carrier for delivery of β sitosterol: effect of formulation variables on physicochemical properties. *Food Chemistry, 260*, 97–105.

Soria, A. C., & Villamiel, M. (2010). Effect of ultrasound on the technological properties and bioactivity of food: a review. *Trends in Food Science & Technology, 21*(7), 323–331.

Sotomayor-Gerding, D., Oomah, B. D., Acevedo, F., Morales, E., Bustamante, M., Shene, C., et al. (2016). High carotenoid bioaccessibility through linseed oil nanoemulsions with enhanced physical and oxidative stability. *Food Chemistry, 199*, 463–470.

Spada, J. C., Marczak, L. D., Tessaro, I. C., & Cardozo, N. S. (2015). Interactions between soy protein from water-soluble soy extract and polysaccharides in solutions with polydextrose. *Carbohydrate Polymers, 134*, 119–127.

Stanic-Vucinic, D., Prodic, I., Apostolovic, D., Nikolic, M., & Velickovic, T. C. (2013). Structure and antioxidant activity of β-lactoglobulin-glycoconjugates obtained by high-intensity-ultrasound-induced maillard reaction in aqueous model systems under neutral conditions. *Food Chemistry, 138*(1), 590–599.

Steed, J. W., Turner, D. R., & Wallace, K. (2007). *Core concepts in supramolecular chemistry and nanochemistry*. UK: John Wiley & Sonss.

Sugumar, S., Ghosh, V., Nirmala, M. J., Mukherjee, A., & Chandrasekaran, N. (2014). Ultrasonic emulsification of eucalyptus oil nanoemulsion: Antibacterial activity against *Staphylococcus aureus* and wound healing activity in Wistar rats. *Ultrasonics Sonochemistry*, *21*(3), 1044–1049.

Sun, D. W., & Li, B. (2003). Microstructural change of potato tissues frozen by ultrasound-assisted immersion freezing. *Journal of Food Engineering*, *57*(4), 337–345.

Suslick, K. S., & Price, G. J. (1999). Applications of ultrasound to materials chemistry. *Annual Review of Materials Science*, *29*(1), 295–326.

Tamjidi, F., Shahedi, M., Varshosaz, J., & Nasirpour, A. (2013). Nanostructured lipid carriers (NLC): a potential delivery system for bioactive food molecules. *Innovative Food Science & Emerging Technologies*, *19*, 29–43.

Tamjidi, F., Shahedi, M., Varshosaz, J., & Nasirpour, A. (2014). Design and characterization of astaxanthin-loaded nanostructured lipid carriers. *Innovative Food Science & Emerging Technologies*, *26*, 366–374.

Tan, C., Xue, J., Abbas, S., Feng, B., Zhang, X., & Xia, S. (2014). Liposome as a delivery system for carotenoids: Comparative antioxidant activity of carotenoids as measured by ferric reducing antioxidant power, DPPH assay and lipid peroxidation. *Journal of Agricultural and Food Chemistry*, *62*(28), 6726–6735.

Tang, S. Y., Shridharan, P., & Sivakumar, M. (2013). Impact of process parameters in the generation of novel aspirin nanoemulsions—Comparative studies between ultrasound cavitation and microfluidizer. *Ultrasonics Sonochemistry*, *20*(1), 485–497.

Tarekegn, A., Joseph, N. M., Palani, S., Zacharia, A., & Ayenew, Z. (2010). Niosomes in targeted drug delivery: some recent advances. *International Journal of Pharmaceutical Science and Research*, *1*(9), 1–8.

Tatar Turan, F., Cengiz, A., Sandıkçı, D., Dervisoglu, M., & Kahyaoglu, T. (2016). Influence of an ultrasonic nozzle in spray-drying and storage on the properties of blueberry powder and microcapsules. *Journal of the Science of Food and Agriculture*, *96*(12), 4062–4076.

Tavakoli, H., Hosseini, O., Jafari, S. M., & Katouzian, I. (2018). Evaluation of physicochemical and antioxidant properties of yogurt enriched by olive leaf phenolics within nanoliposomes. *Journal of Agricultural and Food Chemistry*, *66*(35), 9231–9240.

Tavano, L., Alfano, P., Muzzalupo, R., & de Cindio, B. (2011). Niosomes vs microemulsions: new carriers for topical delivery of capsaicin. *Colloids and Surfaces B: Biointerfaces*, *87*(2), 333–339.

Tavano, L., Muzzalupo, R., Picci, N., & de Cindio, B. (2014). Co-encapsulation of lipophilic antioxidants into niosomal carriers: Percutaneous permeation studies for cosmeceutical applications. *Colloids and Surfaces B: Biointerfaces*, *114*, 144–149.

Taylor, T. M., Weiss, J., Davidson, P. M., & Bruce, B. D. (2005). Liposomal nanocapsules in food science and agriculture. *Critical Reviews in Food Science and Nutrition*, *45*(7–8), 587–605.

Thompson, A. K., Mozafari, M. R., & Singh, H. (2007). The properties of liposomes produced from milk fat globule membrane material using different techniques. *Le Lait*, *87* (4–5), 349–360.

Thorat, A. A., & Dalvi, S. V. (2012). Liquid antisolvent precipitation and stabilization of nanoparticles of poorly water soluble drugs in aqueous suspensions: recent developments and future perspective. *Chemical Engineering Journal*, *181*, 1–34.

Tischer, P. C. F., Sierakowski, M. R., Westfahl, H., Jr., & Tischer, C. A. (2010). Nanostructural reorganization of bacterial cellulose by ultrasonic treatment. *Biomacromolecules*, *11*(5), 1217–1224.

Tomalia, D. A., & Frechet, J. M. (2001). Introduction to the dendritic state. In *Dendrimers and other dendritic polymers* (pp. 1–44). John Wiley & Sons Ltd.

Torchilin, V., & Weissig, V. (Eds.), (2003). *Liposomes: A practical approach*. In Vol. 264. Oxford University Press.

Tsai, M. L., Bai, S. W., & Chen, R. H. (2008). Cavitation effects versus stretch effects resulted in different size and polydispersity of ionotropic gelation chitosan–sodium tripolyphosphate nanoparticle. *Carbohydrate Polymers*, *71*(3), 448–457.

Tu, Y. S., Fu, J. W., Sun, D. M., Zhang, J. J., Yao, N., Huang, D. E., et al. (2014). Preparation, characterisation and evaluation of curcumin with piperine-loaded cubosome nanoparticles. *Journal of Microencapsulation*, *31*(6), 551–559.

Ubbink, J., Burbidge, A., & Mezzenga, R. (2008). Food structure and functionality: a soft matter perspective. *Soft Matter*, *4*(8), 1569–1581.

Uchegbu, I. F., & Vyas, S. P. (1998). Non-ionic surfactant based vesicles (niosomes) in drug delivery. *International Journal of Pharmaceutics*, *172*(1–2), 33–70.

Urandur, S., Marwaha, D., Gautam, S., Banala, V. T., Sharma, M., & Mishra, P. R. (2018). Nonlamellar liquid crystals: a new paradigm for the delivery of small molecules and bio-macromolecules. *Therapeutic Delivery*, *9*(9), 667–689.

Van der Sman, R. G. M. (2012). Soft matter approaches to food structuring. *Advances in Colloid and Interface Science*, *176*, 18–30.

Verma, S., Singh, S. K., Syan, N., Mathur, P., & Valecha, V. (2010). Nanoparticle vesicular systems: a versatile tool for drug delivery. *Journal of Chemical and Pharmaceutical Research*, *2*(2), 496–509.

Walia, N., Dasgupta, N., Ranjan, S., Chen, L., & Ramalingam, C. (2017). Fish oil based vitamin D nanoencapsulation by ultrasonication and bioaccessibility analysis in simulated gastro-intestinal tract. *Ultrasonics Sonochemistry*, *39*, 623–635.

Wang, M., Yuan, Y., Gao, Y., Ma, H. M., Xu, H. T., Zhang, X. N., et al. (2012). Preparation and characterization of 5-fluorouracil pH-sensitive niosome and its tumor-targeted evaluation: in vitro and in vivo. *Drug Development and Industrial Pharmacy*, *38*(9), 1134–1141.

Wei, L., Li, X., Guo, F., Liu, X., & Wang, Z. (2018). Structural properties, in vitro release and radical scavenging activity of lecithin based curcumin-encapsulated inverse hexagonal (H II) liquid crystals. *Colloids and Surfaces A: Physicochemical and Engineering Aspects*, *539*, 124–131.

Wörle, G., Drechsler, M., Koch, M. H. J., Siekmann, B., Westesen, K., & Bunjes, H. (2007). Influence of composition and preparation parameters on the properties of aqueous monoolein dispersions. *International Journal of Pharmaceutics*, *329*(1–2), 150–157.

Wu, T., Zivanovic, S., Hayes, D. G., & Weiss, J. (2008). Efficient reduction of chitosan molecular weight by high-intensity ultrasound: underlying mechanism and effect of process parameters. *Journal of Agricultural and Food Chemistry*, *56*(13), 5112–5119.

Yaghmur, A., & Glatter, O. (2009). Characterization and potential applications of nanostructured aqueous dispersions. *Advances in Colloid and Interface Science*, *147*, 333–342.

Yi, J., Li, Y., Zhong, F., & Yokoyama, W. (2014). The physicochemical stability and in vitro bioaccessibility of beta-carotene in oil-in-water sodium caseinate emulsions. *Food Hydrocolloids*, *35*, 19–27.

Yu, F., & Tang, X. (2016). Novel long-circulating liposomes consisting of PEG modified β-sitosterol for gambogic acid delivery. *Journal of Nanoscience and Nanotechnology*, *16*(3), 3115–3121.

Zardini, A. A., Mohebbi, M., Farhoosh, R., & Bolurian, S. (2018). Production and characterization of nanostructured lipid carriers and solid lipid nanoparticles containing lycopene for food fortification. *Journal of Food Science and Technology*, *55*(1), 287–298.

Zhang, Y., Chan, J. W., Moretti, A., & Uhrich, K. E. (2015). Designing polymers with sugar-based advantages for bioactive delivery applications. *Journal of Controlled Release*, *219*, 355–368.

Zhao, X. L., Yang, C. R., Yang, K. L., Li, K. X., Hu, H. Y., & Chen, D. W. (2010). Preparation and characterization of nanostructured lipid carriers loaded traditional Chinese medicine, zedoary turmeric oil. *Drug Development and Industrial Pharmacy*, *36*(7), 773–780.
Zimet, P., & Livney, Y. D. (2009). Beta-lactoglobulin and its nanocomplexes with pectin as vehicles for ω-3 polyunsaturated fatty acids. *Food Hydrocolloids*, *23*(4), 1120–1126.

Further reading

Acosta, E. (2009). Bioavailability of nanoparticles in nutrient and nutraceutical delivery. *Current Opinion in Colloid & Interface Science*, *14*(1), 3–15.
Bhavsar, R. S. (2011). Overview of homogenizer processes. *Pharmacy Times*, *43*(10), 37–41.
Corzo-Martínez, M., Montilla, A., Megías-Pérez, R., Olano, A., Moreno, F. J., & Villamiel, M. (2014). Impact of high-intensity ultrasound on the formation of lactulose and Maillard reaction glycoconjugates. *Food Chemistry*, *157*, 186–192.
Esposito, E., Sticozzi, C., Ravani, L., Drechsler, M., Muresan, X. M., Cervellati, F., et al. (2015). Effect of new curcumin-containing nanostructured lipid dispersions on human keratinocytes proliferative responses. *Experimental Dermatology*, *24*(6), 449–454.
Kim, H. Y., Han, J. A., Kweon, D. K., Park, J. D., & Lim, S. T. (2013). Effect of ultrasonic treatments on nanoparticle preparation of acid-hydrolyzed waxy maize starch. *Carbohydrate Polymers*, *93*(2), 582–588.
Lacatusu, I., Mitrea, E., Badea, N., Stan, R., Oprea, O., & Meghea, A. (2013). Lipid nanoparticles based on omega-3 fatty acids as effective carriers for lutein delivery. Preparation and in vitro characterization studies. *Journal of Functional Foods*, *5*(3), 1260–1269.
Linkevičiūtė, A., Misiūnas, A., Naujalis, E., & Barauskas, J. (2015). Preparation and characterization of quercetin-loaded lipid liquid crystalline systems. *Colloids and Surfaces B: Biointerfaces*, *128*, 296–303.
Liu, T., & Guo, R. (2005). Preparation of a highly stable niosome and its hydrotrope-solubilization action to drugs. *Langmuir*, *21*(24), 11034–11039.
Mason, T. J. (2002). High powered ultrasound in physical and chemical processing. In *New acoustics. Selected topics*. Madrid: CSIC.
Narsaiah, K., Jha, S. N., Wilson, R. A., Mandge, H. M., Manikantan, M. R., Malik, R. K., et al. (2013). Pediocin-loaded nanoliposomes and hybrid alginate–nanoliposome delivery systems for slow release of pediocin. *BioNanoScience*, *3*(1), 37–42.
Neethirajan, S., & Jayas, D. S. (2011). Nanotechnology for the food and bioprocessing industries. *Food and Bioprocess Technology*, *4*(1), 39–47.
Shilpa, S., Srinivasan, B. P., & Chauhan, M. (2011). Niosomes as vesicular carriers for delivery of proteins and biologicals. *International Journal of Drug Delivery*, *3*(1), 14–24.
Shishir, M. R. I., Taip, F. S., Aziz, N. A., Talib, R. A., & Sarker, M. S. H. (2016). Optimization of spray drying parameters for pink guava powder using RSM. *Food Science and Biotechnology*, *25*(2), 461–468.
Suslick, K. S. (1995). Applications of ultrasound to materials chemistry. *MRS Bulletin*, *20*(4), 29–34.
Weiss, J., Takhistov, P., & McClements, D. J. (2006). Functional materials in food nanotechnology. *Journal of Food Science*, *71*(9), R107–R116.
Yaghmur, A., Kriechbaum, M., Amenitsch, H., Steinhart, M., Laggner, P., & Rappolt, M. (2009). Effects of pressure and temperature on the self-assembled fully hydrated nanostructures of monoolein – oil systems. *Langmuir*, *26*(2), 1177–1185.

Index

Note: Page numbers followed by *f* indicate figures and *t* indicate tables.

T

U

MIX
Papier | Fördert gute Waldnutzung
FSC® C083411

Kolibri 360 GmbH
Ferdinand-Jühlke-Straße 7
99095 Erfurt, Deutschland
produktsicherheit@kolibri360.de